Fast Algorithms for Signal Processing

Efficient algorithms for signal processing are critical to very large scale future applications such as video processing and four-dimensional medical imaging. Similarly, efficient algorithms are important for embedded and power-limited applications since, by reducing the number of computations, power consumption can be reduced considerably. This unique textbook presents a broad range of computationally-efficient algorithms, describes their structure and implementation, and compares their relative strengths. All the necessary background mathematics is presented, and theorems are rigorously proved. The book is suitable for researchers and practitioners in electrical engineering, applied mathematics, and computer science.

Richard E. Blahut is a Professor of Electrical and Computer Engineering at the University of Illinois, Urbana-Champaign. He is Life Fellow of the IEEE and the recipient of many awards including the IEEE Alexander Graham Bell Medal (1998) and Claude E. Shannon Award (2005), the Tau Beta Pi Daniel C. Drucker Eminent Faculty Award, and the IEEE Millennium Medal. He was named a Fellow of the IBM Corporation in 1980, where he worked for over 30 years, and was elected to the National Academy of Engineering in 1990.

Fast Algorithms for Signal Processing

Richard E. Blahut

Henry Magnuski Professor in Electrical and Computer Engineering,
University of Illinois, Urbana-Champaign

CAMBRIDGE UNIVERSITY PRESS
Cambridge, New York, Melbourne, Madrid, Cape Town, Singapore, São Paulo, Delhi

Cambridge University Press
The Edinburgh Building, Cambridge CB2 8RU, UK

Published in the United States of America by Cambridge University Press, New York

www.cambridge.org
Information on this title: www.cambridge.org/9780521190497

© Cambridge University Press 2010

This publication is in copyright. Subject to statutory exception
and to the provisions of relevant collective licensing agreements,
no reproduction of any part may take place without the written
permission of Cambridge University Press.

First published 2010

Printed in the United Kingdom at the University Press, Cambridge

A catalogue record for this publication is available from the British Library

Library of Congress Cataloguing in Publication data
Blahut, Richard E.
Fast algorithms for signal processing / Richard E. Blahut.
    p.   cm.
Includes bibliographical references and index.
ISBN 978-0-521-19049-7 (hardback)
1. Signal processing – Digital techniques.   2. Algorithms.   I. Title.
TK5102.9.A378   1995
621.382′2 – dc22   2010007111

ISBN 978-0-521-19049-7 Hardback

Cambridge University Press has no responsibility for the persistence or
accuracy of URLs for external or third-party internet websites referred to
in this publication, and does not guarantee that any content on such
websites is, or will remain, accurate or appropriate.

In loving memory of
    Jeffrey Paul Blahut
    May 2, 1968 – June 13, 2004

Many small make a great.
— **Chaucer**

Contents

Preface xi
Acknowledgments xiii

1 Introduction 1

1.1 Introduction to fast algorithms 1
1.2 Applications of fast algorithms 6
1.3 Number systems for computation 8
1.4 Digital signal processing 9
1.5 History of fast signal-processing algorithms 17

2 Introduction to abstract algebra 21

2.1 Groups 21
2.2 Rings 26
2.3 Fields 30
2.4 Vector space 34
2.5 Matrix algebra 37
2.6 The integer ring 44
2.7 Polynomial rings 48
2.8 The Chinese remainder theorem 58

3 Fast algorithms for the discrete Fourier transform 68

3.1 The Cooley–Tukey fast Fourier transform 68
3.2 Small-radix Cooley–Tukey algorithms 72
3.3 The Good–Thomas fast Fourier transform 80

3.4	The Goertzel algorithm	83
3.5	The discrete cosine transform	85
3.6	Fourier transforms computed by using convolutions	91
3.7	The Rader–Winograd algorithm	97
3.8	The Winograd small fast Fourier transform	102

4 Fast algorithms based on doubling strategies — 115

4.1	Halving and doubling strategies	115
4.2	Data structures	119
4.3	Fast algorithms for sorting	120
4.4	Fast transposition	122
4.5	Matrix multiplication	124
4.6	Computation of trigonometric functions	127
4.7	An accelerated euclidean algorithm for polynomials	130
4.8	A recursive radix-two fast Fourier transform	139

5 Fast algorithms for short convolutions — 145

5.1	Cyclic convolution and linear convolution	145
5.2	The Cook–Toom algorithm	148
5.3	Winograd short convolution algorithms	155
5.4	Design of short linear convolution algorithms	164
5.5	Polynomial products modulo a polynomial	168
5.6	Design of short cyclic convolution algorithms	171
5.7	Convolution in general fields and rings	176
5.8	Complexity of convolution algorithms	178

6 Architecture of filters and transforms — 194

6.1	Convolution by sections	194
6.2	Algorithms for short filter sections	199
6.3	Iterated filter sections	202
6.4	Symmetric and skew-symmetric filters	207
6.5	Decimating and interpolating filters	213
6.6	Construction of transform computers	216
6.7	Limited-range Fourier transforms	221
6.8	Autocorrelation and crosscorrelation	222

7 Fast algorithms for solving Toeplitz systems — 231

- 7.1 The Levinson and Durbin algorithms — 231
- 7.2 The Trench algorithm — 239
- 7.3 Methods based on the euclidean algorithm — 245
- 7.4 The Berlekamp–Massey algorithm — 249
- 7.5 An accelerated Berlekamp–Massey algorithm — 255

8 Fast algorithms for trellis search — 262

- 8.1 Trellis and tree searching — 262
- 8.2 The Viterbi algorithm — 267
- 8.3 Sequential algorithms — 270
- 8.4 The Fano algorithm — 274
- 8.5 The stack algorithm — 278
- 8.6 The Bahl algorithm — 280

9 Numbers and fields — 286

- 9.1 Elementary number theory — 286
- 9.2 Fields based on the integer ring — 293
- 9.3 Fields based on polynomial rings — 296
- 9.4 Minimal polynomials and conjugates — 299
- 9.5 Cyclotomic polynomials — 300
- 9.6 Primitive elements — 304
- 9.7 Algebraic integers — 306

10 Computation in finite fields and rings — 311

- 10.1 Convolution in surrogate fields — 311
- 10.2 Fermat number transforms — 314
- 10.3 Mersenne number transforms — 317
- 10.4 Arithmetic in a modular integer ring — 320
- 10.5 Convolution algorithms in finite fields — 324
- 10.6 Fourier transform algorithms in finite fields — 328
- 10.7 Complex convolution in surrogate fields — 331

	10.8	Integer ring transforms	336
	10.9	Chevillat number transforms	339
	10.10	The Preparata–Sarwate algorithm	339

11 Fast algorithms and multidimensional convolutions — 345

	11.1	Nested convolution algorithms	345
	11.2	The Agarwal–Cooley convolution algorithm	350
	11.3	Splitting algorithms	357
	11.4	Iterated algorithms	362
	11.5	Polynomial representation of extension fields	368
	11.6	Convolution with polynomial transforms	371
	11.7	The Nussbaumer polynomial transforms	372
	11.8	Fast convolution of polynomials	376

12 Fast algorithms and multidimensional transforms — 384

	12.1	Small-radix Cooley–Tukey algorithms	384
	12.2	The two-dimensional discrete cosine transform	389
	12.3	Nested transform algorithms	391
	12.4	The Winograd large fast Fourier transform	395
	12.5	The Johnson–Burrus fast Fourier transform	399
	12.6	Splitting algorithms	403
	12.7	An improved Winograd fast Fourier transform	410
	12.8	The Nussbaumer–Quandalle permutation algorithm	411

A A collection of cyclic convolution algorithms — 427

B A collection of Winograd small FFT algorithms — 435

Bibliography — 442
Index — 449

Preface

A quarter of a century has passed since the previous version[1] of this book was published, and signal processing continues to be a very important part of electrical engineering. It forms an essential part of systems for telecommunications, radar and sonar, image formation systems such as medical imaging, and other large computational problems, such as in electromagnetics or fluid dynamics, geophysical exploration, and so on. Fast computational algorithms are necessary in large problems of signal processing, and the study of such algorithms is the subject of this book. Over those several decades, however, the nature of the need for fast algorithms has shifted both to much larger systems on the one hand and to embedded power-limited applications on the other.

Because many processors and many problems are much larger now than they were when the original version of this book was written, and the relative cost of addition and multiplication now may appear to be less dramatic, some of the topics of twenty years ago may be seen by some to be of less importance today. I take exactly the opposite point of view for several reasons. Very large three-dimensional or four-dimensional problems now under consideration require massive amounts of computation and this computation can be reduced by orders of magnitude in many cases by the choice of algorithm. Indeed, these very large problems can be especially suitable for the benefits of fast algorithms. At the same time, smaller signal processing problems now appear frequently in handheld or remote applications where power may be scarce or nonrenewable. The designer's care in treating an embedded application, such as a digital television, can repay itself many times by significantly reducing the power expenditure. Moreover, the unfamiliar algorithms of this book now can often be handled automatically by computerized design tools, and in embedded applications where power dissipation must be minimized, a search for the algorithm with the fewest operations may be essential.

Because the book has changed in its details and the title has been slightly modernized, it is more than a second edition, although most of the topics of the original book have been retained in nearly the same form, but usually with the presentation rewritten. Possibly, in time, some of these topics will re-emerge in a new form, but that time

---

[1] *Fast Algorithms for Digital Signal Processing*, Addison-Wesley, Reading, MA, 1985.

is not now. A newly written book might look different in its choice of topics and its balance between topics than does this one. To accommodate this consideration here, the chapters have been rearranged and revised, even those whose content has not changed substantially. Some new sections have been added, and all of the book has been polished, revised, and re-edited. Most of the touch and feel of the original book is still evident in this new version.

The heart of the book is in the Fourier transform algorithms of Chapters 3 and 12 and the convolution algorithms of Chapters 5 and 11. Chapters 12 and 11 are the multidimensional continuations of Chapters 3 and 4, respectively, and can be partially read immediately thereafter if desired. The study of one-dimensional convolution algorithms and Fourier transform algorithms is only completed in the context of the multidimensional problems. Chapters 2 and 9 are mathematical interludes; some readers may prefer to treat them as appendices, consulting them only as needed. The remainder, Chapters 4, 7, and 8, are in large part independent of the rest of the book. Each can be read independently with little difficulty.

This book uses branches of mathematics that the typical reader with an engineering education will not know. Therefore these topics are developed in Chapters 2 and 9, and all theorems are rigorously proved. I believe that if the subject is to continue to mature and stand on its own, the necessary mathematics must be a part of such a book; appeal to a distant authority will not do. Engineers cannot confidently advance through the subject if they are frequently asked to accept an assertion or to visit their mathematics library.

Acknowledgments

My major debt in writing this book is to Shmuel Winograd. Without his many contributions to the subject, the book would be shapeless and much shorter. He was also generous with his time in clarifying many points to me, and in reviewing early drafts of the original book. The papers of Winograd and also the book of Nussbaumer were a source for much of the material discussed in this book.

The original version of this book could not have reached maturity without being tested, critiqued, and rewritten repeatedly. I remain indebted to Professor B. W. Dickinson, Professor Toby Berger, Professor C. S. Burrus, Professor J. Gibson, Professor J. G. Proakis, Professor T. W. Parks, Dr B. Rice, Professor Y. Sugiyama, Dr W. Vanderkulk, and Professor G. Verghese for their gracious criticisms of the original 1985 manuscript. That book could not have been written without the support that was given by the International Business Machines Corporation. I am deeply grateful to IBM for this support and also to Cornell University for giving me the opportunity to teach several times from the preliminary manuscript of the earlier book. The revised book was written in the wonderful collaborative environment of the Department of Electrical and Computer Engineering and the Coordinated Science Laboratory of the University of Illinois. The quality of the book has much to with the composition skills of Mrs Francie Bridges and the editing skills of Mrs Helen Metzinger. And, as always, Barbara made it possible.

1 Introduction

Algorithms for computation are found everywhere, and efficient versions of these algorithms are highly valued by those who use them. We are mainly concerned with certain types of computation, primarily those related to signal processing, including the computations found in digital filters, discrete Fourier transforms, correlations, and spectral analysis. Our purpose is to present the advanced techniques for fast digital implementation of these computations. We are not concerned with the function of a digital filter or with how it should be designed to perform a certain task; our concern is only with the computational organization of its implementation. Nor are we concerned with why one should want to compute, for example, a discrete Fourier transform; our concern is only with how it can be computed efficiently. Surprisingly, there is an extensive body of theory dealing with this specialized topic – the topic of fast algorithms.

1.1 Introduction to fast algorithms

An algorithm, like most other engineering devices, can be described either by an input/output relationship or by a detailed explanation of its internal construction. When one applies the techniques of signal processing to a new problem one is concerned only with the input/output aspects of the algorithm. Given a signal, or a data record of some kind, one is concerned with what should be done to this data, that is, with what the output of the algorithm should be when such and such a data record is the input. Perhaps the output is a filtered version of the input, or the output is the Fourier transform of the input. The relationship between the input and the output of a computational task can be expressed mathematically without prescribing in detail all of the steps by which the calculation is to be performed.

Devising such an algorithm for an information processing problem, from this input/output point of view, may be a formidable and sophisticated task, but this is not our concern in this book. We will assume that we are given a specification of a relationship between input and output, described in terms of filters, Fourier transforms, interpolations, decimations, correlations, modulations, histograms, matrix operations, and so forth. All of these can be expressed with mathematical formulas and so can be computed just as written. This will be referred to as the obvious implementation.

One may be content with the obvious implementation, and it might not be apparent that the obvious implementation need not be the most efficient. But once people began to compute such things, other people began to look for more efficient ways to compute them. This is the story we aim to tell, the story of fast algorithms for signal processing. By a fast algorithm, we mean a detailed description of a computational procedure that is not the obvious way to compute the required output from the input. A fast algorithm usually gives up a conceptually clear computation in favor of one that is computationally efficient.

Suppose we need to compute a number A, given by

$$A = ac + ad + bc + bd.$$

As written, this requires four multiplications and three additions to compute. If we need to compute A many times with different sets of data, we will quickly notice that

$$A = (a+b)(c+d)$$

is an equivalent form that requires only one multiplication and two additions, and so it is to be preferred. This simple example is quite obvious, but really illustrates most of what we shall talk about. Everything we do can be thought of in terms of the clever insertion of parentheses in a computational problem. But in a big problem, the fast algorithms cannot be found by inspection. It will require a considerable amount of theory to find them.

A nontrivial yet simple example of a fast algorithm is an algorithm for complex multiplication. The complex product[1]

$$(e + jf) = (a + jb) \cdot (c + jd)$$

can be defined in terms of real multiplications and real additions as

$$e = ac - bd$$
$$f = ad + bc.$$

We see that these formulas require four real multiplications and two real additions. A more efficient "algorithm" is

$$e = (a-b)d + a(c-d)$$
$$f = (a-b)d + b(c+d)$$

whenever multiplication is harder than addition. This form requires three real multiplications and five real additions. If c and d are constants for a series of complex

---

[1] The letter j is used for $\sqrt{-1}$ and j is used as an index throughout the book. This should not cause any confusion.

1.1 Introduction to fast algorithms

multiplications, then the terms $c + d$ and $c - d$ are constants also and can be computed off-line. It then requires three real multiplications and three real additions to do one complex multiplication.

We have traded one multiplication for an addition. This can be a worthwhile saving, but only if the signal processor is designed to take advantage of it. Most signal processors, however, have been designed with a prejudice for a complex multiplication that uses four multiplications. Then the advantage of the improved algorithm has no value. The storage and movement of data between additions and multiplications are also important considerations in determining the speed of a computation and of some importance in determining power dissipation.

We can dwell further on this example as a foretaste of things to come. The complex multiplication above can be rewritten as a matrix product

$$\begin{bmatrix} e \\ f \end{bmatrix} = \begin{bmatrix} c & -d \\ d & c \end{bmatrix} \begin{bmatrix} a \\ b \end{bmatrix},$$

where the vector $\begin{bmatrix} a \\ b \end{bmatrix}$ represents the complex number $a + jb$, the matrix $\begin{bmatrix} c & -d \\ d & c \end{bmatrix}$ represents the complex number $c + jd$, and the vector $\begin{bmatrix} e \\ f \end{bmatrix}$ represents the complex number $e + jf$. The matrix–vector product is an unconventional way to represent complex multiplication. The alternative computational algorithm can be written in matrix form as

$$\begin{bmatrix} e \\ f \end{bmatrix} = \begin{bmatrix} 1 & 0 & 1 \\ 0 & 1 & 1 \end{bmatrix} \begin{bmatrix} (c-d) & 0 & 0 \\ 0 & (c+d) & 0 \\ 0 & 0 & d \end{bmatrix} \begin{bmatrix} 1 & 0 \\ 0 & 1 \\ 1 & -1 \end{bmatrix} \begin{bmatrix} a \\ b \end{bmatrix}.$$

The algorithm, then, can be thought of as nothing more than the unusual matrix factorization:

$$\begin{bmatrix} c & -d \\ d & c \end{bmatrix} = \begin{bmatrix} 1 & 0 & 1 \\ 0 & 1 & 1 \end{bmatrix} \begin{bmatrix} (c-d) & 0 & 0 \\ 0 & (c+d) & 0 \\ 0 & 0 & d \end{bmatrix} \begin{bmatrix} 1 & 0 \\ 0 & 1 \\ 1 & -1 \end{bmatrix}.$$

We can abbreviate the algorithm as

$$\begin{bmatrix} e \\ f \end{bmatrix} = BDA \begin{bmatrix} a \\ b \end{bmatrix},$$

where A is a three by two matrix that we call a matrix of preadditions; D is a three by three diagonal matrix that is responsible for all of the general multiplications; and B is a two by three matrix that we call a matrix of postadditions.

We shall find that many fast computational procedures for convolution and for the discrete Fourier transform can be put into this factored form of a diagonal matrix in the center, and on each side of which is a matrix whose elements are 1, 0, and -1. Multiplication by a matrix whose elements are 0 and ± 1 requires only additions and subtractions. Fast algorithms in this form will have the structure of a batch of additions, followed by a batch of multiplications, followed by another batch of additions.

The final example of this introductory section is a fast algorithm for multiplying two arbitrary matrices. Let

$$C = AB,$$

where A and B are any ℓ by n, and n by m, matrices, respectively. The standard method for computing the matrix C is

$$c_{ij} = \sum_{k=1}^{n} a_{ik} b_{kj} \qquad \begin{array}{l} i = 1, \ldots, \ell \\ j = 1, \ldots, m, \end{array}$$

which, as it is written, requires $m\ell n$ multiplications and $(n-1)\ell m$ additions. We shall give an algorithm that reduces the number of multiplications by almost a factor of two but increases the number of additions. The total number of operations increases slightly.

We use the identity

$$a_1 b_1 + a_2 b_2 = (a_1 + b_2)(a_2 + b_1) - a_1 a_2 - b_1 b_2$$

on the elements of A and B. Suppose that n is even (otherwise append a column of zeros to A and a row of zeros to B, which does not change the product C). Apply the above identity to pairs of columns of A and pairs of rows of B to write

$$c_{ij} = \sum_{i=1}^{n/2} (a_{i,2k-1} b_{2k-1,j} + a_{i,2k} b_{2k,j})$$

$$= \sum_{k=1}^{n/2} (a_{i,2k-1} + b_{2k,j})(a_{i,2k} + b_{2k-1,j}) - \sum_{k=1}^{n/2} a_{i,2k-1} a_{i,2k} - \sum_{k=1}^{n/2} b_{2k-1,j} b_{2k,j}$$

for $i = 1, \ldots, \ell$ and $j = 1, \ldots, m$.

This results in computational savings because the second term depends only on i and need not be recomputed for each j, and the third term depends only on j and need not be recomputed for each i. The total number of multiplications used to compute matrix C is $\frac{1}{2} n\ell m + \frac{1}{2} n(\ell + m)$, and the total number of additions is $\frac{3}{2} n\ell m + \ell m + \left(\frac{1}{2} n - 1\right)(\ell + m)$. For large matrices the number of multiplications is about half the direct method.

This last example may be a good place for a word of caution about numerical accuracy. Although the number of multiplications is reduced, this algorithm is more sensitive to roundoff error unless it is used with care. By proper scaling of intermediate steps,

1.1 Introduction to fast algorithms

Algorithm	Multiplications/pixel*	Additions/pixel
Direct computation of discrete Fourier transform 1000 x 1000	8000	4000
Basic Cooley–Tukey FFT 1024 x 1024	40	60
Hybrid Cooley–Tukey/Winograd FFT 1000 x 1000	40	72.8
Winograd FFT 1008 x 1008	6.2	91.6
Nussbaumer–Quandalle FFT 1008 x 1008	4.1	79

*1 pixel – 1 output grid point

Figure 1.1   Relative performance of some two-dimensional Fourier transform algorithms

however, one can obtain computational accuracy that is nearly the same as the direct method. Consideration of computational noise is always a practical factor in judging a fast algorithm, although we shall usually ignore it. Sometimes when the number of operations is reduced, the computational noise is reduced because fewer computations mean that there are fewer sources of noise. In other algorithms, though there are fewer sources of computational noise, the result of the computation may be more sensitive to one or more of them, and so the computational noise in the result may be increased.

Most of this book will be spent studying only a few problems: the problems of linear convolution, cyclic convolution, multidimensional linear convolution, multidimensional cyclic convolution, the discrete Fourier transform, the multidimensional discrete Fourier transforms, the solution of Toeplitz systems, and finding paths in a trellis. Some of the techniques we shall study deserve to be more widely used – multidimensional Fourier transform algorithms can be especially good if one takes the pains to understand the most efficient ones. For example, Figure 1.1 compares some methods of computing a two-dimensional Fourier transform. The improvements in performance come more slowly toward the end of the list. It may not seem very important to reduce the number of multiplications per output cell from six to four after the reduction has already gone from forty to six, but this can be a shortsighted view. It is an additional savings and may be well worth the design time in a large application. In power-limited applications, a potential of a significant reduction in power may itself justify the effort.

There is another important lesson contained in Figure 1.1. An entry, labeled the *hybrid Cooley–Tukey/Winograd FFT*, can be designed to compute a 1000 by 1000-point two-dimensional Fourier transform with forty real multiplications per grid point. This example may help to dispel an unfortunate myth that the discrete Fourier transform is practical only if the blocklength is a power of two. In fact, there is no need to insist that one should use only a power of two blocklength; good algorithms are available for many values of the blocklength.

1.2 Applications of fast algorithms

Very large scale integrated circuits, or chips, are now widely available. A modern chip can easily contain many millions of logic gates and memory cells, and it is not surprising that the theory of algorithms is looked to as a way to efficiently organize these gates on special-purpose chips. Sometimes a considerable performance improvement, either in speed or in power dissipation, can be realized by the choice of algorithm. Of course, a performance improvement in speed can also be realized by increasing the size or the speed of the chip. These latter approaches are more widely understood and easier to design, but they are not the only way to reduce power or chip size.

For example, suppose one devises an algorithm for a Fourier transform that has only one-fifth of the computation of another Fourier transform algorithm. By using the new algorithm, one might realize a performance improvement that can be as real as if one increased the speed or the size of the chip by a factor of five. To realize this improvement, however, the chip designer must reflect the architecture of the algorithm in the architecture of the chip. A naive design can dissipate the advantages by increasing the complexity of indexing, for example, or of data flow between computational steps. An understanding of the fast algorithms described in this book will be required to obtain the best system designs in the era of very large-scale integrated circuits.

At first glance, it might appear that the two kinds of development – fast circuits and fast algorithms – are in competition. If one can build the chip big enough or fast enough, then it seemingly does not matter if one uses inefficient algorithms. No doubt this view is sound in some cases, but in other cases one can also make exactly the opposite argument. Large digital signal processors often create a need for fast algorithms. This is because one begins to deal with signal-processing problems that are much larger than before. Whether competing algorithms for some problem of interest have running times proportional to n^2 or n^3 may be of minor importance when n equals three or four; but when n equals 1000, it becomes critical.

The fast algorithms we shall develop are concerned with digital signal processing, and the applications of the algorithms are as broad as the application of digital signal processing itself. Now that it is practical to build a sophisticated algorithm for signal processing onto a chip, we would like to be able to choose such an algorithm to maximize the performance of the chip. But to do this for a large chip involves a considerable amount of theory. In its totality the theory goes well beyond the material that will be discussed in this book. Advanced topics in logic design and computer architecture, such as parallelism and pipelining, must also be studied before one can determine all aspects of practical complexity.

1.2 Applications of fast algorithms

We usually measure the performance of an algorithm by the number of multiplications and additions it uses. These performance measures are about as deep as one can go at the level of the computational algorithm. At a lower level, we would want to know the area of the chip or the number of gates on it and the time required to complete a computation. Often one judges a circuit by the area–time product. We will not give performance measures at this level because this is beyond the province of the algorithm designer, and entering the province of the chip architecture.

The significance of the topics in this book cannot be appreciated without understanding the massive needs of some processing applications of the near future and the power limitations of other embedded applications now in widespread use. At the present time, applications are easy to foresee that require orders of magnitude more signal processing than current technology can satisfy.

Sonar systems have now become almost completely digital. Though they process only a few kilohertz of signal bandwidth, these systems can use hundreds of millions of multiplications per second and beyond, and even more additions. Extensive racks of digital equipment may be needed for such systems, and yet reasons for even more processing in sonar systems are routinely conceived.

Radar systems also have become digital, but many of the front-end functions are still done by conventional microwave or analog circuitry. In principle, radar and sonar are quite similar, but radar has more than one thousand times as much bandwidth. Thus, one can see the enormous potential for digital signal processing in radar systems.

Seismic processing provides the principal method for exploration deep below the Earth's surface. This is an important method of searching for petroleum reserves. Many computers are already busy processing the large stacks of seismic data, but there is no end to the seismic computations remaining to be done.

Computerized tomography is now widely used to synthetically form images of internal organs of the human body by using X-ray data from multiple projections. Improved algorithms are under study that will reduce considerably the X-ray dosage, or provide motion or function to the imagery, but the signal-processing requirements will be very demanding. Other forms of medical imaging continue to advance, such as those using ultrasonic data, nuclear magnetic resonance data, or particle decay data. These also use massive amounts of digital signal processing.

It is also possible, in principle, to enhance poor-quality photographs. Pictures blurred by camera motion or out-of-focus pictures can be corrected by signal processing. However, to do this digitally takes large amounts of signal-processing computations. Satellite photographs can be processed digitally to merge several images or enhance features, or combine information received on different wavelengths, or create stereoscopic images synthetically. For example, for meteorological research, one can create a moving three-dimensional image of the cloud patterns moving above the Earth's surface based on a sequence of satellite photographs from several aspects. The nondestructive testing of manufactured articles, such as castings, is possible by means of computer-generated internal images based on the response to induced acoustic vibrations.

Other applications for the fast algorithms of signal processing could be given, but these should suffice to prove the point that a need exists and continues to grow for fast signal-processing algorithms.

All of these applications are characterized by computations that are massive but are fairly straightforward and have an orderly structure. In addition, in such applications, once a hardware module or a software subroutine is designed to do a certain task, it is permanently dedicated to this task. One is willing to make a substantial design effort because the design cost is not what matters; the operational performance, both speed and power dissipation, is far more important.

At the same time, there are embedded applications for which power reduction is of critical importance. Wireless handheld and desktop devices and untethered remote sensors must operate from batteries or locally generated power. Chips for these devices may be produced in the millions. Nonrecurring design time to reduce the computations needed by the required algorithm is one way to reduce the power requirements.

1.3 Number systems for computation

Throughout the book, when we speak of the complexity of an algorithm, we will cite the number of multiplications and additions, as if multiplications and additions were fundamental units for measuring complexity. Sometimes one may want to go a little deeper than this and look at how the multiplier is built so that the number of bit operations can be counted. The structure of a multiplier or adder critically depends on how the data is represented. Though we will not study such issues of number representation, a few words are warranted here in the introduction.

To take an extreme example, if a computation involves mostly multiplication, the complexity may be less if the data is provided in the form of logarithms. The additions will now be more complicated; but if there are not too many additions, a savings will result. This is rarely the case, so we will generally assume that the input data is given in its natural form either as real numbers, as complex numbers, or as integers.

There are even finer points to consider in practical digital signal processors. A number is represented by a binary pattern with a finite number of bits; both floating-point numbers and fixed-point numbers are in use. Fixed-point arithmetic suffices for most signal-processing tasks, and so it should be chosen for reasons of economy. This point cannot be stressed too strongly. There is always a temptation to sweep away many design concerns by using only floating-point arithmetic. But if a chip or an algorithm is to be dedicated to a single application for its lifetime – for example, a digital-processing chip to be used in a digital radio or television for the consumer market – it is not the design cost that matters; it is the performance of the equipment, the power dissapation, and the recurring manufacturing costs that matter. Money spent on features to ease the designer's work cannot be spent to increase performance.

A nonnegative integer j smaller than q^m has an m-symbol fixed-point radix-q representation, given by

$$j = j_0 + j_1 q + j_2 q^2 + \cdots + j_{m-1} q^{m-1}, \qquad 0 \le j_i < q.$$

The integer j is represented by the m-tuple of coefficients $(j_0, j_1, \ldots, j_{m-1})$. Several methods are used to handle the sign of a fixed-point number. These are sign-and-magnitude numbers, q-complement numbers, and $(q-1)$-complement numbers. The same techniques can be used for numbers expressed in any base. In a binary notation, q equals two, and the complement representations are called two's-complement numbers and one's-complement numbers.

The sign-and-magnitude convention is easiest to understand. The magnitude of the number is augmented by a special digit called the sign digit; it is zero – indicating a plus sign – for positive numbers and it is one – indicating a minus sign – for negative numbers. The sign digit is treated differently from the magnitude digits during addition and multiplication, in the customary way. The complement notations are a little harder to understand, but often are preferred because the hardware is simpler; an adder can simply add two numbers, treating the sign digit the same as the magnitude digits. The sign-and-magnitude convention and the $(q-1)$-complement convention each leads to the existence of both a positive and a negative zero. These are equal in meaning, but have separate representations. The two's-complement convention in binary arithmetic and the ten's-complement convention in decimal arithmetic have only a single representation for zero.

The $(q-1)$-complement notation represents the negative of a number by replacing digit j, including the sign digit, by $q-1-j$. For example, in nine's-complement notation, the negative of the decimal number $+62$, which is stored as 062, is 937; and the negative of the one's-complement binary number $+011$, which is stored as 0011, is 1100. The $(q-1)$-complement representation has the feature that one can multiply any number by minus one simply by taking the $(q-1)$-complement of each digit.

The q-complement notation represents the negative of a number by adding one to the $(q-1)$-complement notation. The negative of zero is zero. In this convention, the negative of the decimal number $+62$, which is stored as 062, is 938; and the negative of the binary number $+011$, which is stored as 0011, is 1101.

1.4 Digital signal processing

The most important task of digital signal processing is the task of filtering a long sequence of numbers, and the most important device is the digital filter. Normally, the data sequence has an unspecified length and is so long as to appear infinite to the

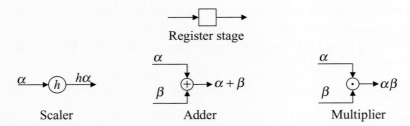

Figure 1.2 Circuit elements

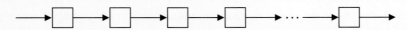

Figure 1.3 A shift register

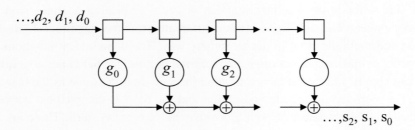

Figure 1.4 A finite-impulse-response filter

processing. The numbers in the sequence are usually either real numbers or complex numbers, but other kinds of number sometimes occur. A digital filter is a device that produces a new sequence of numbers, called the *output sequence*, from the given sequence, now called the *input sequence*. Filters in common use can be constructed out of those circuit elements, illustrated in Figure 1.2, called *shift-register stages*, *adders*, *scalers*, and *multipliers*. A shift-register stage holds a single number, which it displays on its output line. At discrete time instants called *clock times*, the shift-register stage replaces its content with the number appearing on the input line, discarding its previous content. A shift register, illustrated in Figure 1.3, is a number of shift-register stages connected in a chain.

The most important kinds of digital filter that we shall study are those known as *finite-impulse-response* (FIR) *filters* and *autoregressive filters*. A FIR filter is simply a tapped shift register, illustrated in Figure 1.4, in which the output of each stage is multiplied by a fixed constant and all outputs are added together to provide the filter output. The output of the FIR filter is a linear convolution of the input sequence and the sequence describing the filter tap weights. An autoregressive filter is also a tapped shift register, now with the output of the filter fed back to the input, as shown in Figure 1.5.

1.4 Digital signal processing

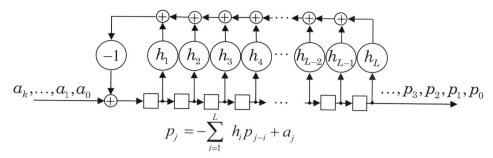

$$p_j = -\sum_{i=1}^{L} h_i p_{j-i} + a_j$$

Figure 1.5 An autoregressive filter

Linear convolution is perhaps the most common computational problem found in signal processing, and we shall spend a great deal of time studying how to implement it efficiently. We shall spend even more time studying ways to compute a cyclic convolution. This may seem a little strange because a cyclic convolution does not often arise naturally in applications. We study it because there are so many good ways to compute a cyclic convolution. Therefore we will develop fast methods of computing long linear convolutions by patching together many cyclic convolutions.

Given the two sequences called the *data sequence*

$$\boldsymbol{d} = \{d_i \mid i = 0, \ldots, N-1\}$$

and the *filter sequence*

$$\boldsymbol{g} = \{g_i \mid i = 0, \ldots, L-1\},$$

where N is the data blocklength and L is the filter blocklength, the linear convolution is a new sequence called the *signal sequence* or the *output sequence*

$$\boldsymbol{s} = \{s_i \mid i = 0, \ldots, L+N-2\},$$

given by the equation

$$s_i = \sum_{k=0}^{N-1} g_{i-k} d_k, \qquad i = 0, \ldots, L+N-2,$$

and $L + N - 1$ is the output blocklength. The convolution is written with the understanding that $g_{i-k} = 0$ if $i - k$ is less than zero. Because each component of $\boldsymbol{d}$ multiplies, in turn, each component of $\boldsymbol{g}$, there are NL multiplications in the obvious implementation of the convolution.

There is a very large body of theory dealing with the design of a FIR filter in the sense of choosing the length L and the tap weights g_i to suit a given application. We are not concerned with this aspect of filter design; our concern is only with fast algorithms for computing the filter output $\boldsymbol{s}$ from the filter $\boldsymbol{g}$ and the input sequence $\boldsymbol{d}$.

A concept closely related to the convolution is the correlation, given by

$$r_i = \sum_{k=0}^{N-1} g_{i+k} d_k, \quad i = 0, \ldots, L+N-2,$$

where $g_{i+k} = 0$ for $i + k \geq L$. The correlation can be computed as a convolution simply by reading one of the two sequences backwards. All of the methods for computing a linear convolution are easily changed into methods for computing the correlation.

We can also express the convolution in the notation of polynomials. Let

$$d(x) = \sum_{i=0}^{N-1} d_i x^i,$$

$$g(x) = \sum_{i=0}^{L-1} g_i x^i.$$

Then

$$s(x) = g(x)d(x),$$

where

$$s(x) = \sum_{i=0}^{L+N-2} s_i x^i.$$

This can be seen by examining the coefficients of the product $g(x)d(x)$. Of course, we can also write

$$s(x) = d(x)g(x),$$

which makes it clear that d and g play symmetric roles in the convolution. Therefore we can also write the linear convolution in the equivalent form

$$s_i = \sum_{k=0}^{L-1} g_k d_{i-k}.$$

Another form of convolution is the *cyclic convolution*, which is closely related to the linear convolution. Given the two sequences d_i for $i = 0, \ldots, n-1$ and g_i for $i = 0, \ldots, n-1$, each of blocklength n, a new sequence s_i' for $i = 0, \ldots, n-1$ of blocklength n now is given by the cyclic convolution

$$s_i' = \sum_{k=0}^{n-1} g_{((i-k))} d_k, \quad i = 0, \ldots, n-1,$$

where the double parentheses denote modulo n arithmetic on the indices (see Section 2.6). That is,

$$((i - k)) = (i - k) \quad \text{modulo } n$$

1.4 Digital signal processing

and

$$0 \leq ((i - k)) < n.$$

Notice that in the cyclic convolution, for every i, every d_k finds itself multiplied by a meaningful value of $g_{((i-k))}$. This is different from the linear convolution where, for some i, d_k will be multiplied by a g_{i-k} whose index is outside the range of definition of $\boldsymbol{g}$, and so is zero.

We can relate the cyclic convolution outputs to the linear convolution as follows. By the definition of the cyclic convolution

$$s'_i = \sum_{k=0}^{n-1} g_{((i-k))} d_k, \qquad i = 0, \ldots, n = 1.$$

We can recognize two kinds of term in the sum: those with $i - k \geq 0$ and those with $i - k < 0$. Those occur when $k \leq i$ and $k > i$, respectively. Hence

$$s'_i = \sum_{k=0}^{i} g_{i-k} d_k + \sum_{k=i+1}^{n-1} g_{n+i-k} d_k.$$

But now, in the first sum, $g_{i-k} = 0$ if $k > i$; and in the second sum, $g_{n+i-k} = 0$ if $k < i$. Hence we can change the limits of the summations as follows:

$$s'_i = \sum_{k=0}^{n-1} g_{i-k} d_k + \sum_{k=0}^{n-1} g_{n+i-k} d_k, \qquad i = 0, \ldots, n-1$$
$$= s_i + s_{n+i}, \qquad i = 0, \ldots, n-1,$$

which relates the cyclic convolution outputs on the left to the linear convolution outputs on the right. We say that coefficients of s with index larger than $n - 1$ are "folded" back into terms with indices smaller than n.

The linear convolution can be computed as a cyclic convolution if the second term above equals zero. This is so if $g_{n+i-k} d_k$ equals zero for all i and k. To ensure this, one can choose n, the blocklength of the cyclic convolution, so that n is larger than $L + N - 2$ (appending zeros to $\boldsymbol{g}$ and $\boldsymbol{d}$ so their blocklength is n). Then one can compute the linear convolution by using an algorithm for computing a cyclic convolution and still get the right answer.

The cyclic convolution can also be expressed as a polynomial product. Let

$$d(x) = \sum_{i=0}^{n-1} d_i x^i,$$

$$g(x) = \sum_{i=0}^{n-1} g_i x^i.$$

Repeat input

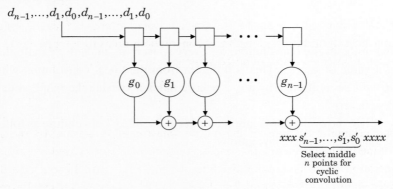

Figure 1.6 Using a FIR filter to form cyclic convolutions

Whereas the linear convolution is represented by

$$s(x) = g(x)d(x),$$

the cyclic convolution is computed by folding back the high-order coefficients of $s(x)$ by writing

$$s'(x) = g(x)d(x) \pmod{x^n - 1}.$$

By the equality modulo $x^n - 1$, we mean that $s'(x)$ is the remainder when $s(x)$ is divided by $x^n - 1$. To reduce $g(x)d(x)$ modulo $x^n - 1$, it suffices to replace x^n by one, or to replace x^{n+i} by x^i wherever a term x^{n+i} with i positive appears. This has the effect of forming the coefficients

$$s'_i = s_i + s_{n+i}, \qquad i = 0, \ldots, n - 1$$

and so gives the coefficients of the cyclic convolution.

From the two forms

$$s'(x) = d(x)g(x) \pmod{x^n - 1}$$
$$= g(x)d(x) \pmod{x^n - 1},$$

it is clear that the roles of d and g are also symmetric in the cyclic convolution. Therefore we have the two expressions for the cyclic convolution

$$s'_i = \sum_{k=0}^{n-1} g_{((i-k))} d_k, \qquad i = 0, \ldots, n - 1$$
$$= \sum_{k=0}^{n-1} d_{((i-k))} g_k, \qquad i = 0, \ldots, n - 1.$$

Figure 1.6 shows a FIR filter that is made to compute a cyclic convolution. To do this, the sequence d is repeated. The FIR filter then produces $3n - 1$ outputs, and within those $3n - 1$ outputs is a consecutive sequence of n outputs that is equal to the cyclic convolution.

A more important technique is to use a cyclic convolution to compute a long linear convolution. Fast algorithms for long linear convolutions break the input datastream into short sections of perhaps a few hundred samples. One section at a time is processed – often as a cyclic convolution – to produce a section of the output datastream. Techniques for doing this are called *overlap* techniques, referring to the fact that nonoverlapping sections of the input datastream cause overlapping sections of the output datastream, while nonoverlapping sections of the output datastream are caused by overlapping sections of the input datastream. Overlap techniques are studied in detail in Chapter 5.

The operation of an autoregressive filter, as was shown in Figure 1.5, also can be described in terms of polynomial arithmetic. Whereas the finite-impulse-response filter computes a polynomial product, an autoregressive filter computes a polynomial division. Specifically, when a finite sequence is filtered by an autoregressive filter (with zero initial conditions), the output sequence corresponds to the quotient polynomial under polynomial division by the polynomial whose coefficients are the tap weights, and at the instant when the input terminates, the register contains the corresponding remainder polynomial. In particular, recall that the output p_j of the autoregressive filter, by appropriate choice of the signs of the tap weights, h_i, is given by

$$p_j = -\sum_{j=1}^{L} h_i p_{j-i} + a_j,$$

where a_j is the jth input symbol and h_i is the weight of the ith tap of the filter. Define the polynomials

$$a(x) = \sum_{j=0}^{n} a_j x^j$$

and

$$h(x) = \sum_{j=0}^{L} h_j x^j,$$

and write

$$a(x) = Q(x)h(x) + r(x),$$

where $Q(x)$ and $r(x)$ are the quotient polynomial and the remainder polynomial under division of polynomials. We conclude that the filter output p_j is equal to the jth coefficient of the quotient polynomial, so $p(x) = Q(x)$. The coefficients of the remainder polynomial $r(x)$ will be left in the stages of the autoregressive filter after the n coefficients a_j are shifted in.

Another computation that is important in signal processing is that of the *discrete Fourier transform* (hereafter called simply the Fourier transform). Let

$\boldsymbol{v} = [v_i \mid i = 0, \ldots, n-1]$ be a vector of complex numbers or a vector of real numbers. The Fourier transform of $\boldsymbol{v}$ is another vector $\boldsymbol{V}$ of length n of complex numbers, given by

$$V_k = \sum_{i=0}^{n-1} \omega^{ik} v_i, \qquad k = 0, \ldots, n-1,$$

where $\omega = e^{-j2\pi/n}$ and $j = \sqrt{-1}$.

Sometimes we write this computation as a matrix–vector product

$$\boldsymbol{V} = \boldsymbol{T}\boldsymbol{v}.$$

When written out, this becomes

$$\begin{bmatrix} V_0 \\ V_1 \\ V_2 \\ \vdots \\ V_{n-1} \end{bmatrix} = \begin{bmatrix} 1 & 1 & 1 & \cdots & 1 \\ 1 & \omega & \omega^2 & \cdots & \omega^{n-1} \\ 1 & \omega^2 & \omega^4 & \cdots & \omega^{2(n-1)} \\ \vdots & \vdots & \vdots & & \vdots \\ 1 & \omega^{n-1} & \omega^{2(n-1)} & \cdots & \omega \end{bmatrix} \begin{bmatrix} v_0 \\ v_1 \\ v_2 \\ \vdots \\ v_{n-1} \end{bmatrix}.$$

If $\boldsymbol{V}$ is the Fourier transform of $\boldsymbol{v}$, then $\boldsymbol{v}$ can be recovered from $\boldsymbol{V}$ by the inverse Fourier transform, which is given by

$$v_i = \frac{1}{n} \sum_{k=0}^{n-1} \omega^{-ik} V_k.$$

The proof is as follows:

$$\sum_{k=0}^{n-1} \omega^{-ik} V_k = \sum_{k=0}^{n-1} \omega^{-ik} \sum_{\ell=0}^{n-1} \omega^{\ell k} v_\ell$$

$$= \sum_{\ell=0}^{n-1} v_\ell \left[\sum_{k=0}^{n-1} \omega^{k(\ell-i)} \right].$$

But the summation on k is clearly equal to n if ℓ is equal to i, while if ℓ is not equal to i the summation becomes

$$\sum_{k=0}^{n-1} (\omega^{(\ell-i)})^k = \frac{1 - \omega^{(\ell-i)n}}{1 - \omega^{(\ell-i)}}.$$

The right side equals zero because $\omega^n = 1$ and the denominator is not zero. Hence

$$\sum_{k=0}^{n-1} \omega^{-ik} V_k = \sum_{\ell=0}^{n-1} v_\ell (n \delta_{i\ell}) = n v_i,$$

where $\delta_{i\ell} = 1$ if $i = \ell$, and otherwise $\delta_{i\ell} = 0$.

There is an important link between the Fourier transform and the cyclic convolution. This link is known as the *convolution theorem* and goes as follows. The vector e is given by the cyclic convolution of the vectors f and g:

$$e_i = \sum_{\ell=0}^{n-1} f_{((i-\ell))} g_\ell, \qquad i = 0, \ldots, n-1,$$

if and only if the Fourier transforms satisfy

$$E_k = F_k G_k, \qquad k = 0, \ldots, n-1.$$

This holds because

$$e_i = \sum_{\ell=0}^{n-1} f_{((i-\ell))} \left[\frac{1}{n} \sum_{k=0}^{n-1} \omega^{-k\ell} G_k \right]$$

$$= \frac{1}{n} \sum_{k=0}^{n-1} \omega^{-ik} G_k \left[\sum_{\ell=0}^{n-1} \omega^{(i-\ell)k} f_{((i-\ell))} \right] = \frac{1}{n} \sum_{k=0}^{n-1} \omega^{-ik} G_k F_k.$$

Because e is the inverse Fourier transform of E, we conclude that $E_k = G_k F_k$.

There are also two-dimensional Fourier transforms, which are useful for processing two-dimensional arrays of data, and multidimensional Fourier transforms, which are useful for processing multidimensional arrays of data. The two-dimensional Fourier transform on an n' by n'' array is

$$V_{k'k''} = \sum_{i'=0}^{n'-1} \sum_{i''=0}^{n''-1} \omega^{i'k'} \mu^{i''k''} v_{i'i''}, \qquad \begin{aligned} k' &= 0, \ldots, n'-1, \\ k'' &= 0, \ldots, n''-1, \end{aligned}$$

where $\omega = e^{-j2\pi/n'}$ and $\mu = e^{-j2\pi/n''}$. Chapter 12 is devoted to the two-dimensional Fourier transforms.

1.5 History of fast signal-processing algorithms

The telling of the history of fast signal-processing algorithms begins with the publication in 1965 of the fast Fourier transform (FFT) algorithm of Cooley and Tukey, although the history itself starts much earlier, indeed, with Gauss. The Cooley–Tukey paper appeared at just the right time and served as a catalyst to bring the techniques of signal processing into a new arrangement. Stockham (1966) soon noted that the FFT led to a good way to compute convolutions. Digital signal processing technology could immediately exploit the FFT, and so there were many applications, and the Cooley–Tukey paper was widely studied. A few years later, it was noted that there was an earlier FFT algorithm, quite different from the Cooley–Tukey FFT, due to Good (1960) and Thomas (1963). The Good–Thomas FFT algorithm had failed to attract much attention at the time it was published. Later, a more efficient though more complicated algorithm was published by Winograd (1976, 1978), who also provided a much deeper understanding of what it means to compute the Fourier transform.

The radix-two Cooley–Tukey FFT is especially elegant and efficient, and so is very popular. This has led some to the belief that the discrete Fourier transform is practical only if the blocklength is a power of two. This belief tends to result in the FFT algorithm dictating the design parameters of an application rather than the application dictating the choice of FFT algorithm. In fact, there are good FFT algorithms for just about any blocklength.

The Cooley–Tukey FFT, in various guises, has appeared independently in other contexts. Essentially the same idea is known as the Butler matrix (1961) when it is used as a method of wiring a multibeam phased-array radar antenna.

Fast convolution algorithms of small blocklength were first constructed by Agarwal and Cooley (1977) using clever insights but without a general technique. Winograd (1978) gave a general method of construction and also proved important theorems concerning the nonexistence of better convolution algorithms in the real field or the complex field. Agarwal and Cooley (1977) also found a method to break long convolutions into short convolutions using the Chinese remainder theorem. Their method works well when combined with the Winograd algorithm for short convolutions.

The earliest idea of modern signal processing that we label as a fast algorithm came much earlier than the FFT. In 1947 the Levinson algorithm was published as an efficient method of solving certain Toeplitz systems of equations. Despite its great importance in the processing of seismic signals, the literature of the Levinson algorithm remained disjoint from the literature of the FFT for many years. Generally, the early literature does not distinguish carefully between the Levinson algorithm as a computational procedure and the filtering problem to which the algorithm might be applied. Similarly, the literature does not always distinguish carefully between the FFT as a computational procedure and the discrete Fourier transform to which the FFT is applied, nor between the Viterbi algorithm as a computational procedure and the minimum-distance pathfinding problem to which the Viterbi algorithm is applied.

Problems for Chapter 1

1.1 Construct an algorithm for the two-point real cyclic convolution

$$(s_1 x + s_0) = (g_1 x + g_0)(d_1 x + d_0) \pmod{x^2 - 1}$$

that uses two multiplications and four additions. Computations involving only g_0 and g_1 need not be counted under the assumption that g_0 and g_1 are constants, and these computations need be done only once off-line.

1.2 Using the result of Problem 1.1, construct an algorithm for the two-point complex cyclic convolution that uses only six real multiplications.

1.3 Construct an algorithm for the three-point Fourier transform

$$V_k = \sum_{i=0}^{2} \omega^{ik} v_i, \qquad k = 0, 1, 2$$

that uses two real multiplications.

1.4 Prove that there does not exist an algorithm for multiplying two complex numbers that uses only two real multiplications. (A throughful "proof" will struggle with the meaning of the term "multiplication.")

1.5 **a** Suppose you are given a device that computes the linear convolution of two fifty-point sequences. Describe how to use it to compute the cyclic convolution of two fifty-point sequences.

b Suppose you are given a device that computes the cyclic convolution of two fifty-point sequences. Describe how to use it to compute the linear convolution of two fifty-point sequences.

1.6 Prove that one can compute a correlation as a convolution by writing one of the sequences backwards, possibly padding a sequence with a string of zeros.

1.7 Show that any algorithm for computing x^{31} that uses only additions, subtractions, and multiplications must use at least seven multiplications, but that if division is allowed, then an algorithm exists that uses a total of six multiplications and divisions.

1.8 Another algorithm for complex multiplication is given by

$$e = ac - bd,$$
$$f = (a+b)(c+d) - ac - bd.$$

Express this algorithm in a matrix form similar to that given in the text. What advantages or disadvantages are there in this algorithm?

1.9 Prove the "translation property" of the Fourier transform. If $\{v_j\} \leftrightarrow \{V_k\}$ is a Fourier transform pair, then the following are Fourier transform pairs:

$$\{\omega^i v_i\} \leftrightarrow \{V_{((k+1))}\},$$
$$\{v_{((i-1))}\} \leftrightarrow \{\omega^k V_k\}.$$

1.10 Prove that the "cyclic correlation" in the real field satisfies the Fourier transform relationship

$$G_k D_k^* \leftrightarrow \sum_{k=0}^{n-1} g_{((i+k))} d_k.$$

1.11 Given two real vectors v' and v'', show how to recover their individual Fourier transforms from the Fourier transform of the sum vector $v = v' + jv''$.

Notes for Chapter 1

A good history of the origins of the fast Fourier transform algorithms is given in a paper by Cooley, Lewis, and Welch (1967). The basic theory of digital signal processing can be found in many books, including the books by Oppenheim and Shafer (1975), Rabiner and Gold (1975), and Proakis and Manolakis (2006).

Algorithms for complex multiplication using three real multiplications became generally known in the late 1950s, but the origin of these algorithms is a little hazy. The matrix multiplication algorithm we have given is due to Winograd (1968).

2 Introduction to abstract algebra

Good algorithms are elegant algebraic identities. To construct these algorithms, we must be familiar with the powerful structures of number theory and of modern algebra. The structures of the set of integers, of polynomial rings, and of Galois fields will play an important role in the design of signal-processing algorithms. This chapter will introduce those mathematical topics of algebra that will be important for later developments but that are not always known to students of signal processing. We will first study the mathematical structures of groups, rings, and fields. We shall see that a discrete Fourier transform can be defined in many fields, though it is most familiar in the complex field. Next, we will discuss the familiar topics of matrix algebra and vector spaces. We shall see that these can be defined satisfactorily in any field. Finally, we will study the integer ring and polynomial rings, with particular attention to the euclidean algorithm and the Chinese remainder theorem in each ring.

2.1 Groups

A group is a mathematical abstraction of an algebraic structure that appears frequently in many concrete forms. The abstract idea is introduced because it is easier to study all mathematical systems with a common structure at once, rather than to study them one by one.

Definition 2.1.1 *A group G is a set together with an operation (denoted by $*$) satisfying four properties.*

1 (***Closure***) *For every a and b in the set, $c = a * b$ is in the set.*
2 (***Associativity***) *For every a, b, and c in the set,*

$$a * (b * c) = (a * b) * c.$$

3 (***Identity***) *There is an element e called the identity element that satisfies*

$$a * e = e * a = a$$

for every a in the set G.

*	e	g_1	g_2	g_3	g_4
e	e	g_1	g_2	g_3	g_4
g_1	g_1	g_2	g_3	g_4	e
g_2	g_2	g_3	g_4	e	g_1
g_3	g_3	g_4	e	g_1	g_2
g_4	g_4	e	g_1	g_2	g_3

+	0	1	2	3	4
0	0	1	2	3	4
1	1	2	3	4	0
2	2	3	4	0	1
3	3	4	0	1	2
4	4	0	1	2	3

Figure 2.1 Example of a finite group

4 (*Inverses*) *If a is in the set, then there is some element b in the set called an inverse of a such that*

$$a * b = b * a = e.$$

A group that has a finite number of elements is called a *finite group*. The number of elements in a finite group G is called the *order* of G. An example of a finite group is shown in Figure 2.1. The same group is shown twice, but represented by two different notations. Whenever two groups have the same structure but a different representation, they are said to be *isomorphic*.[1]

Some groups satisfy the property that for all a and b in the group

$$a * b = b * a.$$

This is called the *commutative property*. Groups with this additional property are called *commutative groups* or *abelian groups*. We shall usually deal with abelian groups.

In an abelian group, the symbol for the group operation is commonly written + and is called addition (even though it might not be the usual arithmetic addition). Then the identity element e is called "zero" and is written 0, and the inverse element of a is written $-a$ so that

$$a + (-a) = (-a) + a = 0.$$

Sometimes the symbol for the group operation is written $*$ and is called multiplication (even though it might not be the usual arithmetic multiplication). In this case, the identity element e is called "one" and is written 1, and the inverse element of a is written a^{-1} so that

$$a * a^{-1} = a^{-1} * a = 1.$$

---

[1] In general, any two algebraic systems that have the same structure but are represented differently are called *isomorphic*.

2.1 Groups

Theorem 2.1.2 *In every group, the identity element is unique. Also, the inverse of each group element is unique, and* $(a^{-1})^{-1} = a$.

Proof Let e and e' be identity elements. Then $e = e * e' = e'$. Next, let b and b' be inverses for element a; then

$$b = b * (a * b') = (b * a) * b' = b'$$

so $b = b'$. Finally, for any a, $a^{-1} * a = a * a^{-1} = e$, so a is an inverse for a^{-1}. But because inverses are unique, $(a^{-1})^{-1} = a$. □

Many common groups have an infinite number of elements. Examples are the set of integers, denoted $\mathbf{Z} = \{0, \pm 1, \pm 2, \pm 3, \ldots\}$, under the operation of addition; the set of positive rationals under the operation of multiplication;[2] and the set of two by two, real-valued matrices under the operation of matrix addition. Many other groups have only a finite number of elements. Finite groups can be quite intricate.

Whenever the group operation is used to combine the same element with itself two or more times, an exponential notation can be used. Thus $a^2 = a * a$ and

$$a^k = a * a * \cdots * a,$$

where there are k copies of a on the right.

A *cyclic group* is a finite group in which every element can be written as a power of some fixed element called a *generator* of the group. Every cyclic group has the form

$$G = \{a^0, a^1, a^2, \ldots, a^{q-1}\},$$

where q is the order of G, a is a generator of G, a^0 is the identity element, and the inverse of a^i is a^{q-i}. To actually form a group in this way, it is necessary that $a^q = a^0$. Because, otherwise, if $a^q = a^i$ with $i \neq 0$, then $a^{q-1} = a^{i-1}$, and there are fewer than q distinct elements, contrary to the definition.

An important cyclic group with q elements is the group denoted by the label $\mathbf{Z}/\langle q \rangle$, by $\mathbf{Z}_q$, or by $\mathbf{Z}/q\mathbf{Z}$, and given by

$$\mathbf{Z}/\langle q \rangle = \{0, 1, 2, \ldots, q - 1\},$$

and the group operation is modulo q addition. In formal mathematics, $\mathbf{Z}/\langle q \rangle$ would be called a *quotient group* because it "divides out" multiples of q from the original group $\mathbf{Z}$.

For example,

$$\mathbf{Z}/\langle 6 \rangle = \{0, 1, 2, \ldots, 5\},$$

and $3 + 4 = 1$.

---

[2] This example is a good place for a word of caution about terminology. In a general abelian group, the group operation is usually called "addition" but is not necessarily ordinary addition. In this example, it is ordinary multiplication.

The group $\mathbf{Z}/\langle q \rangle$ can be chosen as a standard prototype of a cyclic group with q elements. There is really only one cyclic group with q elements; all others are isomorphic copies of it differing in notation but not in structure. Any other cyclic group G with q elements can be mapped into $\mathbf{Z}/\langle q \rangle$ with the group operation in G replaced by modulo q addition. Any properties of the structure in G are also true in $\mathbf{Z}/\langle q \rangle$, and the converse is true as well.

Given two groups G' and G'', it is possible to construct a new group G, called the *direct product*,[3] or, more simply, the *product* of G' and G'', and written $G = G' \times G''$. The elements of G are pairs of elements (a', a''), the first from G' and the second from G''. The group operation in the product group G is defined by

$$(a', a'') * (b', b'') = (a' * b', a'' * b'').$$

In this formula, $*$ is used three times with three meanings. On the left side it is the group operation in G, and on the right side it is the group operation in G' or in G'', respectively.

For example, for $\mathbf{Z}_2 \times \mathbf{Z}_3$, we have the set

$$\mathbf{Z}_2 \times \mathbf{Z}_3 = \{(0,0), (0,1), (0,2), (1,0), (1,1), (1,2)\}.$$

A typical entry in the addition table for $\mathbf{Z}_2 \times \mathbf{Z}_3$ is

$$(1, 2) + (0, 2) = (1, 1)$$

with $\mathbf{Z}_2$ addition in the first position and $\mathbf{Z}_3$ addition in the second position. Notice that $\mathbf{Z}_2 \times \mathbf{Z}_3$ is itself a cyclic group generated by the element $(1, 1)$. Hence $\mathbf{Z}_2 \times \mathbf{Z}_3$ is isomorphic to $\mathbf{Z}_6$. The reason this is so is that two and three have no common integer factor. In contrast, $\mathbf{Z}_3 \times \mathbf{Z}_3$ is not isomorphic to $\mathbf{Z}_9$.

Let G be a group and let H be a subset of G. Then H is called a *subgroup* of G if H is itself a group with respect to the restriction of $*$ to H. As an example, in the set of integers (positive, negative, and zero) under addition, the set of even integers is a subgroup, as is the set of multiples of three.

One way to get a subgroup H of a finite group G is to take any element h from G and let H be the set of elements obtained by multiplying h by itself an arbitrary number of times to form the sequence of elements $h, h^2, h^3, h^4, \ldots$. The sequence must eventually repeat because G is a finite group. The first element repeated must be h itself, and the element in the sequence just before h must be the group identity element because the construction gives a cyclic group. The set H is called the *cyclic subgroup* generated by h. The number q of elements in the subgroup H satisfies $h^q = 1$, and q is called the *order* of the element h. The set of elements $h, h^2, h^3, \ldots, h^q = 1$ is called a *cycle* in the group G, or the *orbit* of h.

---

[3] If the group G is an abelian group, the direct product is often called the *direct sum*, and denoted $\oplus$. For this reason, one may also use the notation $\mathbf{Z}_2 \oplus \mathbf{Z}_3$.

To prove that a nonempty subset H of G is a subgroup of G, it is necessary only to check that $a*b$ is in H whenever a and b are in H and that the inverse of each a in H is also in H. The other properties required of a group will then be inherited from the group G. If the group is finite, then even the inverse property is satisfied automatically, as we shall see later in the discussion of cyclic subgroups.

Given a finite group G and a subgroup H, there is an important construction known as the *coset decomposition* of G that illustrates certain relationships between H and G. Let the elements of H be denoted by $h_1, h_2, h_3, \ldots$, and choose h_1 to be the identity element. Construct the array as follows. The first row consists of the elements of H, with the identity at the left and every other element of H appearing once. Choose any element of G not appearing in the first row. Call it g_2 and use it as the first element of the second row. The rest of the elements of the second row are obtained by multiplying each element in the first row by g_2. Then construct a third, fourth, and fifth row similarly, each time choosing a previously unused group element for the first element in the row. Continue in this way until, at the completion of a row, all the group elements appear somewhere in the array. This occurs when there is no unused element remaining. The process must stop, because G is finite. The final array is

$$
\begin{array}{ccccccc}
h_1 = 1 & h_2 & h_3 & h_4 & \cdots & h_n \\
g_2 * h_1 = g_2 & g_2 * h_2 & g_2 * h_3 & g_2 * h_4 & \cdots & g_2 * h_n \\
g_3 * h_1 = g_3 & g_3 * h_2 & g_3 * h_3 & g_3 * h_4 & \cdots & g_3 * h_n \\
\vdots & \vdots & \vdots & \vdots & \vdots & \vdots \\
g_m * h_1 = g_m & g_m * h_2 & g_m * h_3 & g_m * h_4 & \cdots & g_m * h_n.
\end{array}
$$

The first element on the left of each row is known as a *coset leader*. Each row in the array is known as a *left coset*, or simply as a *coset* when the group is abelian. If the coset decomposition is defined instead with the new elements of G multiplied on the right, the rows are known as *right cosets*. The coset decomposition is always rectangular with all rows completed because it is constructed that way. The next theorem says that every element of G appears exactly once in the final array.

Theorem 2.1.3 *Every element of G appears once and only once in a coset decomposition of G.*

Proof Every element appears at least once because, otherwise, the construction is not halted. We now prove that an element cannot appear twice in the same row and then prove that an element cannot appear in two different rows.

Suppose that two elements in the same row, $g_i * h_j$ and $g_i * h_k$, are equal. Then multiplying each by g_i^{-1} gives $h_j = h_k$. This is a contradiction because each subgroup element appears only once in the first row.

Suppose that two elements in different rows, $g_i * h_j$ and $g_k * h_\ell$, are equal with k less than i. Multiplying on the right by h_j^{-1} gives $g_i = g_k * h_\ell * h_j^{-1}$. Then g_i is in the kth coset, because $h_\ell * h_j^{-1}$ is in the subgroup H. This contradicts the rule of construction that coset leaders must be previously unused. Thus the same element cannot appear twice in the array. □

Corollary 2.1.4 *If H is a subgroup of G, then the number of elements in H divides the number of elements in G. That is,*

(Order of H) (Number of cosets of G with respect to H) = (Order of G).

Proof Follows immediately from the rectangular structure of the coset decomposition. □

Theorem 2.1.5 (Lagrange) *The order of a finite group is divisible by the order of any of its elements.*

Proof The group contains the cyclic subgroup generated by any element; then Corollary 2.1.4 proves the theorem. □

Corollary 2.1.6 *For any a in a group with q elements, $a^q = 1$.*

Proof By Theorem 2.1.5, the order of a divides q, so $a^q = 1$. □

2.2 Rings

The next algebraic structure we will need is that of a ring. A ring is an abstract set that is an abelian group and also has an additional operation.

Definition 2.2.1 *A ring R is a set with two operations defined – the first called addition (denoted by $+$) and the second called multiplication (denoted by juxtaposition) – and the following axioms are satisfied.*
1 *R is an abelian group under addition.*
2 *(Closure) For any a, b in R, the product ab is in R.*
3 *(Associativity)*

$a(bc) = (ab)c.$

4 *(Distributivity)*

$a(b + c) = ab + ac$

$(b + c)a = ba + ca.$

The distributivity property in the definition of a ring links the addition and multiplication operations. The addition operation is always required to be commutative in a ring, but the multiplication operation need not be commutative. A *commutative ring* is one in which multiplication is commutative; that is, $ab = ba$ for all a, b in R.

Some important examples of rings are the ring of integers, denoted $\mathbf{Z}$, and the ring of integers under modulo q arithmetic, denoted $\mathbf{Z}/\langle q \rangle$. The ring $\mathbf{Z}/\langle q \rangle$ is an example of a quotient ring because it uses modulo q arithmetic to "divide out" q from $\mathbf{Z}$. We have already seen the various $\mathbf{Z}/\langle q \rangle$ as examples of groups under addition. Because there is also a multiplication operation that behaves properly on these sets, they are also rings. Another example of a ring is the set of all polynomials in x with rational coefficients. This ring is denoted $\mathbf{Q}[x]$. It is easy to verify that $\mathbf{Q}[x]$ has all the properties required of a ring. Similarly, the set of all polynomials with real coefficients is a ring, denoted $\mathbf{R}[x]$. The ring $\mathbf{Q}[x]$ is a subring of $\mathbf{R}[x]$. The set of all two by two matrices over the reals is an example of a noncommutative ring, as can be easily checked.

Several consequences that are well-known properties of familiar rings are implied by the axioms as can be proved as follows.

Theorem 2.2.2 *For any elements a, b in a ring R,*
(i) $a0 = 0a = 0$;
(ii) $a(-b) = (-a)b = -(ab)$.

Proof
(i) $a0 = a(0+0) = a0 + a0$. Hence adding $-a0$ to both sides gives $0 = a0$. The second half of (i) is proved the same way.
(ii) $0 = a0 = a(b-b) = ab + a(-b)$. Hence

$$a(-b) = -(ab).$$

The second half of (ii) is proved the same way. □

The addition operation in a ring has an identity called "zero." The multiplication operation need not have an identity, but if there is an identity, it is unique. A ring that has an identity under multiplication is called a *ring with identity*. The identity is called "one" and is denoted by 1. Then

$$1a = a1 = a$$

for all a in R.

Every element in a ring has an inverse under the addition operation. Under the multiplication operation, there need not be any inverses, but in a ring with identity, inverses may exist. That is, given an element a, there may exist an element b with $ab = 1$. If so, b is called a *right inverse* for a. Similarly, if there is an element c such that $ca = 1$, then c is called a *left inverse* for a.

Theorem 2.2.3 *In a ring with identity:*
  (i) The identity is unique.
  (ii) If an element a has both a right inverse b and a left inverse c, then $b = c$. In this case the element a is said to have an inverse (denoted a^{-1}). The inverse is unique.
(iii) $(a^{-1})^{-1} = a$.

Proof The proof is similar to that used in Theorem 2.1.2. □

The identity of a ring, if there is one, can be added to itself or subtracted from itself any number of times to form the doubly infinite sequence

$$\ldots, -(1+1+1), -(1+1), -1, 0, 1, (1+1), (1+1+1), \ldots.$$

These elements of the ring are called the *integers* of the ring and are sometimes denoted more simply as $0, \pm 1, \pm 2, \pm 3, \pm 4, \ldots$. There may be a finite number or an infinite number of integers. The number of integers in a ring with identity, if this is finite, is called the *characteristic* of the ring. If the characteristic of a ring is not finite then, by definition, the ring has characteristic zero. If the characteristic is the finite number q, then we can write the integers of the ring as the set

$$\{1, 1+1, 1+1+1, \ldots\},$$

denoting these more simply with the notation of the usual integers

$$\{1, 2, 3, \ldots, q-1, 0\}.$$

This subset is a subgroup of the additive group of the ring; in fact, it is the cyclic subgroup generated by the element one. Hence addition of the integers of a ring is modulo q addition whenever the characteristic of the ring is finite. If the characteristic is infinite, then the integers of the ring add as integers. Hence every ring with identity contains a subset that behaves under addition either as **Z** or as $\mathbf{Z}/\langle q \rangle$. In fact, it also behaves in this way under multiplication, because if α and β are each a finite sum of the ring identity one, then

$$\alpha * \beta = \beta + \beta + \cdots + \beta,$$

where there are α copies of β on the right. Because addition of integers in R behaves like addition in **Z** or in $\mathbf{Z}/\langle q \rangle$, then so does multiplication of integers in R.

Within a ring R, any element α can be raised to an integer power; the notation α^m simply means the product of m copies of α. If the ring has an identity and the number of integers of the ring is a prime, then the following theorem is sometimes useful for simplifying powers of sums.

Theorem 2.2.4 *Let p be a prime, and let R be a commutative ring with p integers. Then for any positive integer m and for any two elements α and β in R,*

$$(a \pm \beta)^{p^m} = \alpha^{p^m} \pm \beta^{p^m},$$

2.2 Rings

and by direct extension,

$$(\Sigma_i \alpha_i)^{p^m} = \Sigma_i \alpha_i^{p^m}$$

for any set of α_i in R.

Proof The binomial theorem,

$$(\alpha \pm \beta)^p = \sum_{i=0}^{p} \binom{p}{i} \alpha^i (\pm \beta)^{p-i}$$

holds in the ring of real numbers, so it must also hold in any commutative ring with identity because the coefficient $\binom{p}{i}$ merely counts how many terms of the expansion have i copies of α and $p - i$ copies of β. This does not depend on the specific ring. However, $\binom{p}{i}$ is interpreted in R as a sum of this many copies of the ring identity. Recall that in a ring with p integers, all integer arithmetic is modulo p. Next, observe that

$$\binom{p}{i} = \frac{p!}{i!(p-i)!} = \frac{p(p-1)!}{i!(p-i)!}$$

is an integer and p is a prime. Hence, as integers, the denominator divides $(p - 1)!$ for $i = 1, \ldots, p - 1$, and so $\binom{p}{i}$ must be a multiple of p. That is, $\binom{p}{i} = 0 \pmod{p}$ for $i = 1, \ldots, p - 1$. Then $(\alpha \pm \beta)^p = \alpha^p + (\pm \beta)^p$. Finally, either $p = 2$ and $-\beta = \beta$ so that $(\pm \beta)^2 = \pm \beta^2$, or p is odd and $(\pm \beta)^p = \pm \beta^p$. Then

$$(\alpha \pm \beta)^p = \alpha^p \pm \beta^p,$$

which proves the theorem for $m = 1$.

This now can again be raised to the pth power,

$$((\alpha \pm \beta)^p)^p = (\alpha^p \pm \beta^p)^p,$$

and because the statement is true for $m = 1$, we have

$$(\alpha^p \pm \beta^p)^p = \alpha^{p^2} \pm \beta^{p^2}.$$

This can be repeated multiple times to get

$$(\alpha \pm \beta)^{p^m} = \alpha^{p^m} \pm \beta^{p^m},$$

which completes the proof of the theorem. □

An element that has an inverse under multiplication in a ring with identity is called a *unit* of the ring. The set of all units is closed under multiplication, because if a and b are units, then $c = ab$ has the inverse $c^{-1} = b^{-1}a^{-1}$.

Theorem 2.2.5
(i) *Under ring multiplication, the set of units of a ring forms a group.*
(ii) *If $c = ab$ and c is a unit, then a has a right inverse and b has a left inverse.*
(iii) *If $c = ab$ and a has no right inverse or b has no left inverse, then c is not a unit.*

Proof Exercise. □

There are many familiar examples of rings such as the following.
1. The set of all real numbers under the usual addition and multiplication is a commutative ring with identity. Every nonzero element is a unit.
2. The set **Z** of all integers under the usual addition and multiplication is a commutative ring with identity. The only units are ± 1.
3. The set of all n by n matrices with real-valued elements under matrix addition and matrix multiplication is a noncommutative ring with identity. The identity is the n by n identity matrix. The units are the nonsingular matrices.
4. The set of all n by n matrices with integer-valued elements under matrix addition and matrix multiplication is a noncommutative ring with identity.
5. The set of all polynomials in x with real-valued coefficients is a commutative ring with identity under polynomial addition and polynomial multiplication. The identity is the zero-degree polynomial $p(x) = 1$. The units are the nonzero polynomials of degree zero.

2.3 Fields

Loosely speaking, an abelian group is a set in which one can "add" and "subtract," and a ring is a set in which one can "add," "subtract," and "multiply." A more powerful algebraic structure, known as a field, is a set in which one can "add," "subtract," "multiply," and "divide."

Definition 2.3.1 *A field F is a set containing at least two elements that has two operations defined on it – addition and multiplication – such that the following axioms are satisfied:*
1. *the set is an abelian group under addition;*
2. *the set is closed under multiplication, and the set of nonzero elements is an abelian group under multiplication;*
3. *the distributive law,*

$$(a+b)c = ac + bc,$$

holds for all a, b, and c in the field.

In a field, it is conventional to denote the identity element under addition by 0 and to call it zero; to denote the additive inverse of a by $-a$; to denote the identity element under multiplication by 1 and to call it one; and to denote the multiplicative inverse of a by a^{-1}. By subtraction $(a - b)$, one means $a + (-b)$; by division (a/b), one means $b^{-1}a$.

2.3 Fields

The following examples of fields are well known.
1. R: the set of real numbers.
2. C: the set of complex numbers.
3. Q: the set of rational numbers.

These fields all have an infinite number of elements. There are many other, less-familiar fields with an infinite number of elements. One that is easy to describe is known as the field of *complex rationals*, denoted $Q(j)$. It is given by

$$Q(j) = \{a + jb\},$$

where a and b are rationals. Addition and multiplication are as complex numbers. With this definition, $Q(j)$ satisfies the requirements of Definition 2.3.1, and so it is a field.

There are also fields with a finite number of elements, and we have uses for these as well. A field with q elements, if there is one, is called a *finite field* or a *Galois field* and is denoted by the label $GF(q)$.

Every field must have an element zero and an element one, thus every field has at least two elements. In fact, with the addition and multiplication tables,

+	0	1
0	0	1
1	1	0

·	0	1
0	0	0
1	0	1

these elements suffice to form a field. This is the field known as $GF(2)$. No other field with two elements exists.

Finite fields can be described by writing out their addition and multiplication tables. Subtraction and division are defined implicitly by the addition and multiplication tables. Later, we shall study finite fields in detail. Here we give three more examples.

The field $GF(3) = \{0, 1, 2\}$ with the operations

+	0	1	2
0	0	1	2
1	1	2	0
2	2	0	1

·	0	1	2
0	0	0	0
1	0	1	2
2	0	2	1

The field $GF(4) = \{0, 1, 2, 3\}$ with the operations

+	0	1	2	3
0	0	1	2	3
1	1	0	3	2
2	2	3	0	1
3	3	2	1	0

·	0	1	2	3
0	0	0	0	0
1	0	1	2	3
2	0	2	3	1
3	0	3	1	2

Notice that multiplication in $GF(4)$ is *not* modulo 4 multiplication, and addition in $GF(4)$ is *not* modulo 4 addition. The field $GF(5) = \{0, 1, 2, 3, 4\}$ with the operations

+	0	1	2	3	4
0	0	1	2	3	4
1	1	2	3	4	0
2	2	3	4	0	1
3	3	4	0	1	2
4	4	0	1	2	3

·	0	1	2	3	4
0	0	0	0	0	0
1	0	1	2	3	4
2	0	2	4	1	3
3	0	3	1	4	2
4	0	4	3	2	1

These examples are very small fields. Much larger fields such as $GF(2^{16}+1)$ can arise in applications.

In any field, whether a finite field or an infinite field, almost all familiar computational procedures are valid. This is because most of the methods in use for the real field or the complex field depend only on the formal structure, given in Definition 2.3.1, and not on any special structure of a particular field.

There is even a Fourier transform in any field F, given by

$$V_k = \sum_{i=0}^{n-1} \omega^{ik} v_i, \qquad k = 0, \ldots, n-1,$$

where ω is now an nth root of unity in F, and v and V are vectors of length n in F. A Fourier transform of blocklength n exists in F only if there is an nth root of unity ω in F. As long as the Fourier transform exists, it will behave as expected. In particular, there will be an inverse Fourier transform, and the convolution theorem will hold because, if we look to the proof of these facts, we will see that no property of F is used except that F satisfies the formal structure given in Definition 2.3.1.

Similarly, a two-dimensional Fourier transform in the field F

$$V_{k'k''} = \sum_{i'=0}^{n'-1} \sum_{i''=0}^{n''-1} \omega^{i'k'} \mu^{i''k''} v_{i'i''}, \qquad \begin{array}{l} k' = 0, \ldots, n'-1 \\ k'' = 0, \ldots, n''-1 \end{array}$$

exists whenever the field F contains both an element ω of order n' and an element μ of order n'', possibly with $n' = n''$ and $\omega = \mu$. If such elements do not exist, then this two-dimensional Fourier transform does not exist.

As an example of the Fourier transform, in $GF(5)$ the element two has order four. Therefore we have the four-point Fourier transform in $GF(5)$

$$V_k = \sum_{i=0}^{3} 2^{ik} v_i, \qquad k = 0, \ldots, 3.$$

We also have the four by four, two-dimensional Fourier transform

$$V_{k'k''} = \sum_{i'=0}^{3} \sum_{i''=0}^{3} 2^{i'k'} 2^{i''k''} v_{i'i''}, \qquad \begin{array}{l} k' = 0, \ldots, 3 \\ k'' = 0, \ldots, 3 \end{array}.$$

2.3 Fields

Because the components of v and V are elements of $GF(5)$, all arithmetic is the arithmetic of $GF(5)$. If

$$v = \begin{bmatrix} 4 \\ 3 \\ 2 \\ 1 \end{bmatrix},$$

then the one-dimensional Fourier transform of v in $GF(5)$ is

$$\begin{bmatrix} V_0 \\ V_1 \\ V_2 \\ V_3 \end{bmatrix} = \begin{bmatrix} 1 & 1 & 1 & 1 \\ 1 & 2 & 4 & 3 \\ 1 & 4 & 1 & 4 \\ 1 & 3 & 4 & 2 \end{bmatrix} \begin{bmatrix} 4 \\ 3 \\ 2 \\ 1 \end{bmatrix} = \begin{bmatrix} 0 \\ 1 \\ 2 \\ 3 \end{bmatrix}.$$

Similarly, the array

$$v = \begin{bmatrix} 4 & 3 & 2 & 1 \\ 3 & 1 & 1 & 4 \\ 2 & 4 & 3 & 1 \\ 1 & 3 & 2 & 2 \end{bmatrix}$$

has a two-dimensional Fourier transform obtained by taking the one-dimensional Fourier transform of each row, then the one-dimensional Fourier transform of each column.

Definition 2.3.2 *Let F be a field. A subset of F is called a subfield when the subset is a field under the inherited addition and multiplication. The original field F is then called an extension field of the subfield.*

The field of rationals is a subfield of the field of reals, which, in turn, is a subfield of the complex field. The field of complex rationals is not a subfield of the real field, but it is a subfield of the complex field. The finite field $GF(2)$ is readily seen to be a subfield of $GF(4)$ because in $GF(4)$, the elements zero and one add and multiply just as they do in $GF(2)$. However, $GF(2)$ is not a subfield of $GF(3)$ nor of $GF(5)$.

To prove that a subset of a finite field is a subfield, it is necessary to prove only that it contains a nonzero element and that it is closed under addition and multiplication. All other necessary properties are inherited from F. Inverses under addition or multiplication of an element β are contained in the cyclic group generated by β under the operation of addition or multiplication.

Each field contains as a subset the set of its integers

$$\{\ldots, -(1+1+1), -(1+1), -1, 0, 1, (1+1), (1+1+1), \ldots\}.$$

The integers of the field are usually written simply as $0, \pm 1, \pm 2, \pm 3, \pm 4, \ldots$. There may be only a finite number of integers in the field, in which case the number of integers, if finite, is called the *characteristic* of the field. (It will then always be a prime p.) If not finite, the characteristic, of the field is defined to be zero. All elements of $GF(3)$ or $GF(5)$ are integers, so these fields have characteristics three or five, respectively. Only the elements zero and one are integers of $GF(4)$, so this field has characteristic two. Both the real field and the complex field have characteristic zero. Every field of characteristic zero will contain the field of rationals (or an isomorphic copy of the field of rationals).

If a field has a finite characteristic p, then the integers of the field form a cyclic group under addition; hence addition of integers is modulo p addition where p is the number of integers. The sum of n copies of the identity element one is written $((n))$, where the double parentheses denote modulo p.

A finite field forms a group in two ways. The elements form a group under the operation of addition, and the nonzero elements form a group under the operation of multiplication. In fact, the nonzero elements form a cyclic group under the operation of multiplication. The fact that the nonzero elements form a cyclic group is difficult to prove, and we defer the proof until Chapter 5, although we will make use of this cyclic structure earlier. Because the set of nonzero elements forms a cyclic group under multiplication, it must be generated by a single element.

Definition 2.3.3 *A primitive element of the Galois field $GF(q)$ is an element of order $q-1$ under multiplication. It generates the multiplicative group of the field.*

For example, in $GF(5)$, the powers of the element 3 are $3^1 = 3$, $3^2 = 4$, $3^3 = 2$, $3^4 = 1$. Hence 3 is a primitive element of $GF(5)$. Likewise, in $GF(7)$, the powers of the element 3 are $3^1 = 3$, $3^2 = 2$, $3^3 = 6$, $3^4 = 4$, $3^5 = 5$, $3^6 = 1$. Hence 3 is a primitive element of $GF(7)$. However, 3 is not a primitive element of $GF(11)$.

2.4 Vector space

Given a field F, the n-tuple of field elements $(v_0, v_1, \ldots, v_{n-1})$ is called a *vector* of length n over the field F. The set of all such vectors of length n, together with two operations called *vector addition* and *scalar multiplication*, is called a *vector space* over the field F. In discussions of vector spaces, the elements of the underlying field F are called *scalars*. Scalar multiplication is an operation that multiplies a vector, denoted v, by a field element, denoted c. Scalar multiplication is defined as

$$c(v_0, v_1, \ldots, v_{n-1}) = (cv_0, cv_1, \ldots, cv_{n-1}).$$

2.4 Vector space

Vector addition is an operation that adds two vectors $\boldsymbol{v} = \boldsymbol{v}' + \boldsymbol{v}''$ according to the following definition:

$$(v'_0, v'_1, \ldots, v'_{n-1}) + (v''_0, v''_1, \ldots, v''_{n-1}) = (v'_0 + v''_0, v'_1 + v''_1, \ldots, v'_{n-1} + v''_{n-1}).$$

A vector space of n-tuples is one example of a vector space. A vector space over a field F can be defined abstractly as a set V of elements called vectors together with two operations. The first operation is called *vector addition* (denoted by $+$) on pairs of elements from V. The second operation is called *scalar multiplication* (denoted by juxtaposition) on an element from F and an element from V to produce an element from V. For V to be a vector space, the operations must satisfy the following axioms.

1 V is an abelian group under vector addition.
2 (*Distributivity*) For any vectors $\boldsymbol{v}_1$ and $\boldsymbol{v}_2$, and for any scalar c,

$$c(\boldsymbol{v}_1 + \boldsymbol{v}_2) = c\boldsymbol{v}_1 + c\boldsymbol{v}_2.$$

3 (*Distributivity*) For any vector $\boldsymbol{v}$, and for any scalars c_1 and c_2, $1\boldsymbol{v} = \boldsymbol{v}$ and

$$(c_1 + c_2)\boldsymbol{v} = c_1\boldsymbol{v} + c_2\boldsymbol{v}.$$

4 (*Associativity*) For any vector $\boldsymbol{v}$, and any scalars c_1 and c_2,

$$(c_1 c_2)\boldsymbol{v} = c_1(c_2\boldsymbol{v}).$$

The zero element of V is called the *origin* of the vector space V and is denoted $\boldsymbol{0}$. Notice that there are two different uses for the symbol $+$: vector addition and addition within the field. Furthermore, the symbol $\boldsymbol{0}$ is used for the origin of the vector space, and the symbol 0 is used for the zero of the field. In practice, these ambiguities cause no confusion.

A subset of a vector space is called a *vector subspace* when it is also a vector space under the original vector addition and scalar multiplication. Under the operation of vector addition, a vector space is a group, and a vector subspace is a subgroup. In order to check whether a nonempty subset of a vector space is a subspace, it is necessary only to check for closure under vector addition and under scalar multiplication. Closure under scalar multiplication ensures that the zero vector is in the subset. All other required properties are then inherited from the original space.

The n-dimensional vector space of n-tuples over F is denoted by F^n. In the n-tuple space F^n, vector addition and scalar multiplication are defined componentwise. In the n-tuple space F^n, there is another operation called the *componentwise product* of two n-tuples. If $\boldsymbol{u} = (a_0, \ldots, a_{n-1})$ and $\boldsymbol{v} = (b_0, \ldots, b_{n-1})$, then the componentwise product is the vector defined componentwise as

$$\boldsymbol{uv} = (a_0 b_0, a_1 b_1, \ldots, a_{n-1} b_{n-1}).$$

Another operation is the *inner product* of two n-tuples. The inner product results in a scalar, defined as

$$\boldsymbol{u} \cdot \boldsymbol{v} = (a_0, \ldots, a_{n-1}) \cdot (b_0, \ldots, b_{n-1})$$
$$= a_0 b_0 + \cdots + a_{n-1} b_{n-1}.$$

It is immediately verified that $\boldsymbol{u} \cdot \boldsymbol{v} = \boldsymbol{v} \cdot \boldsymbol{u}$, that $(c\boldsymbol{u}) \cdot \boldsymbol{v} = c(\boldsymbol{u} \cdot \boldsymbol{v})$, and that $\boldsymbol{w} \cdot (\boldsymbol{u} + \boldsymbol{v}) = (\boldsymbol{w} \cdot \boldsymbol{u}) + (\boldsymbol{w} \cdot \boldsymbol{v})$. If the inner product of two vectors is zero, the vectors are said to be *orthogonal*. There are some fields over which it is possible for a nonzero vector to be orthogonal to itself, but this cannot happen in the real field. A vector orthogonal to every vector in a set is said to be orthogonal to the set.

Theorem 2.4.1 *Let V be the vector space of n-tuples over a field F, and let W be a subspace. The set of vectors orthogonal to W is itself a subspace.*

Proof Let U be the set of all vectors orthogonal to W. Because $\boldsymbol{0}$ is in U, U is not empty. Let $\boldsymbol{w}$ be any vector in W, and let $\boldsymbol{u}_1$ and $\boldsymbol{u}_2$ be any vectors in U. Then $\boldsymbol{w} \cdot \boldsymbol{u}_1 = \boldsymbol{w} \cdot \boldsymbol{u}_2 = 0$, and

$$\boldsymbol{w} \cdot \boldsymbol{u}_1 + \boldsymbol{w} \cdot \boldsymbol{u}_2 = 0 = \boldsymbol{w} \cdot (\boldsymbol{u}_1 + \boldsymbol{u}_2),$$

so $\boldsymbol{u}_1 + \boldsymbol{u}_2$ is in U. Also $\boldsymbol{w} \cdot (c\boldsymbol{u}_1) = c(\boldsymbol{w} \cdot \boldsymbol{u}_1) = 0$, so $c\boldsymbol{u}_1$ is in U. Therefore U is a subspace. □

The set of vectors orthogonal to a subspace W is called the *orthogonal complement* of W and is denoted by $W^\perp$. In the vector space of n-tuples over the real numbers, $\boldsymbol{R}^n$, the intersection of the subspaces W and $W^\perp$ contains only the all-zero vector; but in a vector space over $GF(q)$, $W^\perp$ may have a nontrivial intersection with W or may even lie within W, contain W, or equal W. For example, in $GF(2)^2$, the subspace, consisting of the two vectors 00 and 11, is its own orthogonal complement.

In a vector space V, a sum of the form

$$\boldsymbol{u} = a_1 \boldsymbol{v}_1 + a_2 \boldsymbol{v}_2 + \cdots + a_k \boldsymbol{v}_k,$$

where the a_i are scalars, is called a *linear combination* of the vectors $\boldsymbol{v}_1, \ldots, \boldsymbol{v}_k$. A set of vectors is said to *span* a vector space if every vector in the space equals at least one linear combination of the vectors in the set. A vector space that is spanned by a finite set of vectors is called a *finite-dimensional vector space*. The number of vectors in a smallest set that spans the space is called the *dimension* of the space. The space of n-tuples over F is an example of a finite-dimensional vector space of dimension n.

A set of vectors $\{\boldsymbol{v}_1, \ldots, \boldsymbol{v}_k\}$ is called *linearly dependent* if there is a set of scalars $\{a_1, \ldots, a_k\}$, not all zero, such that

$$a_1 \boldsymbol{v}_1 + a_2 \boldsymbol{v}_2 + \cdots + a_k \boldsymbol{v}_k = 0.$$

A set of vectors that is not linearly dependent is called *linearly independent*. No vector in a linearly independent set can be expressed as a linear combination of the others. Note that the all-zero vector **0** cannot belong to a linearly independent set; every set containing **0** is linearly dependent. A set of k linearly independent vectors that spans a vector space is called a *basis* of the space.

2.5 Matrix algebra

The methods of matrix algebra are often studied only for the field of real numbers and the field of complex numbers, but most of the operations remain valid in an arbitrary field (sometimes even in an arbitrary ring).

Definition 2.5.1 *An n by m matrix **A** over a field F consists of nm elements from F arranged in a rectangular array of n rows and m columns.*

A matrix is written as:

$$\boldsymbol{A} = \begin{bmatrix} a_{11} & a_{12} & \cdots & a_{1m} \\ a_{21} & a_{22} & \cdots & a_{2m} \\ \vdots & \vdots & & \vdots \\ a_{n1} & a_{n2} & \cdots & a_{nm} \end{bmatrix}.$$

If $n = m$, the matrix is called a *square matrix*. A square matrix for which $a_{ij} = a_{i'j'}$ whenever $i - j = i' - j'$ is called a *Toeplitz matrix*. A Toeplitz matrix has the form

$$\boldsymbol{A} = \begin{bmatrix} a_0 & a_1 & a_2 & \cdots & a_{n-1} \\ a_{-1} & a_0 & a_1 & \cdots & a_{n-2} \\ a_{-2} & a_{-1} & a_0 & & \\ \vdots & & & & \vdots \\ a_{-(n-1)} & & & \cdots & a_0 \end{bmatrix}$$

with the same elements along any diagonal.

Two n by m matrices, **A** and **B**, over a field F are added by the rule

$$\boldsymbol{A} + \boldsymbol{B} = \begin{bmatrix} a_{11} + b_{11} & a_{12} + b_{12} & \cdots & a_{1m} + b_{1m} \\ \vdots & & & \vdots \\ a_{n1} + b_{n1} & a_{n2} + b_{n2} & \cdots & a_{nm} + b_{nm} \end{bmatrix}.$$

An ℓ by n matrix **A** and an n by m matrix **B** can be multiplied to produce an ℓ by m matrix **C** by using the following rule:

$$c_{ij} = \sum_{k=1}^{n} a_{ik} b_{kj}, \qquad \begin{array}{l} i = 1, \ldots, \ell \\ j = 1, \ldots, m \end{array}.$$

In the matrix A, usually a square matrix, the set of elements a_{ii}, for which the column number and row number are equal, is called the (main) *diagonal* of the matrix. An *identity matrix*, denoted by I, is an n by n matrix with the field element one in every entry of the diagonal and the field element zero in every other matrix entry. An *exchange matrix*, denoted by J, is a matrix with the field element one in every entry of the antidiagonal (the entries where $j = n + 1 - i$) and with the field element zero in every other matrix entry. Notice that $J^2 = I$. Examples of a three by three identity matrix and a three by three exchange matrix are

$$I = \begin{bmatrix} 1 & 0 & 0 \\ 0 & 1 & 0 \\ 0 & 0 & 1 \end{bmatrix}, \qquad J = \begin{bmatrix} 0 & 0 & 1 \\ 0 & 1 & 0 \\ 1 & 0 & 0 \end{bmatrix}.$$

With the above definitions of matrix multiplication and matrix addition, the set of n by n square matrices over any field F forms a ring, as can easily be verified. It is a noncommutative ring but does have an identity, namely, the n by n identity matrix.

The *transpose* of an n by m matrix A is an m by n matrix, denoted A^T, such that $a_{ij}^T = a_{ji}$. That is, the rows of A^T are the columns of A, and the columns of A^T are the rows of A. It is easy to verify that if $C = AB$, then $C^T = B^T A^T$.

The inverse of the square matrix A is the square matrix A^{-1}, if it exists, such that $A^{-1}A = AA^{-1} = I$. The set of all square n by n matrices for which an inverse exists is a group under matrix multiplication. Therefore, whenever a matrix has an inverse, it is unique, because we saw in Theorem 2.1.2 that this uniqueness property holds in any group. A matrix that has an inverse is called *nonsingular*; otherwise, it is called *singular*. Let $C = AB$. We see from part (iii) of Theorem 2.2.5 that if the inverse of either A or B does not exist, then neither does the inverse of C. If the inverses of A and B both exist, then $C^{-1} = B^{-1}A^{-1}$ because $(B^{-1}A^{-1})C = I = C(B^{-1}A^{-1})$.

Definition 2.5.2 *For any field F and for each n, the determinant, $\det(A)$, is a function from the set of n by n matrices over F into the field F, given by*

$$\det(A) = \sum \xi_{i_k \cdots i_n} a_{1i_1} a_{2i_2} a_{3i_3} \cdots a_{ni_n},$$

where the sum is over all permutations $i_1, i_2, \ldots, i_n$ of the integers $1, 2, \ldots, n$; $\xi_{i_1 \ldots i_n}$, is equal to 1 if the permutation can be obtained by an even number of transpositions; otherwise, it is equal to -1. A transposition is an interchange of two terms.

If a matrix A' is obtained from A by interchanging two rows, then every permutation of rows of the new matrix A' that can be obtained by an even (odd) number of transpositions looks like a permutation of rows of A that can be obtained by an odd (even) number of transpositions. From this it follows that if two rows of a matrix are interchanged, the determinant is replaced by its negative. Similar reasoning shows that if two rows of a real matrix are equal, the determinant is equal to zero.

2.5 Matrix algebra

The following theorem gives, without proof, properties of the determinant that follow easily from the definition.

Theorem 2.5.3
(i) If all elements of any row of a square matrix are zero, the determinant of the matrix is zero.
(ii) The determinant of a matrix equals the determinant of its transpose.
(iii) If two rows of a square matrix are interchanged, the determinant is replaced by its negative.
(iv) If two rows of a square matrix are equal, the determinant is zero.
(v) If all elements of one row of a square matrix are multiplied by a field element c, the value of the determinant is multiplied by c.
(vi) If two matrices $\mathbf{A}$ and $\mathbf{B}$ differ only in row i, the sum of their determinants equals the determinant of a matrix $\mathbf{C}$ whose ith row is the sum of the ith rows of $\mathbf{A}$ and $\mathbf{B}$ and whose other rows equal the corresponding rows of $\mathbf{A}$ or $\mathbf{B}$.
(vii) If a scaler multiple of any row is added to any other row, the determinant is unchanged.
(viii) The determinant of a square matrix is nonzero if and only if its rows (or columns) are linearly independent.

If the row and column containing an element a_{ij} in a square matrix are deleted, then the determinant of the remaining square array, denoted here by M_{ij}, is called the *minor* of a_{ij}. The *cofactor* of a_{ij}, denoted here by C_{ij}, is defined by

$$C_{ij} = (-1)^{i+j} M_{ij}.$$

By examination of the definition of the determinant, it is seen that the cofactor of a_{ij} is the coefficient of a_{ij} in the expansion of $\det(\mathbf{A})$. Therefore the determinant can be written

$$\det(\mathbf{A}) = \sum_{k=1}^{n} a_{ik} C_{ik}.$$

This is known as the *Laplace expansion formula* for determinants. The Laplace expansion formula is used as a recursive method of computing the determinant. It gives the determinant of an n by n matrix in terms of n determinants of $(n-1)$ by $(n-1)$ submatrices.

If a_{ik} is replaced by a_{jk}, then $\sum_{k=1}^{n} a_{jk} C_{ik}$ is the determinant of a new matrix in which the elements of the ith row are replaced by the elements of the jth row; hence it is zero if $j \neq i$. Thus

$$\sum_{k=1}^{n} a_{jk} C_{ik} = \begin{cases} \det(\mathbf{A}), & i = j \\ 0, & i \neq j. \end{cases}$$

Therefore the matrix $A = [a_{ij}]$ has the inverse

$$A^{-1} = \left[\frac{C_{ji}}{\det(A)} \right],$$

provided that $\det(A) \neq 0$. When $\det(A) = 0$, an inverse does not exist.

A matrix can be broken into pieces as follows

$$A = \left[\begin{array}{c|c} A_{11} & A_{12} \\ \hline A_{21} & A_{22} \end{array} \right],$$

where A_{11}, A_{12}, A_{21}, and A_{22} are smaller matrices whose dimensions add up to the dimension of A. That is, the number of rows of A_{11} (or A_{12}) plus the number of rows of A_{21} (or A_{22}) equals the number of rows of A. A similar statement holds for columns. Matrices can be multiplied in blocks. That is, if

$$A = \left[\begin{array}{c|c} A_{11} & A_{12} \\ \hline A_{21} & A_{22} \end{array} \right], \qquad B = \left[\begin{array}{c|c} B_{11} & B_{12} \\ \hline B_{21} & B_{22} \end{array} \right]$$

and $C = AB$, then

$$C = \left[\begin{array}{c|c} A_{11}B_{11} + A_{12}B_{21} & A_{11}B_{12} + A_{12}B_{22} \\ \hline A_{21}B_{11} + A_{22}B_{21} & A_{21}B_{12} + A_{22}B_{22} \end{array} \right],$$

provided that the dimensions of the blocks are compatible in the sense that all matrix products and additions are defined. This decomposition can be readily verified as a simple consequence of the associativity and distributivity properties of the underlying field.

Definition 2.5.4 Let $A = [a_{ik}]$ be an I by K matrix, and let $B = [b_{j\ell}]$ be a J by L matrix. Then the Kronecker product of A and B, denoted $A \times B$, is a matrix with IJ rows and KL columns whose entry in row $(i-1)J + j$ and column $(k-1)L + \ell$ is given by

$$c_{ij,k\ell} = a_{ik} b_{j\ell}.$$

The Kronecker product, $A \times B$, is an I by K array of J by L blocks, with the (i, k)th such block being $a_{ik}B$. It is apparent from the definition that the Kronecker product is not commutative, but it is associative:

$$A \times B \neq B \times A$$
$$(A \times B) \times C = A \times (B \times C).$$

The elements of $B \times A$ are the same as those of $A \times B$, but they are arranged differently. It is also clear that the Kronecker product distributes over ordinary matrix addition.

2.5 Matrix algebra

The most familiar example of a Kronecker product is the *outer product* of two vectors. Suppose that both A and B are column vectors, say $a = (a_1, \ldots, a_I)^T$ and $b = (b_1, \ldots, b_J)^T$, respectively. Then $K = L = 1$ and $a \times b^T$ is an I by J matrix with entry $a_i b_j$ in row i and column j. This is denoted simply as ab^T because, in this simple case, the Kronecker product coincides with the ordinary matrix product.

The following useful theorem says that the Kronecker product of the matrix product of matrices is the matrix product of the Kronecker products.

Theorem 2.5.5 *The Kronecker product satisfies* $(A \times B)(C \times D) = (AC) \times (BD)$, *provided that the matrix products all exist.*

Proof Let the matrices A, B, C, and D have dimensions $I \times K$, $J \times L$, $K \times M$, and $L \times N$, respectively. Because $A \times B$ has KL columns and $C \times D$ has KL rows, the matrix product $(A \times B)(C \times D)$ is defined. It has IJ rows, which we doubly index by (i, j), and MN columns, which we doubly index by (m, n). The entry in row (i, j) and column (m, n) is $\sum_{k\ell} a_{ik} b_{j\ell} c_{km} d_{\ell n}$. Because AC has I rows and M columns and BD has J rows and L columns, $(AC) \times (BD)$ also is an IJ by MN matrix. Its entry in row (i, j) and column (m, n) is

$$\sum_k a_{ik} c_{km} \sum_\ell b_{j\ell} d_{\ell n} = \sum_{k\ell} a_{ik} b_{j\ell} c_{km} d_{\ell n},$$

which completes the proof. □

The rows of an n by m matrix A over a field F comprise a set of vectors in F^m having m components. The *row space* of the matrix A is the set of all linear combinations of the row vectors of A. The row space is a subspace of F^m. The dimension of the row space is called the *row rank*. Similarly, the columns of A may be thought of as a set of vectors in F^n having n components. The *column space* of A is the set of all linear combinations of column vectors of A, and the dimension of the column space is called the *column rank*. The set of vectors v such that $Av^T = 0$ is called the *null space* of the matrix A. It is clear that the null space is a vector subspace of F^n. In particular, the null space of A is the orthogonal complement of the row space of A because the null space can be described as the set of all vectors orthogonal to all vectors of the row space.

The *elementary row operations* on a matrix are as follows:
1 interchange of any two rows;
2 multiplication of any row by a nonzero field element;
3 replacement of any row by the sum of itself and a multiple of any other row.

Each elementary row operation on an n by m matrix can be effected by multiplying the matrix on the left by an appropriate n by n matrix E, called an *elementary matrix*. The elementary matrices are of the form of the following modifications of an identity matrix:

$$\begin{bmatrix} 1 & & & & & \\ & \ddots & & & & \\ & & 0 & 1 & & \\ & & & \ddots & & \\ & & 1 & 0 & & \\ & & & & \ddots & \\ & & & & & 1 \end{bmatrix}, \begin{bmatrix} 1 & & & & \\ & \ddots & & & \\ & & a & & \\ & & & \ddots & \\ & & & & 1 \end{bmatrix},$$

or

$$\begin{bmatrix} 1 & & & & & \\ & \ddots & & & & \\ & & 1 & & & \\ & & & \ddots & & \\ & & a & & 1 & \\ & & & & & \ddots \\ & & & & & & 1 \end{bmatrix}$$

Each elementary row operation is inverted by an elementary row operation of the same kind.

Elementary row operations can be used to put a matrix into a standard form, known as the *row-echelon form*, with the same row space. The row-echelon form is as follows.
1 The leading nonzero term of every nonzero row is one.
2 Every column containing such a leading term has all its other entries equal to zero.
3 The leading term of any row is to the right of the leading term in every higher row. Every zero row is below every nonzero row.

An example of a matrix in row-echelon form is the matrix

$$A = \begin{bmatrix} 1 & 1 & 0 & 1 & 3 & 0 \\ 0 & 0 & 1 & 1 & 0 & 0 \\ 0 & 0 & 0 & 0 & 0 & 1 \\ 0 & 0 & 0 & 0 & 0 & 0 \end{bmatrix}.$$

Notice the all-zero row at the bottom. Also, notice that if the all-zero row is deleted, then all columns of a three by three identity matrix appear as columns of the matrix, but not as consecutive columns. In general, if there are k rows, none of them all zero, and at least this many columns, then all columns of a k by k identity matrix will appear within a row-echelon matrix.

Theorem 2.5.6 *If two matrices A and A' are related by a succession of elementary row operations, both matrices have the same row space.*

Proof Each row of A' is a linear combination of rows of A; therefore any linear combination of rows of A' is a linear combination of rows of A also, so the row space of A' contains the row space of A'. But A can be obtained from A' by the inverse succession of elementary row operations, so by the same argument the row space of A' contains the row space of A. Therefore A and A' have equal row spaces. □

Theorem 2.5.7 *If two matrices A and A' are related by a succession of elementary row operations, any set of columns that is linearly independent in A is also linearly independent in A'.*

Proof It suffices to prove the theorem for a single elementary row operation, and the theorem is obvious if it is the first or second kind of elementary row operation. Hence suppose A' is formed from A by adding a multiple of row α to row β. Choose any linearly dependent combination of columns of A'. The column sum of the elements in row α must combine to give zero and so must also sum to zero within row β. That is, this set of columns is also linearly dependent in A. □

Theorem 2.5.8 *A k by n matrix A whose k rows are linearly independent also has k linearly independent columns.*

Proof Put A in row-echelon form A'. Because the rows of A are linearly independent, the rows of A' are also linearly independent, so it has no all-zero row. Hence, for each row of A', there is a column in which that row has a one and every other row has a zero. This set of k columns of A' is linearly independent, so by Theorem 2.5.7 this same set of columns of A is linearly independent. □

Theorem 2.5.9 *The row rank of a matrix A equals its column rank, and both are equal to the dimension of any largest square submatrix with determinant not equal to zero. (Hence this value is called simply the rank of the matrix.)*

Proof It is necessary only to show that the row rank of A is equal to the dimension of a largest square submatrix with nonzero determinant. The same proof applied to the transpose of A then proves the same for the column rank of A, and so proves that the row rank equals the column rank.

A submatrix of A is a matrix obtained by deleting any number of rows and columns from A. Let M be a nonsingular square submatrix of A of largest dimension. Because M is nonsingular, the rows of M are linearly independent by part (viii) of Theorem 2.5.3, and so the rows of A that give rise to these rows of M must be linearly independent. Therefore the row rank of A is at least as large as the dimension of M.

On the other hand, choose any set of k linearly independent rows. A matrix of these rows, by Theorem 2.5.8, has k linearly independent columns. Hence choosing these k columns from these k rows gives a matrix with nonzero determinant. Therefore the dimension of a largest nonsingular submatrix of A is at least as large as the row rank of A. Hence the row rank of A is equal to the dimension of M. This completes the proof. $\square$

2.6 The integer ring

The integers (positive, negative, and zero) form a deceptively simple mathematical set. Nothing could appear more uniform and regular than the integers, yet, upon close examination one can see very complex patterns and interrelationships in this set, starting with the notions of factors and primes. Clever designers have built efficient signal-processing algorithms upon these properties of the set of integers.

Under the usual operations of addition and multiplication, the integers form a ring which is conventionally denoted by the label $\mathbf{Z}$. Within the ring of integers, while subtraction is always possible, division is not always possible. This limitation of the division operation is one thing that makes the integer ring so interesting and rich in structure.

We say that the integer s is divisible by the integer r or that r divides s or is a factor of s if $ra = s$ for some integer a. In symbols this is written $r|s$, which is read "r divides s." Whenever r both divides s and is divisible by s, then $r = \pm s$. This is because $r = sa$ and $s = rb$ for some a and b. Therefore $r = rab$, so ab must equal one. Because a and b are integers, a and b must each be either plus one or minus one.

A positive integer p larger than one, divisible only by $\pm p$ or ± 1, is called a *prime integer* or a *prime*. The smallest primes are 2, 3, 5, 7, 11, 13, ...; the integer 1 is not a prime. A positive integer larger than one, not a prime, is called *composite*. The *greatest common divisor* of two integers r and s, denoted by GCD$[r, s]$, is the largest positive integer that divides both of them. The *least common multiple* of two integers r and s, denoted by LCM$[r, s]$, is the smallest positive integer that is divisible by both of them. Two integers are said to be *coprime*, or *relatively prime*, if their greatest common divisor is one. Thus, every positive integer n is coprime with one.

It is always possible to cancel in the integer ring; if $ca = cb$ and c is nonzero, then $a = b$. The integer ring also has a weak form of division known as *division with remainder* or as the *division algorithm*. We state it as a self-evident theorem.

Theorem 2.6.1 (Division algorithm) *For every integer c and positive integer d, there is a unique pair of integers Q, the quotient, and s, the remainder, such that $c = dQ + s$, where $0 \leq s < d$.*

2.6 The integer ring

The quotient is sometimes denoted by

$$Q = \left\lfloor \frac{c}{d} \right\rfloor.$$

Usually, we will be more interested in the remainder than in the quotient. When s and c have the same remainder under division by d, we write

$$s \equiv c \pmod{d}.$$

In this form, the expression is called a *congruence* and is read: s is congruent to c modulo d. In a congruence, neither s nor c is necessarily smaller than d. The remainder will also be written

$$s = R_d[c],$$

which is read: s is the remainder of c when divided by d, or s is the residue of c modulo d. We will also use double parentheses $((c))$ to denote the same thing; in this case, d is understood from the context. Yet another notation is

$$s = c \pmod{d},$$

where now an equal sign is used. This is not quite the same as a congruence. Now s is the remainder of c when divided by d.

The computation of the remainder of a complicated expression is facilitated by the following theorem, which says that the process of computing a remainder can be interchanged with addition and multiplication.

Theorem 2.6.2 *With the modulus d fixed,*
(i) $R_d[a + b] = R_d[R_d[a] + R_d[b]]$;
(ii) $R_d[a \cdot b] = R_d[R_d[a] \cdot R_d[b]]$.

Proof Exercise. □

Given two positive integers s and t, their greatest common divisor can be computed by an iterative application of the division algorithm. This procedure is known as the *euclidean algorithm*. Suppose that t is less than s; the euclidean algorithm consists of the steps

$$s = Q^{(1)}t + t^{(1)},$$
$$t = Q^{(2)}t^{(1)} + t^{(2)},$$
$$t^{(1)} = Q^{(3)}t^{(2)} + t^{(3)},$$
$$\vdots$$
$$t^{(n-2)} = Q^{(n)}t^{(n-1)} + t^{(n)},$$
$$t^{(n-1)} = Q^{(n+1)}t^{(n)},$$

where the process stops when a remainder of zero is obtained. The steps of the euclidean algorithm can be expressed concisely in matrix notation as

$$\begin{bmatrix} s^{(r)} \\ t^{(r)} \end{bmatrix} = \begin{bmatrix} 0 & 1 \\ 1 & -Q^{(r)} \end{bmatrix} \begin{bmatrix} s^{(r-1)} \\ t^{(r-1)} \end{bmatrix}.$$

The last nonzero remainder $t^{(n)}$ is the greatest common divisor. This will be proved in the next theorem.

Theorem 2.6.3 (Euclidean algorithm) *Given two positive integers s and t, with s larger than t, let $s^{(0)} = s$ and $t^{(0)} = t$, the following recursive equations for $r = 1, \ldots, n$:*

$$Q^{(r)} = \left\lfloor \frac{s^{(r-1)}}{t^{(r-1)}} \right\rfloor,$$

$$\begin{bmatrix} s^{(r)} \\ t^{(r)} \end{bmatrix} = \begin{bmatrix} 0 & 1 \\ 1 & -Q^{(r)} \end{bmatrix} \begin{bmatrix} s^{(r-1)} \\ t^{(r-1)} \end{bmatrix}$$

satisfy

$$s^{(n)} = \text{GCD}[s, t],$$

where n is the integer for which $t^{(n)} = 0$.

Proof Because $t^{(r+1)}$ is less than $t^{(r)}$ and all remainders are nonnegative, eventually for some n, $t^{(n)} = 0$, so the termination is well-defined. The following matrix inverse is readily verified:

$$\begin{bmatrix} 0 & 1 \\ 1 & -Q^{(r)} \end{bmatrix}^{-1} = \begin{bmatrix} Q^{(r)} & 1 \\ 1 & 0 \end{bmatrix}.$$

Therefore

$$\begin{bmatrix} s \\ t \end{bmatrix} = \left\{ \prod_{\ell=1}^{n} \begin{bmatrix} Q^{(\ell)} & 1 \\ 1 & 0 \end{bmatrix} \right\} \begin{bmatrix} s^{(n)} \\ 0 \end{bmatrix},$$

so $s^{(n)}$ must divide both s and t and hence divides $\text{GCD}[s, t]$. Inverting this equation gives

$$\begin{bmatrix} s^{(n)} \\ 0 \end{bmatrix} = \left\{ \prod_{\ell=n}^{1} \begin{bmatrix} 0 & 1 \\ 1 & -Q^{(\ell)} \end{bmatrix} \right\} \begin{bmatrix} s \\ t \end{bmatrix},$$

so that any divisor of both s and t divides $s^{(n)}$. Hence $\text{GCD}[s, t]$ divides $s^{(n)}$ and is divisible by $s^{(n)}$. Thus

$$s^{(n)} = \text{GCD}[s, t].$$

This completes the proof of the theorem. □

2.6 The integer ring

There are several important corollaries to this theorem. Let

$$A^{(r)} = \prod_{\ell=r}^{1} \begin{bmatrix} 0 & 1 \\ 1 & -Q^{(\ell)} \end{bmatrix}$$
$$= \begin{bmatrix} 0 & 1 \\ 1 & -Q^{(r)} \end{bmatrix} A^{(r-1)}.$$

We then have the following corollary, an important and nonintuitive result of number theory. It says that the greatest common divisor of two integers is an integer combination of them.

Corollary 2.6.4 *For any integers s and t, there exist integers a and b such that*

$\text{GCD}[s, t] = as + bt.$

Proof It suffices to prove the corollary for s and t positive. Then, because

$$\begin{bmatrix} s^{(n)} \\ 0 \end{bmatrix} = A^{(n)} \begin{bmatrix} s \\ t \end{bmatrix}$$

and

$s^{(n)} = \text{GCD}[s, t],$

the theorem follows with $a = A_{11}^{(n)}$ and $b = A_{12}^{(n)}$. $\square$

The integers solving Corollary 2.6.4 are not unique, because we can write

$\text{GCD}[s, t] = (a - \ell t)s + (b + \ell s)t$

for any integer ℓ.

Corollary 2.6.5 *For any positive coprime integers s and t, there exist integers a and b such that*

$as + bt = 1.$

Proof This is an immediate consequence of Corollary 2.6.4. $\square$

The proof of the corollary tells how to compute the integers a and b as elements of the matrix A. This procedure is referred to as the *extended euclidian algorithm*. The other two elements of the matrix also have a direct interpretation. To interpret those elements, we will need the inverse of the matrix $A^{(r)}$. Recall that

$$A^{(r)} = \prod_{\ell=r}^{1} \begin{bmatrix} 0 & 1 \\ 1 & -Q^{(\ell)} \end{bmatrix}.$$

From this it is clear that the determinant of $A^{(r)}$ is $(-1)^r$. The inverse is

$$\begin{bmatrix} A_{11}^{(r)} & A_{12}^{(r)} \\ A_{21}^{(r)} & A_{22}^{(r)} \end{bmatrix}^{-1} = (-1)^r \begin{bmatrix} A_{22}^{(r)} & -A_{12}^{(r)} \\ -A_{21}^{(r)} & A_{11}^{(r)} \end{bmatrix}.$$

Corollary 2.6.6 *The matrix elements $A_{21}^{(n)}$ and $A_{22}^{(n)}$ produced by the euclidean algorithm satisfy*

$$s = (-1)^n A_{22}^{(n)} \mathrm{GCD}[s, t],$$
$$t = (-1)^n A_{21}^{(n)} \mathrm{GCD}[s, t].$$

Proof Using the above expression for the inverse gives

$$\begin{bmatrix} s \\ t \end{bmatrix} = (-1)^n \begin{bmatrix} A_{22}^{(n)} & -A_{12}^{(n)} \\ -A_{21}^{(n)} & A_{11}^{(n)} \end{bmatrix} \begin{bmatrix} s^{(n)} \\ 0 \end{bmatrix}$$

from which the corollary follows. □

Using the division algorithm, we can find the greatest common divisor of two integers. As an example, GCD[814, 187] is found as follows:

$$\begin{bmatrix} s^{(n)} \\ 0 \end{bmatrix} = \begin{bmatrix} 0 & 1 \\ 1 & -5 \end{bmatrix} \begin{bmatrix} 0 & 1 \\ 1 & -1 \end{bmatrix} \begin{bmatrix} 0 & 1 \\ 1 & -2 \end{bmatrix} \begin{bmatrix} 0 & 1 \\ 1 & -4 \end{bmatrix} \begin{bmatrix} 814 \\ 187 \end{bmatrix}$$
$$= \begin{bmatrix} 3 & -13 \\ -17 & 74 \end{bmatrix} \begin{bmatrix} 814 \\ 187 \end{bmatrix} = \begin{bmatrix} 11 \\ 0 \end{bmatrix}.$$

From this calculation, we immediately have that GCD[814, 187] is 11, and also that

$$\mathrm{GCD}[814, 187] = 3 \times 814 - 13 \times 187,$$

as given by Corollary 2.6.6.

2.7 Polynomial rings

For each field F, there is a ring $F[x]$ called the ring of polynomials over F. A polynomial ring is analogous in many ways to the ring of integers. To make this evident, this section will closely follow Section 2.6.

A *polynomial* over a field F is a mathematical expression

$$f(x) = f_n x^n + f_{n-1} x^{n-1} + \cdots + f_1 x + f_0$$
$$= \sum_{i=0}^{n} f_i x^i,$$

2.7 Polynomial rings

where the symbol x is an *indeterminate* and the *coefficients* $f_0, \ldots, f_n$ are elements of the field. The *zero polynomial* is

$$f(x) = 0.$$

The *degree* of a polynomial $f(x)$, denoted deg $f(x)$, is the largest index of a nonzero coefficient. The degree of a nonzero polynomial is always finite. By convention, the degree of the zero polynomial is negative infinity ($-\infty$). A *monic polynomial* is a polynomial whose coefficient f_n with largest index is equal to one. Two polynomials are equal if all coefficients f_i are equal.

To form a ring from the set of all polynomials over a given field, addition and multiplication are defined as the usual addition and multiplication of polynomials. For each field F, we define such a polynomial ring, denoted by the label $F[x]$, and consisting of all polynomials with coefficients in F. In discussions about the ring $F[x]$, the polynomials of degree zero are elements of the field F. They are sometimes called *scalars*.

In the usual way, the sum of two polynomials in $F[x]$ is another polynomial in $F[x]$, defined by

$$f(x) + g(x) = \sum_{i=0}^{\infty} (f_i + g_i) x^i,$$

where, of course, terms higher than the larger of the degrees of $f(x)$ and $g(x)$ are all zero. The degree of the sum is not greater than the larger of these two degrees. The product of two polynomials in $F[x]$ is another polynomial in $F[x]$, defined by

$$f(x)g(x) = \sum_{i} \left(\sum_{j=0}^{i} f_j g_{i-j} \right) x^i.$$

The degree of the product of two polynomials is equal to the sum of the degrees of the two factors. If $f(x) \neq 0$ and $g(x) \neq 0$, then $f(x)g(x) \neq 0$ because deg $p(x)$ equals negative infinity if and only if $p(x) = 0$.

Within a ring of polynomials, while subtraction is always possible, division is not always possible. We write $r(x)|s(x)$ and say that the polynomial $s(x)$ is *divisible* by the polynomial $r(x)$, or that $r(x)$ *divides* $s(x)$, or $r(x)$ is a factor of $s(x)$, if there is a polynomial $a(x)$ such that $r(x)a(x) = s(x)$. A nonzero polynomial $p(x)$ that is divisible only by $p(x)$ or by α, where α is an arbitrary field element, is called an *irreducible polynomial*. A monic irreducible polynomial is called a *prime polynomial*.

To say that a polynomial is a prime polynomial, it is necessary to know within which field the polynomial is to be regarded. The polynomial $p(x) = x^4 - 2$ is a prime polynomial over the field of rationals, but it is not a prime polynomial over the real field. Over the real field, $p(x) = (x^2 - \sqrt{2})(x^2 + \sqrt{2})$ is a product of two prime polynomials. Over the complex field those two polynomials are not prime because they can be factored further.

Whenever $r(x)$ both divides $s(x)$ and is divisible by $s(x)$, then $r(x) = \alpha s(x)$ where α is an element of the field F. This is proved as follows. There must exist polynomials $a(x)$ and $b(x)$ such that $r(x) = s(x)a(x)$ and $s(x) = r(x)b(x)$. Therefore $r(x) = r(x)b(x)a(x)$. But the degree of the right side is the sum of the degrees of $r(x)$, $b(x)$, and $a(x)$. Because this must equal the degree of the left side, $a(x)$ and $b(x)$ must have zero degree; that is, they are scalars.

The *greatest common divisor* of two polynomials $r(x)$ and $s(x)$, denoted by $\text{GCD}[r(x), s(x)]$, is the monic polynomial of largest degree that divides both of them. If the greatest common divisor of two polynomials is one, then they are said to be *coprime* (or *relatively prime*).

The *least common multiple* of two polynomials $r(x)$ and $s(x)$, denoted by $\text{LCM}[r(x), s(x)]$, is the monic polynomial of smallest degree divisible by both of them. We shall see that the greatest common divisor and the least common multiple of $r(x)$ and $s(x)$ are unique.

Differentiation is defined in the real field in terms of limits. This definition does not work in all fields because, in some fields, there is no notion of an arbitrary small number. In such fields it is convenient simply to define an operation on polynomials that behaves the way we want derivatives to behave. This is called the *formal derivative* of a polynomial.

Definition 2.7.1 *Let $r(x) = r_n x^n + r_{n-1} x^{n-1} + \cdots + r_1 x + r_0$ be a polynomial over the field F. The formal derivative of $r(x)$ is a polynomial $r'(x)$, given by*

$$r'(x) = n r_n x^{n-1} + (n-1) r_{n-1} x^{n-2} + \cdots + 2 r_2 x + r_1,$$

where the new coefficients $i r_i$ are computed in the field F as the sum of i copies of r_i:

$$i r_i = r_i + r_i + \cdots + r_i.$$

It is easy to verify many of the usual properties of derivatives, namely that

$$[r(x)s(x)]' = r'(x)s(x) + r(x)s'(x)$$

and that if $a(x)^2$ divides $r(x)$, then $a(x)$ divides $r'(x)$.

Cancellation is valid in a ring of polynomials over a field; if $c(x)a(x) = c(x)b(x)$ and $c(x)$ is nonzero, then $a(x) = b(x)$. A ring of polynomials also has a weak form of division known as *division with remainder* or as the *division algorithm*.

Theorem 2.7.2 (Division algorithm for polynomials) *For every polynomial $c(x)$ and nonzero polynomial $d(x)$, there is a unique pair of polynomials $Q(x)$, the quotient* polynomial, and $s(x)$, the remainder polynomial, such that

$$c(x) = d(x)Q(x) + s(x)$$

and

$\deg s(x) < \deg d(x)$.

Proof The quotient polynomial and the remainder polynomial can be found by elementary long division of polynomials. They are unique because if

$$c(x) = d(x)Q_1(x) + s_1(x) = d(x)Q_2(x) + s_2(x),$$

then

$$d(x)[Q_1(x) - Q_2(x)] = s_1(x) - s_2(x).$$

If the right side is nonzero, it has degree less than $\deg d(x)$, while if the left side is nonzero, it has degree at least as large as $\deg d(x)$. Hence both are zero, and the representation is unique. □

In practice, one can compute the quotient polynomial and the remainder polynomial by simple long division of polynomials. The quotient polynomial is sometimes denoted by

$$Q(x) = \left\lfloor \frac{c(x)}{d(x)} \right\rfloor.$$

Usually, we will be more interested in the remainder polynomial than in the quotient polynomial. The remainder polynomial will also be written

$$s(x) = R_{d(x)}[c(x)]$$

or

$$s(x) = c(x) \pmod{d(x)}.$$

One also can write the congruence

$$s(x) \equiv c(x) \pmod{d(x)},$$

which only means that $s(x)$ and $c(x)$ have the same remainder under division by $d(x)$.

To find $R_{d(x)}[c(x)]$, it seems that we must carry through a polynomial division. Actually, there are several shortcuts that simplify the work. First, notice that

$$R_{d(x)}[c(x)] = R_{d(x)}[c(x) + a(x)d(x)],$$

so we can add any multiple of $d(x)$ to $c(x)$ without changing the remainder. Hence, without changing the remainder, one can cancel the largest-index nonzero coefficient of $c(x)$ by adding a multiple of $d(x)$. By using this principle, reduction of $c(x)$ modulo the monic polynomial

$$d(x) = x^n + \sum_{i=0}^{n-1} d_i x^i$$

can be simplified by replacing x^n with the polynomial $-\sum_{i=0}^{n-1} d_i x^i$ whenever it is convenient to do so. In this way, finding the remainder polynomial is simplified.

Another method to simplify the computation of a remainder is given in the following theorem.

Theorem 2.7.3 *Let $d(x)$ be a multiple of $g(x)$. Then for any $a(x)$,*

$$R_{g(x)}[a(x)] = R_{g(x)}[R_{d(x)}[a(x)]].$$

Proof Let $d(x) = g(x)h(x)$ for some $h(x)$. Expanding the meaning of the right side gives

$$\begin{aligned} a(x) &= Q_1(x)d(x) + R_{d(x)}[a(x)] \\ &= Q_1(x)h(x)g(x) + Q_2(x)g(x) + R_{g(x)}[R_{d(x)}[a(x)]], \end{aligned}$$

where the remainder polynomial has a degree less than $\deg g(x)$. Expanding the meaning of the left side gives

$$a(x) = Q(x)g(x) + R_{g(x)}[a(x)],$$

and the division algorithm says that there is only one such expansion with the degree of $R_{g(x)}[a(x)]$ smaller than the degree of $g(x)$. The theorem follows by identifying the like terms in the two expansions. $\square$

As an example of the use of Theorem 2.7.3, we will divide $x^7 + x + 1$ by $x^4 + x^3 + x^2 + x + 1$. Long division would be tedious, but if we remember that

$$(x - 1)(x^4 + x^3 + x^2 + x + 1) = x^5 - 1,$$

we can first divide by $x^5 - 1$, then by $x^4 + x^3 + x^2 + x + 1$. But then

$$R_{x^5-1}[x^7 + x + 1] = x^2 + x + 1,$$

which is now trivial to divide by $x^4 + x^3 + x^2 + x + 1$. Thus

$$R_{x^4+x^3+x^2+x+1}[x^7 + x + 1] = x^2 + x + 1.$$

Another handy reduction is given in the next theorem.

2.7 Polynomial rings

Theorem 2.7.4
(i) $R_{d(x)}[a(x) + b(x)] = R_{d(x)}[a(x)] + R_{d(x)}[b(x)]$.
(ii) $R_{d(x)}[a(x) \cdot b(x)] = R_{d(x)}\{R_{d(x)}[a(x)] \cdot R_{d(x)}[b(x)]\}$.

Proof Use the division algorithm to interpret the left side of the first equation as

$$a(x) + b(x) = Q(x)d(x) + R_{d(x)}[a(x) + b(x)].$$

Use the division algorithm to interpret the right side of the first equation as

$$a(x) + b(x) = Q'(x)d(x) + R_{d(x)}[a(x)] + Q''(x)d(x) + R_{d(x)}[b(x)].$$

Equating these, part (i) follows from the uniqueness of the division algorithm. Part (ii) is proved in the same way. □

Just as it is often useful to express an integer as a product of primes, it is also often useful to express polynomials as products of irreducible polynomials. To make the factoring of integers into primes unique, one adopts the convention that only positive integers can be primes. Similarly, to make the factoring of polynomials into irreducible polynomials unique, one adopts the convention that the irreducible polynomials used as factors must be monic polynomials.

Theorem 2.7.5 (Unique factorization theorem) *A polynomial over a field has a unique factorization into a field element times a product of prime polynomials over the field, each polynomial with degree at least one.*

Proof Clearly, the field element must be the coefficient of p_n, where n is the degree of the polynomial $p(x)$. We can factor out this field element and prove the theorem for monic polynomials.

Suppose the theorem is false. Let $p(x)$ be a polynomial of the lowest degree for which the theorem fails. Then there are two factorizations:

$$p(x) = a_1(x)a_2(x) \cdots a_k(x) = b_1(x)b_2(x) \cdots b_j(x),$$

where the $a_k(x)$ and $b_j(x)$ are prime polynomials.

All of the $a_k(x)$ must be different from all of the $b_j(x)$ because, otherwise, the common terms could be canceled to give a polynomial of lower degree that can be factored in two different ways.

Without loss of generality, suppose that $b_1(x)$ has a degree not larger than that of $a_1(x)$. Then

$$a_1(x) = b_1(x)Q(x) + s(x),$$

where $\deg s(x) < \deg b_1(x) \leq \deg a_1(x)$. Then

$$s(x)a_2(x)a_3(x) \cdots a_k(x) = b_1(x)[b_2(x) \cdots b_j(x) - Q(x)a_2(x) \cdots a_k(x)].$$

Factor both $s(x)$ and the bracketed term on the right into their prime polynomial factors and, if necessary, divide by a field element to make all factors monic. Because $b_1(x)$ does not appear on the left side, we have two different factorizations of another monic polynomial whose degree is smaller than the degree of $p(x)$, contrary to the choice of $p(x)$. The contradiction proves the theorem. □

Now, from the unique factorization theorem, for any polynomials $s(x)$ and $t(x)$, it is clear that both $\text{GCD}[s(x), t(x)]$ and $\text{LCM}[s(x), t(x)]$ are unique because the greatest common divisor is the product of all prime factors common to both $s(x)$ and $t(x)$, each factor raised to the smallest power with which it appears in either $s(x)$ or $t(x)$, and because the least common multiple is the product of all prime factors that appear in either $s(x)$ or $t(x)$, each factor raised to the largest power that appears in either $s(x)$ or $t(x)$.

The division algorithm for polynomials has an important consequence known as the *euclidean algorithm for polynomials*. Given two polynomials $s(x)$ and $t(x)$, their greatest common divisor can be computed by an iterative application of the division algorithm. Without loss of generality, we can suppose that $\deg s(x) \geq \deg t(x)$; the computation is

$$s(x) = Q^{(1)}(x)t(x) + t^{(1)}(x)$$
$$t(x) = Q^{(2)}(x)t^{(1)}(x) + t^{(2)}(x)$$
$$t^{(1)}(x) = Q^{(3)}(x)t^{(2)}(x) + t^{(3)}(x)$$
$$\vdots$$
$$t^{(n-2)}(x) = Q^{(n)}(x)t^{(n-1)}(x) + t^{(n)}(x)$$
$$t^{(n-1)}(x) = Q^{(n+1)}(x)t^{(n)}(x),$$

where the process stops when a remainder of zero is obtained. We claim that the last nonzero remainder $t^{(n)}(x)$ is a scalar multiple of the greatest common divisor. The proof is given in the following theorem.

Theorem 2.7.6 (Euclidean algorithm for polynomials) *Given two polynomials $s(x)$ and $t(x)$ with $\deg s(x) \geq \deg t(x)$, let $s^{(0)}(x) = s(x)$ and $t^{(0)}(x) = t(x)$. The following recursive equations for $r = 1, \ldots, n$:*

$$Q^{(r)}(x) = \left\lfloor \frac{s^{(r-1)}(x)}{t^{(r-1)}(x)} \right\rfloor$$

$$\begin{bmatrix} s^{(r)}(x) \\ t^{(r)}(x) \end{bmatrix} = \begin{bmatrix} 0 & 1 \\ 1 & -Q^{(r)}(x) \end{bmatrix} \begin{bmatrix} s^{(r-1)}(x) \\ t^{(r-1)}(x) \end{bmatrix}$$

2.7 Polynomial rings

satisfy

$$s^{(n)}(x) = \alpha \text{GCD}[s(x), t(x)],$$

where n is the smallest integer for which $t^{(n)}(x) = 0$, and α is a field element.

Proof Because $\deg t^{(r+1)}(x)$ is strictly decreasing, eventually $t^{(n)}(x) = 0$ for some n, so the termination is well-defined. The following matrix inverse is readily verified:

$$\begin{bmatrix} 0 & 1 \\ 1 & -Q^{(r)}(x) \end{bmatrix}^{-1} = \begin{bmatrix} Q^{(r)}(x) & 1 \\ 1 & 0 \end{bmatrix}.$$

Therefore

$$\begin{bmatrix} s(x) \\ t(x) \end{bmatrix} = \left\{ \prod_{\ell=1}^{n} \begin{bmatrix} Q^{(\ell)}(x) & 1 \\ 1 & 0 \end{bmatrix} \right\} \begin{bmatrix} s^{(n)}(x) \\ 0 \end{bmatrix},$$

so $s^{(n)}(x)$ must divide both $s(x)$ and $t(x)$, and hence divides $\text{GCD}[s(x), t(x)]$. Further,

$$\begin{bmatrix} s^{(n)}(x) \\ 0 \end{bmatrix} = \left\{ \prod_{\ell=n}^{1} \begin{bmatrix} 0 & 1 \\ 1 & -Q^{(\ell)}(x) \end{bmatrix} \right\} \begin{bmatrix} s(x) \\ t(x) \end{bmatrix}$$

so that any divisor of both $s(x)$ and $t(x)$ divides $s^{(n)}(x)$. Hence $\text{GCD}[s(x), t(x)]$ both divides $s^{(n)}(x)$ and is divisible by $s^{(n)}(x)$. Thus

$$s^{(n)}(x) = \alpha \text{GCD}[s(x), t(x)],$$

where α is a nonzero field element. This completes the proof of the theorem. □

Again, as in the case of the integer ring, there are two important corollaries. Define the matrix of polynomials

$$A^{(r)}(x) = \prod_{\ell=r}^{1} \begin{bmatrix} 0 & 1 \\ 1 & -Q^{(\ell)}(x) \end{bmatrix}$$

$$= \begin{bmatrix} 0 & 1 \\ 1 & -Q^{(r)}(x) \end{bmatrix} A^{(r-1)}(x),$$

where $A^{(0)}(x)$ is the identity matrix. We then have the following corollary.

Corollary 2.7.7 (Bézout's identity) *For any polynomials $s(x)$ and $t(x)$ over the field F, there exist two other polynomials $a(x)$ and $b(x)$ over the same field such that*

$$\text{GCD}[s(x), t(x)] = a(x)s(x) + b(x)t(x).$$

Proof Because

$$\begin{bmatrix} s^{(n)}(x) \\ 0 \end{bmatrix} = A^{(n)}(x) \begin{bmatrix} s(x) \\ t(x) \end{bmatrix}$$

and $s^{(n)}(x) = \alpha \text{GCD}[s(x), t(x)]$, the corollary follows with $a(x) = \alpha^{-1} A_{11}^{(n)}(x)$ and $b(x) = \alpha^{-1} A_{12}^{(n)}(x)$. □

The polynomials $a(x)$ and $b(x)$ are not unique because, given any $a(x)$ and $b(x)$ satisfying the statement of the corollary, we can also write

$$\text{GCD}[s(x), t(x)] = [a(x) + t(x)]s(x) + [b(x) - s(x)]t(x).$$

Corollary 2.7.8 *For any two coprime polynomials $s(x)$ and $t(x)$ over the field F, there exist two other polynomials, $a(x)$ and $b(x)$, over the same field such that*

$$a(x)s(x) + b(x)t(x) = 1.$$

Proof This is an immediate consequence of Corollary 2.7.7. □

The polynomials $a(x)$ and $b(x)$ are known as the *Bézout polynomials*. They can be obtained as two elements of the matrix $A(x)$ normalized by α. The other two elements of $A(x)$ also have a direct interpretation. We shall need the inverse of the matrix $A^{(r)}(x)$. Because

$$A^{(r)}(x) = \prod_{\ell=r}^{1} \begin{bmatrix} 0 & 1 \\ 1 & -Q^{(\ell)}(x) \end{bmatrix},$$

it is clear that the determinant of $A^{(r)}(x)$ is $(-1)^r$. The inverse is

$$\begin{bmatrix} A_{11}^{(r)}(x) & A_{12}^{(r)}(x) \\ A_{21}^{(r)}(x) & A_{22}^{(r)}(x) \end{bmatrix}^{-1} = (-1)^r \begin{bmatrix} A_{22}^{(r)}(x) & -A_{12}^{(r)}(x) \\ -A_{21}^{(r)}(x) & A_{11}^{(r)}(x) \end{bmatrix}.$$

Corollary 2.7.9 *The $A_{21}^{(n)}(x)$ and $A_{22}^{(n)}(x)$ produced by the euclidean algorithm satisfy*

$$s(x) = (-1)^n A_{22}^{(n)}(x) \alpha \text{GCD}[s(x), t(x)],$$
$$t(x) = -(-1)^n A_{21}^{(n)}(x) \alpha \text{GCD}[s(x), t(x)].$$

Proof Using the above expression for the matrix inverse gives

$$\begin{bmatrix} s(x) \\ t(x) \end{bmatrix} = (-1)^n \begin{bmatrix} A_{22}^{(n)}(x) & -A_{12}^{(n)}(x) \\ -A_{21}^{(n)}(x) & A_{11}^{(n)}(x) \end{bmatrix} \begin{bmatrix} s^{(n)}(x) \\ 0 \end{bmatrix}$$

from which the corollary follows. □

2.7 Polynomial rings

As an example of the euclidean algorithm for polynomials, let $s(x) = x^4 - 1$, and $t(x) = x^3 + 2x^2 + 2x + 1$. Then

$$\begin{bmatrix} s^{(n)}(x) \\ 0 \end{bmatrix} = \begin{bmatrix} 0 & 1 \\ 1 & -\frac{8}{3}x - \frac{4}{3} \end{bmatrix} \begin{bmatrix} 0 & 1 \\ 1 & -\frac{1}{2}x - \frac{1}{4} \end{bmatrix} \begin{bmatrix} 0 & 1 \\ 1 & -x + 2 \end{bmatrix} \begin{bmatrix} s(x) \\ t(x) \end{bmatrix}$$

$$= \begin{bmatrix} -\frac{1}{2}x - \frac{1}{4} & \frac{1}{2}x^2 - \frac{3}{4}x + \frac{1}{2} \\ \frac{4}{3}x^2 + \frac{4}{3}x + \frac{4}{3} & -\frac{4}{3}x^3 + \frac{4}{3}x^2 - \frac{4}{3}x + \frac{4}{3} \end{bmatrix} \begin{bmatrix} x^4 - 1 \\ x^3 + 2x^2 + 2x + 1 \end{bmatrix}$$

$$= \begin{bmatrix} \frac{3}{4}(x+1) \\ 0 \end{bmatrix}.$$

Hence $\text{GCD}[x^4 - 1, x^3 + 2x^2 + 2x + 1] = x + 1$. In addition,

$$x + 1 = \left(-\tfrac{2}{3}x - \tfrac{1}{3}\right)s(x) + \left(\tfrac{2}{3}x^2 - x + \tfrac{2}{3}\right)t(x),$$

as promised by Corollary 2.7.7.

A polynomial $p(x)$ over the field F can be evaluated at any element β of the field F. This is done by substituting the field element β for the indeterminate x to obtain the field element $p(\beta)$, given by

$$p(\beta) = \sum_{i=0}^{\deg p(x)} p_i \beta^i.$$

A polynomial over the field F also can be evaluated at an element of any larger field that contains F. This is done by substituting the element of the extension field for the indeterminate x. When F is the real field, evaluation of a polynomial in an extension field is a familiar concept. For example, polynomials with real coefficients are commonly evaluated over the complex field.

A field element β is called a *zero* of the polynomial $p(x)$ if $p(\beta) = 0$. A polynomial does not necessarily have zeros in its own field. The polynomial $p(x) = x^2 + 1$ has no zeros in the real field.

Theorem 2.7.10 *A nonzero polynomial $p(x)$ has the field element β as a zero if and only if $(x - \beta)$ is a factor of $p(x)$. Furthermore, there are at most n field elements that are zeros of a polynomial of degree n.*

Proof From the division algorithm,

$$p(x) = (x - \beta)Q(x) + s(x),$$

where $s(x)$ has degree less than one. That is, $s(x)$ is a field element s_0. Hence

$$0 = p(\beta) = (\beta - \beta)Q(\beta) + s_0,$$

so that $s_0 = 0$ and hence $p(x) = (x - \beta)Q(x)$. Conversely, if $(x - \beta)$ is a factor of $p(x)$, then

$$p(x) = (x - \beta)Q(x)$$

and $p(\beta) = (\beta - \beta)Q(\beta) = 0$, so that β is a zero of $p(x)$.

Next factor $p(x)$ into a field element times a product of prime polynomials. The degree of $p(x)$ equals the sum of the degrees of the prime polynomial factors, and one such prime polynomial factor exists for each zero. Hence there are at most n zeros. □

Theorem 2.7.11 (Lagrange interpolation) *Let $\beta_0, \ldots, \beta_n$ be a set of $n + 1$ distinct points, and let $p(\beta_k)$ for $k = 0, \ldots, n$ be given. There is exactly one polynomial $p(x)$ of degree n or less that has value $p(\beta_k)$ when evaluated at β_k for $k = 0, \ldots, n$. It is given by*

$$p(x) = \sum_{i=0}^{n} p(\beta_i) \frac{\prod_{j \neq i}(x - \beta_j)}{\prod_{j \neq i}(\beta_i - \beta_j)}.$$

Proof The stated polynomial $p(x)$ passes through the given points, as can be verified by substituting β_k for x. Uniqueness follows because if $p'(x)$ and $p''(x)$ both satisfy the requirements and $P(x) = p'(x) - p''(x)$, then $P(x)$ has degree at most n and has $n + 1$ zeros at β_k for $k = 0, \ldots, n$. Hence $P(x)$ equals the zero polynomial. □

2.8 The Chinese remainder theorem

It is possible to uniquely determine a nonnegative integer given only its moduli with respect to each of several integers, provided that the integer is known to be smaller than the product of the moduli. This is known as the *Chinese remainder theorem*. The Chinese remainder theorem, summarized in Figure 2.2, is proved in two parts. First, we prove the uniqueness of a solution. Then we prove the existence of a solution by giving a procedure for finding it.

Before we develop the theory formally, we will give a simple example. Choose the moduli $m_0 = 3$, $m_1 = 4$, and $m_2 = 5$, and let $M = m_0 m_1 m_2 = 60$. Given the integer c satisfying $0 \leq c < 60$, let $c_i = R_{m_i}[c]$. The Chinese remainder theorem says that there is a one-to-one map between the sixty values that c is allowed to take on and the sixty values that the vector of residues (c_0, c_1, c_2) can take on. Suppose that $c_0 = 2$, $c_1 = 1$, and $c_2 = 2$. These three conditions imply, in turn, that

$c \in \{2, 5, 8, 11, 14, 17, 20, 23, 26, 29, \ldots\}$,

$c \in \{1, 5, 9, 13, 17, 21, 25, 29, 33, \ldots\}$,

$c \in \{2, 7, 12, 17, 22, 27, 32, 37, \ldots\}$.

2.8 The Chinese remainder theorem

Direct equations
$$c_i = R_{m_i}[c] \quad i = 0,\ldots,k$$
where the m_i are coprime.

Inverse equations
$$c = \sum_{i=0}^{k} c_i N_i M_i \pmod{M}$$
where $M = \prod_{i=0}^{k} m_i \quad M_i = M/m_i$
and N_i is the solution of $N_i M_i + n_i m_i = 1$.

Figure 2.2 The Chinese remainder theorem

The unique solution for c is seventeen. Later, we shall give a simple algorithm for finding c from its residues.

The example suggests that the residues uniquely determine the original integer. The following theorem proves this in the general case.

Theorem 2.8.1 *Given a set of integers* $m_0, m_1, \ldots, m_k$ *that are pairwise coprime and a set of integers* $c_0, c_1, \ldots, c_k$ *with* $c_i < m_i$, *then the system of equations*

$$c_i \equiv c \pmod{m_i}, \quad i = 0, \ldots, k$$

has at most one solution for c in the interval

$$0 \le c < \prod_{i=0}^{k} m_i.$$

Proof Suppose that c and c' are solutions in this interval. Then, for each i,
$$c = Q_i m_i + c_i,$$
$$c' = Q'_i m_i + c_i,$$
so $c - c'$ is a multiple of m_i for each i. Then $c - c'$ is a multiple of $\prod_{i=0}^{k} m_i$ because the m_i are pairwise coprime. But $c - c'$ satisfies
$$-\prod_{i=0}^{k} m_i < c - c' < \prod_{i=0}^{k} m_i.$$
The only possibility is $c - c' = 0$. Hence $c = c'$. □

There is a simple way to find the solution to the system of congruences of Theorem 2.8.1, which is based on the corollary to the euclidean algorithm. Corollary 2.6.4 says that, for each s and t, there exist integers a and b that satisfy

$$\text{GCD}[s, t] = as + bt.$$

Therefore, using the set of pairwise coprime integers $m_0, m_1, \ldots, m_k$ as moduli, define $M = \prod_{r=0}^{k} m_r$ and $M_i = M/m_i$. Then $\text{GCD}[M_i, m_i] = 1$, so for each i there exist integers N_i and n_i with

$$N_i M_i + n_i m_i = 1, \qquad i = 0, \ldots, k.$$

We are now ready to prove the following theorem.

Theorem 2.8.2 *Let $M = \prod_{r=0}^{k} m_r$ be a product of pairwise coprime integers; let $M_i = M/m_i$ and for each i, let N_i satisfy $N_i M_i + n_i m_i = 1$. Then the system of congruences*

$$c_i \equiv c \pmod{m_i}, \qquad i = 0, \ldots, k$$

is uniquely solved by

$$c = \sum_{i=0}^{k} c_i N_i M_i \pmod{M}.$$

Proof We need only show that this c solves the specified system of congruences because we already know that the solution is unique. But for this c,

$$c = \sum_{r=0}^{k} c_r N_r M_r \equiv c_i N_i M_i \pmod{m_i}$$

because m_i divides M_r if $r \neq i$. Finally, because

$$N_i M_i + n_i m_i = 1,$$

we have

$$N_i M_i \equiv 1 \pmod{m_i}$$

and

$$c \equiv c_i \pmod{m_i},$$

which completes the proof. $\square$

The earlier example can be continued to illustrate Theorem 2.8.2. Notice that $M = 60$, $M_0 = 20$, $M_1 = 15$, and $M_2 = 12$. Further,

$$1 = (-1)M_0 + 7m_0,$$
$$1 = (-1)M_1 + 4m_1,$$
$$1 = (-2)M_2 + 5m_2,$$

as can be computed from the euclidean algorithm or simply written down by inspection. Therefore

$$N_0 M_0 = -20, \quad N_1 M_1 = -15, \quad N_2 M_2 = -24,$$

and the inverse operation is

$$c = -20c_0 - 15c_1 - 24c_2 \pmod{60}.$$

In particular, if $c_0 = 2$, $c_1 = 1$, and $c_2 = 2$, then

$$c = -103 \pmod{60}$$
$$= 17,$$

as we saw earlier.

On the basis of the Chinese remainder theorem, one can form an alternative system for representing integers, a representation in which multiplication is easy. Suppose that we need to perform the multiplication

$$c = ab.$$

For each i, let $a_i = R_{m_i}[a]$, $b_i = R_{m_i}[b]$, and $c_i = R_{m_i}[c]$. Then for $i = 0, \ldots, k$,

$$c_i = a_i b_i \pmod{m_i}.$$

This can be an easy computation because a_i and b_i are now small integers. Similarly, if instead we had the addition

$$c = a + b,$$

then for $i = 0, \ldots, k$,

$$c_i = a_i + b_i \pmod{m_i}.$$

In either case, the final answer c can be recovered from its residues by using the Chinese remainder theorem.

In this way, by taking residues, large integers are broken down into small pieces that are easy to add, subtract, and multiply. As long as a computation involves only these operations, this representation provides an alternative system of arithmetic. If the computation is simple, then the mapping from the natural representation of the data into the residue representation and the mapping of the answer back into the natural representation more than offset any possible computational advantage. If, however, the computation is lengthy, savings may be found because intermediate values can be left in the residue form. Only the final answer needs to be converted to a conventional integer form.

In a ring of polynomials over any field, there again is a Chinese remainder theorem, summarized in Figure 2.3, which is developed in the same way as for the case of integers.

> **Direct equations**
>
> $$c^{(i)}(x) = R_{m^{(i)}(x)}[c(x)], \quad i = 0,\ldots,k,$$
>
> where the $m^{(i)}(x)$ are coprime.
>
> **Inverse equations**
>
> $$c(x) = \sum_{i=0}^{k} c^{(i)}(x) N^{(i)}(x) M^{(i)}(x) \pmod{M(x)},$$
>
> where $M(x) = \prod_{i=0}^{k} m^{(i)}(x)$, $M^{(i)}(x) = M(x)/m^{(i)}(x)$,
>
> and $N^{(i)}(x)$ is the solution of $N^{(i)}(x)M^{(i)}(x) + n^{(i)}(x)m^{(i)}(x) = 1$.

Figure 2.3 The Chinese remainder theorem for polynomials

Theorem 2.8.3 *Given a set of polynomials $m^{(0)}(x), m^{(1)}(x), \ldots, m^{(k)}(x)$ that are pairwise coprime and a set of polynomials $c^{(0)}(x), c^{(1)}(x), \ldots, c^{(k)}(x)$ with $\deg c^{(i)}(x) < \deg m^{(i)}(x)$, then the system of equations*

$$c^{(i)}(x) \equiv c(x) \pmod{m^{(i)}(x)}, \qquad i = 0, \ldots, k$$

has at most one solution for $c(x)$, satisfying

$$\deg c(x) < \sum_{i=0}^{k} \deg m^{(i)}(x).$$

Proof The proof is similar to the proof of Theorem 2.8.1. Suppose that $c(x)$ and $c'(x)$ are solutions:

$$c(x) = Q^{(i)}(x) m^{(i)}(x) + c^{(i)}(x),$$
$$c'(x) = Q'^{(i)}(x) m^{(i)}(x) + c^{(i)}(x),$$

so $c(x) - c'(x)$ is a multiple of $m^{(i)}(x)$ for each i. Because the $m^{(i)}(x)$ are pairwise coprime, $c(x) - c'(x)$ is a multiple of $\prod_{i=0}^{k} m^{(i)}(x)$, and the degree of $c(x) - c'(x)$ is less than the degree of $\prod_{i=0}^{k} m^{(i)}(x)$. That is, $c(x) - c'(x) = 0$, and the proof is complete. □

The system of congruences can be solved in a way similar to the case of the integer ring. Corollary 2.7.7 states that in a ring of polynomials over a field, given any $s(x)$ and $t(x)$, there exist polynomials $a(x)$ and $b(x)$ that satisfy

$$\text{GCD}[s(x), t(x)] = a(x)s(x) + b(x)t(x).$$

Hence let $M(x) = \prod_{r=0}^{k} m^{(r)}(x)$ and $M^{(i)}(x) = M(x)/m^{(i)}(x)$. Then $\text{GCD}[M^{(i)}(x), m^{(i)}(x)] = 1$. Let $N^{(i)}(x)$ and $n^{(i)}(x)$ be the polynomials that satisfy

$$N^{(i)}(x)M^{(i)}(x) + n^{(i)}(x)m^{(i)}(x) = 1.$$

Theorem 2.8.4 *Let $M(x) = \prod_{r=0}^{k} m^{(r)}(x)$ be a product of pairwise coprime polynomials. Let $M^{(i)}(x) = M(x)/m^{(i)}(x)$ and $N^{(i)}(x)$ satisfy $N^{(i)}(x)M^{(i)}(x) + n^{(i)}(x)m^{(i)}(x) = 1$. Then the system of congruences*

$$c^{(i)}(x) \equiv c(x) \pmod{m^{(i)}(x)}, \qquad i = 0, \ldots, k$$

is uniquely solved by

$$c(x) = \sum_{i=0}^{k} c^{(i)}(x) N^{(i)}(x) M^{(i)}(x) \pmod{M(x)}.$$

Proof We need only show that this $c(x)$ satisfies every congruence in the system of congruences. But

$$c(x) \equiv c^{(i)}(x) N^{(i)}(x) M^{(i)}(x) \pmod{m^{(i)}(x)}$$

because $M^{(r)}(x)$ has $m^{(i)}(x)$ as a factor if $r \neq i$. Then, because

$$N^{(i)}(x)M^{(i)}(x) + n^{(i)}(x)m^{(i)}(x) = 1,$$

we have

$$N^{(i)}(x)M^{(i)}(x) \equiv 1 \pmod{m^{(i)}(x)}$$

and

$$c(x) \equiv c^{(i)}(x) \pmod{m^{(i)}(x)},$$

which completes the proof of the theorem. □

Problems for Chapter 2

2.1   **a** Show that only one group with three elements exists. Construct it and show that it is abelian.

b Show that only two groups with four elements exist. Construct them and show that they are abelian. Show that one of the two groups with four elements has no element of order four. This group is called the *Klein four-group*.

2.2 Let the group operation in the groups of Problem 2.1 be called addition.

a Define multiplication to make the three-element group a ring. Is it unique?

b For each of the two four-element groups, define multiplication to make it a ring. Is each definition unique?

2.3 Which of the three rings in Problem 2.2 are also fields? Can multiplication be defined differently to get a field?

2.4 Prove that, in a cyclic group with q elements, $a^q = a^0$ and $(a^i)^{-1} = a^{q-i}$ for any element a.

2.5 **a** Show that $\mathbf{Z}_2 \times \mathbf{Z}_3$ is isomorphic to $\mathbf{Z}_6$.
b Show that $\mathbf{Z}_2 \times \mathbf{Z}_4$ is not isomorphic to $\mathbf{Z}_8$.

2.6 Give an example of a ring without identity.

2.7 Prove the following standard properties of the discrete Fourier transform, starting with the Fourier transform pair $\{v_i\} \leftrightarrow \{V_k\}$:
a Linearity $\{av_i + bv'_i\} \leftrightarrow \{aV_k + bV'_k\}$;
b Cyclic shift $\{v_{((i-1))}\} \leftrightarrow \{\omega^k V_k\}$;
c Modulation $\{\omega^i v_i\} \leftrightarrow \{V_{((k+1))}\}$.

2.8 Show that the Fourier transform of the vector with components $v_i = \omega^{ri}$, r an integer, has a single nonzero spectral component. Which component is it if $r = 0$? Show that a vector that is nonzero in only a single component has a nonzero spectrum everywhere.

2.9 Prove that if A is a Toeplitz matrix and J is an exchange matrix of the same size, then

$$A^T = JAJ.$$

2.10 **a** Use the euclidean algorithm to find GCD[1573, 308].
b Find integers A and B, satisfying

$$\text{GCD}[1573, 308] = A1573 + B308.$$

2.11 The set of powers of 3 modulo 2^m is a cyclic group. By enumerating the elements of the set for $m = 3, 4, 5, 6$, and 7, show that the order of the group is 2^{m-2} for these values of m. Is the statement true for all values of m not smaller than 3?

2.12 Consider the set $S = \{0, 1, 2, 3\}$ with the operations

+	0	1	2	3		·	0	1	2	3
0	0	1	2	3		0	0	0	0	0
1	1	2	3	0		1	0	1	2	3
2	2	3	0	1		2	0	2	3	1
3	3	0	1	2		3	0	3	1	2

Is this a field?

2.13 Prove that the complex-valued discrete Fourier transform of a data sequence satisfies the symmetry condition

$$V_k = V^*_{n-k}, \quad k = 0, \ldots, n-1$$

if and only if the data sequence is real.

2.14 Let G be an arbitrary group (not necessarily finite). For convenience, call the group operation "multiplication" and call the identity "one." Let g be any element and suppose that ν is the smallest integer, if there is one, such that $g^\nu = 1$, where g^ν means $g * g * \cdots * g$, ν times. Then ν is called the *order* of g. Prove that the subset $\{g, g^2, g^3, \ldots, g^{\nu-1}, g^\nu\}$ is a subgroup of G. Prove that the subgroup is abelian even if G is not.

2.15 Prove that the set of real numbers of the form $\{a + b\sqrt{2}\}$, where a and b are rational, is a field under the conventional arithmetic operations.

2.16 The ring of quaternions consists of all expressions of the form

$$a = a_0 + a_1 i + a_2 j + a_3 k,$$

where a_0, a_1, a_2, and a_3 are real numbers, and i, j, and k are indeterminates. Addition and multiplication are defined by

$$a + b = (a_0 + b_0) + (a_1 + b_1)i + (a_2 + b_2)j + (a_3 + b_3)k$$
$$ab = (a_0 b_0 - a_1 b_1 - a_2 b_2 - a_3 b_3)$$
$$+ (a_1 b_0 + a_0 b_1 - a_3 b_2 + a_2 b_3)i$$
$$+ (a_2 b_0 + a_3 b_1 + a_0 b_2 - a_1 b_3)j$$
$$+ (a_3 b_0 - a_2 b_1 + a_1 b_2 + a_0 b_3)k.$$

Prove that the ring of quaternions is indeed a ring but is not a field. What field property is lacking?

2.17 Prove the following.
  (i) Under ring multiplication, the set of units of a ring forms a group.
  (ii) If $c = ab$ and c is a unit, then a has a right inverse and b has a left inverse.
  (iii) If $c = ab$ and a has no right inverse or b has no left inverse, then c is not a unit.

2.18 The field with three elements, $GF(3)$, is given by the arithmetic tables

+	0	1	2
0	0	1	2
1	1	2	0
2	2	0	1

·	0	1	2
0	0	0	0
1	0	1	2
2	0	2	1

Calculate the determinant of the following matrix and show that its rank is three:

$$M = \begin{bmatrix} 2 & 1 & 2 \\ 1 & 1 & 2 \\ 1 & 0 & 1 \end{bmatrix}.$$

2.19 (Discrete Fourier transform of a permuted sequence) Given the discrete Fourier transform

$$V_k = \sum_{i=0}^{n-1} \omega^{ik} v_i, \qquad i = 0, \ldots, n-1,$$

suppose that a and n are coprime. Let

$$v'_i = v_{((ai))}$$

define a permutation of the components of v. Prove that

$$V'_k = \sum_{i=0}^{n-1} \omega^{ik} v'_i$$

is a permutation of the components of V, given by $V'_k = V_{((bk))}$ for some b coprime to n.

2.20 A year has at most 366 days. Suppose that all months except the last have 31 days.
  a Is it possible to uniquely determine the day of the year, given the day of the month and the day of the week?
  b Suppose next that a month has 31 days and a week has twelve days. Is it now possible to uniquely determine the day of the year, given the day of the month and the day of the week?
  c Using a 31-day month and a twelve-day week, give a formula for the day of the year when given the day of month and day of week.
  d Work through some numerical examples.

2.21 How many vectors are there in the vector space $GF(2)^n$?

2.22 Is it true that if x, y, and z are linearly independent vectors over $GF(q)$, then so also are $x + y$, $y + z$, and $z + x$?

2.23 If S and T are distinct two-dimensional subspaces of a three-dimensional vector space, show that their intersection is a one-dimensional subspace.

2.24 Let S be any finite set. Let G be the set of subsets of S. If A and B are two subsets, let $A \cup B$ denote the set of elements in either A or B, let $A \cap B$ denote the set of elements in both A and B, and let $A - B$ denote the set of elements in A but not in B.
  a Show that G with the operation $*$ as set union $\cup$ is not a group.
  b The set operation of symmetric difference $\triangle$ is given by

  $$A \triangle B = (A - B) \cup (B - A).$$

  Show that G with $*$ as the operation of symmetric difference does give a group. Is it abelian?
  c Show that G, together with the operations $\triangle$ and $\cap$, gives a ring. Is it a commutative ring? Is there a unit?

Notes for Chapter 2

This chapter deals with standard topics in modern algebra. Many textbooks can be found that cover the material more thoroughly. The book by Birkhoff and MacLane (1941) is intended as an introductory text and is easily understood at the level of this book. The two-volume work by Van der Waerden (1949, 1953) is a more advanced work, addressed primarily to mathematicians, and goes more deeply into many topics. The material on linear algebra and matrix theory can also be found in textbooks written specifically for these topics. The book by Thrall and Tornheim (1957) is especially suitable because it does not presuppose that the underlying field is the real field or the complex field as do many other books. Pollard (1971) explicitly put forth the notion of a Fourier transform in an arbitrary field.

The Galois fields are named for Évariste Galois (1811–1832). The abelian groups are named for Niels Henrik Abel (1802–1829).

3 Fast algorithms for the discrete Fourier transform

One of our major goals is the development of a collection of techniques for computing the discrete Fourier transform. We shall find many such techniques, each with different advantages and each best-used in different circumstances. There are two basic strategies. One strategy is to change a one-dimensional Fourier transform into a two-dimensional Fourier transform of a form that is easier to compute. The second strategy is to change a one-dimensional Fourier transform into a small convolution, which is then computed by using the techniques described in Chapter 5. Good algorithms for computing the discrete Fourier transform will use either or both of these strategies to minimize the computational load. In Chapter 6, we shall describe how the fast Fourier transform algorithms are used to perform, in conjunction with the convolution theorem, the cyclic convolutions that are used to compute the long linear convolutions forming the output of a digital filter.

Throughout the chapter, we shall usually regard the complex field as the field of the computation, or perhaps the real field. However, most of the algorithms we study do not depend on the particular field over which the Fourier transform is defined. In such cases, the algorithms are valid in an arbitrary field. In some cases, the general idea behind an algorithm does not depend on the field over which the Fourier transform is defined, but some small detail of the algorithm may depend on the field. Then the algorithm would need to be derived or verified for each field of interest.

3.1 The Cooley–Tukey fast Fourier transform

The Fourier transform of a vector v,

$$V_k = \sum_{i=0}^{n-1} \omega^{ik} v_i,$$

as it is written, requires on the order of n^2 multiplications and n^2 additions. If n is composite, there are several ways to change this one-dimensional Fourier transform into a two-dimensional Fourier transform, or something similar to it. This changes the

3.1 The Cooley–Tukey fast Fourier transform

Cooley–Tukey FFT (1965)

$$n = n'n''$$

$$i = i' + n'i'' \qquad \begin{matrix} i' = 0,\ldots,n'-1; \\ i'' = 0,\ldots,n''-1 \end{matrix}$$

$$k = n''k' + k'' \qquad \begin{matrix} k' = 0,\ldots,n'-1; \\ k'' = 0,\ldots,n''-1 \end{matrix}$$

$$V_{k'k''} = \sum_{i'=0}^{n'-1} \beta^{i'k'} \left[\omega^{i'k''} \sum_{i''=0}^{n''-1} \gamma^{i''k''} v_{i'i''} \right]$$

Number of multiplications $\approx n(n' + n'') + n$

Figure 3.1 Cooley–Tukey FFT

computation to a form that is much more efficient; but the price is an increased difficulty of understanding. Algorithms of this kind are known collectively as *fast Fourier transform (FFT) algorithms*. Figure 3.1 summarizes the structure of the Cooley–Tukey FFT algorithm – the algorithm that will be studied in this section. Figure 3.1 should be compared with Figure 3.8 of Section 3.3, which summarizes the structure of the Good–Thomas FFT algorithm.

To derive the general form of the Cooley–Tukey FFT algorithm, suppose that $n = n'n''$. Replace each of the indices in the above expression for the Fourier transform by a coarse index and a vernier index as follows:

$$i = i' + n'i'', \qquad \begin{matrix} i' = 0, \ldots, n'-1, \\ i'' = 0, \ldots, n''-1; \end{matrix}$$

$$k = n''k' + k'', \qquad \begin{matrix} k' = 0, \ldots, n'-1, \\ k'' = 0, \ldots, n''-1. \end{matrix}$$

Then

$$V_{n''k'+k''} = \sum_{i''=0}^{n''-1} \sum_{i'=0}^{n'-1} \omega^{(i'+n'i'')(n''k'+k'')} v_{i'+n'i''}.$$

Expand the product in the exponent and let $\omega^{n'} = \gamma$ and $\omega^{n''} = \beta$. Because ω has order $n'n''$, the term $\omega^{n'n''i''k'} = 1$ and can be dropped. Now define the two-dimensional variables, which we also call v and V, given by

$$v_{i'i''} = v_{i'+n'i''}, \qquad \begin{matrix} i' = 0, \ldots, n'-1, \\ i'' = 0, \ldots, n''-1; \end{matrix}$$

$$V_{k'k''} = V_{n''k'+k''}, \qquad \begin{matrix} k' = 0, \ldots, n'-1, \\ k'' = 0, \ldots, n''-1. \end{matrix}$$

Fast algorithms for the discrete Fourier transform

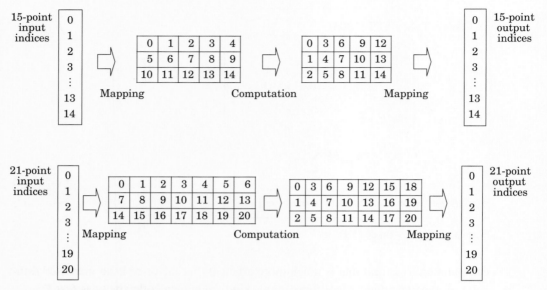

Figure 3.2 Examples of Cooley–Tukey address shuffling

In this way, the input and output data vectors are mapped into two-dimensional arrays. Observe that the components of the transform V are found arranged differently in the array than are the components of the signal v. This is known as *address shuffling*. In terms of the two-dimensional variables, the formula becomes

$$V_{k'k''} = \sum_{i'=0}^{n'-1} \beta^{i'k'} \left[\omega^{i'k''} \sum_{i''=0}^{n''-1} \gamma^{i''k''} v_{i'i''} \right].$$

Although this form is more difficult to understand than the original, the number of multiplications and additions is much less. In fact, at most $n(n' + n'' + 1)$ complex multiplications and $n(n' + n'' - 2)$ complex additions are required, compared to about n^2 complex multiplications and n^2 complex additions previously.

The computations of the Cooley–Tukey FFT can be visualized as mapping a two-dimensional signal-domain array into a two-dimensional transform-domain array, as shown in Figure 3.2 for $n = 15$ and for $n = 21$. The computation consists of an n''-point discrete Fourier transform on each column, followed by an element-by-element complex multiplication throughout the new array by $\omega^{i'k''}$, followed by an n'-point discrete Fourier transform on each row.

To compute the number of complex multiplications and the number of complex additions when the input vector v is complex, suppose that the inner Fourier transform and the outer Fourier transform are each computed in the obvious way with n''^2 and n'^2 complex multiplications, respectively, and with $n''(n'' - 1)$ and $n'(n' - 1)$ complex additions, respectively. Each of these Fourier transforms is computed n' times and n'' times, respectively. Besides these computations, there are $n'n''$ additional complex multiplications needed for the element-by-element adjustment terms $\omega^{i'k''}$. Therefore the number of complex multiplications $M_C(n)$ and the number of complex additions $A_C(n)$ are given by

$$M_C(n) = n'(n'')^2 + n''(n')^2 + n'n''$$
$$= n(n' + n'' + 1),$$
$$A_C(n) = n'n''(n'' - 1) + n''n'(n' - 1)$$
$$= n(n' + n'' - 2),$$

as we have already asserted. Among the multiplications are some trivial multiplications by one. These occur in the adjustment terms whenever i' or k'' equals zero. If one wants to take the care, the algorithm can skip these multiplications. Then the number of complex multiplications is

$$M_C(n) = n(n' + n'') + (n' - 1)(n'' - 1)$$
$$= (n - 1)(n' + n'') + (n + 1).$$

The inner Fourier transform and the outer Fourier transform can themselves be computed by a fast algorithm, not necessarily the Cooley–Tukey FFT. Then the number of complex multiplications $M_C(n)$ and the number of complex additions $A_C(n)$ needed by the Cooley–Tukey FFT will satisfy

$$M_C(n) = n'M_C(n'') + n''M_C(n') + n,$$
$$A_C(n) = n'A_C(n'') + n''A_C(n'),$$

where the new $M_C(n')$, $M_C(n'')$, $A_C(n')$, and $A_C(n'')$ on the right side denote the number of complex multiplications and complex additions needed by the selected n'-point fast Fourier transform algorithm and the selected n''-point fast Fourier transform algorithm, respectively. Of course, if n' or n'' is itself composite, the smaller transforms may themselves be computed by the Cooley–Tukey FFT. In this way, a transform whose size n equals $\prod_\ell n_\ell$ can be broken down into a form requiring about $n \sum_\ell n_\ell$ complex multiplications. Figure 3.3 shows one way of several that a 75-point Fourier transform can be broken down. We may want to think of the intermediate nodes in that figure as representing calls to a subroutine and the terminal nodes as the actual subroutines. If the computations are to be so broken into subroutines, then one way to do it is shown in Figure 3.4. At the bottom level are a three-point Fourier transform and a five-point Fourier transform that are computed explicitly. Later, we shall study the Winograd small FFT, which can be used instead for these routines at the bottom level. The Winograd small FFTs are highly optimized routines (by one criterion of optimality) designed to compute a small Fourier transform whose blocklength is a small prime or a power of a small prime.

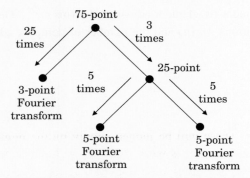

Figure 3.3 Structure of a 75-point Cooley–Tukey FFT

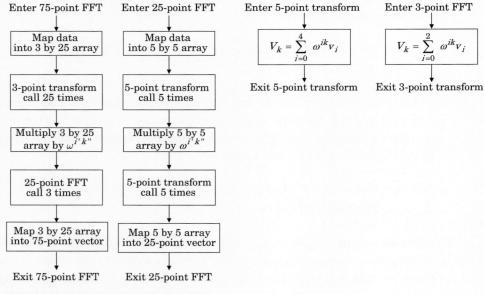

Figure 3.4 Subroutines for a 75-point FFT

3.2 Small-radix Cooley–Tukey algorithms

Many applications of the Cooley–Tukey FFT use a blocklength n that is a power of two or of four. The blocklength 2^m is factored either as $2 \cdot 2^{m-1}$ or as $2^{m-1} \cdot 2$ to form the FFT. The FFT then is called a *radix-two*[1] Cooley–Tukey FFT. Similarly, the blocklength 4^m is factored either as $4 \cdot 4^{m-1}$ or as $4^{m-1} \cdot 4$. The FFT then is called a *radix-four* Cooley–Tukey FFT. In this section, we shall describe many variations of

---

[1] The term radix-two refers to the fact that the indices are represented to the base two. The data components may be in any number representation, including a base-two representation.

3.2 Small-radix Cooley–Tukey algorithms

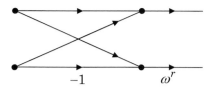

Figure 3.5 Decimation-in-time butterfly

the radix-two and radix-four FFTs. These will be judged simply by the number of multiplications and additions. However, it takes work to avoid work, and one must choose an algorithm only after all considerations are examined.

The 2^m-point radix-two Cooley–Tukey FFT with n' equal to 2 and n'' equal to 2^{m-1} is known as a *decimation-in-time* radix-two Cooley–Tukey FFT. By setting $n' = 2$ and $n'' = n/2$ in the expression of Figure 3.1, the equations of the FFT then can be put into the simple form

$$V_k = \sum_{i=0}^{(n/2)-1} \omega^{2ik} v_{2i} + \omega^k \sum_{i=0}^{(n/2)-1} \omega^{2ik} v_{2i+1},$$

$$V_{k+n/2} = \sum_{i=0}^{(n/2)-1} \omega^{2ik} v_{2i} - \omega^k \sum_{i=0}^{(n/2)-1} \omega^{2ik} v_{2i+1}$$

for $k = 0, \ldots, (n/2) - 1$, where we have used the fact that $\beta = \omega^{n/2} = -1$. The decimation-in-time FFT breaks the input data vector v into the set of components with odd index and the set of components with even index. The output transform vector V is broken into the set containing the first $n/2$ components and the set containing the second $n/2$ components. The decimation-in-time Cooley–Tukey FFT calls for the computation of two Fourier transforms, given by

$$V'_k = \sum_{i=0}^{(n/2)-1} \mu^{ik} v_{2i}$$

and

$$V''_k = \sum_{i=0}^{(n/2)-1} \mu^{ik} v_{2i+1},$$

where $\mu = \omega^2$. Each of these expressions is a Fourier transform of blocklength $n/2$. Then V is obtained by the equations

$$V_k = V'_k + \omega^k V''_k,$$
$$V_{k+n/2} = V'_k - \omega^k V''_k$$

for $k = 0, \ldots, (n/2) - 1$. This is illustrated by the so-called decimation-in-time "butterfly" in Figure 3.5.

Fast algorithms for the discrete Fourier transform

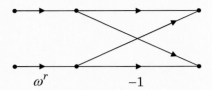

Figure 3.6 Decimation-in-frequency butterfly

The 2^m-point radix-two Cooley–Tukey algorithm, with n' equal to $n/2$ and n'' equal to 2 in the expression of Figure 3.1, is known as a *decimation-in-frequency* radix-two Cooley–Tukey FFT. The equations of this FFT are

$$V_{2k'} = \sum_{i'=0}^{(n/2)-1} (v_{i'} + v_{i'+n/2})\omega^{2i'k'},$$

$$V_{2k'+1} = \sum_{i'=0}^{(n/2)-1} (v_{i'} - v_{i'+n/2})\omega^{i}\omega^{2i'k'}$$

for $k' = 0, \ldots, (n/2) - 1$. This is illustrated by the so-called decimation-in-frequency "butterfly" in Figure 3.6. The decimation-in-frequency FFT breaks the input data vector v into the first $n/2$ components and the second $n/2$ components. The output Fourier transform vector V is broken into the set of components with odd index and the set of components with even index. By taking the sum and difference of two half data vectors, the decimation-in-frequency FFT sets up two Fourier transforms of blocklength $n/2$.

The decimation-in-time algorithm and the decimation-in-frequency algorithm are different in structure and in the sequence of the computations, but the number of computations is the same. Also, the performance is the same, but the user may prefer one of them because of some unique implementation considerations. We shall study in detail only the performance of the decimation-in-time algorithm.

The decimation-in-time algorithm changes an n-point Fourier transform into two $n/2$-point Fourier transforms plus some extra additions and multiplications. Some of the multiplications are multiplications by one or by j. These are trivial and need no actual computation. To by-pass the trivial multiplications, however, does require that such special cases be handled separately in the implementation. Sometimes, to make the implementation clean, the designer will elect to execute all multiplications, even the trivial ones. We begin with this case.

The decimation-in-time algorithm is used recursively, at each level replacing an n-point Fourier transform by two $n/2$-point Fourier transforms, which, in turn, are broken down in the same way. By examining the equations, it is easy to see that the number of complex multiplications for an n-point FFT, $M_C(n)$, satisfies the recursion

$$M_C(n) = 2M_C(n/2) + n/2$$

3.2 Small-radix Cooley–Tukey algorithms

and the number of complex additions satisfies

$$A_C(n) = 2A_C(n/2) + n,$$

where n is a power of two. These recursions are satisfied by

$$M_C(n) = \tfrac{1}{2}n \log_2 n,$$
$$A_C(n) = n \log_2 n.$$

Each complex multiplication can be implemented with four real multiplications and two real additions. The performance of the radix-two FFT is then measured by

$$M_R(n) = 2n \log_2 n,$$
$$A_R(n) = 3n \log_2 n.$$

Alternatively, each complex multiplication can be implemented with three real multiplications and three real additions. Then the performance is described by

$$M_R(n) = \tfrac{3}{2}n \log_2 n,$$
$$A_R(n) = \tfrac{7}{2}n \log_2 n.$$

Suppose, instead, that we are willing to build into the algorithm a provision to skip the trivial multiplications. Then these performance measures will go down. A careful analysis will show that all multiplications in the innermost stage are trivial, given by $(-1)^k$ for $k = 0, 1$; all multiplications in the next stage are trivial, given by j^k for $k = 0, 1, 2, 3$; and in subsequent stages the numbers of trivial multiplications are $n/4$, $n/8$, and so forth. Therefore the number of complex multiplications is

$$M_C(n) = \tfrac{1}{2}n(-3 + \log_2 n) + 2.$$

When using four real multiplications and two real additions to compute a complex multiplication, the performance of the radix-two FFT is described by

$$M_R(n) = 2n(-3 + \log_2 n) + 8,$$
$$A_R(n) = 3n(-1 + \log_2 n) + 4$$

real multiplications and real additions. When using three real multiplications and three real additions to compute a complex multiplication, the performance is described by

$$M_R(n) = \tfrac{3}{2}n(-3 + \log_2 n) + 6,$$
$$A_R(n) = \tfrac{1}{2}n(-9 + 7\log_2 n) + 6$$

real multiplications and real additions.

There is still one more symmetry within the trigonometric functions that can be used to perform even a little better. Notice that

$$\omega^{n/8} = (1 - j)/\sqrt{2}.$$

Fast algorithms for the discrete Fourier transform

Table 3.1 *Performance of some Cooley–Tukey FFT algorithms*

Blocklength n	Basic radix-two complex FFT		Fully optimized* radix-two complex FFT		Rader–Brenner radix-two complex FFT	
	Number of real multiplications	Number of real additions	Number of real multiplications	Number of real additions	Number of real multiplications	Number of real additions
8	48	72	4	52	4	64
16	128	192	24	152	20	192
32	320	480	88	408	68	512
64	768	1152	264	1032	196	1280
128	1792	2688	712	2504	516	3072
256	4096	6144	1800	5896	1284	7168
512	9216	13824	4360	13576	3076	16384
1024	20480	30720	10248	30728	7172	36864
2048	45056	67584	23560	68616	16388	81920
4096	98304	147456	59256	151560	36868	180224

** • Complex multiplication using three real multiplications and three real additions
 • Trivial multiplications (by ±1 or ±j) not counted
 • Symmetries of trigonometric functions fully used

Multiplication by this complex constant requires only two real multiplications and two real additions. There are $n/4$, $n/8$, ... such multiplications in stages 3, 4, ... These can be handled by a special multiplication procedure. In such an implementation, the performance is measured by

$$M_R(n) = n(-7 + 2\log_2 n) + 12,$$
$$A_R(n) = 3n(-1 + \log_2 n) + 4$$

real multiplications and real additions, or

$$M_R(n) = \tfrac{1}{2}n(-10 + 3\log_2 n) + 8,$$
$$A_R(n) = \tfrac{1}{2}n(-10 + 7\log_2 n) + 8,$$

depending on which complex multiplication rule is used.

We can see that the number of variations of the radix-two Cooley–Tukey algorithm is quite large, but we are still not finished. There are even more options available. Table 3.1 summarizes the performance of some Cooley–Tukey FFTs. In addition to the ordinary form of the Cooley–Tukey FFT, the figure also includes the *Rader–Brenner fast Fourier transform*. This FFT algorithm is a variation of the Cooley–Tukey FFT algorithm and is based on the observation that some of the multiplications by complex constants can be made into multiplications by real constants by rearranging the equations.

3.2 Small-radix Cooley–Tukey algorithms

The Rader–Brenner FFT can be developed by starting from the decimation-in-time equations or from the decimation-in-frequency equations. We choose to start with the decimation-in-frequency equations:

$$V_{2k} = \sum_{i=0}^{(n/2)-1} (v_i + v_{i+n/2})\omega^{2ik},$$

$$V_{2k+1} = \sum_{i=0}^{(n/2)-1} (v_i - v_{i+n/2})\omega^i \omega^{2ik}$$

for $k = 0, \ldots, (n/2) - 1$. Define a working vector a, given by

$$a_i = \begin{cases} 0, & i = 0, \\ (v_i - v_{i+n/2})/[2j \sin(2\pi i/n)], & i = 1, \ldots, (n/2) - 1, \end{cases}$$

and let

$$A_k = \sum_{i=0}^{(n/2)-1} \omega^{2ik} a_i, \qquad k = 0, \ldots, \frac{n}{2} - 1.$$

We can relate V_{2k+1} to A_k as follows:

$$A_{k+1} - A_k = \sum_{i=0}^{(n/2)-1} \omega^{2ik} a_i (\omega^{2i} - 1)$$

$$= \sum_{i=0}^{(n/2)-1} \omega^{2ik} \omega^i a_i 2j \sin(2\pi i/n)$$

$$= \sum_{i=1}^{(n/2)-1} (v_i - v_{i+n/2})\omega^i \omega^{2ik}.$$

Therefore

$$V_{2k+1} = A_{k+1} - A_k + (V_0 - V_{n/2})$$

for $k = 0, \ldots, (n/2) - 1$. Thus we have replaced the multiplication by the complex constant ω^i with multiplication by the imaginary constant $[2j \sin(2\pi i/n)]^{-1}$, which reduces the computational complexity. Care must be taken, however, because when n is large and i is small, the new constants can become quite large, so wordlength problems may arise.

A summary of the Rader–Brenner FFT algorithm is shown in Figure 3.7. Each of the two $n/2$-point discrete Fourier transforms can, in turn, be broken down in the same way. At each stage the algorithm requires $(n/2) - 2$ multiplications of a complex number by an imaginary number, for a total of $n - 4$ real multiplications. (We choose not to count the multiplication by $\frac{1}{2}$ when i equals $n/4$.) There is a total of $2n$ complex additions per stage, or $4n$ real multiplications. The performance of the Rader–Brenner

$$a_i = \begin{cases} 0, & i = 0, \\ j(v_{i+n/2} - v_i)/[2\sin(2\pi i/n)], & i = 1,\ldots,(n/2)-1, \end{cases}$$

$$A_k = \sum_{i=0}^{(n/2)-1} a_i \omega^{2ik}, \quad k = 0,\ldots,(n/2)-1,$$

$$V_{2k} = \sum_{i=0}^{(n/2)-1} (v_i + v_{i+n/2}) \omega^{2ik}, \quad k = 0,\ldots,(n/2)-1,$$

$$V_{2k+1} = A_{k+1} - A_k + (V_0 - V_{n/2}), \quad k = 0,\ldots,(n/2)-1,$$

Figure 3.7 Rader–Brenner FFT algorithm

algorithm is measured by the recursive equations

$M_R(n) = n - 4 + 2M_R(n/2),$
$A_R(n) = 4n + 2A_R(n/2),$

with initial conditions

$M_R(4) = 0,$
$A_R(4) = 16.$

We have not counted the multiplications by ± 1 or $\pm j$ that occur in the innermost two stages.

The performance of the Rader–Brenner FFT is shown in Table 3.1 in comparison with other forms of the Cooley–Tukey FFT. Notice that the fully optimized radix-two FFT for small blocklength has fewer additions. This suggests a hybrid approach, breaking down the Fourier transform with the Rader–Brenner algorithm until a blocklength of sixteen is reached; then continuing with the fully optimized radix-two algorithm. Even better performance can be obtained by using a Winograd sixteen-point FFT, discussed in Section 3.6, at the innermost stage. The recursive equations given above still apply, but now with initial conditions

$M_R(16) = 20,$
$A_R(16) = 148.$

The radix-four Cooley–Tukey FFT algorithms are also popular. They can be used when the blocklength n is a power of four and so factorable as $4 \cdot 4^{m-1}$ or as $4^{m-1} \cdot 4$. We will discuss only the decimation-in-time radix-four Cooley–Tukey FFT. The equations of this FFT can be obtained in a simple form by setting $n' = 4$ and $n'' = n/4$ in the general equation for the Cooley–Tukey FFT, given in Figure 3.1. Then for $k = 0, \ldots, (n/4) - 1$,

$$\begin{bmatrix} V_k \\ V_{k+n/4} \\ V_{k+n/2} \\ V_{k+3n/4} \end{bmatrix} = \begin{bmatrix} 1 & 1 & 1 & 1 \\ 1 & -j & -1 & j \\ 1 & -1 & 1 & -1 \\ 1 & j & -1 & -j \end{bmatrix} \begin{bmatrix} \sum_{i=0}^{(n/4)-1} \omega^{4ik} v_{4i} \\ \omega^k \sum_{i=0}^{(n/4)-1} \omega^{4ik} v_{4i+1} \\ \omega^{2k} \sum_{i=0}^{(n/4)-1} \omega^{4ik} v_{4i+2} \\ \omega^{3k} \sum_{i=0}^{(n/4)-1} \omega^{4ik} v_{4i+3} \end{bmatrix}.$$

For each of $n/4$ values of k, there is such a matrix equation giving four values of the transform. In this way, the n-point Fourier transform is replaced with four $n/4$-point Fourier transforms plus some supporting computations. As written, the matrix equation has only three distinct complex multiplications and twelve complex additions for each k as supporting computations. The innermost FTT stage – a four-point Fourier transform – has no multiplications.

The number of additions can be reduced further. Rewrite the equation for $k = 0, \ldots, (n/4) - 1$ as

$$\begin{bmatrix} V_k \\ V_{k+n/4} \\ V_{k+n/2} \\ V_{k+3n/4} \end{bmatrix} = \begin{bmatrix} 1 & 0 & 1 & 0 \\ 0 & 1 & 0 & -j \\ 1 & 0 & -1 & 0 \\ 0 & 1 & 0 & j \end{bmatrix} \begin{bmatrix} 1 & 0 & 1 & 0 \\ 1 & 0 & -1 & 0 \\ 0 & 1 & 0 & 1 \\ 0 & 1 & 0 & -1 \end{bmatrix} \begin{bmatrix} \sum_{i=0}^{(n/4)-1} \omega^{4ik} v_{4i} \\ \omega^k \sum_{i=0}^{(n/4)-1} \omega^{4ik} v_{4i+1} \\ \omega^{2k} \sum_{i=0}^{(n/4)-1} \omega^{4ik} v_{4i+2} \\ \omega^{3k} \sum_{i=0}^{(n/4)-1} \omega^{4ik} v_{4i+3} \end{bmatrix}.$$

Now only eight additions are needed to execute the factored form of the four by four matrix, compared to twelve previously. This is the final form of the radix-four Cooley–Tukey FFT. The performance is described by

$$M_C(n) = \tfrac{3}{4} n (\log_4 n - 1) = \tfrac{3}{8} n (\log_2 n - 2),$$
$$A_C(n) = 2n \log_4 n = n \log_2 n$$

complex multiplications and complex additions, respectively.

If the complex multiplications are computed by using three real multiplications and three real additions, the performance of the basic algorithm is measured by the equations

$$M_R(n) = \tfrac{9}{8} n (\log_2 n - 2),$$
$$A_R(n) = \tfrac{25}{8} n \log_2 n - \tfrac{9}{4} n.$$

It is possible to do even better if one wishes to design enough logic in the program to catch all multiplications by ± 1, by $\pm j$, or by odd powers of $(1/\sqrt{2})(1 - j)$ because these multiplications do not require three real multiplications. The performance

Table 3.2 *Performance of some radix-four FFT algorithms*

Blocklength n	Basic radix-four complex FFT		Fully optimized* radix-four complex FFT	
	Number of real multiplications	Number of real additions	Number of real multiplications	Number of real additions
4	0	16	0	16
16	36	164	20	148
64	288	1128	208	976
256	1728	5824	1392	5488
1024	9216	29696	7856	28336
4096	46080	144384	40642	138928

*• Complex multiplication using 3 real multiplications and 3 real additions
• Trival multiplications (by ±1 or ±j) not counted

equations then become

$$M_R(n) = \tfrac{9}{8} n \log_2 n - \tfrac{43}{12} n + \tfrac{16}{3},$$
$$A_R(n) = \tfrac{25}{8} n \log_2 n - \tfrac{43}{12} n + \tfrac{16}{3}.$$

The performance of the radix-four Cooley–Tukey FFT is summarized in Table 3.2.

3.3 The Good–Thomas fast Fourier transform

The second type of FFT algorithm is the Good–Thomas fast Fourier transform. It is based on factorization of the blocklength into distinct prime powers. This algorithm is a little more complicated conceptually than the Cooley–Tukey algorithm, but is a little simpler computationally. The Good–Thomas algorithm, summarized in Figure 3.8, is another way of organizing a linear array of $n = n'n''$ numbers into an n' by n'' array, but in such a way that a one-dimensional Fourier transform can be turned into a true two-dimensional Fourier transform. The idea is very different from the idea of the Cooley–Tukey algorithm. Now n' and n'' must be coprime, and the mapping is based on the Chinese remainder theorem. Refer to Figure 3.9 to see how the input data is arranged. It is stored in the two-dimensional array by starting in the upper left corner and listing the components down the "extended diagonal." Because the number of rows and the number of columns are coprime, the extended diagonal passes through every element of the array. After a true two-dimensional Fourier transform, the transform appears in another two-dimensional array. The order of the output components in the

3.3 The Good–Thomas fast Fourier transform

Good–Thomas FFT (1960–1963)

$n = n'n''$ coprime

Scramble input indices

$$\left.\begin{array}{l} i' = i \pmod{n'} \\ i'' = i \pmod{n''} \end{array}\right\} \leftrightarrow \begin{cases} i = i'N''n'' + i''N'n' \pmod{n} \\ \text{where} \\ N'n' + N''n'' = 1 \end{cases}$$

Scramble output indices

$$\left.\begin{array}{l} k' = N''k \pmod{n'} \\ k'' = N'k \pmod{n''} \end{array}\right\} \leftrightarrow k = n''k' + n'k'' \pmod{n}$$

$$V_{k'k''} = \sum_{i'=0}^{n'-1} \beta^{i'k'} \left[\sum_{i''=0}^{n''-1} \gamma^{i''k''} v_{i'i''} \right]$$

Number of multiplications $\approx n(n' + n'')$

Figure 3.8 Good–Thomas FFT

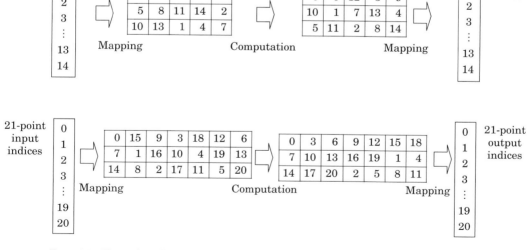

Figure 3.9 Examples of Good–Thomas shuffling

output array, however, is different from the order of input components in the input array. The ordering of the input and output arrays is described below.

The derivation of the Good–Thomas FFT algorithm is based on the Chinese remainder theorem for integers. The input index is described by its residues as follows:

$$i' = i \pmod{n'},$$
$$i'' = i \pmod{n''}.$$

This is the map of the input index i down the extended diagonal of a two-dimensional array indexed by (i', i''). By the Chinese remainder theorem, there exist integers N' and N'' such that the input index can be recovered as follows:

$$i = i'N''n'' + i''N'n' \pmod{n},$$

where N' and N'' are the integers that satisfy

$$N'n' + N''n'' = 1.$$

The output index is described somewhat differently. Define

$$k' = N''k \pmod{n'},$$
$$k'' = N'k \pmod{n''}.$$

For this purpose, N' and N'' can be reduced modulo n' and modulo n'', respectively. The output index k is recovered as follows:

$$k = n''k' + n'k'' \pmod{n}.$$

To verify this, write it out:

$$k = n''(N''k + Q_1 n') + n'(N'k + Q_2 n'') \pmod{n'n''}$$
$$= k(n''N'' + n'N') \pmod{n}$$
$$= k.$$

Now, with these new indices, we convert the Fourier transform

$$V_k = \sum_{i=0}^{n-1} \omega^{ik} v_i$$

into the formula

$$V_{n''k' + n'k''} = \sum_{i''=0}^{n''-1} \sum_{i'=0}^{n'-1} \omega^{(i'N''n'' + i''N'n')(n''k' + n'k'')} v_{i'N''n'' + i''N'n'}.$$

Multiply out the exponent. Because ω has order $n'n''$, terms in the exponent of ω involving $n'n''$ can be dropped. Treat the input and output vectors as two-dimensional arrays by using the index transformations given above to replace $n''k' + n'k''$ by (k', k'')

and to replace $v_{i'N''n''+i''N'n'}$ by (i', i''). Then

$$V_{k'k''} = \sum_{i'=0}^{n'-1} \sum_{i''=0}^{n''-1} \omega^{N''(n'')^2 i'k'} \omega^{N'(n')^2 i''k''} v_{i'i''}$$

$$= \sum_{i'=0}^{n'-1} \sum_{i''=0}^{n''-1} \beta^{i'k'} \gamma^{i''k''} v_{i'i''},$$

where $\beta = \omega^{N''(n'')^2}$ and $\gamma = \omega^{N'(n')^2}$. The terms β and γ are an n'th root of unity and an n''th root of unity, respectively, which are needed for the n'-point Fourier transform and the n''-point Fourier transform. To see this for β, notice that $\beta = (\omega^{n''})^{N''n''}$. Because $\omega^{n''} = e^{-j2\pi/n'}$, and $N''n'' = 1$ modulo n', we see that $\beta = e^{-j2\pi/n'}$. A similar analysis shows that $\gamma = e^{-j2\pi/n''}$.

The equation is now in the form of an n' by n'' two-dimensional Fourier transform. The number of multiplications is about $n(n' + n'')$, and the number of additions is about the same. The Fourier transform on the rows or on the columns can, in turn, be simplified by another application of the fast Fourier transform if the size is composite. In this way, a transform whose size n has coprime factors n_ℓ can be broken down into a form requiring about $n \sum_\ell n_\ell$ multiplications and $n \sum_\ell n_\ell$ additions.

One can choose either the Cooley–Tukey algorithm or the Good–Thomas algorithm to do Fourier transforms. It is even possible to build a Fourier transform algorithm by using both the Cooley–Tukey FFT and the Good–Thomas FFT. For example, a 63-point transform can be broken into a seven-point transform and a nine-point transform by using the Good–Thomas FFT; the nine-point transform can then be broken into two three-point transforms by using the Cooley–Tukey FFT. One then has a computation in a form similar to a three by three by seven three-dimensional Fourier transform. Figure 3.10 shows some ways that a 1000-point discrete Fourier transform could be built out of smaller pieces. Each example uses a two-point module three times and a five-point module three times. They are, however, distinctly different procedures; the small modules are used in a different order. The number of multiplications and additions, the sensitivity to computational noise, and the ease of implementation will be different.

3.4 The Goertzel algorithm

A single component of a Fourier transform can be computed by *Horner's rule*. This is a way of evaluating a polynomial

$$v(x) = v_{n-1} x^{n-1} + v_{n-2} x^{n-2} + \cdots + v_1 x + v_0$$

at some point β. Horner's rule is the arrangement

$$v(\beta) = (\cdots ((v_{n-1}\beta + v_{n-2})\beta + v_{n-3})\beta + \cdots + v_1)\beta + v_0,$$

Fast algorithms for the discrete Fourier transform

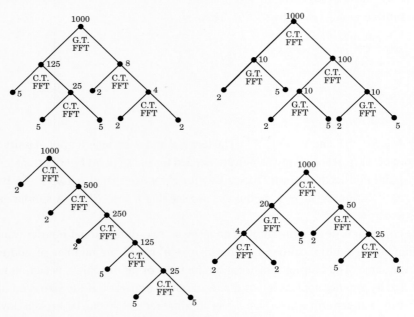

Figure 3.10 Some ways to build a 1000-point Fourier transform

which requires $n - 1$ multiplications and $n - 1$ additions in the field of β. If the various powers of β are prestored, then Horner's rule has no advantage over a direct computation. The advantage of Horner's rule is that prestorage is not necessary.

If $\beta = \omega^k$, then Horner's rule computes the kth component of the Fourier transform in $n - 1$ complex multiplications and $n - 1$ complex additions. A more efficient algorithm is the *Goertzel algorithm*, which is another procedure for computing a discrete Fourier transform. It reduces the number of multiplications by only a small factor. It is not an FFT because the complexity is still proportional to n^2. The Goertzel algorithm is useful when only a few components of the discrete Fourier transform are to be computed – typically not more than $\log_2 n$ of the n components. This is because, if more than $\log_2 n$ components were to be computed, an FFT algorithm would compute all the components with about $n \log_2 n$ operations; then those that are not needed could be discarded.

To compute one component of the Fourier transform

$$V_k = \sum_{i=0}^{n-1} \omega^{ik} v_i$$

introduce the polynomial $p(x)$, given by

$$p(x) = (x - \omega^k)(x - \omega^{-k}).$$

This polynomial is the smallest-degree polynomial with real coefficients having ω^k as a zero. More succinctly, it is the minimal polynomial of ω^k over the reals. It is

$$p(x) = x^2 - 2\cos\left(\frac{2\pi}{n}k\right)x + 1.$$

Let

$$v(x) = \sum_{i=0}^{n-1} v_i x^i$$

and write

$$v(x) = p(x)Q(x) + r(x).$$

The quotient polynomial $Q(x)$ and the remainder polynomial $r(x)$ can be found by long division of polynomials. Then V_k is computed from the remainder polynomial by

$$V_k = v(\omega^k) = r(\omega^k),$$

because, by construction, $p(\omega^k)$ is equal to zero. Most of the work of the Goertzel algorithm is in the long division of polynomials. If $v(x)$ has complex coefficients, then the division by $p(x)$ requires $2(n-2)$ real multiplications; if $v(x)$ has real coefficients, then $n-2$ real multiplications are needed. Likewise, $4(n-2)$ real additions are needed in the complex case, and $2(n-2)$ real additions are needed in the real case.

Because $r(x)$ has degree one, to compute $r(\omega^k)$ after the division takes only one more complex multiplication and one more complex addition. Hence, when the input data is complex, the Goertzel algorithm has $2n-1$ real multiplications and $4n-1$ real additions for each output component V_k computed.

A circuit diagram implementing the Goertzel algorithm is shown in Figure 3.11. This circuit has the form of an autoregressive filter. This is a consequence of the fact that a circuit for dividing one polynomial by another has the form of an autoregressive filter. After the polynomial $v(x)$ is shifted in, the circuit of Figure 3.11 will contain the remainder polynomial $r(x)$ when $v(x)$ is divided by $p(x)$. The quotient polynomial $Q(x)$ is of no importance to the computation, and is lost.

3.5 The discrete cosine transform

The Fourier transform of a real-valued vector is a complex-valued vector. This means that every component of the transform, in general, has both a real part and an imaginary part, which results in a doubling of the data storage when converting a real time-domain signal into the frequency domain. Moreover, the computation of the Fourier transform of a real-valued vector may have nearly as many operations as the computation of

Shift in $v(x)$

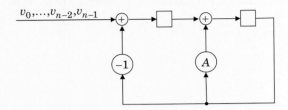

Notes: Real or complex input data
$A = 2\cos\dfrac{2\pi}{n}k$
$r(x)$ remains in shift register after division is complete
One divide circuit for each V_k
$V_k = \omega^k r_1 + r_0$

Figure 3.11 Flow diagram for Goertzel algorithm

the Fourier transform of a complex-valued vector, while it may seem that it should have only half as many. One way to effectively halve the number of computations is to combine two real time-domain vectors into one complex time-domain vector so that both can be transformed simultaneously. Then the resulting Fourier transform can be disentangled in the standard way (see Problem 1.11 and Section 5.1) into the two Fourier transforms of the two real vectors. Those two transforms, however, will each still be complex and must be stored as such. Another method – one that yields a real-valued transform – is to use a *discrete cosine transform* as described in this section.

The discrete cosine transform is an alternative transform with the property that the transform of a real-valued vector is another real-valued vector. The discrete cosine transform of blocklength n is defined as

$$V_k = \sum_{i=0}^{n-1} v_i \cos\frac{\pi(2i+1)k}{2n}, \qquad k = 0, \ldots, n-1.$$

The inverse discrete cosine transform is

$$v_i = \frac{1}{n}\sum_{k=0}^{n-1} V_k \left(1 - \tfrac{1}{2}\delta_k\right) \cos\frac{\pi(2i+1)k}{2n}, \qquad i = 0, \ldots, n-1,$$

where $\delta_k = 1$ if $k = 0$ and, otherwise, $\delta_k = 0$. The inverse discrete cosine transform can be verified directly. The summand in the inverse discrete cosine transform differs from the summand in the direct discrete cosine transform due to the curious factor of $\left(1 - \tfrac{1}{2}\delta_k\right)$, whose only purpose is to change V_0 to $V_0/2$.

3.5 The discrete cosine transform

An example may be helpful to make the structure of the discrete cosine transform explicit. To this end, notice that the four-point discrete cosine transform can be written out as

$$\begin{bmatrix} V_0 \\ V_1 \\ V_2 \\ V_3 \end{bmatrix} = \begin{bmatrix} 1 & 1 & 1 & 1 \\ \cos\frac{\pi}{8} & \cos\frac{3\pi}{8} & -\cos\frac{3\pi}{8} & -\cos\frac{\pi}{8} \\ \cos\frac{2\pi}{8} & -\cos\frac{2\pi}{8} & -\cos\frac{2\pi}{8} & \cos\frac{2\pi}{8} \\ \cos\frac{3\pi}{8} & -\cos\frac{\pi}{8} & \cos\frac{\pi}{8} & -\cos\frac{3\pi}{8} \end{bmatrix} \begin{bmatrix} v_0 \\ v_1 \\ v_2 \\ v_3 \end{bmatrix}.$$

As written, this requires twelve real multiplications, although there are some obvious groupings of terms that immediately reduce this to five real multiplications.

Mathematically, the discrete cosine transform is closely related to the discrete Fourier transform. Indeed, the discrete cosine transform of the vector v is equal to the discrete Fourier transform of a vector u of blocklength $4n$ that is formed from v in two steps. First, double the length of v by following it with its own time reversal and divide the resulting vector by two. Then double the length again by inserting a zero between every two elements. The new vector u is a vector of length $4n$ given by

$$u_{2i+1} = \begin{cases} \frac{1}{2}v_i, & i = 0, \ldots, n-1, \\ \frac{1}{2}v_{2n-i-1}, & i = n, \ldots, 2n-1, \end{cases}$$

$$u_{2i} = 0, \qquad i = 0, \ldots, 2n-1.$$

Theorem 3.5.1 *The discrete cosine transform of blocklength n of the vector v is equal to the first n components of the discrete Fourier transform of blocklength $4n$ of the vector u given by $u_{2i+1} = v_i$ for $i = 0, \ldots, n-1$, and equal to v_{2n-i-1} for $i = n, \ldots, 2n-1$, and otherwise, the components of u are zero.*

Proof To compute the $4n$-point Fourier transform of the vector u, let $\omega = e^{-j2\pi/4n}$ be a $4n$th root of unity, and let the $4n$ components of u be indexed by ℓ. Then the first n components of the Fourier transform of u are given by

$$V_k = \sum_{\ell=0}^{4n-1} \omega^{\ell k} u_\ell, \qquad k = 0, \ldots, n-1$$

$$= \frac{1}{2}\sum_{i=0}^{n-1} \omega^{(2i+1)k} v_i + \frac{1}{2}\sum_{i=n}^{2n-1} \omega^{(2i+1)k} v_{2n-i-1}.$$

Set $i = 2n - i' - 1$ in the second term and note that $\omega^{4n} = 1$. Then

$$V_k = \frac{1}{2}\sum_{i=0}^{n-1} \omega^{(2i+1)k} v_i + \frac{1}{2}\sum_{i'=0}^{n-1} \omega^{-(2i'+1)k} v_{i'}.$$

But $\omega = e^{-j2\pi/4n}$, so, with i' changed to i in the second term,

$$V_k = \frac{1}{2}\sum_{i=0}^{n-1} e^{-j2\pi(2i+1)k/4n} v_i + \frac{1}{2}\sum_{i=0}^{n-1} e^{j2\pi(2i+1)k/4n} v_i$$

$$= \sum_{i=0}^{n-1} v_i \cos\frac{\pi(2i+1)k}{2n}, \qquad k = 0, \ldots, n-1,$$

as was to be proved. □

From Theorem 3.5.1 we can conclude that any fast algorithm for computing a $4n$-point discrete Fourier transform could be used to compute an n-point discrete cosine transform by first rearranging the components of $\boldsymbol{v}$ into the vector $\boldsymbol{u}$ of blocklength $4n$. The resulting algorithm, however, can be simplified because of the alternating zeros in the vector $\boldsymbol{u}$ of blocklength $4n$. In this way, any of the algorithms for computing the discrete Fourier transform can be restructured to provide an algorithm for computing the discrete cosine transform. One must examine the Fourier transform algorithm to eliminate from that algorithm those operations that become vacuous or redundant when applied to the discrete cosine transform. A first step in this direction is the following corollary.

Corollary 3.5.2 *The discrete cosine transform of the vector $\boldsymbol{v}$ of blocklength n is given by*

$$V_k = \mathrm{Re}\left[e^{-j2\pi k/4n} \sum_{i=0}^{2n-1} r_i e^{-j2\pi ik/2n} \right], \qquad k = 0, \ldots, n-1,$$

where

$$r_i = \begin{cases} v_i, & i = 0, \ldots, n-1, \\ v_{2n-i-1}, & i = n, \ldots, 2n-1. \end{cases}$$

Proof The proof applies the decimation-in-time equations of the radix-two Cooley–Tukey fast Fourier transform to the expression of Theorem 3.5.1. This decimation leads to one expression for all of the even values of the index k in the equation of the theorem, and another expression for all of the odd values of that index. The expression involving the even indices can be dropped because those terms are all zero. Then, with $\omega = e^{-j2\pi/4n}$, we have

$$V_k = \omega^k \sum_{i=0}^{2n-1} \omega^{2ik} u_i, \qquad k = 0, \ldots, n-1$$

$$= \omega^k \sum_{i=0}^{n-1} \omega^{2ik} v_i + \omega^k \sum_{i=n}^{2n-1} \omega^{2ik} v_{4n-4i-1}.$$

3.5 The discrete cosine transform

Replace $4n - 4i - 1$ by i in the second term:

$$V_k = \omega^k \sum_{i=0}^{n-1} r_i \omega^{2ik} + \omega^{-k} \sum_{i=0}^{n-1} r_i \omega^{-2ik}.$$

Because r_i is real, this becomes

$$V_k = \operatorname{Re}\left[\sum_{i=0}^{n-1} r_i \omega^{(2i+1)k}\right],$$

which completes the proof. $\square$

Of course, the statement in Corollary 3.5.2 could instead be obtained directly and more simply from the definition of the discrete cosine transform.

Theorem 3.5.1 and Corollary 3.5.2 show that the discrete cosine transform of blocklength n can be computed with a fast Fourier transform algorithm of blocklength either $4n$ or $2n$. In particular, a radix-two discrete cosine transform of blocklength n can be computed with complexity on the order of $n \log n$. However, such algorithms do use complex numbers, and may be inconvenient for small or moderate values of n. A more direct algorithm may be preferred for small blocklengths when n is equal to a power of two. As an instructive alternative, we will develop an algorithm directly from the definition of the discrete cosine transform.

Theorem 3.5.3 (Decimation of the DCT) *The discrete cosine transform of even blocklength can be written for even k as*

$$V_{2k'} = \sum_{i=0}^{(n/2)-1} (v_i \mp v_{i+n/2}) \cos \frac{\pi(2i+1)k'}{n}, \qquad k' = 0, \ldots, (n/2) - 1,$$

the sign depending on whether k' is odd or even, and for odd k as

$$V_{2k'+1} = \sum_{i=0}^{(n/2)-1} (v_i - v_{n-1-i}) \cos \frac{\pi(2i+1)(2k'+1)}{2n}, \qquad k' = 0, \ldots, (n/2) - 1.$$

Proof The discrete cosine transform is given by

$$V_k = \sum_{i=0}^{n-1} v_i \cos \frac{\pi(2i+1)k}{2n}.$$

This will be decimated by treating odd indices and even indices separately.

For the decimation step for even indices, replace k by $2k'$ and write

$$V_{2k'} = \sum_{i=0}^{n-1} v_i \cos \frac{\pi(2i+1)2k'}{2n}$$

$$= \sum_{i=0}^{(n/2)-1} v_i \cos \frac{\pi(2i+1)k'}{n} + \sum_{i=n/2}^{n-1} v_i \cos \frac{\pi(2i+1)k'}{n}.$$

Now replace i by $i' + n/2$ in the second sum, noting that $\cos \pi(2i'+n+1)k'/n = \mp \cos \pi(2i'+1)k'/n$. Therefore, replacing, in turn, i' by i, we have

$$V_{2k'} = \sum_{i=0}^{(n/2)-1} (v_i \mp v_{i+n/2}) \cos \frac{\pi(2i+1)k'}{n}, \qquad k' = 0, \ldots, (n/2) - 1.$$

This has the same form as the original equation except that n is replaced by $n/2$.

For the decimation step for odd integers, replace k by $2k' + 1$ and write

$$V_{2k'+1} = \sum_{i=0}^{n-1} v_i \cos \frac{\pi(2i+1)(2k'+1)}{2n}$$

$$= \sum_{i=0}^{(n/2)-1} v_i \cos \frac{\pi(2i+1)(2k'+1)}{2n} + \sum_{i=n/2}^{n-1} v_i \cos \frac{\pi(2i+1)(2k'+1)}{2n}.$$

Now replace i by $n - 1 - i'$ in the second term. Then

$$V_{2k'+1} = \sum_{i=0}^{(n/2)-1} v_i \cos \frac{\pi(2i+1)(2k'+1)}{2n} + \sum_{i'=0}^{(n/2)-1} v_{n-1-i'} \frac{\cos \pi(2n - 2i' - 1)(2k'+1)}{2n}$$

$$= \sum_{i=0}^{(n/2)-1} (v_i - v_{n-1-i}) \cos \frac{\pi(2i+1)(2k'+1)}{2n},$$

which completes the proof of the theorem. □

The first decimation equation in Theorem 3.5.3 has the same form as did the original equation for the discrete cosine transform, and so, in turn, that expression can be decimated in the same way provided $n/2$ is even. The second decimation equation in Theorem 3.5.3 does not have this same form and so cannot be decimated in the same way. It can be computed as written using $(n/2)^2$ multiplications. If $M_R(n)$ denotes the number of real multiplications needed to compute the n-point discrete cosine transform, then this decimation procedure leads to the recursion

$$M_R(n) = M_R(n/2) + (n/2)^2.$$

By starting with $M(4) = 5$, this gives $M(8) = 21$, and $M(16) = 85$. In general, this recursion shows that it requires $\frac{1}{3}(n^2 - 1)$ multiplications to compute a discrete cosine transform of blocklength n with this procedure.

It is possible to do even better if $n = 2^m$. Then the indices i and k' are elements of $\mathbf{Z}_{2^m}$, and so $2i + 1$ and $2k' + 1$ are elements of $\mathbf{Z}_{2^m}^*$. This means that the methods to be studied in Section 3.7 can be used to represent this computation as a two-dimensional cyclic convolution for which fast algorithms exist. For this purpose, the structure of $\mathbf{Z}_{2^m}^*$ is described in Chapter 9. In particular, Theorem 9.1.8 states that $\mathbf{Z}_{2^m}^*$ is isomorphic to the direct sum $\mathbf{Z}_2 \oplus \mathbf{Z}_{2^{m-2}}$. This means that the elements of $\mathbf{Z}_{2^m}^*$ can be written as $\sigma^{\ell'}\eta^{\ell''}$ or as $\sigma^{-r'}\eta^{-r''}$ for some integers σ and η. In particular, we can write $(2i+1)(2k+1) = \sigma^{\ell'}\eta^{\ell''}\sigma^{-r'}\eta^{-r''} = \sigma^{\ell'-r'}\eta^{\ell''-r''}$ for some ℓ', r', ℓ'', and r''. In Section 3.7, it is explained how, with this representation, the decimation equation can be put in the form

$$V'_{r'r''} = \sum_{\ell'=0}^{1} \sum_{\ell''=0}^{n} g_{\ell'-r',\ell''-r''} v_{\ell',\ell''}$$

which is a two-dimensional cyclic convolution. Thus, the terms in the second of the decimation equations of Theorem 3.5.3 can be computed by any fast algorithm for computing a two-dimensional cyclic convolution.

3.6 Fourier transforms computed by using convolutions

The discrete Fourier transform

$$V_k = \sum_{i=0}^{n-1} \omega^{ik} v_i$$

can be computed efficiently by first changing it into a convolution. This may seem like a peculiar thing to consider, because we have already suggested that a good way to compute a convolution is to make use of the convolution theorem and a fast Fourier transform algorithm. However, sometimes there can be an advantage in turning a Fourier transform into a convolution, and, conversely, sometimes there can be an advantage in using a Fourier transform to do a convolution. Even more surprising, one can gain an advantage by using a Fourier transform to do a convolution, while at the same time implementing that very Fourier transform by turning it back into a convolution, albeit one of a different blocklength than the original.

The *Bluestein chirp algorithm* and the *Rader prime algorithm* are two different ways of turning a Fourier transform into a convolution. The Bluestein algorithm, shown in Figure 3.12, turns an n-point Fourier transform into an n-point convolution plus $2n$ side multiplications. The Rader algorithm, shown in Figure 3.13, turns an n-point Fourier transform into an $(n-1)$-point convolution but it requires that n be a prime. Indeed, an attractive feature of the Rader algorithm is that it can be used when n is a prime, while the usual FFT algorithms cannot.

92   Fast algorithms for the discrete Fourier transform

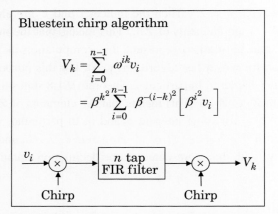

Figure 3.12 The Bluestein algorithm for computing a Fourier transform

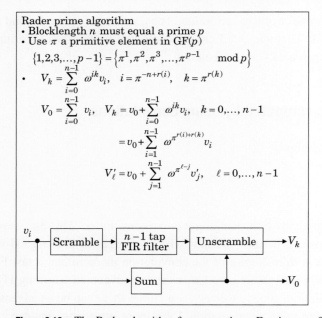

Figure 3.13 The Rader algorithm for computing a Fourier transform

The Bluestein chirp algorithm is less useful but is easy to describe, so we will begin with it. It is given by the expression

$$V_k = \beta^{-k^2} \sum_{i=0}^{n-1} \beta^{(i-k)^2} (\beta^{-i^2} v_i),$$

where β is a square root of ω. This variation of the Fourier transform is based on the calculation

$$\beta^{-k^2} \sum_{i=0}^{n-1} \beta^{(i-k)^2} (\beta^{-i^2} v_i) = \sum_{i=0}^{n-1} \beta^{2ik} v_i = \sum_{i=0}^{n-1} \omega^{ik} v_i = V_k.$$

3.6 Fourier transforms computed by using convolutions

The Bluestein chirp algorithm requires n multiplications for the pointwise product of v_i by β^{-i^2}, an n-tap finite impulse response filter for the cyclic convolution with β^{i^2}, followed by n multiplications for the pointwise product with β^{-k^2}. The number of operations is still on the order of n^2, so the Bluestein chirp algorithm is not asymptotically more efficient than the direct Fourier transform. However, it can be easier to implement in hardware in some applications. Further, it is possible to replace the direct convolution by the fast convolution algorithms of Chapter 5.

The Bluestein chirp algorithm requires $2n$ multiplications plus a convolution of length n. The next algorithm – the Rader prime algorithm – is generally preferred in situations where it can be used because the $2n$ multiplications are not needed. However, the Rader algorithm can be used only when the blocklength is a prime p. The Rader algorithm requires only some indexing operations plus the computations of a cyclic convolution – now, however, a cyclic convolution of blocklength $p - 1$, which is not a prime.

The Rader prime algorithm can be used to compute a Fourier transform with a blocklength equal to a prime p in any field F. Because p is a prime, we can make use of the structure of $GF(p)$ to reindex the vector components. The index field $GF(p)$ is defined as modulo p integer arithmetic. It should not be confused with F, the field over which the Fourier transform of v is to be computed.

Choose a primitive element π in the field $GF(p)$. Then each integer less than p can be expressed as a unique power of π. The Fourier transform

$$V_k = \sum_{i=0}^{p-1} \omega^{ik} v_i, \qquad k = 0, \ldots, p-1$$

will be rewritten with i and k expressed as powers of the primitive element π. Because i and k each take on the value zero, and zero is not a power of π, the zero frequency component (with $k = 0$) and the zero time component (with $i = 0$) must be treated specially. To this end, write

$$V_0 = \sum_{i=0}^{p-1} v_i,$$

$$V_k = v_0 + \sum_{i=1}^{p-1} \omega^{ik} v_i, \qquad k = 1, \ldots, p-1.$$

For each i from 1 to $p - 1$, let $r(i)$ be the unique integer from 1 to $p - 1$ such that in $GF(p)$, $\pi^{r(i)} = i$. The function $r(i)$ is a map from the set $\{1, 2, \ldots, p-1\}$ onto the set $\{1, 2, \ldots, p-1\}$; it is a permutation of $\{1, 2, \ldots, p-1\}$. Then V_k can be written

$$V_{\pi^{r(k)}} = v_0 + \sum_{i=1}^{p-1} \omega^{\pi^{r(i)+r(k)}} v_{\pi^{r(i)}}, \qquad k = 1, \ldots, p-2.$$

Because $r(i)$ is a permutation, we can set $r(k) = \ell$, and set $r(i) = p - 1 - j$. Because a sum is unaffected by the order of the summations, we can use j as the index of summation to get

$$V_{\pi^\ell} = v_0 + \sum_{j=1}^{p-1} \omega^{\pi^{\ell-j}} v_{\pi^{p-1-j}}, \qquad \ell = 0, \ldots, p-2,$$

recalling that $\pi^{p-1} = 1$. Finally, we write this equation in this slightly modified form as

$$V'_\ell - V_0 = \sum_{j=0}^{p-2} (\omega^{\pi^{\ell-j}} - 1) v'_j, \qquad \ell = 0, \ldots, p-2,$$

where $V'_\ell = V_{\pi^\ell}$ and $v'_j = v_{\pi^{p-1-j}}$ are scrambled input and output data sequences. This is now the equation of a cyclic convolution between the scrambled input vector $\boldsymbol{v} = [v'_j]$ and the vector $\boldsymbol{g} = [\omega^{\pi^j} - 1]$. Accordingly, define the *Rader polynomial* as

$$g(x) = \sum_{j=0}^{p-2} (\omega^{\pi^j} - 1) x^j.$$

By scrambling the input and output indices, we have turned the Fourier transform of blocklength p into a cyclic convolution of blocklength $p - 1$. As it is written, the number of operations needed to implement the convolution is still on the order of p^2, but a fast Fourier transform can be used to reduce the number of multiplications. In Section 3.8 we shall combine the Rader prime algorithm with the Winograd convolution algorithm to obtain the Winograd small FFT algorithm in which the number of multiplications is so reduced.

Using the Rader prime algorithm, let us construct a binary five-point Fourier transform algorithm as an example. This algorithm will compute

$$V_k = \sum_{i=0}^{4} \omega^{ik} v_i, \qquad k = 0, \ldots, 4,$$

where $\omega = e^{-j2\pi/5}$. First, rewrite this expression as

$$V_0 = \sum_{i=0}^{4} v_i,$$

$$V_k = V_0 + \sum_{i=1}^{4} (\omega^{ik} - 1) v_i, \qquad k = 1, \ldots, 4,$$

3.6 Fourier transforms computed by using convolutions

and work only with the terms $\sum_{i=1}^{4}(\omega^{ik} - 1)v_i$. The element two is easily found to be primitive in $GF(5)$, and so in $GF(5)$ we have

$$2^0 = 1, \quad 2^0 = 1,$$
$$2^1 = 2, \quad 2^{-1} = 3,$$
$$2^2 = 4, \quad 2^{-2} = 4,$$
$$2^3 = 3, \quad 2^{-3} = 2.$$

Hence

$$V'_\ell - V_0 = \sum_{j=0}^{3}(\omega^{2^{i-j}} - 1)v'_j.$$

The summation is recognized as a four-point cyclic convolution. We now write the Rader polynomial $g(x)$ of blocklength four over the field F as

$$g(x) = (\omega^3 - 1)x^3 + (\omega^4 - 1)x^2 + (\omega^2 - 1)x + (\omega - 1),$$

where $g_j = \omega^{2^j} - 1$. The input to the filter and the output from the filter can be expressed as polynomials whose coefficients are scrambled coefficients of v and V. The Rader five-point Fourier transform algorithm is now summarized by the following equations:

$$d(x) = v_2 x^3 + v_4 x^2 + v_3 x + v_1,$$
$$s(x) = (V_3 - V_0)x^3 + (V_4 - V_0)x^2 + (V_2 - V_0)x + (V_1 - V_0),$$

and

$$s(x) = g(x)d(x) \pmod{x^4 - 1}.$$

The Rader polynomial $g(x)$ is a fixed polynomial with precomputed coefficients. The polynomial $d(x)$ is formed by scrambling the coefficients of $v(x)$. The polynomial $V(x)$ is obtained by unscrambling the coefficients of the polynomial $s(x)$. The algorithm is summarized in Figure 3.14.

It is instructive to reexamine the Rader algorithm in a matrix formulation. The five-point Fourier transform is

$$\begin{bmatrix} V_0 \\ V_1 \\ V_2 \\ V_3 \\ V_4 \end{bmatrix} = \begin{bmatrix} 1 & 1 & 1 & 1 & 1 \\ 1 & \omega & \omega^2 & \omega^3 & \omega^4 \\ 1 & \omega^2 & \omega^4 & \omega & \omega^3 \\ 1 & \omega^3 & \omega & \omega^4 & \omega^2 \\ 1 & \omega^4 & \omega^3 & \omega^2 & \omega^1 \end{bmatrix} \begin{bmatrix} v_0 \\ v_1 \\ v_2 \\ v_3 \\ v_4 \end{bmatrix}.$$

From this we obtain

$$V_0 = v_0 + v_1 + v_2 + v_3 + v_4,$$

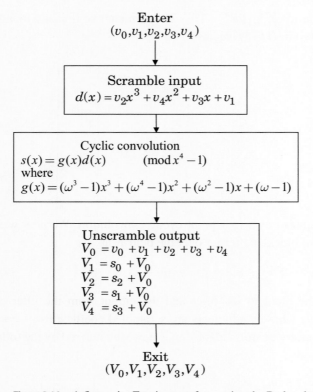

Figure 3.14 A five-point Fourier transform using the Rader algorithm

and

$$\begin{bmatrix} V_1 - V_0 \\ V_2 - V_0 \\ V_3 - V_0 \\ V_4 - V_0 \end{bmatrix} = \begin{bmatrix} \omega^1 - 1 & \omega^2 - 1 & \omega^3 - 1 & \omega^4 - 1 \\ \omega^2 - 1 & \omega^4 - 1 & \omega^1 - 1 & \omega^3 - 1 \\ \omega^3 - 1 & \omega^1 - 1 & \omega^4 - 1 & \omega^2 - 1 \\ \omega^4 - 1 & \omega^3 - 1 & \omega^2 - 1 & \omega^1 - 1 \end{bmatrix} \begin{bmatrix} v_1 \\ v_2 \\ v_3 \\ v_4 \end{bmatrix}.$$

By the scrambling rules of the Rader algorithm, this becomes

$$\begin{bmatrix} V_1 - V_0 \\ V_2 - V_0 \\ V_4 - V_0 \\ V_3 - V_0 \end{bmatrix} = \begin{bmatrix} \omega^1 - 1 & \omega^3 - 1 & \omega^4 - 1 & \omega^2 - 1 \\ \omega^2 - 1 & \omega^1 - 1 & \omega^3 - 1 & \omega^4 - 1 \\ \omega^4 - 1 & \omega^2 - 1 & \omega^1 - 1 & \omega^3 - 1 \\ \omega^3 - 1 & \omega^4 - 1 & \omega^2 - 1 & \omega^1 - 1 \end{bmatrix} \begin{bmatrix} v_1 \\ v_3 \\ v_4 \\ v_2 \end{bmatrix},$$

which can be recognized as the matrix representation of a cyclic convolution

$$\begin{bmatrix} V_1 - V_0 \\ V_2 - V_0 \\ V_4 - V_0 \\ V_3 - V_0 \end{bmatrix} = \begin{bmatrix} g_0 & g_3 & g_2 & g_1 \\ g_1 & g_0 & g_3 & g_2 \\ g_2 & g_1 & g_0 & g_3 \\ g_3 & g_2 & g_1 & g_0 \end{bmatrix} \begin{bmatrix} v_1 \\ v_3 \\ v_4 \\ v_2 \end{bmatrix},$$

where $g_0 = \omega - 1$, $g_1 = \omega^2 - 1$, $g_3 = \omega^4 - 1$, and $g_3 = \omega^3 - 1$. The coefficients of the polynomial $g(x)$ can be specified once the field F is specified. For example, if the field is the complex field, then $\omega = e^{-j2\pi/5}$ and the coefficients of $g(x)$ are complex constants.

3.7 The Rader–Winograd algorithm

The idea of the Rader algorithm can still be used when the blocklength n is a power of an odd prime. In this case, not only must the zero time component and the zero frequency component be treated specially, but certain other time components and other frequency components must be treated specially as well. This is because the nonzero elements of $\mathbf{Z}/\langle p^m \rangle$ do not form a cyclic group, as is explained within the discussion on number theory in Chapter 9. We shall develop the ideas of this section by referring forward to the discussion of Section 9.1. Theorem 9.1.8 of Chapter 9 promises that, when p is an odd prime, there is an element of order $p^{m-1}(p-1)$, and we use this element to construct an algorithm. To this point, all multiples of p must be deleted to find a cyclic group, denoted $\mathbf{Z}^*_{p^m}$, that is contained in $\mathbf{Z}/\langle p^m \rangle$.

To compute

$$V_k = \sum_{i=0}^{p^m-1} \omega^{ik} v_i, \qquad k = 0, \ldots, p^m - 1,$$

we shall make use of the cyclic structure of $\mathbf{Z}^*_{p^m}$, portrayed by Theorem 9.1.8, to reindex the components of the vector. When q is a power of an odd prime, $\mathbf{Z}/\langle p^m \rangle$ contains a cyclic group with $p^{m-1}(p-1)$ elements. From the p^m by p^m matrix

$$\mathbf{W} = [\omega^{ik}]$$

simply strike out all troublesome rows and columns to get a $p^{m-1}(p-1)$ by $p^{m-1}(p-1)$ matrix for which Rader's idea can be used. The indices of the remaining rows and columns are all of the elements of $\mathbf{Z}^*_{p^m}$ and can be written as powers of a generator π so that ω^{ij} can be replaced by $\omega^{\pi^{\ell-r}}$. With this procedure, the notion of the Rader polynomial is then replaced by the notion of the generalized Rader polynomial. An example is given in Section 3.8.

The troublesome rows and columns are those whose index is divisible by p. They are then handled separately. Even these troublesome rows and columns can themselves be arranged into yet smaller cyclic convolutions, as we shall see in the next section. Hence the p^m-point Fourier transform, though too irregular to be swallowed whole, can be handled with just a few judicious bites.

The case where the blocklength is a power of two is a little more complicated and requires one more layer of manipulation. This is because the set of indices coprime to

2^m – the set of odd indices – does not form a cyclic group under multiplication. Rather, as will be shown in Theorem 9.1.8, this group is isomorphic to $Z_2 \times Z_{2^{m-2}}$. The idea of the procedure is as follows. From the 2^m by 2^m matrix

$$W = [\omega^{ik}],$$

strike out all the rows and columns with even indices to get a 2^{m-1} by 2^{m-1} matrix, denoted W'. Similarly, strike out all the components of v and V with even indices to form the reduced vectors v' and V'. The surviving indices are all odd and, under multiplication, the set of these indices is isomorphic to $Z_2 \times Z_{2^{m-2}}$. The isomorphism can be used to define a permutation of the rows of the reduced matrix W' and a similar permutation of the columns of that matrix that will now put the reduced matrix W' into the form

$$W'' = \begin{bmatrix} W_1 & W_2 \\ W_2 & W_1 \end{bmatrix},$$

where W_1 and W_2 are 2^{m-2} by 2^{m-2} matrices, each of which has the structure of a cyclic convolution and with indices in $Z_{2^{m-2}}$. Moreover, the index designating W_1 or W_2 is an element of Z_2. Hence, the full matrix W'' is indexed by $Z_2 \times Z_{2^{m-2}}$. Similar permutations on the components of v' and V' based on $Z_2 \times Z_{2^{m-2}}$ put these vectors into the form

$$v'' = \begin{bmatrix} v_1 \\ v_2 \end{bmatrix}$$

and

$$V'' = \begin{bmatrix} V_1 \\ V_2 \end{bmatrix}.$$

In keeping with this background, we will require two indices. One index designates the top half versus the bottom half (or left half versus the right half), and the other index designates a component within that half. These are described explicitly in Corollary 9.1.9. Specifically, let $\pi = 3$ and let $\sigma = 2^m - 1$. The group of odd integers under multiplication modulo 2^m is generated by σ and π; every element of $Z_{2^m}^*$ can be expressed as $\sigma^{\ell'} \pi^{\ell''}$, where $\ell' = 0, 1$ and $\ell'' = 0, \ldots, 2^{m-2}$. Hence we can write

$$\omega^{ik} = \omega^{\sigma^{\ell'} \pi^{\ell''} \sigma^{r'} \pi^{r''}},$$

and by the indicated permutations

$$W'' = \begin{bmatrix} [\omega^{\pi^{\ell''+r''}}] & [\omega^{\sigma \pi^{\ell''+r''}}] \\ [\omega^{\sigma \pi^{\ell''+r''}}] & [\omega^{\pi^{\ell''+r''}}] \end{bmatrix},$$

3.7 The Rader–Winograd algorithm

where ℓ'' and r'' index the rows and columns, respectively, in each of the four submatrices. Each of the four submatrices now corresponds to a cyclic convolution of blocklength 2^{m-2}.

Moreover, the matrix computation, denoted by

$$\begin{bmatrix} V_1 \\ V_2 \end{bmatrix} = \begin{bmatrix} W_1 & W_2 \\ W_2 & W_1 \end{bmatrix} \begin{bmatrix} v_1 \\ v_2 \end{bmatrix},$$

has the form of a two-point cyclic convolution of matrices. One way to compute it is as follows:

$$\begin{bmatrix} V_1 \\ V_2 \end{bmatrix} = \begin{bmatrix} 1 & 1 \\ 1 & -1 \end{bmatrix} \begin{bmatrix} \frac{1}{2}(W_1 + W_2) & 0 \\ 0 & \frac{1}{2}(W_1 - W_2) \end{bmatrix} \begin{bmatrix} 1 & 1 \\ 1 & -1 \end{bmatrix} \begin{bmatrix} v_1 \\ v_2 \end{bmatrix}.$$

The diagonal elements each lead to a complex cyclic convolution of blocklength 2^{m-2}, which is a reduction from the four cyclic convolutions in the previous expression. However, we shall find that we can still do a little better. For one thing, we will see a little later that $W_1 + W_2$ is a purely real matrix and $W_1 - W_2$ is a purely imaginary matrix.

As an example, we shall look at the two four-point cyclic convolutions that arise in this way at the core of a sixteen-point Fourier transform. Let

$$V_k = \sum_{i=0}^{15} \omega^{ik} v_i, \quad k = 0, \ldots, 15,$$

where $\omega^{16} = 1$. Consider the matrix obtained by striking out even indices from the range of i and k. The subcomputation is

$$\begin{bmatrix} V_1' \\ V_3' \\ V_5' \\ V_7' \\ V_9' \\ V_{11}' \\ V_{13}' \\ V_{15}' \end{bmatrix} = \begin{bmatrix} \omega^1 & \omega^3 & \omega^5 & \omega^7 & \omega^9 & \omega^{11} & \omega^{13} & \omega^{15} \\ \omega^3 & \omega^9 & \omega^{15} & \omega^5 & \omega^{11} & \omega^1 & \omega^7 & \omega^{13} \\ \omega^5 & \omega^{15} & \omega^9 & \omega^3 & \omega^{13} & \omega^7 & \omega^1 & \omega^{11} \\ \omega^7 & \omega^5 & \omega^3 & \omega^1 & \omega^{15} & \omega^{13} & \omega^{11} & \omega^9 \\ \omega^9 & \omega^{11} & \omega^{13} & \omega^{15} & \omega^1 & \omega^3 & \omega^5 & \omega^7 \\ \omega^{11} & \omega^1 & \omega^7 & \omega^{13} & \omega^3 & \omega^9 & \omega^{15} & \omega^5 \\ \omega^{13} & \omega^7 & \omega^1 & \omega^{11} & \omega^5 & \omega^{15} & \omega^9 & \omega^3 \\ \omega^{15} & \omega^{13} & \omega^{11} & \omega^9 & \omega^7 & \omega^5 & \omega^3 & \omega^1 \end{bmatrix} \begin{bmatrix} v_1 \\ v_3 \\ v_5 \\ v_7 \\ v_9 \\ v_{11} \\ v_{13} \\ v_{15} \end{bmatrix}.$$

To find the permutation, we write the indices as $15^{\ell'} 3^{\ell''}$ for $\ell' = 0, 1$ and $\ell'' = 0, 1, 2, 3$. The powers of three (modulo 16) are

$3^0 = 1,$ $\quad 3^0 = 1,$
$3^1 = 3,$ $\quad 3^{-1} = 11,$
$3^2 = 9,$ $\quad 3^{-2} = 9,$
$3^3 = 11,$ $\quad 3^{-3} = 3,$

and

$$15 \cdot 3^0 = 15, \qquad 15 \cdot 3^0 = 15,$$
$$15 \cdot 3^1 = 13, \qquad 15 \cdot 3^{-1} = 5,$$
$$15 \cdot 3^2 = 7, \qquad 15 \cdot 3^{-2} = 7,$$
$$15 \cdot 3^3 = 5, \qquad 15 \cdot 3^{-3} = 13.$$

The input indices are scrambled by using $15^{-\ell'}3^{-\ell''}$ (mod 16), and the output indices are scrambled by using $15^{\ell'}3^{\ell''}$ (mod 16). Then

$$\begin{bmatrix} V'_1 \\ V'_{11} \\ V'_9 \\ V'_3 \\ V'_{15} \\ V'_5 \\ V'_7 \\ V'_{13} \end{bmatrix} = \begin{bmatrix} \begin{bmatrix} \omega^1 & \omega^3 & \omega^9 & \omega^{11} \\ \omega^{11} & \omega^1 & \omega^3 & \omega^9 \\ \omega^9 & \omega^{11} & \omega^1 & \omega^3 \\ \omega^3 & \omega^9 & \omega^{11} & \omega^1 \end{bmatrix} & \begin{bmatrix} \omega^{15} & \omega^{13} & \omega^7 & \omega^5 \\ \omega^5 & \omega^{15} & \omega^{13} & \omega^7 \\ \omega^7 & \omega^5 & \omega^{15} & \omega^{13} \\ \omega^{13} & \omega^7 & \omega^5 & \omega^{15} \end{bmatrix} \\ \begin{bmatrix} \omega^{15} & \omega^{13} & \omega^7 & \omega^5 \\ \omega^5 & \omega^{15} & \omega^{13} & \omega^7 \\ \omega^7 & \omega^5 & \omega^{15} & \omega^{13} \\ \omega^{13} & \omega^7 & \omega^5 & \omega^{15} \end{bmatrix} & \begin{bmatrix} \omega^1 & \omega^3 & \omega^9 & \omega^{11} \\ \omega^{11} & \omega^1 & \omega^3 & \omega^9 \\ \omega^9 & \omega^{11} & \omega^1 & \omega^3 \\ \omega^3 & \omega^9 & \omega^{11} & \omega^1 \end{bmatrix} \end{bmatrix} \begin{bmatrix} v_1 \\ v_3 \\ v_9 \\ v_{11} \\ v_{15} \\ v_{13} \\ v_7 \\ v_5 \end{bmatrix}.$$

The matrix has been partitioned to show the four cyclic convolutions that have formed. If the Fourier transform is in the complex field, the blocks are related as complex conjugates. This is more evident if the matrix equation is rewritten as

$$\begin{bmatrix} V'_1 \\ V'_{11} \\ V'_9 \\ V'_3 \\ V'_{15} \\ V'_5 \\ V'_7 \\ V'_{13} \end{bmatrix} = \begin{bmatrix} \begin{bmatrix} \omega^1 & \omega^3 & \omega^9 & \omega^{11} \\ \omega^{11} & \omega^1 & \omega^3 & \omega^9 \\ \omega^9 & \omega^{11} & \omega^1 & \omega^3 \\ \omega^3 & \omega^9 & \omega^{11} & \omega^1 \end{bmatrix} & \begin{bmatrix} \omega^{-1} & \omega^{-3} & \omega^{-9} & \omega^{-11} \\ \omega^{-11} & \omega^{-1} & \omega^{-3} & \omega^{-9} \\ \omega^{-9} & \omega^{-11} & \omega^{-1} & \omega^{-3} \\ \omega^{-3} & \omega^{-9} & \omega^{-11} & \omega^{-1} \end{bmatrix} \\ \begin{bmatrix} \omega^{-1} & \omega^{-3} & \omega^{-9} & \omega^{-11} \\ \omega^{-11} & \omega^{-1} & \omega^{-3} & \omega^{-9} \\ \omega^{-9} & \omega^{-11} & \omega^{-1} & \omega^{-3} \\ \omega^{-3} & \omega^{-9} & \omega^{-11} & \omega^{-1} \end{bmatrix} & \begin{bmatrix} \omega^1 & \omega^3 & \omega^9 & \omega^{11} \\ \omega^{11} & \omega^1 & \omega^3 & \omega^9 \\ \omega^9 & \omega^{11} & \omega^1 & \omega^3 \\ \omega^3 & \omega^9 & \omega^{11} & \omega^1 \end{bmatrix} \end{bmatrix} \begin{bmatrix} v_1 \\ v_3 \\ v_9 \\ v_{11} \\ v_{15} \\ v_{13} \\ v_7 \\ v_5 \end{bmatrix}.$$

Notice that the bottom four rows are the complex conjugates of the top four rows. If the input v is real, then only computations associated with the first four rows need be performed. Hence one way to proceed is to write the computation of the first four rows as the sum of a pair of cyclic convolutions as follows:

$$V'_3 x^3 + V'_9 x^2 + V'_{11} x + V'_1 = (\omega^{11} x^3 + \omega^9 x^2 + \omega^3 x + \omega)(v_{11} x^3 + v_9 x^2 + v_3 x + v_1)$$
$$+ (\omega^{-11} x^3 + \omega^{-9} x^2 + \omega^{-3} x + \omega^{-1})(v_5 x^3 + v_7 x^2$$
$$+ v_{13} x + v_{15}) \pmod{x^4 - 1}.$$

3.7 The Rader–Winograd algorithm

We shall develop an alternative method for the complex field by viewing the above matrix equation as a two-point cyclic convolution of blocks written in the form

$$\begin{bmatrix} \overline{V}'_1 \\ \overline{V}'_2 \end{bmatrix} = \begin{bmatrix} 1 & 1 \\ 1 & -1 \end{bmatrix} \begin{bmatrix} \frac{1}{2}(W_1 + W_2) & 0 \\ 0 & \frac{1}{2}(W_1 - W_2) \end{bmatrix} \begin{bmatrix} 1 & 1 \\ 1 & -1 \end{bmatrix} \begin{bmatrix} \overline{v}_1 \\ \overline{v}_2 \end{bmatrix},$$

where

$$\tfrac{1}{2}(W_1 + W_2) = \begin{bmatrix} \cos\theta & \cos 3\theta & \cos 9\theta & \cos 11\theta \\ \cos 11\theta & \cos\theta & \cos 3\theta & \cos 9\theta \\ \cos 9\theta & \cos 11\theta & \cos\theta & \cos 3\theta \\ \cos 3\theta & \cos 9\theta & \cos 11\theta & \cos\theta \end{bmatrix}$$

and

$$\tfrac{1}{2}(W_1 - W_2) = \begin{bmatrix} j\sin\theta & j\sin 3\theta & j\sin 9\theta & j\sin 11\theta \\ j\sin 11\theta & j\sin\theta & j\sin 3\theta & j\sin 9\theta \\ j\sin 9\theta & j\sin 11\theta & j\sin\theta & j\sin 3\theta \\ j\sin 3\theta & j\sin 9\theta & j\sin 11\theta & j\sin\theta \end{bmatrix},$$

and where $\theta = 2\pi/16$. Notice that of the two cyclic convolutions now indicated, one is purely real and one is purely imaginary. For real input data, we need to compute the two real cyclic convolutions of the form

$$s(x) = [\cos 3\theta x^3 + \cos 9\theta x^2 + \cos 11\theta x + \cos\theta] d(x) \pmod{x^4 - 1}$$

and

$$s'(x) = [\sin 3\theta x^3 + \sin 9\theta x^2 + \sin 11\theta x + \sin\theta] d'(x) \pmod{x^4 - 1}.$$

To reduce this computation, we use the Chinese remainder theorem for polynomials with the factorization

$$x^4 - 1 = (x^2 - 1)(x^2 + 1)$$

and note that $\theta = \pi/8$, so $\cos 11\theta = -\cos 3\theta$, $\sin 11\theta = -\sin 3\theta$, $\cos 9\theta = -\cos\theta$, and $\sin 9\theta = -\sin\theta$. Therefore some terms drop out because

$$\cos 3\theta x^3 + \cos 9\theta x^2 + \cos 11\theta x + \cos\theta = 0 \pmod{x^2 - 1}$$

and

$$\sin 3\theta x^3 + \sin 9\theta x^2 + \sin 11\theta x + \sin\theta = 0 \pmod{x^2 - 1}.$$

This means that the multiplications associated with the residue modulo $x^2 - 1$ are not needed. The residues modulo $x^2 + 1$ will each use three multiplications. Hence the two cyclic convolutions require a total of only six multiplications.

In the next section, we shall see that four more multiplications are needed to process the rows and columns with even indices. Altogether, by incorporating all these devices,

Table 3.3 *Performance of Winograd small FFT algorithms*

Blocklength n	Number of real multiplications*	Number of nontrivial real multiplications	Number of real additions
2	2	0	2
3	3	2	6
4	4	0	8
5	6	5	17
7	9	8	36
8	8	2	26
9	11	10	44
11	21	20	84
13	21	20	94
16	18	10	74
17	36	35	157
19	39	38	186

* Including multiplications by ± 1 or $\pm j$

the sixteen-point Fourier transform of a real vector v requires a total of ten nontrivial real multiplications. If the input vector v is complex, the algorithm can be applied separately to the real part and the imaginary part of v and the results then added. This requires twenty real multiplications.

3.8 The Winograd small fast Fourier transform

The *Winograd small fast Fourier transform* is a method of efficiently computing the discrete Fourier transform for small blocklengths. It is built from three ideas: the Rader prime algorithm of Section 3.6, the Rader–Winograd algorithm of Section 3.7, and the Winograd convolution algorithm to be given in Chapter 5. We have already touched on this merger of algorithms in the previous section and this section is an elaboration of that section.

There are three cases that must be treated: blocklength equal to a prime; blocklength equal to a power of an odd prime; and blocklength equal to a power of two. The most suitable blocklengths for the Winograd small FFT are 2, 3, 4, 5, 7, 8, 9, and 16. The performance of the Winograd small FFT algorithms of these blocklengths is given in Table 3.3. (The algorithms themselves can be found in Appendix B.) The number of real multiplications is given in two ways in Table 3.3. Only the number of nontrivial multiplications matters if the small FFT itself is to be computed. However, if the small FFT is to be used as a building block in the nesting algorithms of Chapter 12, then the multiplications by ± 1 and $\pm j$ will propagate into nontrivial multiplications (and also into additions) in the larger algorithm. This is why we always record the total number of multiplications as well.

There are also some trivial additions that are included in Table 3.3. These are additions of a purely real number and a purely imaginary number, which are not really executed as additions. We choose not to distinguish between trivial additions and nontrivial additions in the bookkeeping. When the real input vector is replaced by a complex input vector, all additions are nontrivial complex additions because the purely real numbers and the purely imaginary numbers both become nontrivial complex numbers.

Blocklength a prime The first step is to change the Fourier transform to a convolution. If n is a small prime, use the Rader prime algorithm to express the transform as a convolution, which is computed using a Winograd small convolution algorithm. Generally, one writes out the equations longhand, so it is not practical to take n too large. The Rader prime algorithm changes the discrete Fourier transform into a convolution using only scrambling of the indices; no additions or multiplications are needed in that step. The convolution algorithm itself has the structure of a set of additions, followed by a set of multiplications, followed by a set of additions.

A five-point Winograd FFT will be constructed that computes

$$V_k = \sum_{i=0}^{4} \omega^{ik} v_i, \qquad k = 0, \ldots, 4,$$

where $\omega = e^{-j2\pi/5}$. First, use the Rader prime algorithm that was discussed in Section 3.6. This changes the Fourier transform into a cyclic convolution

$$s(x) = g(x)d(x) \pmod{x^4 - 1},$$

where the Rader polynomial

$$g(x) = (\omega^3 - 1)x^3 + (\omega^4 - 1)x^2 + (\omega^2 - 1)x + (\omega - 1)$$

has fixed coefficients. The input to the filter and the output from the filter are the polynomials

$$d(x) = v_2 x^3 + v_4 x^2 + v_3 x + v_1,$$
$$s(x) = (V_3 - V_0)x^3 + (V_4 - V_0)x^2 + (V_2 - V_0)x + (V_1 - V_0)$$

whose coefficients are scrambled coefficients of v and V. The polynomial $d(x)$ is formed by scrambling the coefficients of $v(x)$. The polynomial $V(x)$ is obtained by unscrambling the coefficients of the polynomial $s(x)$ and adding V_0 to each coefficient.

$$V_k = \sum_{i=0}^{4} \omega^{ik} v_i, \quad \begin{matrix} i = 0,\ldots,4, \\ \omega = e^{-j2\pi/5}, \end{matrix}$$

$$V = Wv$$
$$= CBAv.$$

$$\begin{bmatrix} V_0 \\ V_1 \\ V_2 \\ V_3 \\ V_4 \end{bmatrix} = \begin{bmatrix} 1 & 0 & 0 & 0 & 0 & 0 \\ 1 & 1 & 1 & 1 & 0 & -1 \\ 1 & 1 & -1 & 1 & 1 & 0 \\ 1 & 1 & -1 & -1 & -1 & 0 \\ 1 & 1 & 1 & -1 & 0 & 1 \end{bmatrix} \begin{bmatrix} B_0 & & & & & \\ & B_1 & & & & \\ & & B_2 & & & \\ & & & B_3 & & \\ & & & & B_4 & \\ & & & & & B_5 \end{bmatrix} \begin{bmatrix} 1 & 1 & 1 & 1 & 1 \\ 0 & 1 & 1 & 1 & 1 \\ 0 & 1 & -1 & -1 & 1 \\ 0 & 1 & -1 & 1 & -1 \\ 0 & 1 & 0 & 0 & -1 \\ 0 & 0 & -1 & 1 & 0 \end{bmatrix} \begin{bmatrix} v_0 \\ v_1 \\ v_2 \\ v_3 \\ v_4 \end{bmatrix}$$

where $B_0 = 1$,

$$\begin{bmatrix} B_1 \\ B_2 \\ B_3 \\ B_4 \\ B_5 \end{bmatrix} = \frac{1}{4} \begin{bmatrix} 1 & 1 & 1 & 1 \\ 1 & -1 & 1 & -1 \\ 2 & 0 & -2 & 0 \\ -2 & 2 & 2 & -2 \\ 2 & 2 & -2 & -2 \end{bmatrix} \begin{bmatrix} \omega - 1 \\ \omega^2 - 1 \\ \omega^4 - 1 \\ \omega^3 - 1 \end{bmatrix} = \begin{bmatrix} \frac{1}{2}(\cos\theta + \cos 2\theta) - 1 \\ \frac{1}{2}(\cos\theta - \cos 2\theta) \\ j\sin\theta \\ j(-\sin\theta + \sin 2\theta) \\ j(\sin\theta + \sin 2\theta) \end{bmatrix}$$

and $\theta = 2\pi/5$.

Figure 3.15 A five-point Winograd small FFT algorithm

The five-point Winograd small FFT is obtained by computing the product $g(x)d(x)$ by using a small convolution algorithm. Refer to Table 5.4 of Chapter 5, which gives a four-point cyclic convolution algorithm with five multiplications. We can rewrite this to do the Fourier transform. Incorporate the scrambling and unscrambling operations into the matrices of the convolution by scrambling the appropriate rows and columns. Also, the coefficients of $g(x)$ are fixed complex numbers, so it is possible to precompute the product of g and its matrix of preadditions. When these changes are made to the four-point convolution algorithm, and the terms V_0 and v_0 are included, it becomes the five-point Winograd small FFT. Figure 3.15 shows the five-point FFT algorithm in a standard matrix form. This standard form will prove useful for the nesting techniques studied in Chapter 12. Notice that, in Figure 3.15, the matrix of preadditions and the matrix of postadditions are not square. The five-point input vector is expanded to a six-point vector, and this is where the multiplications occur. The top two rows inside the braces have to do with v_0 and V_0, and have no multiplications. The other five rows come from the four-point cyclic convolution algorithm. One of the multiplying constants turns out to be a one, so there are really only five multiplications in the algorithm. The algorithm has five nontrivial multiplications and one trivial multiplication.

3.8 The Winograd small fast Fourier transform

Figure 3.15 illustrates another important point. Although we have given no reason to expect it, the diagonal elements turn out to be purely real or purely imaginary.[2] This is important because it means that each diagonal element is responsible for only one real multiplication if the input data is real and is responsible for only two real multiplications if the input data is complex. This phenomenon is quite general, as we show next.

Whenever p is odd, we can write

$$x^{p-1} - 1 = (x^{(p-1)/2} - 1)(x^{(p-1)/2} + 1).$$

The factors of $x^{p-1} - 1$ divide one of the two terms on the right. Hence, whenever the Winograd convolution algorithm is built on the factors of $x^{p-1} - 1$, the following theorem describes the situation.

Theorem 3.8.1 *Let $g(x)$ be a Rader polynomial. For every odd prime p, the coefficients of $g(x) \pmod{x^{(p-1)/2} - 1}$ are real numbers and the coefficients of $g(x) \pmod{x^{(p-1)/2} + 1}$ are imaginary numbers.*

Proof The Rader polynomial of blocklength $p - 1$ is given by

$$g(x) = \sum_{k=0}^{p-2} (\omega^{\pi^k} - 1) x^k,$$

First break this sum into two sums

$$g(x) = \sum_{k=0}^{(p-3)/2} (\omega^{\pi^k} - 1) x^k + \sum_{k=(p-1)/2}^{p-2} (\omega^{\pi^k} - 1) x^k$$

and set $k = k' + (p-1)/2$ in the second summation so that

$$g(x) = \sum_{k=0}^{(p-3)/2} (\omega^{\pi^k} - 1) x^k + \sum_{k'=0}^{(p-3)/2} (\omega^{\pi^{k'+(p-1)/2}}) x^{k'+(p-1)/2}.$$

Because π is primitive, $\pi^{p-1} = 1$, which means that $\pi^{(p-1)/2} = -1$. Therefore

$$g(x) = \sum_{k=0}^{(p-3)/2} [(\omega^{\pi^k} - 1) \mp (\omega^{-\pi^k} - 1)] x^k \pmod{x^{(p-1)/2} \pm 1}.$$

The theorem now follows by choosing, in turn, the plus sign and then the minus sign. □

---

[2] If so desired, one can suppress j from the diagonal matrix by introducing j into the matrix of postadditions. Then the elements of the diagonal matrix will be purely real.

We have shown how the Winograd small FFT of blocklength n can be derived whenever n is a prime. A Winograd small FFT also can be derived whenever n is a prime power. This construction requires a method like the Rader algorithm for turning a Fourier transform of size p^m, p a prime, into a convolution. However, the set of integers modulo p^m is not a field, and an element π of order $p^m - 1$ does not exist. Therefore a customized development is needed, as was discussed in Section 3.7.

There are two cases, p equal to two and p equal to an odd prime, that require somewhat different techniques. We first treat the case of p an odd prime.

Blocklength a power of an odd prime The construction is a bit complicated. We first remove all integers that contain the factor p from the set $\{1, 2, \ldots, p^m - 1\}$ to get a cyclic group with $p^{m-1}(p - 1)$ elements. This cyclic group leads to a cyclic convolution of length $p^{m-1}(p - 1)$, which is the core of the Fourier transform. As before, it is computed with a fast convolution algorithm. The output of the convolution must be unscrambled and then augmented by the p^{m-1} rows and columns that were dropped to form the convolution. As we shall see, one can find smaller convolutions in these auxiliary terms because they contain smaller Fourier transforms.

For example, with $N = 9 = 3^2$, we delete the integers 0, 3, and 6 to obtain the set $\{1, 2, 4, 5, 7, 8\}$, which forms a cyclic group under multiplication modulo 9 and is isomorphic to $\mathbf{Z}_6$, the additive group of integers $\{0, 1, 2, 3, 4, 5\}$ under addition modulo 6. The integer 2 generates the multiplicative group, because the powers of 2 modulo 9 are 1, 2, 4, 8, 7, 5. Thus there are six rows and six columns within the nine-point Fourier transform that can be isolated, scrambled, and then computed as a convolution, It is not hard to anticipate that the remaining rows and columns (those with index 0, 3, and 6) will have a structure akin to that of a three-point Fourier transform, and so some of the correction terms can be expressed as a smaller convolution.

The development of the nine-point Winograd FFT proceeds as follows. Write out the matrix equation

$$\begin{bmatrix} V_0 \\ V_1 \\ V_2 \\ V_3 \\ V_4 \\ V_5 \\ V_6 \\ V_7 \\ V_8 \end{bmatrix} = \begin{bmatrix} 1 & 1 & 1 & 1 & 1 & 1 & 1 & 1 & 1 \\ 1 & \omega & \omega^2 & \omega^3 & \omega^4 & \omega^5 & \omega^6 & \omega^7 & \omega^8 \\ 1 & \omega^2 & \omega^4 & \omega^6 & \omega^8 & \omega & \omega^3 & \omega^5 & \omega^7 \\ 1 & \omega^3 & \omega^6 & 1 & \omega^3 & \omega^6 & 1 & \omega^3 & \omega^6 \\ 1 & \omega^4 & \omega^8 & \omega^3 & \omega^7 & \omega^2 & \omega^6 & \omega & \omega^5 \\ 1 & \omega^5 & \omega & \omega^6 & \omega^2 & \omega^7 & \omega^3 & \omega^8 & \omega^4 \\ 1 & \omega^6 & \omega^3 & 1 & \omega^6 & \omega^3 & 1 & \omega^6 & \omega^3 \\ 1 & \omega^7 & \omega^5 & \omega^3 & \omega & \omega^8 & \omega^6 & \omega^4 & \omega^2 \\ 1 & \omega^8 & \omega^7 & \omega^6 & \omega^5 & \omega^4 & \omega^3 & \omega^2 & \omega \end{bmatrix} \begin{bmatrix} v_0 \\ v_1 \\ v_2 \\ v_3 \\ v_4 \\ v_5 \\ v_6 \\ v_7 \\ v_8 \end{bmatrix}.$$

By looking at this matrix we can see that rows and columns with index 0, 3, or 6 are novel because they contain repeated elements. Permute the rows and columns of the matrix to bring these to the top and left and to bring the other rows into the order 1, 2,

3.8 The Winograd small fast Fourier transform

4, 8, 7, 5, the powers of two modulo 9, and the columns into the order 1, 5, 7, 8, 4, 2, the powers of 2^{-1} modulo 9. Then

$$\begin{bmatrix} V_0 \\ V_3 \\ V_6 \\ V_1 \\ V_2 \\ V_4 \\ V_8 \\ V_7 \\ V_5 \end{bmatrix} = \begin{bmatrix} 1 & 1 & 1 & 1 & 1 & 1 & 1 & 1 & 1 \\ 1 & 1 & 1 & \omega^3 & \omega^6 & \omega^3 & \omega^6 & \omega^3 & \omega^6 \\ 1 & 1 & 1 & \omega^6 & \omega^3 & \omega^6 & \omega^3 & \omega^6 & \omega^3 \\ 1 & \omega^3 & \omega^6 & \omega^1 & \omega^5 & \omega^7 & \omega^8 & \omega^4 & \omega^2 \\ 1 & \omega^6 & \omega^3 & \omega^2 & \omega^1 & \omega^5 & \omega^7 & \omega^8 & \omega^4 \\ 1 & \omega^3 & \omega^6 & \omega^4 & \omega^2 & \omega^1 & \omega^5 & \omega^7 & \omega^8 \\ 1 & \omega^6 & \omega^3 & \omega^8 & \omega^4 & \omega^2 & \omega^1 & \omega^5 & \omega^7 \\ 1 & \omega^3 & \omega^6 & \omega^7 & \omega^8 & \omega^4 & \omega^2 & \omega^1 & \omega^5 \\ 1 & \omega^6 & \omega^3 & \omega^5 & \omega^7 & \omega^8 & \omega^4 & \omega^2 & \omega^1 \end{bmatrix} \begin{bmatrix} v_0 \\ v_3 \\ v_6 \\ v_1 \\ v_5 \\ v_7 \\ v_8 \\ v_4 \\ v_2 \end{bmatrix}.$$

The matrix has been partitioned to show the cyclic convolutions that have formed. There is one six-point cyclic convolution, and there are also six two-point cyclic convolutions. However, those two-point cyclic convolutions in the second and third column are a repetition of the same computation, and those in the second and third row can be combined into one by writing

$$\begin{bmatrix} V_3 \\ V_6 \end{bmatrix} = \begin{bmatrix} \omega^3 & \omega^6 \\ \omega^6 & \omega^3 \end{bmatrix} \begin{bmatrix} v_1 + v_7 + v_4 \\ v_5 + v_8 + v_2 \end{bmatrix} + \begin{bmatrix} v_0 + v_3 + v_6 \\ v_0 + v_3 + v_6 \end{bmatrix}.$$

Thus the Fourier transforms can be broken into a six-point cyclic convolution and two two-point cyclic convolutions by appropriately partitioning and scrambling the input data. We should anticipate that all of this will require twelve complex multiplications. However, the following theorem shows that two things will happen.

1 All multiplications will be by a purely real or a purely imaginary number, so the twelve complex multiplications reduce to twelve real multiplications.

2 Two of the coefficients are equal to zero, so the number of multiplications is reduced to ten.

Actually, we will write the final algorithm with eleven multiplications. The extra multiplication is an unneeded multiplication by one and used only to bring row zero into the algorithm in the same form as the others.

Theorem 3.8.2 *Let p be an odd prime and let m be an integer larger than one. Let $b = (p-1)p^{m-1}$. Let $g(x)$ be the generalized Rader polynomial*

$$g(x) = \sum_{k=0}^{b-1} \omega^{\pi^k} x^k,$$

where ω is a p^mth root of unity and π is an integer of order b under multiplication modulo p^m. Then

(i) the coefficients of $g(x)$ (mod $x^{b/2} - 1$) are real numbers;

(ii) the coefficients of $g(x)$ (mod $x^{b/2} + 1$) are imaginary numbers;
(iii) the coefficients of $g(x)$ (mod $x^{b/p} - 1$) are zero.

Proof The proof of the first two parts is similar to the proof of Theorem 3.8.1. Because π has order b, and b is even, we know that $\pi^b = 1$ modulo p^m, and $\pi^{b/2} = -1$ modulo p^m. Then the coefficients of $g(x)$ (mod $x^{b/2} \pm 1$) are

$$g_k = \omega^{\pi^k} \mp \omega^{\pi^{(k+b/2)}}.$$

Because $\pi^{b/2} = -1$, the first two parts of the theorem follow simply by noting that $\omega^{\pi^k} \mp \omega^{-\pi^k}$ is real or imaginary depending on the choice of sign.

To prove part (iii), let

$$g'(x) = g(x) \pmod{x^{b/p} - 1}$$
$$= \sum_{k=0}^{b-1} \omega^{\pi^k} x^k \pmod{x^{b/p} - 1}.$$

The polynomial $g'(x)$ has b/p coefficients, and each of these coefficients is the sum of p coefficients of $g(x)$. For nonzero r, the coefficient g'_r is given by

$$g'_r = \sum_{i=0}^{p-1} \omega^{\pi^{r+ib/p}}, \qquad r = 1, \ldots, (b/p) - 1.$$

We shall first prove that g'_r equals zero for all nonzero r. Rewrite the equation as

$$g'_r = \sum_{i=0}^{p-1} [\omega^{\pi^r}]^{\pi^{ib/p}}, \qquad r = 1, \ldots, (b/p) - 1.$$

Because ω^{π^r} is a p^mth root of unity when r is nonzero, the sum reduces to a sum of p ones, and a sum of p ones is zero modulo p. It only remains to carry out the proof for the case in which r is equal to zero. The coefficient g'_0 is the sum of those g_k for which k is a multiple of b/p. In this case,

$$g'_0 = \sum_{i=0}^{p-1} \omega^{\alpha^i},$$

where $\alpha = \pi^{b/p}$ has order p. Thus, α^i is a permutation of the integers of $\mathbf{Z}_p$, so with $h = \alpha^i$, the expression becomes

$$g'_0 = \sum_{h=0}^{p-1} \omega^h = \frac{1 - \omega^p}{1 - \omega},$$

where ω is a pth root of unity. Hence $g'_0 = 0$, and the proof is complete. □

We can construct a Winograd small FFT in this way for any blocklength equal to a prime power. The p^m by p^m matrix describing that Fourier transform will break down

3.8 The Winograd small fast Fourier transform

into one cyclic convolution of blocklength $(p^m - p^{m-1})$, and $p^{m-1} + 1$ cyclic convolutions of blocklength $(p - 1)$. Theorem 3.8.2 guarantees that all these convolutions can be computed by using the Winograd cyclic convolution algorithms with only purely real or purely imaginary multiplications.

Blocklength a power of two The Fourier transforms of blocklength equal to a power of two must be studied separately because only half of the integers less than 2^m are coprime to 2. These, of course, are the odd integers. By rearranging the matrix of the Fourier transform to put the rows with even indices first followed by the rows with odd indices, and putting the columns with even indices first followed by the columns with odd indices, the matrix is partitioned into four $n/2$ by $n/2$ subarrays.

The submatrix in which the rows and columns both have even indices can be expressed separately in terms of several Fourier transforms of size 2^{m-1}. This portion of the computation can use any 2^{m-1}-point Fourier transform algorithm. The partitioning into even and odd indices has some similarity to a single stage of a radix-two Cooley–Tukey algorithm.

Starting with the elementary expression,

$$V_k = \sum_{i=0}^{n-1} \omega^{ik} v_i, \qquad k = 0, \ldots, n-1,$$

the components of V with even values of k can be rewritten, using $2k'$ in place of k, as

$$V_{2k'} = \sum_{i'=0}^{(n/2)-1} (v_{i'} + v_{i'+n/2}) \omega^{2i'k'}, \qquad k' = 0, \ldots, (n/2) - 1.$$

This is a 2^{m-1}-point Fourier transform that can be computed by an FFT algorithm of that blocklength, 2^{m-1}. In this way, the components of V with an even index are computed with a smaller Fourier transform of smaller blocklength.

The components with odd values of k can be rewritten, using $2k' + 1$ in place of k, as

$$V_{2k'+1} = \sum_{i'=0}^{(n/2)-1} v_{2i'} \omega^{2i'(2k'+1)} + \sum_{i'=0}^{(n/2)-1} v_{2i'+1} \omega^{(2i'+1)(2k'+1)}.$$

With a little manipulation, the first term can be made into a Fourier transform of blocklength 2^{m-2} because ω^4 is a 2^{m-2}th root of unity. Thus

$$V_{2k'+1} = \sum_{i=0}^{(n/4)-1} [\omega^{2i'}(v_{2i'} - v_{2i'+n/4})] \omega^{4i'k'}$$
$$+ \sum_{i'=0}^{(n/2)-1} v_{2i'+1} \omega^{(2i'+1)(2k'+1)}, \qquad k' = 0, \ldots, (n/2) - 1,$$

where the first term needs only to be computed for $k' = 0, \ldots, (n/4) - 1$, because it then repeats. To compute the first term, if the input data is real, requires $(n/2) - 6$ real multiplications plus an $n/4$-point Fourier transform and, if the input data is complex, requires $(3/4)n - 8$ real multiplications plus an $n/4$-point Fourier transform.

The second term in the equation for $V_{2k'+1}$ can be computed by a generalized form of the Rader algorithm, as was discussed in the previous section. The odd integers do not form a cyclic group under multiplication modulo 2^m (except when $m = 1$ or $m = 2$). Rather, the set of odd integers is isomorphic to $\mathbf{Z}_2 \times \mathbf{Z}_{2^{m-2}}$. Because of this, the construction of a Winograd FFT of blocklength 2^m requires one more embellishment. First, we pick out the 2^{m-1} rows and the 2^{m-1} columns whose index is odd, as was done in Section 3.7. This portion of the matrix is rearranged – not into a single cyclic convolution – but into four cyclic convolutions. This will require two cyclic convolutions of blocklength 2^{m-2}. The following theorem shows that when these cyclic convolutions are computed by using the Chinese remainder theorem, some of the multiplications will be multiplications by zeros and can be dropped.

Theorem 3.8.3 *Let m be an integer larger than 2. Let $g(x, y)$ be the two-dimensional generalized Rader polynomial*

$$g(x, y) = \sum_{i=0}^{n'-1} \sum_{i'=0}^{1} \omega^{3^i(-1)^{i'}} x^i y^{i'},$$

where ω is a 2^nth root of unity. Then
  (i) the coefficients of $g(x, y)$ (mod $y - 1$) are real numbers;
  (ii) the coefficients of $g(x, y)$ (mod $y + 1$) are imaginary numbers;
  (iii) the coefficients of $g(x, y)$ (mod $x^{n'/2} - 1$) are zero.

Proof The proof is similar to the proof of Theorem 3.8.2. □

The first two parts of the theorem tell us that the cyclic convolutions (mod $x^{n/4} - 1$) are real cyclic convolutions (more precisely, one convolution is purely real and one is purely imaginary). The second part of the theorem tells us that, when computing the cyclic convolution, only the residue modulo $x^{n/8} + 1$ needs to be computed. The remaining residues are zero. Because $x^{n/8} + 1$ is irreducible, this requires $2(n/8) - 1$ real multiplications.

We now have decomposed the 2^m-point Fourier transform into the following pieces: (1) a 2^{m-1}-point Fourier transform, (2) a 2^{m-2}-point Fourier transform preceded by $2^{m-2} - 1$ complex multiplications, and (3) two products of real polynomials modulo the irreducible polynomial $x^{n/8} + 1$.

3.8 The Winograd small fast Fourier transform

The eight-point Fourier transform breaks into the two parts

$$\begin{bmatrix} V_0 \\ V_2 \\ V_4 \\ V_6 \end{bmatrix} = \begin{bmatrix} 1 & 1 & 1 & 1 \\ 1 & \omega^2 & \omega^4 & \omega^6 \\ 1 & \omega^4 & 1 & \omega^4 \\ 1 & \omega^6 & \omega^4 & \omega^2 \end{bmatrix} \begin{bmatrix} v_0 + v_4 \\ v_1 + v_5 \\ v_2 + v_6 \\ v_3 + v_7 \end{bmatrix},$$

which is a four-point Fourier transform, and

$$\begin{bmatrix} V_1 \\ V_3 \\ V_5 \\ V_7 \end{bmatrix} = \begin{bmatrix} 1 & 1 \\ 1 & \omega^4 \\ 1 & 1 \\ 1 & \omega^4 \end{bmatrix} \begin{bmatrix} \omega^0(v_0 - v_4) \\ \omega^2(v_2 - v_6) \end{bmatrix} + \begin{bmatrix} \omega^1 & \omega^3 & \omega^5 & \omega^7 \\ \omega^3 & \omega^1 & \omega^7 & \omega^5 \\ \omega^5 & \omega^7 & \omega^1 & \omega^3 \\ \omega^7 & \omega^5 & \omega^3 & \omega^1 \end{bmatrix} \begin{bmatrix} v_1 \\ v_3 \\ v_5 \\ v_7 \end{bmatrix}.$$

The second term is rearranged first as

$$\begin{bmatrix} V_1' \\ V_3' \\ V_7' \\ V_5' \end{bmatrix} = \begin{bmatrix} \omega^1 & \omega^3 & \omega^7 & \omega^5 \\ \omega^3 & \omega^1 & \omega^5 & \omega^7 \\ \omega^7 & \omega^5 & \omega^1 & \omega^3 \\ \omega^5 & \omega^7 & \omega^3 & \omega^1 \end{bmatrix} \begin{bmatrix} v_1 \\ v_3 \\ v_7 \\ v_5 \end{bmatrix}.$$

We now proceed more directly than in the general case. Using $\omega^4 = -1$, rewrite the equation as follows:

$$\begin{bmatrix} V_1' \\ V_3' \\ V_7' \\ V_5' \end{bmatrix} = \begin{bmatrix} \omega^1 & -\omega^7 & \omega^7 & -\omega^1 \\ -\omega^7 & \omega^1 & -\omega^1 & \omega^7 \\ \omega^7 & -\omega^1 & \omega^1 & -\omega^7 \\ -\omega^1 & \omega^7 & -\omega^7 & \omega^1 \end{bmatrix} \begin{bmatrix} v_1 \\ v_3 \\ v_7 \\ v_5 \end{bmatrix}.$$

Then

$$\begin{bmatrix} V_1' \\ V_7' \end{bmatrix} = \begin{bmatrix} \omega^1 & \omega^7 \\ \omega^7 & \omega^1 \end{bmatrix} \begin{bmatrix} v_1 - v_5 \\ v_7 - v_3 \end{bmatrix},$$

$$\begin{bmatrix} V_5' \\ V_3' \end{bmatrix} = -\begin{bmatrix} \omega^1 & \omega^7 \\ \omega^7 & \omega^1 \end{bmatrix} \begin{bmatrix} v_1 - v_5 \\ v_7 - v_3 \end{bmatrix} = -\begin{bmatrix} V_1' \\ V_7' \end{bmatrix}.$$

The cyclic convolution can be computed by

$$\begin{bmatrix} V_1' \\ V_7' \end{bmatrix} = \begin{bmatrix} 1 & 1 \\ 1 & -1 \end{bmatrix} \begin{bmatrix} \cos\theta & 0 \\ 0 & j\sin\theta \end{bmatrix} \begin{bmatrix} 1 & 1 \\ 1 & -1 \end{bmatrix} \begin{bmatrix} v_1 - v_5 \\ v_7 - v_3 \end{bmatrix}.$$

These are the only two nontrivial multiplications used by the eight-point Winograd FFT. In addition, there are six trivial multiplications.

For blocklengths equal to a large power of two, it becomes clumsy to explicitly list all of the equations needed to compute the Winograd FFT. It is better to express the radix-two FFT in recursive form. This we will postpone until Chapter 4. In Section 4.8 we will give an efficient radix-two FFT based on putting Winograd's ideas into recursive form.

Problems for Chapter 3

3.1 Given a device that computes an n-point Fourier transform in the complex field, describe how it can be used to simultaneously compute the Fourier transforms of two real vectors of length n.

3.2 Given a device that computes an n-point Fourier transform, describe how it can be used to compute an n-point inverse Fourier transform.

3.3 Sketch a circuit that computes a five-point Fourier transform using the Bluestein chirp algorithm.

3.4 **a** Prove that two is a primitive element in $GF(11)$.
   b Use the Rader prime algorithm to express the eleven-point Fourier transform over the complex numbers
$$V_k = \sum_{i=0}^{10} \omega^{ik} v_i$$
   in terms of a convolution
$$s(x) = g(x)d(x),$$
   writing out the polynomials $d(x)$, $s(x)$, and $g(x)$ and in terms of V, v, and ω.

3.5 Find prime integers n and n' with n less than n' such that the n-point Winograd small FFT uses more multiplications than the n'-point Winograd small FFT.

3.6 How many real multiplications will there be in a 25-point Winograd small FFT if all convolutions are optimal? How does this compare with a procedure that uses the Cooley–Tukey algorithm to combine two five-point Winograd FFT algorithms?

3.7 Show that by conjugating the input and output, one can use any FFT algorithm to compute an inverse Fourier transform.

3.8 State the inverse discrete cosine transform and prove its correctness.

3.9 Suppose that v is a vector of blocklength n of real numbers. Convert v to a vector u of blocklength $2n$ by following v by its time reversal
$$\mu_i = \begin{cases} v_i, & i = 0, \ldots, n-1, \\ v_{2n-1-i}, & i = n, \ldots, 2n-1. \end{cases}$$

How does the discrete cosine transform of v relate to the discrete Fourier transform of u?

Problems

3.10 Using only elementary algebraic and trigonometric properties, restructure the four-point discrete cosine transform to minimize the number of real multiplications.

3.11 The Good–Thomas algorithm reindexes the input and output differently. Show that one can interchange the input and output indexing scheme without any basic change in the algorithm.

3.12 Given a two-point and a five-point Fourier transform computed in the straightforward way (with four multiplications and 25 multiplications, respectively),

a Count the number of multiplications used by each of the following schemes for computing a 100-point Fourier transform:

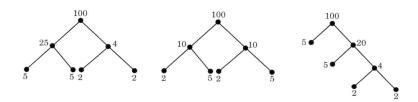

Use the Good–Thomas algorithm when possible; otherwise, use the Cooley–Tukey algorithm. Count all multiplications including trivial ones (those by $\pm 1, \pm j$).

b Repeat, but do not count trivial multiplications (those by $\pm 1, \pm j$).

c Now suppose you are given a subroutine that computes a two-point Fourier transform with no (nontrivial) multiplications and another subroutine that computes a five-point Fourier transform with five (nontrivial) real multiplications. Repeat part b.

3.13 The optimality of the Winograd small FFT does not follow immediately from the choice of an optimal convolution algorithm to compute the cyclic convolution generated by the Rader algorithm. This is because the coefficients of the Rader filter are not arbitrary; they are definite powers of ω, and conceivably there may be dependencies that can be exploited to reduce the number of multiplications. Specifically, we have seen that complex multiplications degenerate to real multiplications, and sometimes coefficients are equal to zero. Prove that the Winograd five-point FFT is optimal as measured by the number of multiplications.

3.14 A sixteen-point Winograd FFT uses eighteen multiplications, eight of which are trivial, and 74 additions.

a Describe how to build a 256-point FFT by using the Cooley–Tukey algorithm to build up from the sixteen-point Winograd FFT.

b How many multiplications are needed if the input data is real?

c How many multiplications are needed if the input data is complex?

Notes for Chapter 3

Fast Fourier transform algorithms came into widespread use in digital signal processing as a result of the well-known paper of Cooley and Tukey (1965) and the companion paper by Singleton (1969). The Cooley–Tukey work became widely known and had an immense impact on the field of signal processing. It was later realized by the community that the essential idea of this algorithm was known privately to Gauss (1866). The history of this algorithm was discussed by Heideman, Johnson, and Burrus (1984), and by Huang (1971). A different FFT, using the Chinese remainder theorem, appeared earlier in the papers of Good (1960) and Thomas (1963). The differences between these FFT algorithms were discussed by Good (1971). An efficient organization of the Cooley–Tukey algorithm was given by Rader and Brenner (1976) and modified by Preuss (1982). Rader (1968) and Bluestein (1970) gave methods for turning a discrete Fourier transform into a convolution. Rader's purpose was to compute a discrete Fourier transform with blocklength equal to a large prime; but ironically, his method turned out to be important when the blocklength is equal to a small prime. Winograd (1978) generalized Rader's prime algorithm to blocklengths that are a power of a prime.

The Winograd FFT was announced in summary form in 1976 and published in detail in 1978. Our treatment diffuses the presentation of the original work by integrating the development into other topics. We have tabulated the small FFT modules that seem to be the most useful. Larger FFT modules have been constructed by Johnson and Burrus (1981). Other methods of computing the Fourier transform were studied by Goertzel (1968) and by Sarwate (1978). Fast algorithms for the discrete cosine transform were studied by Chen, Smith, and Fralick (1977); and by Narasimha and Peterson (1978), by Makhoul (1980), and by Feig and Winograd (1992).

4 Fast algorithms based on doubling strategies

Many good algorithms can be derived by strategies that double an algorithm for half of the problem. Given a problem of size n in some parameter, split the problem in half, if possible, to obtain two problems of size $n/2$, but each with the same structure as the original problem. If algorithms for the half problems can be easily combined into an algorithm for the original problem, then one may have succeeded in finding an algorithm that is efficient.

A radix-two Cooley–Tukey FFT algorithm can be thought of as an algorithm constructed by halving and doubling because an n-point Cooley–Tukey FFT is built out of two $n/2$-point FFTs. In Chapter 5, we will describe the iterated filter sections that also have this doubling structure whereby an n-point filter section can be built out of two $n/2$-point filter sections. This chapter will develop other fast algorithms based on halving and doubling. These algorithms are important in their own right for signal processing. They illustrate a way of thinking that can be used to construct algorithms for many kinds of processing tasks.

4.1 Halving and doubling strategies

Consider the task of computing a polynomial $p(x)$ of degree n, given its set of zeros $\beta_0, \beta_1, \ldots, \beta_{n-1}$. This means that the polynomial $p(x)$ can be written as

$$p(x) = (x - \beta_{n-1})(x - \beta_{n-2}) \cdots (x - \beta_0),$$

and from this expression on the right, the coefficients of $p(x)$ can be computed. The most natural way to do this computation is to start at one end, say the right end, and to multiply new factors one at a time by using the following procedure:

$$p^{(i)}(x) = (x - \beta_i)p^{(i-1)}(x), \qquad i = 1, \ldots, n-1,$$

and starting with $p^{(0)}(x) = (x - \beta_0)$. This procedure requires i multiplications and i additions at iteration i, a total of $\frac{1}{2}n(n-1)$ multiplications and the same number of additions.

A better algorithm is obtained by halving and doubling. Suppose that n is a power of two given by 2^m. (It is easy to modify the procedure for other values of n by jumping over some of the steps.) Now let

$$p'(x) = \prod_{i=0}^{(n/2)-1} (x - \beta_i),$$

$$p''(x) = \prod_{i=0}^{(n/2)-1} (x - \beta_{(n/2)+i}),$$

and

$$p(x) = p''(x)p'(x).$$

This last equation requires $(n/2)^2$ multiplications as it is written. If $p'(x)$ and $p''(x)$ are each computed in the direct way, they each require $\frac{1}{2}(n/2)((n/2) - 1)$ multiplications. The total number of multiplications is

$$\left(\frac{n}{2}\right)^2 + 2\frac{1}{2}\left(\frac{n}{2}\right)\left(\frac{n}{2} - 1\right) = \frac{1}{2}n(n - 1),$$

which is no better than the direct method.

In order to gain any benefit from the doubling strategy, we need a better method to combine the two parts in the computation

$$p(x) = p''(x)p'(x).$$

But the set of coefficients of $p(x)$ is a linear convolution of the sets of coefficients of $p'(x)$ and $p''(x)$. By using a fast Fourier transform, a linear convolution can be done in fewer than $An \log_2 n$ operations for some small constant A. Hence the total number of multiplications is fewer than

$$M(n) = An \log_2 n + \frac{n}{2}\left(\frac{n}{2} - 1\right),$$

which is an improvement for large n. The number can be reduced further by using the same idea again to compute $p'(x)$ and $p''(x)$. Each of these can be split, in turn, and computed from two half solutions using fewer than $A(n/2)\log_2(n/2)$ operations – a total of fewer than $An \log_2(n/2)$ operations to compute both. By continuing to halve the problem in this way, the total number of multiplications is reduced to

$$M(n) = A\sum_{i=1}^{m} n \log_2 \frac{n}{2^i}$$

$$= A\frac{n}{2}(\log_2^2 n - \log_2 n).$$

4.1 Halving and doubling strategies

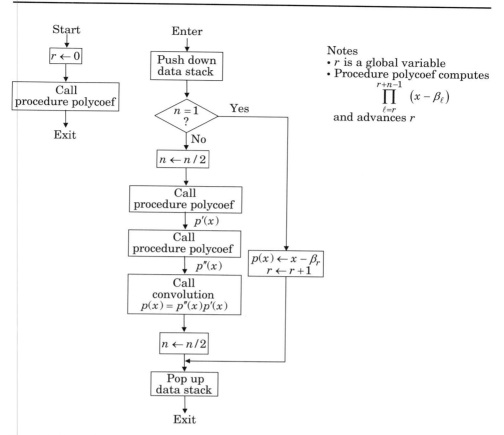

Figure 4.1 Procedure *polycoef*

This is less than $\frac{1}{2}n(n-1)$ except for very small n, so the halving and doubling strategy has yielded an improved algorithm.

Figure 4.1 shows an organization of the computation of the polynomial product using halving and doubling. This procedure is a good example of a kind of procedure known as a *recursive procedure*. Recursion is a sophisticated computational principle. It does not explicitly describe every level of the computation; it describes one level, but that level contains a copy of the same procedure. The procedure calls itself. This requires that the temporary data be organized in the form of a push-down stack, as described in the next section. Each time the procedure is called, it pushes down the existing stack of data to open a clear workspace.

Similar doubling strategies can be used for many problems. If a computational problem depends on an integer n that is a power of two, one tries to obtain an answer for $n = 2^m$ from the answers to two half-problems with $n = 2^{m-1}$. The radix-two Cooley–Tukey FFT is shown in the recursive form of a doubling algorithm in Figure 4.2. It can be instructive to reflect on how the sequence of operations differs in Figure 4.2

Fast algorithms based on doubling strategies

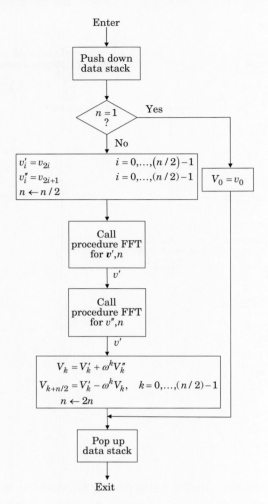

Figure 4.2 Procedure FFT

from the organization to be given in Figure 6.5 of Chapter 6. The recursive form may require more temporary memory because it will create a stack of temporary Fourier transforms of various sizes. On the other hand, the recursive form may be easier to use if a single program is required to be able to compute any radix-two Fourier transform. It also may be a convenient organization for a mixed-radix FFT because it is easy to branch to subroutines for other blocklengths.

Doubling strategies usually can be extended to problems where n is not a power of two. One way is to append enough dummy iterations to make n the next larger power of two, though in some cases, as for the Fourier transform, this will not work. One can also split a problem into pieces of some other size, such as thirds or fifths, but splitting it into halves is usually best if the problem permits it.

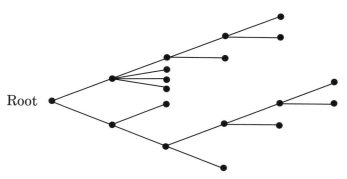

Figure 4.3 A tree

4.2 Data structures

Any collection of data in a computation must be arranged in some way before it can be processed. Similarly, intermediate data must be stored in a convenient way. The two basic methods of organizing data are *lists* and *trees*. A list of length L is an ordered set of L data items; each data item may itself be a complicated collection of data, perhaps even containing lists of its own.

When the elements of a list are numbers, the list is sometimes called a *vector*. We prefer to reserve this term only for an element of a vector space. When the elements of a list are elements from a finite alphabet, the list may be called a *string*. A string need not have a fixed length, but a vector usually does. The difference between a list, a vector, and a string is a matter of the context of the application.

A variable-length list is one whose length is not fixed, but grows or shrinks with time. A variable-length list can grow or shrink by adding or deleting items from any point in the list, but, in many cases, items are added or deleted only at either or both of the two ends. A list that has additions and deletions only from one end – say, the top – is called a *stack*, or a *push-down stack*, or a *last-in first-out (LIFO) buffer*. A list that has insertions only at one end and deletions from the other is called a *queue*, or a *first-in first-out (FIFO) buffer*.

A tree is a data structure in which each data item can be followed by one or more data items called *descendants*; a graphical representation is given in Figure 4.3. Each node represents a data item, and each data item may have several other descending data items to which it refers. Possibly several nodes in a tree are identical copies of the same data item. A tree can be compared to a list; in a list, each data item has only one descendant.

The data items in a list or a tree might be quite complex, perhaps involving text from a natural language, but for the purpose of studying the list structure, each data item is treated as a single unit. Lists or trees can be stored in a memory conveniently by using names for the items; a good name is the memory address where the data item begins.

Data record

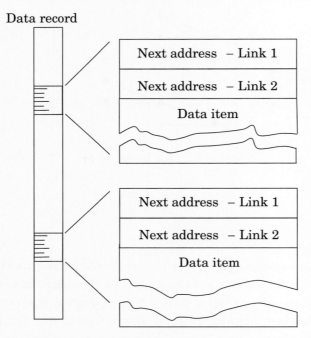

Figure 4.4 A doubly-linked list

The list (or the tree) is stored by listing in sequence the names of the items. The list need not be stored in close proximity to the data items.

Another method, which is sometimes better, is an indirect addressing method called a *linked list*, as shown in Figure 4.4. The data items appear in arbitrary order, and each entry begins with the starting address of the next data item on the list. Figure 4.4 shows a *doubly-linked list*. In a doubly-linked list, the data items are listed in two orders – say, alphabetical and chronological – but need not be stored twice. If the list is frequently revised, a linked list is convenient because the data need not be moved. Only the link addresses need to be changed.

A stack can be constructed out of complex data items simply by giving the address of the first data item and attaching to each data item the address of the next data item in the stack. To push down the stack with a new data item on top, simply attach the address of the previous top entry to the new entry, and change the address of the first data item in the list to the address of the new item. To pop up the stack, reverse this procedure.

4.3 Fast algorithms for sorting

The sorting problem is formulated as follows. We are given an arbitrary sequence of n elements, taken from a set on which we have some notion of a natural order. We

4.3 Fast algorithms for sorting

are to rearrange these data items into their natural order. The sorting algorithm may handle the data item itself and relocate that item in memory. Alternatively, the sorting algorithm handles not the data item itself, but only some parameters attached to it. The data is sorted by rearranging a list of indirect addresses. Sorting by rearranging addresses is useful when the data items are themselves large.

Suppose that each data item has a numerical parameter associated with it. We are to order the data items in the numerical order of this parameter from largest to smallest. Any sorting problem can be viewed in this way.

A naive sorting procedure looks at the items one by one to find and tag the largest. Then it looks at the reduced list to find the new largest item. It continues in this way until all items are sorted. In everyday life, where n is small, this is a satisfactory procedure. However, the average number of steps is proportional to n^2. For sorting large lists, we can do much better.

Good sorting algorithms are based on halving and doubling. *Mergesort* is a sorting algorithm with a complexity proportional to $n \log n$. Split a list of n data items into two halves, and sort each half. From the two sorted lists, produce a single sorted list by merging them. Merging works as follows. Look at the top element of the two lists and choose the largest as the next entry of the composite list. Delete it from the half-list when it is moved to the composite list. Because a data item is placed on the composite list after every comparison, there can be at most n comparisons. Therefore the complexity $C(n)$ of sorting n data items satisfies the recursion

$$C(n) \leq 2C\left(\frac{n}{2}\right) + n.$$

This implies that the complexity of mergesort is bounded by

$$C(n) \leq n \log_2 n.$$

There is only one place in the argument where an inequality occurs. A few comparisons may not be needed in the merge, so we might save a little. However, we expect the bound to be rather tight.

There are other ways to split the sorting problem. *Quicksort* randomly chooses one of the data parameters as a number used to split the list. The performance of quicksort is a random variable, quite good on the average, but slow on worst-case data sets.

Quicksort works as follows. If n equals zero or one, then the list is already sorted. Otherwise, randomly choose an element of the list, then move each other element of the list above or below that chosen element, according to whether that other element is larger or smaller than the chosen element with ties going either way. This results in two half-lists of random length, those above the chosen element and those below. Then sort each half-list in the same way.

In the worst case, the randomly chosen element at each step is the largest (or the smallest). Then the two new lists have length zero and $n - 1$, respectively. If this occurs at each step, the complexity at step i is $n - i$, and the total complexity is proportional to n^2.

The two new lists have $n - 1 - i$ and i elements, respectively, where i is a random variable, equiprobable over the set $\{0, \ldots, n - 1\}$. Forming these two sets has a complexity proportional to n. The expected complexity of sorting the original set of n data points is

$$C(n) = An + \frac{1}{n}\sum_{i=0}^{n-1} C(i) + \frac{1}{n}\sum_{i=0}^{n-1} C(n - 1 - i)$$

for some constant, A. This gives the recursive formula for $C(n)$ when n is larger than two:

$$C(n) = An + \frac{2}{n}\sum_{i=0}^{n-1} C(i).$$

This is initialized with $C(0) = C(1)$ equal to zero, or perhaps to some small number, but this detail changes little and has no effect on the asymptotic complexity. We shall show that for n greater than two, $C(n)$ is less than $2An \log n$. The argument is by induction. Assume that for all i less than n, $C(i)$ is less than $2Ai \log_e i$. This holds for n equal to two. Then

$$C(n) < An + \frac{4A}{n}\sum_{i=2}^{n-1} i \log_e i.$$

Because $i \log_e i$ is a convex function, the right side can be bounded as follows:

$$C(n) < An + \frac{4A}{n}\int_2^n x \log_e x \, dx.$$

Hence

$$C(n) < An + \frac{4A}{n}\left[\frac{n^2 \log_e n}{2} - \frac{n^2}{4}\right] = 2An \log_e n.$$

The average performance of quicksort can be asymptotically better than mergesort, but the worst case performance is poorer.

4.4 Fast transposition

Whenever one processes a large two-dimensional array, such as a digitized image, it may be that only a small part of the array is directly accessible within the processor.

4.4 Fast transposition

For example, to store a 1024 by 1024 array requires more than a million words of memory and more than two million words if the data is complex. Most of the array is held in a bulk memory system and is transferred in sections into local memory within the processor. Most commonly, the n by n array is stored by columns (or by rows), and it may be convenient to transfer one column at a time between local memory and bulk memory to select one element of a row. Hence n columns must be read to form a row. To transfer two elements from two different columns may be as difficult as transferring two entire columns. To form each row in turn, n^2 columns must be read; direct matrix transposition reads n^2 columns.

A fast transposition algorithm interchanges rows and columns in the bulk memory in applications where the local memory is small. We shall study the case in which the local memory can hold two columns of the square array. This is the most interesting case, and it illustrates all of the main ideas. If the local memory cannot hold two columns, then the doubling algorithm loses its efficiency; if it can hold more, then further improvements by a small constant are possible.

Transposing a matrix that is in random-access local memory is trivial, because it is simply a matter of modifying addresses when calling data. We will assume that taking a transpose of a small matrix amounts to nothing more than reading an array into local memory, then reading it out. Let A, a 2^m by 2^m matrix, be partitioned into blocks of size 2^{m-1} by 2^{m-1} as

$$A = \begin{bmatrix} A_{11} & A_{12} \\ A_{21} & A_{22} \end{bmatrix}.$$

Suppose we have a method of computing

$$\begin{bmatrix} A_{11}^T & A_{12}^T \\ A_{21}^T & A_{22}^T \end{bmatrix}.$$

Then to compute

$$A^T = \begin{bmatrix} A_{11}^T & A_{21}^T \\ A_{12}^T & A_{22}^T \end{bmatrix}$$

it suffices to interchange A_{21}^T and A_{12}^T. This can be done by reading two columns at a time to interchange one column of A_{12}^T with one column of A_{21}^T. It requires that $n/2$ pairs of columns be transferred to complete the interchange.

Now we apply the same idea recursively to compute A_{11}^T, A_{21}^T, A_{12}^T, and A_{22}^T by partitioning each of these blocks. Because A_{11} and A_{21} appear in the same columns of the array, both of these transpositions can be done with the same set of column transfers. Hence, at each level of the recursion, n columns need to be transferred, and there are $\log_2 n$ levels in the recursion. Fast matrix transposition uses $n \log_2 n$ column transfers, as compared to n^2 column transfers used by the direct method.

4.5 Matrix multiplication

The pairwise product of two by two matrices

$$\begin{bmatrix} c_{11} & c_{12} \\ c_{21} & c_{22} \end{bmatrix} = \begin{bmatrix} a_{11} & a_{12} \\ a_{21} & a_{22} \end{bmatrix} \begin{bmatrix} b_{11} & b_{12} \\ b_{21} & b_{22} \end{bmatrix}$$

can be written out as

$$c_{11} = a_{11}b_{11} + a_{12}b_{21},$$
$$c_{12} = a_{11}b_{12} + a_{12}b_{22},$$
$$c_{21} = a_{21}b_{11} + a_{22}b_{21},$$
$$c_{22} = a_{21}b_{12} + a_{22}b_{22},$$

from which we see that the computation in this form requires eight multiplications and four additions. The *Strassen algorithm* is a way to do the computation in seven multiplications.

The Strassen algorithm first computes the following products:

$$m_1 = (a_{12} - a_{22})(b_{21} + b_{22}),$$
$$m_2 = (a_{11} + a_{22})(b_{11} + b_{22}),$$
$$m_3 = (a_{11} - a_{21})(b_{11} + b_{12}),$$
$$m_4 = (a_{11} + a_{12})b_{22},$$
$$m_5 = a_{11}(b_{12} - b_{22}),$$
$$m_6 = a_{22}(b_{21} - b_{11}),$$
$$m_7 = (a_{21} + a_{22})b_{11}.$$

The following equations:

$$c_{11} = m_1 + m_2 - m_4 + m_6,$$
$$c_{12} = m_4 + m_5,$$
$$c_{21} = m_6 + m_7,$$
$$c_{22} = m_2 - m_3 + m_5 - m_7$$

then give the elements of the matrix.

4.5 Matrix multiplication

The Strassen algorithm can be expressed in a matrix form as

$$\begin{bmatrix} c_{11} \\ c_{12} \\ c_{21} \\ c_{22} \end{bmatrix} = \begin{bmatrix} 1 & 1 & 0 & -1 & 0 & 1 & 0 \\ 0 & 0 & 0 & 1 & 1 & 0 & 0 \\ 0 & 0 & 0 & 0 & 0 & 1 & 1 \\ 0 & 1 & -1 & 0 & 1 & 0 & -1 \end{bmatrix} \begin{bmatrix} G_1 \\ G_2 \\ G_3 \\ G_4 \\ G_5 \\ G_6 \\ G_7 \end{bmatrix} \begin{bmatrix} 0 & 0 & 1 & 1 \\ 1 & 0 & 0 & 1 \\ 1 & 1 & 0 & 0 \\ 0 & 0 & 0 & 1 \\ 0 & 1 & 0 & -1 \\ -1 & 0 & 1 & 0 \\ 1 & 0 & 0 & 0 \end{bmatrix} \begin{bmatrix} b_{11} \\ b_{12} \\ b_{21} \\ b_{22} \end{bmatrix},$$

where the center matrix is a diagonal matrix whose diagonal elements are given by

$$\begin{bmatrix} G_1 \\ G_2 \\ G_3 \\ G_4 \\ G_5 \\ G_6 \\ G_7 \end{bmatrix} = \begin{bmatrix} 0 & 1 & 0 & -1 \\ 1 & 0 & 0 & 1 \\ 1 & 0 & -1 & 0 \\ 1 & 1 & 0 & 0 \\ 1 & 0 & 0 & 0 \\ 0 & 0 & 0 & 1 \\ 0 & 0 & 1 & 1 \end{bmatrix} \begin{bmatrix} a_{11} \\ a_{12} \\ a_{21} \\ a_{22} \end{bmatrix}.$$

The Strassen algorithm uses seven multiplications and eighteen additions. If one of the two matrices in the product is a constant and is to be used many times, then some of the additions can be done once off-line, and so only thirteen additions are required.

In the best case, the Strassen algorithm trades one multiplication for nine additions as compared to straightforward matrix multiplication. It has no practical advantage for multiplying two by two matrices.

Now consider the problem of multiplying n by n matrices. The direct method of matrix multiplication uses n^3 multiplications and $(n-1)n^2$ additions. We suppose that n is a power of two and can be written as 2^m for some m; otherwise, append columns of zeros on the right and append rows of zeros on the bottom to make n into a power of two.

The matrix product $C = AB$ can be partitioned as

$$\begin{bmatrix} C_{11} & C_{12} \\ C_{21} & C_{22} \end{bmatrix} = \begin{bmatrix} A_{11} & A_{12} \\ A_{21} & A_{22} \end{bmatrix} \begin{bmatrix} B_{11} & B_{12} \\ B_{21} & B_{22} \end{bmatrix},$$

where now each block is 2^{m-1} by 2^{m-1} matrix. If we multiply out these blocks as written, then there are eight matrix multiplications of $n/2$ by $n/2$ matrices and four matrix additions of $n/2$ by $n/2$ matrices. If these are computed in the direct way, then the total number of multiplications is

$$M(n) = 8\left(\frac{n}{2}\right)^3 = n^3,$$

$$A(n) = 8\left(\frac{n}{2} - 1\right)\left(\frac{n}{2}\right)^2 + 4\left(\frac{n}{2}\right)^2$$
$$= (n-1)n^2.$$

This is the same as before, so halving and doubling has no advantage unless some other improvement is made. The Strassen algorithm, when applied to matrix blocks, is such an improvement.

The Strassen algorithm applies when the arguments are matrix blocks because it does not depend on the commutative property of arithmetic. The Strassen algorithm first computes the following matrix products:

$$M_1 = (A_{12} - A_{22})(B_{21} + B_{22}),$$
$$M_2 = (A_{11} + A_{22})(B_{11} + B_{22}),$$
$$M_3 = (A_{11} - A_{21})(B_{11} + B_{12}),$$
$$M_4 = (A_{11} + A_{12})B_{22},$$
$$M_5 = A_{11}(B_{12} - B_{22}),$$
$$M_6 = A_{22}(B_{21} - B_{11}),$$
$$M_7 = (A_{21} + A_{22})B_{11}.$$

The blocks of C are then computed by

$$C_{11} = M_1 + M_2 - M_4 + M_6,$$
$$C_{12} = M_4 + M_5,$$
$$C_{21} = M_6 + M_7,$$
$$C_{22} = M_2 - M_3 + M_5 - M_7.$$

In the case in which A is a matrix of constants, there are seven $n/2$ by $n/2$ matrix multiplications and thirteen $n/2$ by $n/2$ matrix additions here, as compared to eight such matrix multiplications and four such matrix additions in the usual procedure. If these are computed in the direct way, the total number of multiplications is

$$M(n) = 7\left(\frac{n}{2}\right)^3 = \frac{7}{8}n^3,$$

which is less than n^3. The total number of additions is

$$A(n) = 7\left(\frac{n}{2} - 1\right)\left(\frac{n}{2}\right)^2 + 13\left(\frac{n}{2}\right)^2$$
$$= \left(\frac{7}{8}n + \frac{3}{2}\right)n^2,$$

which is less than $(n-1)n^2$ if n is larger than twenty.

The Strassen algorithm gives even better performance if it is applied recursively, breaking each matrix product into smaller pieces using the same equations. Then the number of multiplications is

$$M(n) = 7^m = 7^{\log_2 n} = n^{\log_2 7}$$
$$= n^{2.81}.$$

The number of additions is more difficult to state in simple terms. It satisfies the recursion

$$A(n) = 7A\left(\frac{n}{2}\right) + 13\left(\frac{n}{2}\right)^2.$$

The number of additions is larger than the number of multiplications, but, for large n, it also grows as $n^{2.81}$. For large enough n, it will also be less than the direct method. For n equal to 1024, the number of additions is about the same as the direct method, but there are only about one-fourth as many multiplications.

4.6 Computation of trigonometric functions

Trigonometric functions are commonly computed using some form of a power series. Power series methods are very different from the doubling methods we have been studying. However, there are several other methods, useful in special applications, that have the flavor of a doubling strategy, and these we shall describe.

The first algorithm we describe is a method for simultaneously computing $\sin\theta$ and $\cos\theta$ when given the angle θ. The computational process is based on the trigonometric double-angle identities:

$$\sin 2\theta = 2\sin\theta\cos\theta$$
$$\cos 2\theta = 1 - 2\sin^2\theta.$$

It will be helpful to write these in the form

$$\sin 2\theta = 2\sin\theta - 2\sin\theta \text{ vers }\theta$$
$$\text{vers } 2\theta = 2\sin^2\theta,$$

where

$$\text{vers }\theta = 1 - \cos\theta.$$

An initial value is generated by dividing θ by a power of two and using small angle approximations:

$$\sin\frac{\theta}{2^m} = \frac{\theta}{2^m}$$
$$\text{vers}\frac{\theta}{2^m} = 0.$$

The accuracy of the algorithm can be adjusted as desired by specifying m. The angle θ is expressed in radians and is limited to $\pm\pi$. No special quadrant determination is necessary because the algorithm automatically places the results in the correct quadrant. The end results of the computation are $\sin\theta$ and $1 - \cos\theta$.

If the initial numbers are very small, large wordlengths will be required. To avoid numbers that are too small, new variables are defined:

$$X_n = \frac{2^m}{2^n}(\sin\theta)_n,$$
$$Y_n = \frac{2^m}{2^n}(\text{vers}\,\theta)_n.$$

The recursive equations then become

$$X_n = X_{n-1} - 2^{n+1-m}X_{n-1}Y_{n-1},$$
$$Y_n = 2^{n+1-m}X_{n-1}^2,$$

where m is the total number of iterations to be executed, and the initial values are $X_0 = \theta$, $Y_0 = 0$.

The accuracy of the algorithm in the initial iteration is determined by the error term

$$(X_0)_e = -\frac{1}{6}\left(\frac{\theta}{2^m}\right)^3,$$

which is the error in the initial approximation. This initial error propagates into a final error, bounded by

$$(\sin\theta)_e \leq \frac{1}{6}\pi^3 2^{-2m},$$
$$(\cos\theta)_e \leq \frac{1}{6}\pi^3 2^{-2m}.$$

Hence the numerical error of the algorithm itself reduces with the number of iterations at a rate of two bits per iteration. To ensure this accuracy, the multiplications must carry enough bits to support that accuracy.

The second trigonometric algorithm to be described in this section is a method for coordinate rotation. Either it computes

$$\begin{bmatrix} x' \\ y' \end{bmatrix} = \begin{bmatrix} \cos\theta & \sin\theta \\ -\sin\theta & \cos\theta \end{bmatrix} \begin{bmatrix} x \\ y \end{bmatrix},$$

or it computes the polar transformation

$$\theta = \tan^{-1}\frac{x}{y}, \qquad r = \sqrt{x^2 + y^2},$$

depending on the way in which it is used. The algorithm is a combination of computational and look-up techniques. The key to the algorithm is the fact that it is easy to rotate

4.6 Computation of trigonometric functions

a vector by a particular angle of the form $\theta = \tan^{-1} 2^{-k}$, by using the trigonometric identities

$$\sin\theta = \frac{\tan\theta}{\sqrt{1+\tan^2\theta}},$$
$$\cos\theta = \frac{1}{\sqrt{1+\tan^2\theta}}.$$

Hence

$$\sin\left[\tan^{-1} 2^{-k}\right] = \frac{2^{-k}}{\sqrt{1+(2^{-k})^2}},$$
$$\cos\left[\tan^{-1} 2^{-k}\right] = \frac{1}{\sqrt{1+(2^{-k})^2}}.$$

Therefore a rotation of the vector (x, y) by $\theta_k = \tan^{-1} 2^{-k}$ can be written

$$x' = \frac{1}{\sqrt{1+(2^{-k})^2}}[x + 2^{-k}y],$$
$$y' = \frac{1}{\sqrt{1+(2^{-k})^2}}[y - 2^{-k}x].$$

To rotate by the negative of $\tan^{-1} 2^{-k}$, the same equations are used but with the sign of 2^{-k} reversed. The sign reversal does not affect the magnitude term multiplying the bracket. Therefore the magnitude of the vector is increased by a fixed constant. An arbitrary angle θ can be expressed as follows:

$$\theta = \pm 90° + \sum_{k=0}^{\infty}(\pm\tan^{-1} 2^{-k})$$

or

$$\theta = \xi_{-1} 90° + \sum_{k=0}^{\infty} \xi_k \tan^{-1} 2^{-k}$$
$$= \sum_{k=-1}^{\infty} \xi_k \theta_k,$$

where ξ_k for $k = -1, 0, 1, \ldots$ is equal to either 1 or -1, according to the table of arc tangents in Figure 4.5. To rotate by the angle θ, simply rotate by each θ_k, in turn, with the sign of the rotation specified by ξ_k.

After n iterations, the magnification is

$$\prod_{k=1}^{m}\sqrt{1+2^{-2k}}$$

which is independent of the signs of the individual rotations.

Iteration k	Scaling 2^{-k}	Angle increment $\theta_i = \tan^{-\ell} 2^{-k}$
−1	0	90°
0	1	45°
1	1/2	26.565 051°
2	1/4	14.036 243°
3	1/8	7.125 016°
4	1/16	3.576 334°
5	1/32	1.789 911°
6	1/64	0.895 174°
7	1/128	0.447 614°
8	1/256	0.223 811°
9	1/512	0.111 906°
10	1/1024	0.055 953°

Figure 4.5 An arc tangent table

The resulting algorithm is shown in Figure 4.6. During the first iteration, the vector is rotated by ±90° by a slightly different rule. Thereafter, at iteration k, by examining the signs of x and y, a decision is made to rotate either by θ_k or by $-\theta_k$. This is effected by one addition or subtraction with scaling. Because the scale factor is 2^{-k}, it consists of a binary shift. Thus the algorithm is almost free of multiplications. The magnitude expansion produced by the algorithm is a constant independent of θ. It can be canceled by multiplying by its reciprocal after the iterations are completed. Even better, it can be buried by absorbing it into constants, if possible, in other places in the larger application.

To compute $\tan^{-1}(x/y)$, the action of the algorithm is reversed. The vector (x, y) is rotated during each iteration in the direction that will reduce the current magnitude of y, and the signed values of θ_k are added to get θ.

4.7 An accelerated euclidean algorithm for polynomials

The euclidean algorithm for polynomials can be accelerated by means of a doubling strategy to get an algorithm with an asymptotic complexity on the order of $n \log^2 n$. For small n, the overhead seems to make the algorithm unattractive. The euclidean algorithm consists of the repeated application of the division algorithm, with each iteration computing, from the current $s(x)$ and $t(x)$, the new quotient polynomial $Q(x)$ and the new remainder polynomial $r(x)$ that satisfy the equation

$$s(x) = Q(x)t(x) + r(x).$$

4.7 An accelerated euclidean algorithm for polynomials

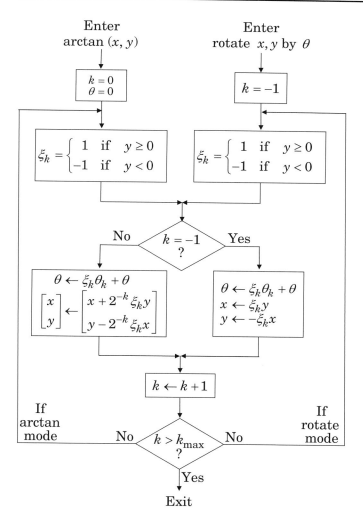

Figure 4.6 Coordinate rotation algorithm

Although all coefficients of $s(x)$ and $t(x)$ are used in computing the remainder polynomial, not all coefficients of $s(x)$ and $t(x)$ are used in computing the quotient polynomial. Only some of the high-order coefficients of $s(x)$ and $t(x)$ are needed to compute $Q(x)$. Moreover, some low-order coefficients of the remainder polynomial may be unneeded for computing the next quotient polynomial, and even may be unneeded in several iterations after that. This means that only a portion of the remainder polynomial needs to be current at each iteration, provided the coefficients of other remainder polynomials that eventually are needed in later iterations can be computed just in time to be used. This observation suggests the formulation of a halving and doubling strategy.

The main computations of the euclidean algorithm are described by the equations

$$Q^{(r)}(x) = \left\lfloor \frac{s^{(r-1)}(x)}{t^{(r-1)}(x)} \right\rfloor$$

and

$$\begin{bmatrix} s^{(r)}(x) \\ t^{(r)}(x) \end{bmatrix} = \begin{bmatrix} 0 & 1 \\ 1 & -Q^{(r)}(x) \end{bmatrix} \begin{bmatrix} s^{(r-1)}(x) \\ t^{(r-1)}(x) \end{bmatrix},$$

where now $t^{(r)}(x)$ denotes the remainder polynomial at iteration r. The iterations stop when $t^{(r+1)}(x)$ equals zero. Computation of the quotient polynomial $Q^{(r)}(x)$ requires only the high-order coefficients of $t^{(r-1)}(x)$. Define

$$A^{(r,r'+1)}(x) = \prod_{r}^{\ell=r'+1} \begin{bmatrix} 0 & 1 \\ 1 & -Q^{(\ell)}(x) \end{bmatrix},$$

with the order of the product written in reverse order to respect the noncommutativity of matrix multiplication. Then

$$\begin{bmatrix} s^{(r)}(x) \\ t^{(r)}(x) \end{bmatrix} = A^{(r,1)}(x) \begin{bmatrix} s(x) \\ t(x) \end{bmatrix},$$

where $A^{(r)}(x) = A^{(r,1)}(x)$.

Clearly, the computation of $A^{(r,1)}(x)$ as a product of matrices has the same structure as a polynomial product, which was shown in Section 4.1 to be amenable to a doubling strategy. We need to partition the terms of the matrix product into two batches of about the same size. If we can do this so that each batch looks like (or can be made to look like) the original problem, then the recursive structure will follow. However, there are several details that must be accommodated. The difficulty is that the number of iterations is not known in advance, so it is not possible to say when half of the iterations have been completed. This difficulty can be accommodated by breaking the computation at a point where approximately half of the iterations will have been completed.

The second difficulty is that the quotient polynomial $Q^{(r)}(x)$ depends on $A^{(r-1)}(x)$ and, in turn, $A^{(r)}(x)$ depends on $Q^{(r)}(x)$. Hence not all of the factors used in computing $A^{(r)}(x)$ are known before the time of the computation. Care must be taken in batching them to ensure that no factor is to be used before it is known. Proposition 4.7.1 will show that if we split the algorithm at the right point, then the two batches will have a structure similar to the original problem, and will not use terms that are not yet known.

Let $A^{(r)}(x)$ be factored as

$$A^{(r)}(x) = A^{(r,r'+1)}(x) A^{(r')}(x).$$

The first batch of computations needed to compute $A^{(r')}(x)$ consists of r' iterations that are nearly identical to the first r' iterations of the original problem except that

4.7 An accelerated euclidean algorithm for polynomials

the polynomial iterates are truncated. The second batch of computations, needed to compute $A^{(r,r'+1)}$ for $r = r'+1, \ldots, R$, is

$$Q^{(r)}(x) = \left\lfloor \frac{s^{(r-1)}(x)}{t^{(r-1)}(x)} \right\rfloor,$$

$$\begin{bmatrix} s^{(r)}(x) \\ t^{(r)}(x) \end{bmatrix} = A^{(r,r'+1)}(x) \begin{bmatrix} s^{(r')}(x) \\ t^{(r')}(x) \end{bmatrix}.$$

The second batch of computations has the same structure as the original computation.

If we compute the two batches in the same way as in the original problem, then the halving and doubling really has not gained much. To reduce the complexity, we will truncate the polynomials that are used in the first batch. The following proposition gives conditions under which it is possible to truncate the divisor and dividend polynomials without changing the quotient polynomial.

Proposition 4.7.1 *Let the two polynomials $f(x)$ and $g(x)$, with $\deg g(x)$ less than $\deg f(x)$, each be expressed in two segments as*

$$f(x) = f'(x)x^k + f''(x),$$
$$g(x) = g'(x)x^k + g''(x),$$

where $f''(x)$ and $g''(x)$ each have a degree smaller than k, and k satisfies

$$k \leq 2 \deg g(x) - \deg f(x).$$

Let

$$f(x) = Q(x)g(x) + r(x)$$

and

$$f'(x) = Q'(x)g'(x) + r'(x)$$

each satisfy the division algorithm. Then
(i) $Q(x) = Q'(x)$,
(ii) $r(x) = r'(x)x^k + r''(x)$,
where the polynomial $r''(x)$ is a polynomial whose degree is smaller than $k + \deg f(x) - \deg g(x)$.

Proof This is an easy consequence of the uniqueness of the division algorithm. Start with

$$f'(x) = Q'(x)g'(x) + r'(x),$$

which leads to

$$f'(x)x^k + f''(x) = Q'(x)g'(x)x^k + r'(x)x^k + f''(x).$$

Now recall the segmentation of $f(x)$ and $g(x)$ to write

$$f(x) = Q'(x)g(x) + r'(x)x^k + f''(x) - Q'(x)g''(x).$$

To show that this has the form of the division algorithm, we must show that

$$\deg[r'(x)x^k + f''(x) - Q'(x)g''(x)] < \deg g(x).$$

Then, from the uniqueness of the division algorithm, we can conclude that

$$Q(x) = Q'(x)$$

and

$$r(x) = r'(x)x^k + f''(x) - Q'(x)g''(x).$$

But, for the conditions of the theorem, the degree condition is easily verified by checking, in turn, that it holds for each of the three terms on the left side:
 (i) $\deg[r'(x)x^k] < \deg g'(x) + k = \deg g(x)$,
 (ii) $\deg f''(x) \leq k - 1 < \deg g(x)$,
 (iii) $\deg[Q'(x)g''(x)] < \deg f'(x) - \deg g'(x) + k$
$$= (\deg f(x) - k) - (\deg g(x) - k) + k$$
$$\leq \deg g(x).$$
The second conclusion of the theorem is obtained by noting in lines *(ii)* and *(iii)* that the polynomial $f''(x) - Q'(x)g''(x)$ has degree smaller than $k + \deg f(x) - \deg g(x)$. □

Corollary 4.7.2 *Let k satisfy*

$$k \leq 2 \deg g(x) - \deg f(x).$$

When dividing $f(x)$ by $g(x)$, the quotient polynomial $Q(x)$ does not depend on the k low-order coefficients of $f(x)$ and $g(x)$, and the k low-order coefficients of $f(x)$ and $g(x)$ affect the remainder polynomial $r(x)$ only in the coefficients of $r(x)$ with index less than $k + \deg f(x) - \deg g(x)$.

Proof This follows directly from the theorem. □

According to Proposition 4.7.1 and its corollary, we can obtain the quotient polynomial $Q(x)$ and a segment of the remainder polynomial $r(x)$ by truncating both $f(x)$ and $g(x)$ to shorter polynomials. We will show next that the missing segment of $r(x)$ will not impair some of the subsequent iterations of the euclidean algorithm. In fact, if k is chosen cleverly, about half of the subsequent iterations can be computed without knowing the missing segment of $r(x)$.

4.7 An accelerated euclidean algorithm for polynomials

The next theorem gives a condition such that the quotients of the remainder sequences, generated by the euclidean algorithm for the unprimed and primed variables, agree at least until the latter reaches a remainder $g''^{(r)}(x)$ whose degree is less than half that of $f'(x)$.

Theorem 4.7.3 Let $f(x) = f'(x)x^k + f''(x)$ and $g(x) = g'(x)x^k + g''(x)$, where $\deg f''(x) < k$ and $\deg g''(x) < k$. Let $\deg f(x) = n$ and $\deg g(x) < \deg f(x)$, and let $\mathbf{A}^{(r)}(x)$ and $\mathbf{A}'^{(r)}(x)$ be the euclidean matrices computed from the unprimed and primed variables, respectively. Then

$$\mathbf{A}^{(r)}(x) = \mathbf{A}'^{(r)}(x)$$

for each r, provided that $\deg g''^{(r)}(x) \geq (n-k)/2$.

Proof The proof consists of applying Corollary 4.7.2 to each iteration of the euclidean algorithm for the unprimed and primed variables initialized with $f^{(0)}(x) = f(x)$, $g^{(0)}(x) = g(x)$ and with $f'^{(0)}(x) = f'(x)$, $g'^{(0)}(x) = g'(x)$.

Step 1 The corollary can be applied to each iteration, provided that the condition

$$k \leq 2 \deg g^{(r)}(x) - \deg f^{(r)}(x)$$

is satisfied. But we are given that

$$\deg g''^{(r)}(x) \geq \frac{n-k}{2}.$$

We shall see that this is equivalent to the desired condition by relating the degrees of the primed polynomials to the degrees of the unprimed polynomials. This follows from the equations

$$\begin{bmatrix} f^{(r)}(x) \\ g^{(r)}(x) \end{bmatrix} = \begin{bmatrix} 0 & 1 \\ 1 & -Q(x) \end{bmatrix} \begin{bmatrix} f^{(r-1)}(x) \\ g^{(r-1)}(x) \end{bmatrix},$$

$$\begin{bmatrix} f'^{(r)}(x) \\ g'^{(r)}(x) \end{bmatrix} = \begin{bmatrix} 0 & 1 \\ 1 & -Q(x) \end{bmatrix} \begin{bmatrix} f'^{(r-1)}(x) \\ g'^{(r-1)}(x) \end{bmatrix},$$

and

$$\deg f'^{(0)}(x) = \deg f^{(0)}(x) - k,$$
$$\deg g'^{(0)}(x) = \deg g^{(0)}(x) - k.$$

Consequently, as long as the quotient polynomial is equal for both the primed and unprimed polynomials, we have

$$\deg f'^{(r)}(x) = \deg f^{(r)}(x) - k,$$
$$\deg g'^{(r)}(x) = \deg g^{(r)}(x) - k.$$

Thus we are given that

$$\deg g^{(r)}(x) - k \geq \frac{n-k}{2} \geq \frac{\deg f^{(r)}(x) - k}{2}$$

because $n = \deg f^{(0)}(x) \geq \deg f^{(r)}(x)$. This immediately reduces to

$$k \leq 2 \deg g^{(r)}(x) - \deg f^{(r)}(x),$$

and so Corollary 4.7.2 can be applied to each iteration, provided $Q(x)$ was correctly computed in every previous iteration.

Step 2 By Corollary 4.7.2, each quotient polynomial will be correct, provided every previous quotient polynomial was correct and there are enough correct coefficients in the most recent remainder polynomial. To verify the latter requirement, we use Corollary 4.7.2 again, this time with k replaced by

$$k^{(r)} = k + \deg f^{(r-1)}(x) - \deg g^{(r-1)}(x)$$

and $k^{(0)} = k$. By Corollary 4.7.2, upon entering the rth iteration, the remainder polynomial will be correct everywhere except possibly in the $k^{(r)}$ low-order coefficients, and by Corollary 4.7.2, this will not affect the quotient polynomial in the rth iteration, provided

$$k^{(r)} \leq 2 \deg g^{(r)}(x) - \deg f^{(r)}(x).$$

We verify that this inequality is satisfied as follows

$$\begin{aligned}
k^{(r)} - 2 \deg g^{(r)}(x) + \deg f^{(r)}(x) &= k + \deg f^{(r-1)}(x) - \deg g^{(r-1)}(x) \\
&\quad - 2 \deg g^{(r)}(x) + \deg f^{(r)}(x) \\
&= k + \deg f^{(r-1)}(x) - 2 \deg g^{(r)}(x) \\
&\leq k + n - 2\left(\frac{n+k}{2}\right) \\
&= 0,
\end{aligned}$$

where we have used the inequalities $\deg f^{(r-1)}(x) \leq n$ and

$$\deg g^{(r)}(x) = \deg g'^{(r)}(x) + k \geq \frac{n+k}{2}.$$

Therefore,

$$k^{(r)} \leq 2 \deg g^{(r)}(x) - \deg f^{(r)}(x),$$

and so in the rth iteration the quotient polynomial does not depend on the unknown coefficients. This completes the proof of the theorem. □

4.7 An accelerated euclidean algorithm for polynomials

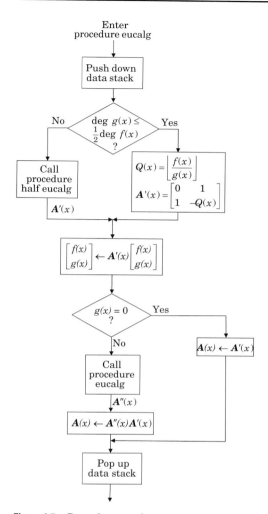

Figure 4.7 Procedure eucalg

The condition of the theorem, that $\deg g'^{(r)}(x) \geq (n-k)/2$, can be stated in terms of the unprimed variables as $\deg g^{(r)}(x) \geq (n+k)/2$ because the degrees differ by k. This equivalence is used in formulating the recursive procedure.

The euclidean algorithm is shown split in halves in the flow diagram of Figure 4.7. The major branch point decides whether to split the problem immediately or to first perform one normal iteration of the euclidean algorithm, because the splitting and doubling procedure cannot yet be applied. This latter path is followed only when the degree of $g(x)$ is smaller than half the degree of $f(x)$. In this case, one normal iteration of the euclidean algorithm cuts the size of the problem in half, so there is nothing lost because of the fact that the doubling procedure could not be used.

138 Fast algorithms based on doubling strategies

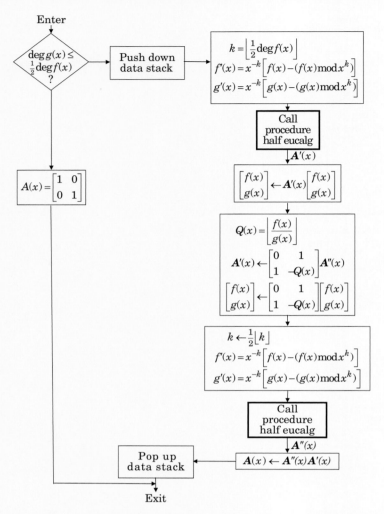

Figure 4.8 Procedure half eucalg

The second half of the split computation has the same form as the original problem and can be computed, in turn, by again calling the same procedure. The first half of the split computation has a similar, but not identical, form because it must terminate when the polynomials become too small. The half computation can itself be split into two halves, and so we can formulate the computations recursively. Figure 4.8 shows this recursive procedure. The validity of Figure 4.8 can be established by induction. We must show that the flow computes those iterations of the euclidean algorithm specified by Theorem 4.7.3. This always happens when n equals two or one because then the left path of Figure 4.8 is followed. When the right path is followed, then by the induction assumption, the first time the procedure of the recursion is called, it terminates with $\deg f(x)$ reduced to $(n + k)/2$, which equals about three-fourths of the original degree of $f(x)$. The second call of the recursion completes those iterations specified by Theorem 4.7.3.

4.8 A recursive radix-two fast Fourier transform

The complex radix-two Fourier transform is

$$V_k = \sum_{i=0}^{n-1} \omega^{ik} v_i, \qquad k = 0, \ldots, n-1,$$

where n is equal to 2^m with m an integer, and $\boldsymbol{v}$ is a complex vector. In Chapter 3, we were interested in constructing a small package of equations that would compute the radix-two Fourier transform by a straight-line computation. This is the right way to do it when n equals eight or sixteen, but it may be cumbersome for larger n. Accordingly, we have described various methods for combining these small algorithms into larger algorithms.

In this section, we shall express the radix-two Fourier transform in recursive form, repeating much of the treatment of the Winograd small FFT of Section 3.8, but now with more emphasis on the general structure and recursive formulation of the algorithm. We shall deal only with complex input data, because for n equal to 32 or greater, a complex Fourier transform can be more efficient than twice using a Fourier transform for real input data.

The Fourier transform can be decomposed, as in Chapter 3, into components with even index and components with odd index. To treat the components with even index, replace k by $2k'$ and write

$$V_{2k'} = \sum_{i=0}^{(n/2)-1} (v_i + v_{i+n/2}) \omega^{2ik'}, \qquad k' = 0, \ldots, (n/2) - 1.$$

This is an $n/2$-point Fourier transform of the new vector that has components $v_i + v_{i+n/2}$ and requires $n/2$ complex additions to prepare. To treat the components with odd index, replace k by $2k' + 1$ and write

$$V_{2k'+1} = \sum_{i=0}^{(n/2)-1} v_{2i} \omega^{2i(2k'+1)} + \sum_{i=0}^{(n/2)-1} v_{2i+1} \omega^{(2i+1)(2k'+1)}, \qquad k' = 0, \ldots, (n/2) - 1.$$

The two terms here will be studied separately. It takes $n/2$ complex additions to combine them. The first term only needs to be computed for $k' = 0, \ldots, (n/4) - 1$ because it then repeats. The first term is an $n/4$-point Fourier transform of the vector with components $v_{2i} \omega^{2i}$. The vector with components $v_{2i} \omega^{2i}$ is computed with $(3/4)n - 8$ real multiplications (because there are $(n/4) - 4$ complex products that will each use three real multiplications, and two complex products that will each use two real multiplications) and the same number of additions.

The bulk of the discussion will deal with the computation of the expression

$$t_{2k'+1} = \sum_{i=0}^{(n/2)-1} v_{2i+1} \omega^{(2i+1)(2k'+1)}, \qquad k' = 0, \ldots, (n/2) - 1.$$

The terms in the exponent of ω are integers taken modulo n because ω has order n. Within the exponents, all operations are multiplications of odd integers modulo n, and n equals 2^m. Under multiplication modulo n, the odd integers form a group that is isomorphic to $\mathbf{Z}_2 \times \mathbf{Z}_2^{m-2}$, as will be shown in Theorem 9.1.8 of Chapter 9. Hence the group has two generators. In fact, 3 and -1 are appropriate generators, and the set of exponents can be written as the multiplicative group

$$\{3^\ell (-1)^{\ell'} : \ell = 0, \ldots, 2^{m-2} - 1; \ell' = 0, 1\}$$

with multiplication modulo 2^m as the group operation. Accordingly, the input indices will be rewritten as

$$2i + 1 = 3^\ell (-1)^{\ell'},$$

and the output indices will be rewritten as

$$2k' + 1 = 3^{-r} (-1)^{-r'}.$$

With the new representation of the indices, the previous equation is changed into

$$t_{rr'} = \sum_{\ell'=0}^{1} \sum_{\ell=0}^{2^{m-2}-1} v_{\ell\ell'} \omega^{3^{\ell-r}(-1)^{\ell'-r'}}, \qquad \begin{array}{l} r = 0, \ldots, 2^{m-2} - 1, \\ r' = 0, 1, \end{array}$$

where the two-dimensional arrays with elements $t_{rr'}$ and $v_{\ell\ell'}$ take their elements from the components of the vectors of $t_{2k'+1}$ and v_{2i+1}, according to the reformulation of the indices. The equation is now a two-dimensional cyclic convolution:

$$t(x, y) = g(x, y)v(x, y) \quad (\bmod\ x^{2^{m-2}} - 1) \ (\bmod\ y^2 - 1),$$

where $g(x, y)$ is the generalized Rader polynomial

$$g(x, y) = \sum_{\ell'=0}^{1} \sum_{\ell=0}^{2^{m-2}-1} \omega^{3^\ell (-1)^{\ell'}} x^\ell y^{\ell'},$$

4.8 A recursive radix-two fast Fourier transform

and $v(x, y)$ and $t(x, y)$ represent the input and output data as two-dimensional polynomials by

$$v(x, y) = \sum_{\ell'=0}^{1} \sum_{\ell=0}^{2^{m-2}-1} v_{\ell\ell'} x^\ell y^{\ell'},$$

$$t(x, y) = \sum_{\ell'=0}^{1} \sum_{\ell=0}^{2^{m-2}-1} t_{\ell\ell'} x^\ell y^{\ell'}.$$

Only permutations are necessary to set up the polynomial $v(x, y)$ from the components v_{2i+1}. Similarly, only permutations are necessary to recover $t_{2k'+1}$ from $t(x, y)$.

The structure may be easier to see if we write this as

$$v(x, y) = v_0(x) + yv_1(x),$$
$$t(x, y) = t_0(x) + yt_1(x),$$

and

$$g(x, y) = \sum_{\ell=0}^{2^{m-2}-1} \omega^{3\ell} x^\ell + \sum_{\ell=0}^{2^{m-2}-1} \omega^{-3\ell} x^\ell y$$
$$= g(x) + yg^*(x),$$

where

$$g(x) = \sum_{\ell=0}^{2^{m-2}-1} \omega^{3\ell} x^\ell.$$

Then

$$\begin{bmatrix} t_0(x) \\ t_1(x) \end{bmatrix} = \begin{bmatrix} g(x) & g^*(x) \\ g^*(x) & g(x) \end{bmatrix} \begin{bmatrix} v_0(x) \\ v_1(x) \end{bmatrix}.$$

If the input data is real, $t_1(x) = t_0^*(x)$, and we need to compute only

$$t_0(x) = g(x)v_0(x) + g^*(x)v_1(x) \pmod{x^{n/4} - 1}$$

to compute this matrix product.

The next step is to use the factorization

$$x^{n/4} - 1 = (x^{n/8} - 1)(x^{n/8} + 1)$$

together with the Chinese remainder theorem. Let

$$g^{(0)}(x) = g(x) \pmod{x^{n/8} - 1},$$
$$g^{(1)}(x) = g(x) \pmod{x^{n/8} + 1},$$

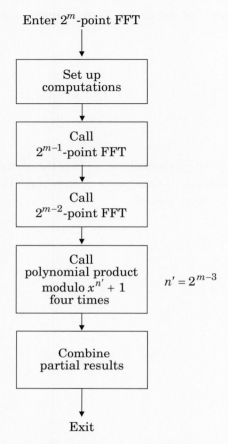

Figure 4.9 Recursive form of the radix-two Winograd FFT

and similarly for the other polynomials. By Theorem 3.8.3, $g^{(0)}(x) = 0$, so we need to deal only with the terms involving $g^{(1)}(x)$:

$$\begin{bmatrix} t_0^{(1)}(x) \\ t_1^{(1)}(x) \end{bmatrix} = \begin{bmatrix} g^{(1)}(x) & g^{(1)*}(x) \\ g^{(1)*}(x) & g^{(1)}(x) \end{bmatrix} \begin{bmatrix} v_0^{(1)}(x) \\ v_1^{(1)}(x) \end{bmatrix}.$$

Next, use a two-point cyclic convolution algorithm to write

$$\begin{bmatrix} t_0^{(1)}(x) \\ t_1^{(1)}(x) \end{bmatrix} = \begin{bmatrix} 1 & 1 \\ 1 & -1 \end{bmatrix} \begin{bmatrix} \frac{1}{2}\left[g^{(1)}(x) + g^{(1)*}(x)\right] \\ \frac{1}{2}\left[g^{(1)}(x) - g^{(1)*}(x)\right] \end{bmatrix}$$
$$\times \begin{bmatrix} 1 & 1 \\ 1 & -1 \end{bmatrix} \begin{bmatrix} v_0^{(1)}(x) \\ v_1^{(1)}(x) \end{bmatrix}.$$

The term $\frac{1}{2}[g^{(1)}(x) + g^{(1)*}(x)]$ is a purely real polynomial, while the term $\frac{1}{2}[g^{(1)}(x) - g^{(1)*}(x)]$ is a purely imaginary polynomial. Therefore we have reduced the computation to four real polynomial products modulo $x^{n/8} + 1$. Each of those, in principle, requires

Problems

Table 4.1 *Performance of some recursive radix-two FFT algorithms*

Blocklength n	Real input data			Complex input data		
	Number of real multiplications		Number of real additions	Number of real multiplications		Number of real additions
	Nontrivial	Total		Nontrivial	Total	
2	0	2	2	0	4	5
4	0	4	8	0	8	16
8	2	8	26	4	16	52
16	10	18	74	20	36	148
32	36	50		68	96	
64	102	124		188	232	
128	258	294		468	540	
256	608	666		1092	1208	
512	1370	1464		2444	2632	
1024	2994	3146		5316	5620	

$(n/4) - 1$ real multiplications to compute, although when n is larger than 32, practical algorithms will use more multiplications.

Figure 4.9 shows an outline for a flow of the FFT computation in recursive form. Table 4.1 tabulates the performance that may be achieved in this way.

Problems for Chapter 4

4.1 Find a doubling algorithm for multiplying a column vector by a Toeplitz matrix.

4.2 Use mergesort to sort the list (3, 1, 5, 2, 7, 3, 9, 8, 2, 6, 1, 4, 9, 2, 5, 1).

4.3 **a** Give a flow diagram for a decimation-in-frequency radix-two Cooley–Tukey FFT expressed as a doubling algorithm.
   b Give a flow diagram of a doubling algorithm in recursive form for computing a filter section whose blocklength is a power of two.
   c Give a flow diagram in recursive form for a doubling algorithm to compute a polynomial product.

4.4 Let n be a power of two, and let a and b be two arbitrary n digit integers. Use doubling to find an efficient algorithm for multiplying a and b.

4.5 Develop a matrix transposition algorithm for a processor that can store four columns of an array at one time in local memory. Give the algorithm in the form of a flow diagram. How many column transfers are needed?

4.6 How many multiplications are needed to compute the following coordinate transformation?

$$\begin{bmatrix} x' \\ y' \end{bmatrix} = \begin{bmatrix} \cos\theta & \sin\theta \\ -\sin\theta & \cos\theta \end{bmatrix} \begin{bmatrix} x \\ y \end{bmatrix}.$$

4.7 Calculate $\tan^{-1}(2/1)$ by using the coordinate rotation algorithm (cordic). By comparing with the true value, estimate the precision of the algorithm.

4.8 Using the Strassen algorithm, compute the matrix product

$$\begin{bmatrix} 1 & 5 \\ 7 & 3 \end{bmatrix} \begin{bmatrix} 4 & 8 \\ 6 & 2 \end{bmatrix}.$$

Notes for Chapter 4

Strategies that use halving and doubling are basic and arose in many places. This short chapter has only touched on this corner of the general theory of algorithms. A broader introduction can be found in the work of Knuth (1968) or Aho, Hopcroft, and Ullman (1974). Common examples of doubling are the sorting algorithm such as mergesort, which we have discussed, as well as the heapsort and quicksort algorithms introduced by Williams (1964) and Hoare (1962), respectively. Many other sorting algorithms are known. The doubling strategy as a view of the Cooley–Tukey FFT was emphasized by Steiglitz (1974). Fiduccia (1972) developed the Cooley–Tukey FFT by taking the polynomial evaluation idea of the Goertzel algorithm, then using halving and doubling. Doubling was used for matrix transposition by Eklundh (1972). The first accelerated form of the euclidean algorithm by doubling appears in the book by Aho, Hopcroft, and Ullman (1974), where it is limited to computing the greatest common denominator.

The Strassen algorithm (1969) is the best known fast matrix multiplication algorithm and is practical for large blocklengths. The Coppersmith–Winograd algorithm (1990) is the fastest known matrix multiplication algorithm, but only for extremely large matrices.

The coordinate rotation algorithm has been in use for many years under the name *cordic algorithm*, and is usually attributed to Volder (1959) with elaborations by Walther (1971). The sine and cosine doubling algorithm was developed by Blahut, and published by Blahut and Waldecker (1970).

5 Fast algorithms for short convolutions

The best-known method for calculating a convolution efficiently is to use the convolution theorem and a fast Fourier transform algorithm. This is usually quite convenient; and while better methods exist, the performance is often satisfactory. However, in applications in which it is worth the trouble to make the computational load even less, one can turn to other methods. When the blocklength is small, the best convolution algorithms, as measured by the number of multiplications and additions, are the Winograd convolution algorithms. The Winograd algorithms form the major topic of this chapter. Later chapters show how to build large convolution algorithms by piecing together small convolution algorithms. The large algorithms will be good only if the small algorithms are good. This is why we are interested in finding the best possible small convolution algorithms.

The details of the convolution algorithms that we derive may depend on the field in which the convolution takes place, but the general idea behind the algorithms does not depend on the field. We shall give methods to derive convolution algorithms for any field of interest, but, of course, the most important applications are in the real field and the complex field.

5.1 Cyclic convolution and linear convolution

A linear convolution can be written compactly as a polynomial product:

$$s(x) = g(x)d(x).$$

The coefficients of $s(x)$ are given by

$$s_i = \sum_{k=0}^{N-1} g_{i-k} d_k, \quad i = 0, \ldots, L + N - 2,$$

where $\deg g(x) = L - 1$ and $\deg d(x) = N - 1$.

The obvious way to compute the polynomial product involves a number of multiplications and additions that are each approximately equal to LN, the product of the degrees of $g(x)$ and $d(x)$, but other ways to compute it involve fewer computations.

The cyclic convolution

$$s(x) = g(x)d(x) \pmod{x^n - 1},$$

where $\deg g(x) = n - 1$ and $\deg d(x) = n - 1$, has coefficients given by

$$s_i = \sum_{k=0}^{n-1} g_{((i-k))} d_k, \qquad i = 0, \ldots, n - 1,$$

where the double parentheses denote modulo n, and involves n^2 multiplications and $n(n - 1)$ additions when computed in the obvious way. The cyclic convolution can be computed by first finding the linear convolution, then reducing it modulo $x^n - 1$. Hence efficient ways of computing a linear convolution lead to efficient ways of computing a cyclic convolution. Conversely, efficient algorithms for computing a cyclic convolution are easily turned into efficient algorithms for computing a linear convolution.

A popular method of computing a cyclic convolution uses the convolution theorem with the discrete Fourier transform. The convolution theorem says that in the frequency domain,

$$S_k = G_k D_k, \qquad k = 0, \ldots, n - 1,$$

so we can compute the cyclic convolution with a Fourier transform, a point-by-point product, and an inverse Fourier transform. This is illustrated in Figure 5.1. The middle block contains n complex multiplications, which is small in comparison with n^2. If n is highly composite, a fast Fourier transform, described in Chapter 3, may be used to reduce the number of computations in the first and third blocks, also to something small in comparison with n^2.

The procedure of Figure 5.1 will also compute a cyclic convolution of two complex sequences. Thus we should expect that when the sequences are real, the algorithm is stronger than necessary and its efficiency can be improved. It is possible, with some modifications, to do two real convolutions at once with nearly the same algorithm. The discrete Fourier transform of a real data sequence satisfies the symmetry condition

$$V_k = V_{n-k}^*, \qquad k = 0, \ldots, n - 1.$$

Suppose that $\boldsymbol{d'}$ and $\boldsymbol{d''}$ are real vectors of length n, and that $\boldsymbol{d}$ is the complex vector of length n with components

$$d_i = d_i' + \mathrm{j} d_i'', \qquad i = 0, \ldots, n - 1.$$

Then the discrete Fourier transform has components

$$D_k = D_k' + \mathrm{j} D_k'', \qquad k = 0, \ldots, n - 1,$$

5.1 Cyclic convolution and linear convolution

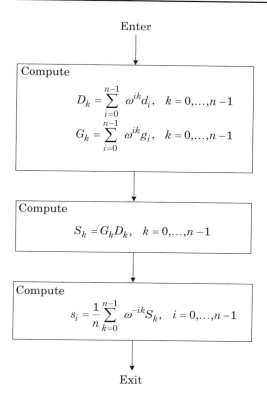

Figure 5.1 Computing a cyclic convolution with the Fourier transform

where, in general, D'_k and D''_k are both complex. Therefore

$$D^*_{n-k} = D'^*_{n-k} - jD''^*_{n-k}, \quad k = 0, \ldots, n-1$$
$$= D'_k - jD''_k.$$

We now conclude that

$$D'_k = \frac{1}{2}[D_k + D^*_{n-k}],$$
$$D''_k = \frac{1}{2j}[D_k - D^*_{n-k}], \quad k = 0, \ldots, n-1.$$

Using these formulas, we can compute the Fourier transforms of two real sequences with one computation of a discrete Fourier transform and some straightforward auxiliary additions.

The idea can be applied in reverse, starting with the two complex transforms of two real data sequences. Given the complex transform-domain vectors $\boldsymbol{D}'$ and $\boldsymbol{D}''$ whose inverse transforms are known to be real, define

$$D_k = D'_k + jD''_k, \quad k = 0, \ldots, n-1,$$

Fast algorithms for short convolutions

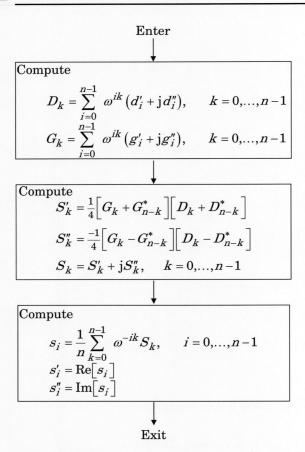

Figure 5.2 Computing two real cyclic convolutions with the Fourier transform

where in general both D'_k and D''_k are complex numbers. Computation of the inverse Fourier transform gives

$$d_i = d'_i + \mathrm{j}d''_i, \qquad i = 0, \ldots, n-1,$$

from which the real sequences are immediately recovered.

By using this idea, we can replace Figure 5.1 with a more efficient procedure for performing two real convolutions. The procedure is summarized in Figure 5.2.

5.2 The Cook–Toom algorithm

The *Cook–Toom algorithm* is an algorithm for linear convolution that is derived as a method of multiplying two polynomials. Write the linear convolution as a polynomial product

$$s(x) = g(x)d(x),$$

5.2 The Cook–Toom algorithm

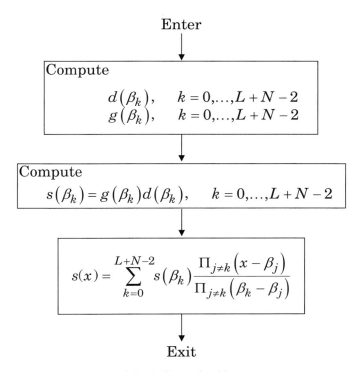

Figure 5.3 Structure of Cook–Toom algorithm

where

$\deg d(x) = N - 1$,

$\deg g(x) = L - 1$.

The output polynomial $s(x)$ has degree $L + N - 2$, and so it is uniquely determined by its value at $L + N - 1$ points. Let $\beta_0, \beta_1, \ldots, \beta_{L+N-2}$ be a set of $L + N - 1$ distinct real numbers. If we know $s(\beta_k)$ for $k = 0, \ldots, L + N - 2$, then we can compute $s(x)$ by Lagrange interpolation. In Theorem 2.7.11, we proved that

$$s(x) = \sum_{i=0}^{n-1} s(\beta_i) \frac{\prod_{j \neq i}(x - \beta_j)}{\prod_{j \neq i}(\beta_i - \beta_j)}$$

is the unique polynomial of degree $n - 1$ that has the value $s(\beta_k)$ when x takes the value β_k for $k = 0, \ldots, n - 1$. The idea of the Cook–Toom algorithm is to first compute $s(\beta_k)$ for $k = 0, \ldots, n - 1$ and then to use Lagrange interpolation.

Figure 5.3 illustrates the Cook–Toom algorithm. The multiplications are given by

$$s(\beta_k) = g(\beta_k) d(\beta_k), \quad k = 0, \ldots, L + N - 2.$$

There are $L + N - 1$ such equations, so there are $L + N - 1$ multiplications here; and if we pick β_k cleverly, those mutiplications will be the only general multiplications.

To evaluate $d(\beta_k)$ and $g(\beta_k)$ and to evaluate the Lagrange interpolation formulas, there will be other multiplications, but these are multiplications by small constants. We do not count them as general multiplications, but we must be careful not to ignore them altogether.

The simplest example is a two by two linear convolution:

$d(x) = d_1 x + d_0,$

$g(x) = g_1 x + g_0,$

and

$s(x) = g(x)d(x).$

This computation consists of passing two data samples through a two-tap FIR filter. The obvious algorithm uses four multiplications and one addition, but we shall find an algorithm with three multiplications and three additions. This might seem to be too small a problem for any practical application, but, in fact, a good algorithm for this small problem is a good building block from which one can construct more elaborate algorithms for larger problems.

Our first attempt will give an algorithm with three multiplications and five additions; then we will improve it. Choose the following points:

$\beta_0 = 0,$
$\beta_1 = 1,$
$\beta_2 = -1.$

Three points are needed because $s(x)$ has degree two. Then

$d(\beta_0) = d_0,$ $\qquad$ $g(\beta_0) = g_0,$
$d(\beta_1) = d_0 + d_1,$ $\qquad$ $g(\beta_1) = g_0 + g_1,$
$d(\beta_2) = d_0 - d_1,$ $\qquad$ $g(\beta_2) = g_0 - g_1,$

and

$s(\beta_0) = g(\beta_0)d(\beta_0),$
$s(\beta_1) = g(\beta_1)d(\beta_1),$
$s(\beta_2) = g(\beta_2)d(\beta_2),$

which requires three multiplications. If the filter represented by $g(x)$ is fixed, then the constants $g(\beta_k)$ need not be recomputed each time the filter is used. They are precomputed once "off-line" and stored. The coefficients of $g(x)$ then do not need to be stored.

Finally, from the Lagrange interpolation formula

$s(x) = s(\beta_0)L_0(x) + s(\beta_1)L_1(x) + s(\beta_2)L_2(x),$

5.2 The Cook–Toom algorithm

where the interpolation polynomials are

$L_0(x) = -x^2 + 1$,
$L_1(x) = \frac{1}{2}(x^2 + x)$,
$L_2(x) = \frac{1}{2}(x^2 - x)$.

This completes the derivation of the Cook–Toom algorithm for computing $s(x)$, but the computations still can be organized more compactly. The factors of one-half can be "buried" so that they do not show up in the computation. Simply replace the $g(\beta_k)$ by new constants to absorb this factor. Let

$G_0 = g_0$,
$G_1 = \frac{1}{2}(g_0 + g_1)$,
$G_2 = \frac{1}{2}(g_0 - g_1)$.

Whenever $g(x)$ is a fixed polynomial, these constants are computed off-line, and it costs nothing to modify them this way. Then

$L_0(x) = -x^2 + 1$,
$L_1(x) = x^2 + x$,
$L_2(x) = x^2 - x$.

Figure 5.4(a) shows the algorithm we have derived, written in the compact matrix–vector notation,

$$s = C\{[Bg] \cdot [Ad]\},$$

where the dot denotes the componentwise product of the vector Bg with the vector Ad. This matrix representation is a convenient way to visualize the algorithm, but, of course, the computation is not performed as a matrix product. Rather, the multiplications by matrices A and C are each computed as a sequence of additions. Figure 5.4(b) shows how the computations might proceed.

The linear convolution itself can be written as a matrix–vector product. In the example, the convolution is

$$\begin{bmatrix} s_0 \\ s_1 \\ s_2 \end{bmatrix} = \begin{bmatrix} g_0 & 0 \\ g_1 & g_0 \\ 0 & g_1 \end{bmatrix} \begin{bmatrix} d_0 \\ d_1 \end{bmatrix},$$

which will be abbreviated as $s = Td$, where

$$T = \begin{bmatrix} g_0 & 0 \\ g_1 & g_0 \\ 0 & g_1 \end{bmatrix}.$$

(a)
$$\begin{bmatrix} s_0 \\ s_1 \\ s_2 \end{bmatrix} = \begin{bmatrix} 1 & 0 & 0 \\ 0 & 1 & -1 \\ -1 & 1 & 1 \end{bmatrix} \left\{ \begin{bmatrix} G_0 \\ G_1 \\ G_2 \end{bmatrix} \bullet \begin{bmatrix} 1 & 0 \\ 1 & 1 \\ 1 & -1 \end{bmatrix} \begin{bmatrix} d_0 \\ d_1 \end{bmatrix} \right\}$$

where

$$\begin{bmatrix} G_0 \\ G_1 \\ G_2 \end{bmatrix} = \begin{bmatrix} 1 & 0 \\ \frac{1}{2} & \frac{1}{2} \\ \frac{1}{2} & -\frac{1}{2} \end{bmatrix} \begin{bmatrix} g_0 \\ g_1 \end{bmatrix}$$

(b) $D_0 = d_0$
$D_1 = d_0 + d_1$
$D_2 = d_0 - d_1$
$S_0 = G_0 D_0$
$S_1 = G_1 D_1$
$S_2 = G_2 D_2$
$s_0 = S_0$
$s_1 = S_1 - S_2$
$s_2 = -S_0 + S_1 + S_2$

(c) $$\begin{bmatrix} s_0 \\ s_1 \\ s_2 \end{bmatrix} = \begin{bmatrix} 1 & 0 & 0 \\ 0 & 1 & -1 \\ -1 & 1 & 1 \end{bmatrix} \begin{bmatrix} G_0 & & \\ & G_1 & \\ & & G_2 \end{bmatrix} \begin{bmatrix} 1 & 0 \\ 1 & 1 \\ 1 & -1 \end{bmatrix} \begin{bmatrix} d_0 \\ d_1 \end{bmatrix}$$

Figure 5.4 A two by two Cook–Toom convolution algorithm

The Cook–Toom algorithm can be understood as a matrix factorization,

$$\begin{bmatrix} s_0 \\ s_1 \\ s_2 \end{bmatrix} = \begin{bmatrix} 1 & 0 & 0 \\ 0 & 1 & -1 \\ -1 & 1 & 1 \end{bmatrix} \begin{bmatrix} g_0 & & \\ & \frac{1}{2}(g_0 + g_1) & \\ & & \frac{1}{2}(g_0 - g_1) \end{bmatrix} \begin{bmatrix} 1 & 0 \\ 1 & 1 \\ 1 & -1 \end{bmatrix} \begin{bmatrix} d_0 \\ d_1 \end{bmatrix}.$$

The algorithm then has the form

$s = CGAd$.

Therefore, as shown in Figure 5.4(c), the Cook–Toom algorithm gives the matrix factorization

$T = CGA$,

5.2 The Cook–Toom algorithm

where A is a matrix of preadditions, C is a matrix of postadditions, and G is a diagonal matrix responsible for all the multiplications. The number of multiplications is equal to the size of the matrix G. This representation often is a useful one for expressing the structure of an algorithm.

In the general case, a linear convolution can be expressed by the relationship

$$s = Td,$$

where the input vector d has length N, the output vector s has length $N + L - 1$, and T is an $N + L - 1$ by N matrix whose elements are the components of the vector g. Then, in the general case, the Cook–Toom algorithm provides the matrix factorization

$$T = CGA,$$

where G is a diagonal matrix and the matrices A and C contain only small integers.

The Cook–Toom algorithm can be modified to give another version with the same number of multiplications but with fewer additions. Notice that $s_{L+N-2} = g_{L-1}d_{N-1}$. This coefficient can be computed with one multiplication. The modified polynomial

$$s(x) - s_{L+N-2}x^{L+N-2} = g(x)d(x) - s_{L+N-2}x^{L+N-2}$$

has degree $L + N - 3$ and can be computed by using the ideas of the Cook–Toom algorithm with $L + N - 2$ multiplications. The extra multiplication $g_{L-1}d_{N-1}$ brings the total back to $L + N - 1$ multiplications just as before, but there will be fewer additions.

We shall derive the modified Cook–Toom algorithm for the two by two linear convolution. We will choose $\beta_0 = 0$ and $\beta_1 = -1$. Then

$$d(\beta_0) = d_0, \qquad g(\beta_0) = g_0,$$
$$d(\beta_1) = d_0 - d_1, \qquad g(\beta_1) = g_0 - g_1,$$

and

$$t(\beta_0) = g(\beta_0)d(\beta_0) - g_1 d_1 \beta_0^2,$$
$$t(\beta_1) = g(\beta_1)d(\beta_1) - g_1 d_1 \beta_1^2,$$

where $t(x) = g(x)d(x) - g_1 d_1 x^2$. From the Lagrange interpolation formula,

$$s(x) - g_1 d_1 x^2 = t(\beta_1)L_1(x) + t(\beta_0)L_0(x),$$

where

$$L_0(x) = x + 1,$$
$$L_1(x) = -x.$$

$$\begin{bmatrix} s_0 \\ s_1 \\ s_2 \end{bmatrix} = \begin{bmatrix} 1 & 0 & 0 \\ 1 & -1 & 1 \\ 0 & 0 & 1 \end{bmatrix} \begin{bmatrix} g_0 & & \\ & (g_0 - g_1) & \\ & & g_1 \end{bmatrix} \begin{bmatrix} 1 & 0 \\ 1 & -1 \\ 0 & 1 \end{bmatrix} \begin{bmatrix} d_0 \\ d_1 \end{bmatrix}$$

Figure 5.5 A modified two by two Cook–Toom convolution algorithm

Combining all the pieces gives

$$s(x) = g_1 d_1 x^2 + [-(d_0 - d_1)(g_0 - g_1) + g_1 d_1 + g_0 d_0]x + g_0 d_0$$

as the desired algorithm. This algorithm has three additions. It is expressed in matrix form in Figure 5.5, which should be compared to the algorithm in Figure 5.4.

The Cook–Toom algorithm is efficient as measured by the number of multiplications, but as the size of the problem increases, the number of additions increases rapidly. This is because the good choices for β_k are 0, 1, and -1, and these are soon used. For larger problems, one must also use ± 2, ± 4, and other small integers. Then the matrices C and A will contain small integers such as two or four. These can still be computed as additions by adding a number to itself several times, but this is reasonable only if the integers are small ones. Because of this, the Cook–Toom algorithm becomes too cumbersome for convolutions larger than three by four, or perhaps four by four. For larger problems, one can use the Winograd convolution algorithms described in the next section, or one can iterate small Cook–Toom convolution algorithms by using the nesting convolution techniques studied in Chapter 11.

There is another way to think of the Cook–Toom algorithm, and this alternative provides a bridge to the next section. Rather than choose the set of numbers $\{\beta_0, \beta_1, \ldots, \beta_{L+N-2}\}$, choose the set of polynomials $[x - \beta_0, x - \beta_1, \ldots, x - \beta_{L+N-2}]$. Then write

$$g(\beta_k) = R_{x-\beta_k}[g(x)],$$
$$d(\beta_k) = R_{x-\beta_k}[d(x)],$$

as shown in Figure 5.6. This is a more complicated way of saying the same thing as before. The advantage is that, in the next section, the polynomials of degree one will be replaced by polynomials of larger degree, thereby greatly increasing the number of available design options. With this more sophisticated viewpoint, and for the special case in which all residue polynomials are polynomials of degree one, the Lagrange interpolation formulas can be seen as the inverse equations associated with the Chinese remainder theorem. In the more general case, Lagrange interpolation is discarded in favor of the Chinese remainder theorem.

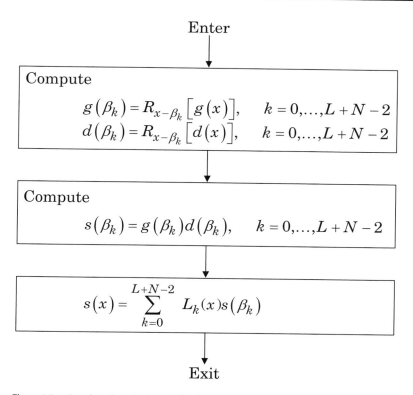

Figure 5.6 Another description of the Cook–Toom algorithm

5.3 Winograd short convolution algorithms

Suppose we want to compute

$$s(x) = g(x)d(x) \pmod{m(x)},$$

where $m(x)$ is a fixed polynomial of degree n in the field F, and $g(x)$ and $d(x)$ are polynomials of degree less than n in the same field. The computational problem of a linear convolution

$$s(x) = g(x)d(x)$$

can be put into this form. Simply let n be an integer larger than the degree of $s(x)$, and choose for $m(x)$ any polynomial of degree n. A trivial restatement of the linear convolution is

$$s(x) = g(x)d(x) \pmod{m(x)}.$$

This is equivalent because the reduction modulo $m(x)$ has no effect on $s(x)$ when the degree of $m(x)$ exceeds that of $s(x)$. In this way, the computation of a linear convolution becomes included in the general method to be discussed in this section. The same general method includes the computation of the cyclic convolution

$$s(x) = g(x)d(x) \pmod{x^n - 1}$$

simply by taking $m(x)$ equal to $x^n - 1$.

To develop the Winograd convolution algorithm, we will replace the computational problem

$$s(x) = g(x)d(x) \pmod{m(x)}$$

by a set of smaller computations. To break the problem into pieces, factor $m(x)$ into pairwise coprime polynomials $m^{(k)}(x)$ over some suitable subfield of F so that

$$m(x) = m^{(0)}(x)m^{(1)}(x) \cdots m^{(K-1)}(x).$$

Usually, if F is the real field or the complex field, one would choose the field of rationals as the subfield for the factorization. For example, if $m(x) = x^6 - 1$ and F is the real field, one would take

$$x^6 - 1 = (x - 1)(x + 1)(x^2 - x + 1)(x^2 + x + 1)$$

as the appropriate factorization. Similarly, if the convolution is in the finite field $GF(p^m)$, then one would usually choose the prime subfield $GF(p)$ for the factorization. The procedure will minimize the number of multiplications in the field F, but will not attempt to minimize the number of multiplications in the subfield. In most cases of interest, these multiplications in the subfield turn out to be multiplications by small integers, usually by $-1, 0$, or 1, and so are trivial. Multiplication by rationals will not be counted as multiplications from now on, but one should check that the rationals are indeed small integers.

The fast convolution algorithm will make use of the residue polynomials

$$s^{(k)}(x) = R_{m^{(k)}(x)}[s(x)], \qquad k = 0, \ldots, K - 1.$$

By the Chinese remainder theorem for polynomials, $s(x)$ can be computed from this set of residue polynomials by

$$s(x) = a^{(0)}(x)s^{(0)}(x) + \cdots + a^{(K-1)}(x)s^{(K-1)}(x) \pmod{m(x)}$$

for appropriate polynomials $a^{(0)}(x), \ldots, a^{(K-1)}(x)$, all with rational coefficients. We divide the computation into three steps. First, compute the residues

$$d^{(k)}(x) = R_{m^{(k)}(x)}[d(x)],$$
$$g^{(k)}(x) = R_{m^{(k)}(x)}[g(x)]$$

5.3 Winograd short convolution algorithms

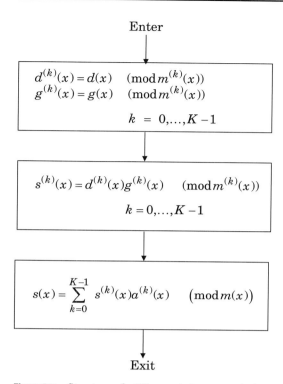

Figure 5.7 Structure of a Winograd short convolution

for $k = 0, \ldots, K - 1$. Computation of the residues $d^{(k)}(x)$ and $g^{(k)}(x)$ requires no multiplications. Next, compute

$$\begin{aligned} s^{(k)}(x) &= g^{(k)}(x)d^{(k)}(x) \pmod{m^{(k)}(x)} \\ &= R_{m^{(k)}(x)}\{R_{m^{(k)}(x)}[g(x)] \cdot R_{m^{(k)}(x)}[d(x)]\} \\ &= R_{m^{(k)}(x)}[g^{(k)}(x)d^{(k)}(x)]. \end{aligned}$$

Finally, compute

$$s(x) = a^{(0)}(x)s^{(0)}(x) + \cdots + a^{(K-1)}(x)s^{(K-1)}(x) \pmod{m(x)}.$$

Because $a^{(k)}(x)$ has only rational coefficients, this last step involves no multiplications.

The structure of a Winograd convolution algorithm is summarized in Figure 5.7. Only the short convolutions represented by the polynomial products $g^{(k)}(x)d^{(k)}(x)$ in the second step require multiplications of numbers. All together, $\sum_{k=0}^{K-1}[\deg m^{(k)}(x)]^2$ multiplications are required if each of the small polynomial products is computed in the obvious way because the number of coefficients in $g^{(k)}(x)$ or in $d^{(k)}(x)$ is equal to the degree of $m^{(k)}(x)$. This can be a considerable reduction in the number of multiplications.

Fast algorithms for short convolutions

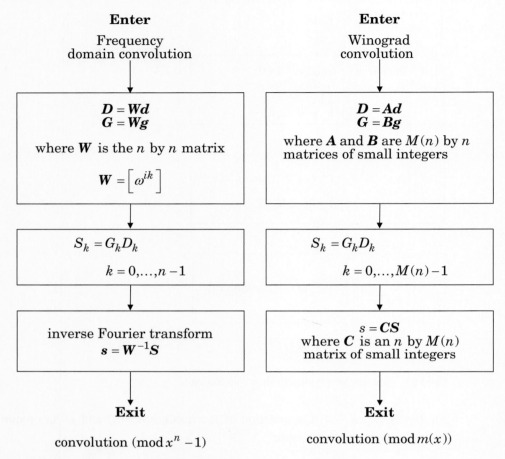

Figure 5.8 Comparison of two convolution methods

We shall see later that the short convolutions can, in turn, be improved upon by using the same idea to break each of them down into yet smaller pieces.

Figure 5.8 gives an instructive comparison of a Winograd convolution algorithm and a cyclic convolution computed by using a discrete Fourier transform. To make this comparison most evident, the Winograd convolution algorithm is expressed in a convenient matrix notation. The terms of the algorithm are gathered together and expressed as the matrix equation

$$s = C\{[Bg] \cdot [Ad]\},$$

where the dot denotes the componentwise product of the vector Bg and the vector Ad. The Winograd convolution algorithm is also written in the more suggestive form

$$s = CGAd,$$

5.3 Winograd short convolution algorithms

where now G is an $M(n)$ by $M(n)$ diagonal matrix with the elements of the vector Bg on the diagonal. In this form the multiplication by matrix A summarizes all the preadditions, and the multiplication by matrix C summarizes all the postadditions. The numerical multiplications are summarized by the multiplication by the diagonal matrix G.

The comparison in Figure 5.8 makes it evident that the Winograd convolution algorithm is a generalization of the method of computing a convolution using the Fourier transform and the convolution theorem.

As an example of a Winograd convolution, we shall take the linear convolution of a three-point vector with a two-point vector. Let

$$g(x) = g_1 x + g_0,$$
$$d(x) = d_2 x^2 + d_1 x + d_0.$$

Direct calculation requires six multiplications and two additions. We will first derive an algorithm that uses five multiplications and twelve additions. It is not a very good algorithm, but the derivation is instructive. Later, we will give a better algorithm, one with four multiplications and seven additions.

The linear convolution $s(x) = g(x)d(x)$ has degree three. Choose

$$m(x) = x(x-1)(x^2+1)$$
$$= m^{(0)}(x) m^{(1)}(x) m^{(2)}(x).$$

The factors are pairwise coprime. There are other polynomials that could be chosen for the polynomial $m(x)$; this choice illustrates the method. The residues are

$$g^{(0)}(x) = g_0, \qquad d^{(0)}(x) = d_0,$$
$$g^{(1)}(x) = g_1 + g_0, \qquad d^{(1)}(x) = d_2 + d_1 + d_0,$$
$$g^{(2)}(x) = g_1 x + g_0, \qquad d^{(2)}(x) = d_1 x + (d_0 - d_2).$$

Therefore

$$s^{(0)}(x) = g_0 d_0,$$
$$s^{(1)}(x) = (g_1 + g_0)(d_2 + d_1 + d_0),$$
$$s^{(2)}(x) = (g_1 x + g_0)(d_1 x + (d_0 - d_2)) \pmod{x^2 + 1}.$$

It takes one multiplication to compute $s^{(0)}(x)$, one multiplication to compute $s^{(1)}(x)$, and, as we shall see, three multiplications to compute $s^{(2)}(x)$. The computation of $s^{(2)}(x)$ involves computing the two terms

$$s_0^{(2)} = g_0^{(2)} d_0^{(2)} - g_1^{(2)} d_1^{(2)},$$
$$s_1^{(2)} = g_0^{(2)} d_1^{(2)} + g_1^{(2)} d_0^{(2)}$$

Table 5.1 *Chinese remainder polynomials*

k	$m^{(k)}$	$M^{(k)}(x)$	$n^{(k)}(x)$	$N^{(k)}(x)$
0	x	$x^3 - x^2 + x - 1$	$x^2 - x + 1$	-1
1	$x - 1$	$x^3 + x$	$-\frac{1}{2}(x^2 + x + 2)$	$\frac{1}{2}$
2	$x^2 + 1$	$x^2 - x$	$-\frac{1}{2}(x^2 - 2)$	$-\frac{1}{2}(x - 1)$

(which, incidentally, have the same structure as complex multiplication). An algorithm for this side computation is

$$\begin{bmatrix} s_0^{(2)} \\ s_1^{(2)} \end{bmatrix} = \begin{bmatrix} 1 & 0 & -1 \\ 1 & 1 & 0 \end{bmatrix} \begin{bmatrix} g_0^{(2)} & & \\ & (g_1^{(2)} - g_0^{(2)}) & \\ & & (g_1^{(2)} + g_0^{(2)}) \end{bmatrix} \begin{bmatrix} 1 & 1 \\ 1 & 0 \\ 0 & 1 \end{bmatrix} \begin{bmatrix} d_0^{(2)} \\ d_1^{(2)} \end{bmatrix},$$

which has three multiplications. The last step of computing $s(x)$ is

$$s(x) = a^{(0)}(x)s^{(0)}(x) + a^{(1)}(x)s^{(1)}(x) + a^{(2)}(x)s^{(2)}(x) \pmod{x^4 - x^3 + x^2 - x},$$

where $a^{(0)}(x)$, $a^{(1)}(x)$, and $a^{(2)}(x)$ are determined by the Chinese remainder theorem for polynomials as follows. Because $m^{(k)}(x)$ and $M^{(k)}(x) = m(x)/m^{(k)}(x)$ are coprime, the relationship

$$n^{(k)}(x)m^{(k)}(x) + N^{(k)}(x)M^{(k)}(x) = 1$$

leads us to construct Table 5.1. Then $a^{(k)}(x) = N^{(k)}(x)M^{(k)}(x)$. Hence

$$s(x) = -(x^3 - x^2 + x - 1)s^{(0)}(x) + \tfrac{1}{2}(x^3 + x)s^{(1)}(x) + \tfrac{1}{2}(x^3 - 2x^2 + x)s^{(2)}(x) \pmod{x^4 - x^3 + x^2 - x}.$$

We can rewrite this as a matrix equation

$$\begin{bmatrix} s_0 \\ s_1 \\ s_2 \\ s_3 \end{bmatrix} = \begin{bmatrix} 1 & 0 & 0 & 0 \\ -1 & 1 & 1 & 1 \\ 1 & 0 & -2 & 0 \\ -1 & 1 & 1 & -1 \end{bmatrix} \begin{bmatrix} s_0^{(0)} \\ \tfrac{1}{2}s_0^{(1)} \\ \tfrac{1}{2}s_0^{(2)} \\ \tfrac{1}{2}s_1^{(2)} \end{bmatrix}.$$

Now we must put all the pieces together to get the desired algorithm. There are five multiplications that we choose to write with a uniform notation as

$$S_k = G_k D_k, \quad k = 0, \ldots, 4.$$

5.3 Winograd short convolution algorithms

We define new notation as needed to fit this chosen form. Combining the steps already developed gives the definition of $\boldsymbol{D}$:

$$\begin{bmatrix} D_0 \\ D_1 \\ D_2 \\ D_3 \\ D_4 \end{bmatrix} = \begin{bmatrix} 1 & 0 & 0 & 0 \\ 0 & 1 & 0 & 0 \\ 0 & 0 & 1 & 1 \\ 0 & 0 & 1 & 0 \\ 0 & 0 & 0 & 1 \end{bmatrix} \begin{bmatrix} d_0^{(0)} \\ d_0^{(1)} \\ d_0^{(2)} \\ d_1^{(2)} \end{bmatrix}$$

$$= \begin{bmatrix} 1 & 0 & 0 & 0 \\ 0 & 1 & 0 & 0 \\ 0 & 0 & 1 & 1 \\ 0 & 0 & 1 & 0 \\ 0 & 0 & 0 & 1 \end{bmatrix} \begin{bmatrix} 1 & 0 & 0 \\ 1 & 1 & 1 \\ 1 & 0 & -1 \\ 0 & 1 & 0 \end{bmatrix} \begin{bmatrix} d_0 \\ d_1 \\ d_2 \end{bmatrix}$$

$$= \begin{bmatrix} 1 & 0 & 0 \\ 1 & 1 & 1 \\ 1 & 1 & -1 \\ 1 & 0 & -1 \\ 0 & 1 & 0 \end{bmatrix} \begin{bmatrix} d_0 \\ d_1 \\ d_2 \end{bmatrix}.$$

We define $\boldsymbol{G}$ similarly, except that we also incorporate into the definition the denominators that would otherwise appear in the matrix of postadditions. Hence, in the following definition of $\boldsymbol{G}$, the leftmost matrix contains factors of one-half, which have been pulled out of the matrix of postadditions and buried here. Then

$$\begin{bmatrix} G_0 \\ G_1 \\ G_2 \\ G_3 \\ G_4 \end{bmatrix} = \begin{bmatrix} 1 & & & & \\ & \frac{1}{2} & & & \\ & & \frac{1}{2} & & \\ & & & \frac{1}{2} & \\ & & & & \frac{1}{2} \end{bmatrix} \begin{bmatrix} 1 & 0 & 0 & 0 \\ 0 & 1 & 0 & 0 \\ 0 & 0 & 1 & 0 \\ 0 & 0 & -1 & 1 \\ 0 & 0 & 1 & 1 \end{bmatrix} \begin{bmatrix} 1 & 0 \\ 1 & 1 \\ 1 & 0 \\ 0 & 1 \end{bmatrix} \begin{bmatrix} g_0 \\ g_1 \end{bmatrix}$$

$$= \begin{bmatrix} 1 & 0 \\ \frac{1}{2} & \frac{1}{2} \\ \frac{1}{2} & 0 \\ -\frac{1}{2} & \frac{1}{2} \\ \frac{1}{2} & \frac{1}{2} \end{bmatrix} \begin{bmatrix} g_0 \\ g_1 \end{bmatrix}.$$

Because the computation of $\boldsymbol{G}$ is an off-line computation, there is no need to keep that computation simple.

$$\begin{bmatrix} s_0 \\ s_1 \\ s_2 \\ s_3 \end{bmatrix} = \begin{bmatrix} 1 & 0 & 0 & 0 & 0 \\ -1 & 1 & 2 & 1 & -1 \\ 1 & 0 & -2 & 0 & 2 \\ -1 & 1 & 0 & -1 & -1 \end{bmatrix} \begin{bmatrix} G_0 \\ G_1 \\ G_2 \\ G_3 \\ G_4 \end{bmatrix} \begin{bmatrix} 1 & 0 & 0 \\ 1 & 1 & 1 \\ 1 & 1 & -1 \\ 1 & 0 & -1 \\ 0 & 1 & 0 \end{bmatrix} \begin{bmatrix} d_0 \\ d_1 \\ d_2 \end{bmatrix}$$

$$\begin{bmatrix} G_0 \\ G_1 \\ G_2 \\ G_3 \\ G_4 \end{bmatrix} = \begin{bmatrix} 1 & 0 \\ \frac{1}{2} & \frac{1}{2} \\ \frac{1}{2} & 0 \\ -\frac{1}{2} & \frac{1}{2} \\ \frac{1}{2} & \frac{1}{2} \end{bmatrix} \begin{bmatrix} g_0 \\ g_1 \end{bmatrix}$$

Figure 5.9 Example of a Winograd convolution algorithm

Finally, the matrix of postadditions is obtained as follows:

$$\begin{bmatrix} s_0 \\ s_1 \\ s_2 \\ s_3 \end{bmatrix} = \begin{bmatrix} 1 & 0 & 0 & 0 \\ -1 & 1 & 1 & 1 \\ 1 & 0 & -2 & 0 \\ -1 & 1 & 1 & -1 \end{bmatrix} \begin{bmatrix} 1 & 0 & 0 & 0 & 0 \\ 0 & 1 & 0 & 0 & 0 \\ 0 & 0 & 1 & 0 & -1 \\ 0 & 0 & 1 & 1 & 0 \end{bmatrix} \begin{bmatrix} S_0 \\ S_1 \\ S_2 \\ S_3 \\ S_4 \end{bmatrix}$$

$$= \begin{bmatrix} 1 & 0 & 0 & 0 & 0 \\ -1 & 1 & 2 & 1 & -1 \\ 1 & 0 & -2 & 0 & 2 \\ -1 & 1 & 0 & -1 & -1 \end{bmatrix} \begin{bmatrix} S_0 \\ S_1 \\ S_2 \\ S_3 \\ S_4 \end{bmatrix}.$$

The algorithm is summarized in matrix form in Figure 5.9. As expressed in the figure, the order in which the additions are done is not really specified. One can experiment with the order of the additions to minimize their number. It is easy to see how to do the preadditions with four real additions. The following sequence of equations does the postadditions with eight real additions:

$s_0 = S_0,$
$c_1 = S_4 - S_2,$
$c_3 = S_3 + S_4,$
$s_2 = c_1 + c_1 + S_0,$
$s_1 = S_1 + c_3 - s_2,$
$s_3 = -c_3 - S_0 + S_1.$

This completes the example of the basic construction of a Winograd convolution algorithm, but we have not yet done our best. A more general form of the Winograd

5.3 Winograd short convolution algorithms

algorithm can be obtained by choosing an $m(x)$ with a somewhat smaller degree. This will produce an incorrect convolution, but one that can be corrected with a few extra computations.

By the division algorithm for polynomials, we can write

$$s(x) = Q(x)m(x) + R_{m(x)}[s(x)].$$

We have already studied the case where $\deg m(x) > \deg s(x)$, in which case the quotient polynomial $Q(x)$ is identically zero. If $\deg m(x) \leq \deg s(x)$, the Winograd algorithm will produce only $s(x)$ modulo $m(x)$. The term $Q(x)m(x)$ is an omitted term that can be determined by a side computation and corrected. The simplest instance is the case in which $\deg m(x) = \deg s(x)$. Then $Q(x)$ must have degree zero and so is a scalar. If $m(x)$ is a monic polynomial of degree n, then, clearly, $Q(x) = s_n$, where s_n is the coefficient of x^n in the polynomial $s(x)$. Consequently,

$$s(x) = s_n m(x) + R_{m(x)}[s(x)]$$

and s_n can be easily computed with one multiplication as the product of the leading coefficients of $g(x)$ and $d(x)$.

This modified procedure can be formally absorbed into the basic Winograd convolution algorithm by replacing $m(x)$ with the formal expression $m(x)(x - \infty)$. The statement

$$s(x) = s(x) \pmod{m(x)(x - \infty)}$$

is only a convenient shorthand for the more proper expression given above.

With this modification, let us return to the earlier example and derive another algorithm for the convolution

$$s(x) = (g_1 x + g_0)(d_2 x^2 + d_1 x + d_0).$$

We now use $x(x - 1)(x + 1)(x - \infty)$ as the modulus polynomial. There will be four multiplications; the factor $(x - \infty)$ symbolizes the product $g_1 d_2$, and the other multiplications are products of the residues modulo x, $x - 1$, and $x + 1$. We immediately have that

$$\begin{bmatrix} S_0 \\ S_1 \\ S_2 \\ S_3 \end{bmatrix} = \left\{ \begin{bmatrix} 1 & 0 \\ 1 & 1 \\ 1 & -1 \\ 0 & 1 \end{bmatrix} \begin{bmatrix} g_0 \\ g_1 \end{bmatrix} \right\} \bullet \left\{ \begin{bmatrix} 1 & 0 & 0 \\ 1 & 1 & 1 \\ 1 & -1 & 1 \\ 0 & 0 & 1 \end{bmatrix} \begin{bmatrix} d_0 \\ d_1 \\ d_2 \end{bmatrix} \right\}.$$

The coefficients of $s(x)$ are recovered by using the Chinese remainder theorem:

$$\begin{aligned} s(x) &= a^{(0)}(x) S_0 + a^{(1)}(x) S_1 + a^{(2)}(x) S_2 + x(x-1)(x+1) S_3 \\ &= (-x^2 + 1) S_0 + \left(\tfrac{1}{2} x^2 + \tfrac{1}{2} x\right) S_1 + \left(\tfrac{1}{2} x^2 - \tfrac{1}{2} x\right) S_2 + (x^3 - x) S_3. \end{aligned}$$

$$\begin{bmatrix} s_0 \\ s_1 \\ s_2 \\ s_3 \end{bmatrix} = \begin{bmatrix} 1 & 0 & 0 & 0 \\ 0 & 1 & -1 & -1 \\ -1 & 1 & 1 & 0 \\ 0 & 0 & 0 & 1 \end{bmatrix} \begin{bmatrix} g_0 \\ \frac{1}{2}(g_0+g_1) \\ \frac{1}{2}(g_0-g_1) \\ g_1 \end{bmatrix} \begin{bmatrix} 1 & 0 & 0 \\ 1 & 1 & 1 \\ 1 & -1 & 1 \\ 0 & 0 & 1 \end{bmatrix} \begin{bmatrix} d_0 \\ d_1 \\ d_2 \end{bmatrix}$$

Figure 5.10 Another Winograd convolution algorithm

Hence

$$\begin{bmatrix} s_0 \\ s_1 \\ s_2 \\ s_3 \end{bmatrix} = \begin{bmatrix} 1 & 0 & 0 & 0 \\ 0 & 1 & -1 & -1 \\ -1 & 1 & 1 & 0 \\ 0 & 0 & 0 & 1 \end{bmatrix} \begin{bmatrix} S_0 \\ \frac{1}{2}S_1 \\ \frac{1}{2}S_2 \\ S_3 \end{bmatrix}.$$

The final form of the algorithm is shown in Figure 5.10. It requires four multiplications and seven additions, provided that the diagonal matrix is precomputed.

5.4 Design of short linear convolution algorithms

The Winograd short convolution algorithm, discussed in the previous section, gives good convolution algorithms. However, that construction does not cover all possible ways of obtaining good algorithms. Sometimes, a good algorithm can be found simply by clever factorization. Consider the following identities:

$$g_0 d_0 = g_0 d_0,$$
$$g_0 d_1 + g_1 d_0 = (g_0 + g_1)(d_0 + d_1) - g_0 d_0 - g_1 d_1,$$
$$g_0 d_2 + g_1 d_1 + g_2 d_0 = (g_0 + g_2)(d_0 + d_2) - g_0 d_0 + g_1 d_1 - g_2 d_2,$$
$$g_1 d_2 + g_2 d_1 = (g_1 + g_2)(d_1 + d_2) - g_1 d_1 - g_2 d_2,$$
$$g_2 d_2 = g_2 d_2.$$

With these factorizations, we can compute the coefficients of the linear convolution $s(x) = (g_0 + g_1 x + g_2 x^2)(d_0 + d_1 x + d_2 x^2)$ using six multiplications and ten additions. In matrix form, this algorithm is

$$\begin{bmatrix} s_0 \\ s_1 \\ s_2 \\ s_3 \\ s_4 \end{bmatrix} = \begin{bmatrix} 1 & 0 & 0 & 0 & 0 & 0 \\ -1 & -1 & 0 & 1 & 0 & 0 \\ -1 & 1 & -1 & 0 & 1 & 0 \\ 0 & -1 & -1 & 0 & 0 & 1 \\ 0 & 0 & 1 & 0 & 0 & 0 \end{bmatrix} \begin{bmatrix} g_0 \\ g_1 \\ g_2 \\ (g_0+g_1) \\ (g_0+g_2) \\ (g_1+g_2) \end{bmatrix} \begin{bmatrix} 1 & 0 & 0 \\ 0 & 1 & 0 \\ 0 & 0 & 1 \\ 1 & 0 & 0 \\ 1 & 0 & 1 \\ 0 & 1 & 1 \end{bmatrix} \begin{bmatrix} d_0 \\ d_1 \\ d_2 \end{bmatrix}.$$

5.4 Design of short linear convolution algorithms

Table 5.2 *Performance of some short linear convolution algorithms*

$$s(x) = g(x)d(x)$$
$$\deg g(x) = L - 1$$
$$\deg d(x) = N - 1$$

Blocklength		Real convolutions	
L	N	Number of real multiplications	Number of real additions
2	2	3	3
2	2	4	7
3	3	5	20
3	3	6	10
3	3	9	4
4	4	7	41
4	4	9	15

		Complex convolutions	
		Number of complex multiplications	Number of complex additions
2	2	3	3
3	3	5	15
4	4	7	

$$\begin{bmatrix} s_0 \\ s_1 \\ s_2 \end{bmatrix} = \begin{bmatrix} 1 & 0 & 0 \\ -1 & 1 & -1 \\ 0 & 0 & 1 \end{bmatrix} \begin{bmatrix} g_0 & & \\ & g_0 + g_1 & \\ & & g_1 \end{bmatrix} \begin{bmatrix} 1 & 0 \\ 1 & 1 \\ 0 & 1 \end{bmatrix} \begin{bmatrix} d_0 \\ d_1 \end{bmatrix}$$

Figure 5.11 A two by two linear convolution algorithm

The algorithm should be compared with the "naive" algorithm, which uses nine multiplications and four additions, and also with the "optimal" algorithm discussed shortly, which uses five multiplications and twenty additions. It is not possible to say which of these algorithms is preferable. It depends on the use.

Table 5.2 tabulates the performance of some linear convolution algorithms for real data sequences developed by the methods of this chapter. Later, in Chapter 11, we shall build large convolution algorithms by combining small convolution algorithms in such a way that both the number of multiplications and the number of additions of the large algorithm depend mostly on the number of multiplications of the small algorithm. In that construction the number of additions of the small convolution algorithm does not much matter. Figure 5.11 gives a two by two linear convolution algorithm with three

$$\begin{bmatrix} s_0 \\ s_1 \\ s_2 \\ s_3 \\ s_4 \end{bmatrix} = \begin{bmatrix} 2 & 0 & 0 & 0 & 0 \\ -1 & 2 & -2 & -1 & 2 \\ -2 & 1 & 3 & 0 & -1 \\ 1 & -1 & -1 & 1 & -2 \\ 0 & 0 & 0 & 0 & 1 \end{bmatrix} \begin{bmatrix} \frac{1}{2}g_0 \\ \frac{1}{2}(g_0+g_1+g_2) \\ \frac{1}{6}(g_0-g_1+g_2) \\ \frac{1}{6}(g_0+2g_1+g_2) \\ g_2 \end{bmatrix} \begin{bmatrix} 1 & 0 & 0 \\ 1 & 1 & 1 \\ 1 & -1 & 1 \\ 1 & 2 & 4 \\ 0 & 0 & 1 \end{bmatrix} \begin{bmatrix} d_0 \\ d_1 \\ d_2 \end{bmatrix}$$

Sequence of preadditions
$t_1 = d_1 + d_2$
$t_2 = d_2 - d_1$
$D_0 = d_0$
$D_1 = d_0 + t_1$
$D_2 = d_0 + t_2$
$D_3 = t_1 + t_1 + t_2 + D_1$
$D_4 = d_2$

Sequence of postadditions
$s_0 = S_0 + S_0$
$T_1 = S_1 + S_1$
$T_2 = S_2 + S_2$
$s_1 = T_1 - T_2 + T_4$
$s_2 = -s_0 + T_2 + T_5 - S_4$
$s_3 = -T_4 - T_5$
$s_4 = S_4$

$T_3 = S_4 + S_4$
$T_4 = T_3 - S_0 - S_3$
$T_5 = S_1 + S_2$

Figure 5.12 A three by three linear convolution algorithm

multiplications and three additions. Figure 5.12 gives a three by three linear convolution algorithm with five multiplications and twenty additions.

The optimal three-point by three-point linear convolution algorithm of Figure 5.12 will be derived as an example. The optimal algorithm, as judged by the number of multiplications, uses five multiplications. This conclusion is suggested by the Cook–Toom algorithm, which uses five multiplications, and is formally proved in Section 5.8. The algorithm we derive is essentially a Cook–Toom algorithm that is obtained by choosing the polynomial

$$m(x) = x(x-1)(x+1)(x-2)(x-\infty).$$

Because all factors are polynomials of degree one, there are five multiplications. The output polynomial is given by

$$s(x) = R_{x(x-1)(x+1)(x-2)}[g(x)d(x)] + g_2 d_2 x(x-1)(x+1)(x-2).$$

Evaluating the residues gives

$$\begin{bmatrix} G_0 \\ G_1 \\ G_2 \\ G_3 \\ G_4 \end{bmatrix} = \begin{bmatrix} 1 & 0 & 0 \\ 1 & 1 & 1 \\ 1 & -1 & 1 \\ 1 & 2 & 4 \\ 0 & 0 & 1 \end{bmatrix} \begin{bmatrix} g_0 \\ g_1 \\ g_2 \end{bmatrix},$$

$$\begin{bmatrix} D_0 \\ D_1 \\ D_2 \\ D_3 \\ D_4 \end{bmatrix} = \begin{bmatrix} 1 & 0 & 0 \\ 1 & 1 & 1 \\ 1 & -1 & 1 \\ 1 & 2 & 4 \\ 0 & 0 & 1 \end{bmatrix} \begin{bmatrix} d_0 \\ d_1 \\ d_2 \end{bmatrix},$$

5.4 Design of short linear convolution algorithms

where G_4 and D_4 are formally defined as residues modulo $(x - \infty)$. Then $S_k = G_k D_k$ for $k = 0, \ldots, 4$. Finally, we use the Chinese remainder theorem (or the Lagrange interpolation formula) to recover $s(x)$ as

$$s(x) = \tfrac{1}{2}(x^3 - 2x^2 - x + 2)S_0 + \tfrac{1}{2}(x^3 - x^2 - 2x)S_1$$
$$+ \tfrac{1}{6}(-x^3 + 3x^2 - 2x)S_2 + \tfrac{1}{6}(x^3 - x)S_3$$
$$+ (x^4 - 2x^3 - x^2 + 2x)S_4.$$

The last step is to express the equation in matrix form:

$$\begin{bmatrix} s_0 \\ s_1 \\ s_2 \\ s_3 \\ s_4 \end{bmatrix} = \begin{bmatrix} 2 & 0 & 0 & 0 & 0 \\ -1 & 2 & -2 & -1 & 2 \\ -2 & 1 & 3 & 0 & -1 \\ 1 & -1 & -1 & 1 & -2 \\ 0 & 0 & 0 & 0 & 1 \end{bmatrix} \begin{bmatrix} \tfrac{1}{2}S_0 \\ \tfrac{1}{2}S_1 \\ \tfrac{1}{6}S_2 \\ \tfrac{1}{6}S_3 \\ S_4 \end{bmatrix}.$$

The constant multipliers can be buried in the constants of the diagonal matrix by redefining S_k and G_k. The convolution requires five multiplications. The matrix of preadditions can be computed with seven additions, and the matrix of postadditions can be computed with thirteen additions.

The final algorithm is given in Figure 5.12. Although this algorithm was derived for real convolutions, it also works perfectly well for complex convolutions. The five multiplications will become five complex multiplications, and the twenty additions will become twenty complex additions.

One may also design an algorithm specifically for the complex field. There will still be five complex multiplications, but there will be fewer complex additions. We will choose the polynomial

$$s(x) = x(x - 1)(x + 1)(x - j)(x + j).$$

Evaluating the residues gives

$$\begin{bmatrix} G_0 \\ G_1 \\ G_2 \\ G_3 \\ G_4 \end{bmatrix} = \begin{bmatrix} 1 & 0 & 0 \\ 1 & 1 & 1 \\ 1 & -1 & 1 \\ 1 & j & -1 \\ 1 & -j & -1 \end{bmatrix} \begin{bmatrix} g_0 \\ g_1 \\ g_2 \end{bmatrix},$$

$$\begin{bmatrix} D_0 \\ D_1 \\ D_2 \\ D_3 \\ D_4 \end{bmatrix} = \begin{bmatrix} 1 & 0 & 0 \\ 1 & 1 & 1 \\ 1 & -1 & 1 \\ 1 & j & -1 \\ 1 & -j & -1 \end{bmatrix} \begin{bmatrix} d_0 \\ d_1 \\ d_2 \end{bmatrix}.$$

The Chinese remainder theorem (or the Lagrange interpolation formula) gives

$$s(x) = (-x^4 + 1)S_0 + \tfrac{1}{4}(x^4 + x^3 + x^2 + x)S_1$$
$$+ \tfrac{1}{4}(x^4 - x^3 + x^2 - x)S_2 + \tfrac{1}{4}(x^4 + jx^3 - x^2 - jx)S_3$$
$$+ \tfrac{1}{4}(x^4 - jx^3 - x^2 + jx)S_4.$$

When expressed in the form of a matrix, this is

$$\begin{bmatrix} s_0 \\ s_1 \\ s_2 \\ s_3 \\ s_4 \end{bmatrix} = \begin{bmatrix} 1 & 0 & 0 & 0 & 0 \\ 0 & 1 & -1 & -j & j \\ 0 & 1 & 1 & -1 & -1 \\ 0 & 1 & -1 & j & -j \\ -1 & 1 & 1 & 1 & 1 \end{bmatrix} \begin{bmatrix} S_0 \\ \tfrac{1}{4}S_1 \\ \tfrac{1}{4}S_2 \\ \tfrac{1}{4}S_3 \\ \tfrac{1}{4}S_4 \end{bmatrix}.$$

The factors of one-fourth can be buried in the diagonal matrix. The algorithm uses five complex multiplications, the equivalent of six complex additions in the matrix of preadditions, and the equivalent of nine complex additions in the matrix of postadditions.

Convolutions bigger than those described in Table 5.2 can also be developed. Large algorithms derived from the basic Winograd procedure, however, tend to be cumbersome and to have too many additions. Instead, one can bind small algorithms together to get large ones. There are several ways that small convolution algorithms can be built into large convolution algorithms. Examples of these, the method of iterated (or nesting) convolution algorithms, and the Agarwal–Cooley algorithm will be studied in Chapter 11. Such algorithms will have a few more multiplications than necessary, but the number of additions will be considerably fewer.

5.5 Polynomial products modulo a polynomial

We have seen how to break a cyclic convolution or a linear convolution into several smaller problems. The smaller problems again have the same form

$$s(x) \equiv g(x)d(x) \pmod{m(x)},$$

where $\deg m(x) = n$, and $g(x)$ and $d(x)$ each has a degree smaller than n. This is a product in the ring of polynomials modulo the polynomial $m(x)$. To get a good algorithm for the original linear convolution, we need to construct a good algorithm for each of the smaller polynomial products. We need to consider only the case in which $m(x)$ is a prime polynomial; otherwise, it would be further factored.

As we shall prove in Section 5.8, the number of general multiplications needed to compute $s(x)$ is $2n - 1$ when $m(x)$ is a prime polynomial. Usually, an algorithm that uses the least possible number of multiplications will use a large number of

5.5 Polynomial products modulo a polynomial

additions. Practical algorithms will attain a reasonable balance between the number of multiplications and the number of additions. This section studies the practical aspects of designing such algorithms.

The most direct method is to compute the linear convolution $g(x)d(x)$ (which uses at least $2n - 1$ multiplications) and then to reduce the result modulo $m(x)$ (which uses no multiplications). This may appear paradoxical at first sight, because we have already proposed that linear convolutions should be broken into convolutions modulo prime polynomials, and now we are turning these back into linear convolutions. However, in this back-and-forth maneuvering, the degree of the linear convolution may be greatly reduced, and this is how the situation improves.

We have already used the polynomial product

$$s(x) = g(x)d(x) \pmod{x^2 + 1},$$

which is formally the same as complex multiplication. At the same time, we can look at the polynomial products

$$s(x) = g(x)d(x) \pmod{x^2}$$

and

$$s(x) = g(x)d(x) \pmod{x^2 + x + 1}.$$

For each of these examples, $g(x) = g_1 x + g_0$ and $d(x) = d_1 x + d_0$. We will modify the two by two linear convolution algorithm that was derived in Section 5.2. Expressed in polynomial form, the linear convolution algorithm is

$$s(x) = g(x)d(x)$$
$$= g_1 d_1 x^2 + [g_1 d_1 + g_0 d_0 - (g_1 - g_0)(d_1 - d_0)]x + g_0 d_0,$$

which requires three multiplications.

The three quantities of interest are the products $g(x)d(x)$ modulo $x^2 + 1$, x^2, and $x^2 + x + 1$, respectively. These are obtained simply by replacing x^2 by -1, 0, and $-x - 1$, respectively. Then

$$s(x) = g(x)d(x) \pmod{x^2 + 1}$$
$$= [g_1 d_1 + g_0 d_0 - (g_1 - g_0)(d_1 - d_0)]x + g_0 d_0 - g_1 d_1,$$
$$s(x) = g(x)d(x) \pmod{x^2}$$
$$= [g_1 d_1 + g_0 d_0 - (g_1 - g_0)(d_1 - d_0)]x + g_0 d_0,$$
$$s(x) = g(x)d(x) \pmod{x^2 + x + 1}$$
$$= [g_0 d_0 - (g_1 - g_0)(d_1 - d_0)]x + g_0 d_0 - g_1 d_1.$$

Fast algorithms for short convolutions

Table 5.3 *Performance of some algorithms for polynomial products modulo a prime polynomial*

Prime polynomial $p(x)$	Number of multiplications	Number of additions
$x^2 + x + 1$	3	3
$x^4 + 1$	9	15
	7	41
$x^4 + x^3 + x^2 + x + 1$	9	16
	7	46
$x^6 + x^3 + 1$	15	39
$x^6 + x^5 + x^4 + x^3 + x^2 + x + 1$	15	53
$x^{18} + x^9 + 1$	75	267

In matrix form, the three algorithms are

$$\begin{bmatrix} s_0 \\ s_1 \end{bmatrix} = \begin{bmatrix} 1 & -1 & 0 \\ 1 & 1 & -1 \end{bmatrix} \begin{bmatrix} g_0 & & \\ & g_1 & \\ & & g_0 - g_1 \end{bmatrix} \begin{bmatrix} 1 & 0 \\ 0 & 1 \\ -1 & 1 \end{bmatrix} \begin{bmatrix} d_0 \\ d_1 \end{bmatrix},$$

$$\begin{bmatrix} s_0 \\ s_1 \end{bmatrix} = \begin{bmatrix} 1 & 0 & 0 \\ 1 & 1 & -1 \end{bmatrix} \begin{bmatrix} g_0 & & \\ & g_1 & \\ & & g_0 - g_1 \end{bmatrix} \begin{bmatrix} 1 & 0 \\ 0 & 1 \\ -1 & 1 \end{bmatrix} \begin{bmatrix} d_0 \\ d_1 \end{bmatrix},$$

and

$$\begin{bmatrix} s_0 \\ s_1 \end{bmatrix} = \begin{bmatrix} 1 & -1 & 0 \\ 1 & 0 & -1 \end{bmatrix} \begin{bmatrix} g_0 & & \\ & g_1 & \\ & & g_0 - g_1 \end{bmatrix} \begin{bmatrix} 1 & 0 \\ 0 & 1 \\ -1 & 1 \end{bmatrix} \begin{bmatrix} d_0 \\ d_1 \end{bmatrix},$$

respectively. The second algorithm is not a good way to compute $g(x)d(x) \pmod{x^2}$. We have allowed the formalism to produce an algorithm that is inferior to the one we would obtain by inspection, namely,

$$\begin{bmatrix} s_0 \\ s_1 \end{bmatrix} = \begin{bmatrix} 1 & 0 & 0 \\ 0 & 1 & 1 \end{bmatrix} \begin{bmatrix} g_0 & & \\ & g_0 & \\ & & g_1 \end{bmatrix} \begin{bmatrix} 1 & 0 \\ 0 & 1 \\ 1 & 0 \end{bmatrix} \begin{bmatrix} d_0 \\ d_1 \end{bmatrix},$$

which uses only one addition.

The performance of some available algorithms for polynomial products modulo a prime polynomial is tabulated in Table 5.3. In some cases, the figure gives the performance of two algorithms for the same problem. For example, using modulo $x^4 + 1$ one can construct a polynomial with nine multiplications and fifteen additions, or an algorithm with seven multiplications and forty-one additions. In selecting between such possibilities, one must use judgment and an understanding of how this computation fits into a larger computation, as discussed in Chapters 11 and 12. It often happens that a slight advantage in the number of multiplications in a small algorithm will project into a large advantage when this small algorithm is embedded into a large algorithm, while a large disadvantage in the number of additions will project into only a small disadvantage in the larger setting. Hence the modulo $x^4 + 1$ algorithm with seven multiplications can actually offer more than it seems.

5.6 Design of short cyclic convolution algorithms

A library of good algorithms for short cyclic convolutions can be constructed as Winograd short convolution algorithms by reducing the polynomial product modulo $x^n - 1$ to many polynomial products, one product modulo each polynomial factor of $x^n - 1$. To get a good Winograd cyclic convolution algorithm, one must ensure that each of the smaller polynomial products itself has a good algorithm. We will refer to the collection of algorithms that was developed in the previous section to do these subcomputations.

In this section, we shall construct a few good algorithms for cyclic convolution as examples. A more extensive collection of short cyclic convolution algorithms, written in the concise form

$$s = CGAd,$$

is given in Appendix A. These short cyclic convolution algorithms can be combined to make long cyclic convolution algorithms by using a method known as the Agarwal–Cooley algorithm, which will be discussed in Chapter 11.

Consider the computation of the cyclic convolution

$$s(x) = g(x)d(x) \pmod{x^n - 1},$$

where $g(x)$ and $d(x)$ each have degree $n - 1$. We could compute this convolution by first computing a linear convolution and then reducing the result modulo $x^n - 1$.

However, we can get the cyclic convolution more directly by choosing the modulus polynomials as the prime polynomial factors of $m(x)$, denoted $m^{(0)}(x), \ldots, m^{(K-1)}(x)$.

For a simple example, consider the cyclic convolution of blocklength four

$$s(x) = g(x)d(x) \pmod{x^4 - 1}.$$

The prime factors of $x^4 - 1$ are given by

$$\begin{aligned} x^4 - 1 &= (x-1)(x+1)(x^2+1) \\ &= m^{(0)}(x)m^{(1)}(x)m^{(2)}(x). \end{aligned}$$

These prime factors are known as cyclotomic polynomials; they will be defined and studied in Section 9.5. In this construction, the cyclotomic polynomials, as the prime factors of $x^n - 1$, are the modulus polynomials that the Winograd algorithm must use. The coefficients of the residues are

$$\begin{bmatrix} d_0^{(0)} \\ d_0^{(1)} \\ d_0^{(2)} \\ d_1^{(2)} \end{bmatrix} = \begin{bmatrix} 1 & 1 & 1 & 1 \\ 1 & -1 & 1 & -1 \\ 1 & 0 & -1 & 0 \\ 0 & 1 & 0 & -1 \end{bmatrix} \begin{bmatrix} d_0 \\ d_1 \\ d_2 \\ d_3 \end{bmatrix},$$

$$\begin{bmatrix} g_0^{(0)} \\ g_0^{(1)} \\ g_0^{(2)} \\ g_1^{(2)} \end{bmatrix} = \begin{bmatrix} 1 & 1 & 1 & 1 \\ 1 & -1 & 1 & -1 \\ 1 & 0 & -1 & 0 \\ 0 & 1 & 0 & -1 \end{bmatrix} \begin{bmatrix} g_0 \\ g_1 \\ g_2 \\ g_3 \end{bmatrix}.$$

We have already seen several algorithms for polynomial multiplication modulo $x^2 + 1$. One is

$$\begin{bmatrix} s_0 \\ s_1 \end{bmatrix} = \begin{bmatrix} 1 & 0 & -1 \\ 1 & 1 & 0 \end{bmatrix} \begin{bmatrix} g_0 & & \\ & g_1 - g_0 & \\ & & g_1 + g_0 \end{bmatrix} \begin{bmatrix} 1 & 1 \\ 1 & 0 \\ 0 & 1 \end{bmatrix} \begin{bmatrix} d_0 \\ d_1 \end{bmatrix}.$$

Define the internal variables

$$\begin{bmatrix} D_0 \\ D_1 \\ D_2 \\ D_3 \\ D_4 \end{bmatrix} = \begin{bmatrix} 1 & 0 & 0 & 0 \\ 0 & 1 & 0 & 0 \\ 0 & 0 & 1 & 1 \\ 0 & 0 & 1 & 0 \\ 0 & 0 & 0 & 1 \end{bmatrix} \begin{bmatrix} 1 & 1 & 1 & 1 \\ 1 & -1 & 1 & -1 \\ 1 & 0 & -1 & 0 \\ 0 & 1 & 0 & -1 \end{bmatrix} \begin{bmatrix} d_0 \\ d_1 \\ d_2 \\ d_3 \end{bmatrix}$$

and

$$\begin{bmatrix} G_0 \\ G_1 \\ G_2 \\ G_3 \\ G_4 \end{bmatrix} = \begin{bmatrix} 1 & 0 & 0 & 0 \\ 0 & 1 & 0 & 0 \\ 0 & 0 & 1 & 0 \\ 0 & 0 & -1 & 1 \\ 0 & 0 & 1 & 1 \end{bmatrix} \begin{bmatrix} 1 & 1 & 1 & 1 \\ 1 & -1 & 1 & -1 \\ 1 & 0 & -1 & 0 \\ 0 & 1 & 0 & -1 \end{bmatrix} \begin{bmatrix} g_0 \\ g_1 \\ g_2 \\ g_3 \end{bmatrix}.$$

Then $S_k = G_k D_k$ for $k = 0, \ldots, 4$. Now we must recover $s(x)$. The Chinese remainder theorem gives

$$s(x) = \tfrac{1}{4}(x^3 + x^2 + x + 1)s^{(0)}(x) - \tfrac{1}{4}(x^3 - x^2 + x - 1)s^{(1)}(x) \\ - \tfrac{1}{2}(x^2 - 1)s^{(2)}(x) \pmod{x^4 - 1}.$$

5.6 Design of short cyclic convolution algorithms

$$\begin{bmatrix} s_0 \\ s_1 \\ s_2 \\ s_3 \end{bmatrix} = \begin{bmatrix} 1 & 1 & 1 & 0 & -1 \\ 1 & -1 & 1 & 1 & 0 \\ 1 & 1 & -1 & 0 & 1 \\ 1 & -1 & -1 & -1 & 0 \end{bmatrix} \begin{bmatrix} G_0 & & & & \\ & G_1 & & & \\ & & G_2 & & \\ & & & G_3 & \\ & & & & G_4 \end{bmatrix} \begin{bmatrix} 1 & 1 & 1 & 1 \\ 1 & -1 & 1 & -1 \\ 1 & 1 & -1 & -1 \\ 1 & 0 & -1 & 0 \\ 0 & 1 & 0 & -1 \end{bmatrix} \begin{bmatrix} d_0 \\ d_1 \\ d_2 \\ d_3 \end{bmatrix}$$

where

$$\begin{bmatrix} G_0 \\ G_1 \\ G_2 \\ G_3 \\ G_4 \end{bmatrix} = \frac{1}{4} \begin{bmatrix} 1 & 1 & 1 & 1 \\ 1 & -1 & 1 & -1 \\ 2 & 0 & -2 & 0 \\ -2 & 2 & 2 & -2 \\ 2 & 2 & -2 & -2 \end{bmatrix} \begin{bmatrix} g_0 \\ g_1 \\ g_2 \\ g_3 \end{bmatrix}$$

Figure 5.13 An algorithm for four-point cyclic convolution

Therefore

$$\begin{bmatrix} s_0 \\ s_1 \\ s_2 \\ s_3 \end{bmatrix} = \begin{bmatrix} 1 & 1 & 1 & 0 \\ 1 & -1 & 0 & 1 \\ 1 & 1 & -1 & 0 \\ 1 & -1 & 0 & -1 \end{bmatrix} \begin{bmatrix} \frac{1}{4}s_0^{(0)} \\ \frac{1}{4}s_0^{(1)} \\ \frac{1}{2}s_0^{(2)} \\ \frac{1}{2}s_1^{(2)} \end{bmatrix}$$

$$= \begin{bmatrix} 1 & 1 & 1 & 0 \\ 1 & -1 & 0 & 1 \\ 1 & 1 & -1 & 0 \\ 1 & -1 & 0 & -1 \end{bmatrix} \begin{bmatrix} 1 & 0 & 0 & 0 & 0 \\ 0 & 1 & 0 & 0 & 0 \\ 0 & 0 & 1 & 0 & -1 \\ 0 & 0 & 1 & 1 & 0 \end{bmatrix} \begin{bmatrix} \frac{1}{4}S_0 \\ \frac{1}{4}S_1 \\ \frac{1}{2}S_2 \\ \frac{1}{2}S_3 \\ \frac{1}{2}S_4 \end{bmatrix}$$

$$= \begin{bmatrix} 1 & 1 & 1 & 0 & -1 \\ 1 & -1 & 1 & 1 & 0 \\ 1 & 1 & -1 & 0 & 1 \\ 1 & -1 & -1 & -1 & 0 \end{bmatrix} \begin{bmatrix} \frac{1}{4}S_0 \\ \frac{1}{4}S_1 \\ \frac{1}{2}S_2 \\ \frac{1}{2}S_3 \\ \frac{1}{2}S_4 \end{bmatrix}.$$

In constructing the algorithm, arbitrary choices were made in arranging the terms. The rows can be permuted in the matrix of preadditions, and the columns can be permuted in the matrix of postadditions without effect. The final algorithm is summarized in Figure 5.13.

Table 5.4 gives the performance of some of the best algorithms known for short cyclic convolutions. These algorithms, some of which are given in Appendix A, are constructed in the way we described. Sometimes, additional techniques, such as are described in the remainder of this section, have been used to reduce the number of additions.

Table 5.4 *Performance of some short cyclic convolution algorithms*

$s(x) = g(x)d(x) \pmod{x^n - 1}$
$\deg g(x) = n - 1$
$\deg s(x) = n - 1$

	Real convolutions	
Blocklength n	Number of real multiplications	Number of real additions
2	2	4
3	4	11
4	5	15
5	8	62
5	10	31
7	16	70
8	12	72
8	14	46
9	19	74
16	33	181
16	35	155

A column for complex convolutions could be appended to Table 5.4. Because every real cyclic convolution algorithm can be used for complex cyclic convolution, entries from the columns for real cyclic convolutions could be entered into the columns for complex cyclic convolutions. However, sometimes the polynomial $x^n - 1$ may factor into more prime polynomials over the complex rationals than it does over the rationals. Then one has additional design options in the complex field.

The Winograd algorithms can be viewed as a method of factoring certain matrices. Let s, g, and d be vectors of length n, whose components are the coefficients of $s(x)$, $g(x)$, and $d(x)$. The cyclic convolution $s(x) = g(x)d(x) \pmod{x^n - 1}$ can be written as a matrix product

$$\begin{bmatrix} s_0 \\ s_1 \\ s_2 \\ \vdots \\ s_{n-1} \end{bmatrix} = \begin{bmatrix} g_0 & g_{n-1} & \cdots & g_2 & g_1 \\ g_1 & g_0 & & g_3 & g_2 \\ \vdots & & & & \vdots \\ g_{n-2} & g_{n-3} & & g_0 & g_{n-1} \\ g_{n-1} & g_{n-2} & \cdots & g_1 & g_0 \end{bmatrix} \begin{bmatrix} d_0 \\ d_1 \\ d_2 \\ \vdots \\ d_{n-1} \end{bmatrix},$$

which is abbreviated as

$$s = Td.$$

The Winograd algorithm written

$$s = C[(Bg) \cdot (Ad)]$$

5.6 Design of short cyclic convolution algorithms

can be written more compactly as

$$s = CGAd,$$

where A is an $M(n)$ by n matrix, G is an $M(n)$ by $M(n)$ diagonal matrix, and C is an n by $M(n)$ matrix. The coefficients of the filter $g(x)$ determine the elements of the diagonal matrix G. Thus the Winograd algorithm can be thought of as the matrix factorization

$$T = CGA,$$

where G is a diagonal matrix and C and A are matrices of small integers.

Whenever $g(x)$ represents a fixed FIR filter, or a FIR filter whose coefficients change infrequently, the computation $G = Bg$ occurs infrequently and can be neglected. Hence the complexity of the matrix B is not significant.

The matrices A and B have symmetric roles, so they can be interchanged just by renaming d and g. The Winograd convolution algorithm may produce matrices A and B that differ only in row permutations and so have the same complexity; but, in general, one would choose the one with the lesser complexity to play the role of A. Surprisingly, it is even possible to interchange the roles of C and A in a cyclic convolution algorithm. The following theorem gives a way to do this.

Theorem 5.6.1 (Matrix exchange theorem) *If a matrix S can be factored as*

$$S = CDE,$$

where D is a diagonal matrix, then it also can be factored as

$$S = (\overline{E})^T D (\underline{C})^T,$$

where $\overline{E}$ is the matrix obtained from E by reversing the order of its columns, and $\underline{C}$ is the matrix obtained from C by reversing the order of its rows.

Proof Let J be an exchange matrix the same size as S. Then

$$\begin{aligned} S^T &= JSJ \\ &= (JC)D(EJ) \\ &= \underline{C}D\overline{E}. \end{aligned}$$

To complete the proof, take the transpose of both sides, noting that D is a diagonal matrix. □

An algorithm

$$\begin{aligned} s &= C[(Bg) \cdot (Ad)] \\ &= C[(Bd) \cdot (Ag)], \end{aligned}$$

which means that either A or B can be chosen as the multiplier of g. But this can be written as

$$s = CDAg,$$

where D is a diagonal matrix with diagonal elements equal to Bd. By Theorem 5.6.1, we can write

$$s = (\overline{A})^T D(\underline{C})^T g.$$

The theorem says that for a cyclic convolution algorithm, the most complex of the three matrices A, B, and C can be chosen as the multiplier of g and buried in G.

5.7 Convolution in general fields and rings

Any convolution algorithm, such as the Cook–Toom algorithm or the Winograd convolution algorithm, is an identity involving the defining properties of the field operations including the properties of associativity, distributivity, and commutativity. An algorithm derived in one field can be used in any extension of that field. Similarly, a convolution algorithm derived for a given field can be used in a commutative ring that contains the field. By using algorithms derived for a subfield, we have a large collection of good algorithms for the extension field. These may not be as good, however, as algorithms derived directly in the extension field.

Specifically, an algorithm for the convolution of sequences of real numbers can be used without change to convolve sequences of complex numbers. If M and A represent the number of real multiplications and real additions to compute the real convolution, then it will take M complex multiplications and A complex additions to compute the complex convolution – a total of $4M$ real multiplications and $2A + 2M$ real additions if the conventional rule for complex multiplication is used. If, instead, the complex multiplication described in Section 1.1 is used and one side of the convolution consists of fixed constants, then only $3M$ real multiplications and $2A + 3M$ real additions are needed.

It is possible to do even better for complex cyclic convolutions whose blocklength is a power of two. To explain this, we begin with a discussion of convolution modulo $x^{2^i} + 1$. Within the ring of polynomials modulo $x^{2^i} + 1$, $i \geq 1$, there is an element that is the square root of minus one. This is because

$$-1 = x^{2^i} \pmod{x^{2^i} + 1},$$

so we have

$$\sqrt{-1} = x^{2^{i-1}}$$

5.7 Convolution in general fields and rings

in this ring. We shall use this element to write a polynomial product modulo $x^{2^i} + 1$, in terms of two real convolutions.

To calculate the polynomial product

$$s(x) = g(x)d(x) \pmod{x^{2^i} + 1},$$

where

$$g(x) = g_R(x) + jg_I(x)$$

and

$$d(x) = d_R(x) + jd_I(x),$$

one can compute the four real convolutions $g_R(x)d_R(x)$, $g_I(x)d_R(x)$, $g_R(x)d_I(x)$, and $g_I(x)d_I(x)$ modulo $x^{2^i} + 1$ and write

$$s_R(x) = g_R(x)d_R(x) - g_I(x)d_I(x),$$
$$s_I(x) = g_R(x)d_I(x) + g_I(x)d_R(x).$$

A better procedure, which has half the number of multiplications, is to define the polynomials

$$a(x) = \tfrac{1}{2}(g_R(x) - x^{2^{i-1}}g_I(x))(d_R(x) - x^{2^{i-1}}d_I(x)) \pmod{x^{2^i} + 1},$$
$$b(x) = \tfrac{1}{2}(g_R(x) + x^{2^{i-1}}g_I(x))(d_R(x) + x^{2^{i-1}}d_I(x)) \pmod{x^{2^i} + 1}.$$

These two polynomials each require one real convolution to compute. From these, the complex polynomial $s(x) \pmod{x^{2^i} + 1}$ then is given by

$$s_R(x) = (a(x) + b(x)),$$
$$s_I(x) = x^{2^{i-1}}(a(x) - b(x)) \pmod{x^{2^i} + 1}.$$

We shall use this relationship to compute the complex cyclic convolution

$$s(x) = g(x)d(x) \pmod{x^n - 1}$$

where n is a power of two. Write

$$x^n - 1 = (x - 1)(x + 1)(x^2 + 1)(x^4 + 1) \cdots (x^{n/2} + 1),$$

so we have multiple short complex convolutions of the form

$$s^{(i)}(x) = g^{(i)}(x)d^{(i)}(x) \pmod{x^{2^i} + 1}.$$

The polynomial products modulo $x - 1$ and modulo $x + 1$ are products of scalars and are computed as products of complex numbers. They contribute very little to the computational complexity. The other polynomial products can each be computed with two real convolutions, as we have seen above. Thus, with this method, a complex cyclic convolution takes about twice as much computation as a real cyclic convolution.

This method is in competition with the method of factoring the polynomial $x^n - 1$ in the complex field to write

$$x^n - 1 = (x - 1)(x + 1)(x - j)(x + j) \cdots (x^{n/4} - j)(x^{n/4} + j).$$

Now the individual subproblems are smaller, but their arithmetic is complex.

5.8 Complexity of convolution algorithms

Given an algorithm for some computation, should one be satisfied with it, or should one attempt to find a better algorithm? This is a difficult question to answer for several reasons. First, it is difficult to decide on criteria for declaring an algorithm to be the best algorithm. Then, even if criteria are agreed on, it is difficult to deduce the performance of the optimal algorithm according to the chosen criteria.

For many problems, one declares the optimal algorithm to be the one with the fewest multiplications. This criterion is simple enough so that we can prove some theorems giving the performance of the optimal algorithm. Of course, after we find the optimal algorithm, we might not like it, perhaps because of the number of additions, or perhaps because of its irregular structure.

In this section, we shall find the performance of the optimal algorithms for convolution as judged by the number of multiplications. To do this, we need to sharpen up our idea of what a multiplication is, and we sharpen it in the way that allows our questions to be answered. Specifically, we want to define the idea of multiplication so that $d \cdot g$ is a multiplication when d and g can take on arbitrary real values, and yet $2g$ is not a multiplication because it can be computed as $g + g$. This distinction is easy to accept, but then what about $3g$ or $\frac{5}{7}g$? For the purpose of this section, we choose a definition that leads to useful results. The computation $d \cdot g$ is a multiplication if both factors can take on arbitrary real values, but it is not a multiplication if only one of the factors can take real values. We shall see that meaningful results follow from the definition, even though it initially appears to be a somewhat hollow one. Notice also that our criterion might appear suspect because in applications all numbers have finite wordlength, and so all numbers in applications are rational. Nevertheless, if, in principle, the variables can take on any real values, the computation is called a multiplication.

The Winograd short cyclic convolution algorithms are optimum in this sense. No n-point cyclic convolution algorithm can have fewer multiplications than a Winograd short convolution algorithm. The proof of this fact is one of the major tasks of this section. The major facts we shall prove are the following.

1. The linear convolution of two sequences of lengths L and N cannot be computed with fewer than $L + N - 1$ general multiplications.
2. The cyclic convolution of two sequences modulo $x^n - 1$ cannot be computed with fewer than $2n - t$ general multiplications, where t is the number of prime factors of $x^n - 1$.

5.8 Complexity of convolution algorithms

3 The polynomial product $g(x)d(x)$ modulo $p(x)$ cannot be computed with fewer than $2n - t$ general multiplications, where t is the number of prime factors of $p(x)$.

Of course, the second statement is a special case of the third, but it is important enough to be stated separately.

The ideas can be developed for any field. Let F be a field, called the *field of the computation*, and let E be a subfield of F, called the *field of constants* or the *ground field*. The elements of E are called *scalars*. Let $\mathbf{d} = (d_0, \ldots, d_{n-1})$ and $\mathbf{g} = (g_0, \ldots, g_{r-1})$ be arbitrary vectors of elements of F of fixed lengths n and r. The components of $\mathbf{d}$ and $\mathbf{g}$ will be referred to as *indeterminates* or as *variables*. These $n + r$ indeterminates are independent; there are no fixed relationships between them. An algorithm is a rule for computing a sequence of elements $f_1, \ldots, f_t$ of F that satisfies the following properties: each element f_i of the sequence is equal to either (1) a component of $\mathbf{d}$ or of $\mathbf{g}$ or the sum, difference, or product of two such components; (2) the sum, difference, or product of a component of $\mathbf{d}$ or of $\mathbf{g}$ and an element f_j of the sequence with j less than i; (3) the sum, difference, or product of two elements, f_j and f_k, of the sequence with both j and k less than i; or (4) an element of the ground field E.

We say that the algorithm computes an output vector $\mathbf{s}$ if the components of $\mathbf{s}$ are included in the sequence $f_1, \ldots, f_t$. It must be emphasized that $\mathbf{s}$ is a vector whose components are variables that have some functional relationship to the variables that make up the components of $\mathbf{d}$ and $\mathbf{g}$. An algorithm that computes $\mathbf{s}$ is a fixed procedure that computes the correct $\mathbf{s}$ for every possible assignment of values from the field F to $\mathbf{d}$ and $\mathbf{g}$.

Notice that the definition includes no provision for division or for branching. For the algorithms we deal with, the operations of division and branching are unnecessary, and they cannot be used to reduce the number of multiplications.

The definition of an algorithm can be illustrated by the problem of complex multiplication. First, write it in the form

$$\begin{bmatrix} e \\ f \end{bmatrix} = \begin{bmatrix} c & -d \\ d & c \end{bmatrix} \begin{bmatrix} a \\ b \end{bmatrix}.$$

Then

$f_1 = ca,$
$f_2 = db,$
$f_3 = f_1 - f_2,$
$f_4 = da,$
$f_5 = cb,$
$f_6 = f_4 + f_5$

is a description of the algorithm in the form of a sequence. We can also write complex multiplication as

$$\begin{bmatrix} e \\ f \end{bmatrix} = \begin{bmatrix} 1 & 0 & 1 \\ 0 & 1 & 1 \end{bmatrix} \begin{bmatrix} (c-d) & 0 & 0 \\ 0 & (c+d) & 0 \\ 0 & 0 & d \end{bmatrix} \begin{bmatrix} 1 & 0 \\ 0 & 1 \\ 1 & -1 \end{bmatrix} \begin{bmatrix} a \\ b \end{bmatrix}.$$

Then

$$f_1 = a - b,$$
$$f_2 = c - d,$$
$$f_3 = c + d,$$
$$f_4 = f_2 a,$$
$$f_5 = f_3 b,$$
$$f_6 = d f_1,$$
$$f_7 = f_4 + f_6,$$
$$f_8 = f_5 + f_6$$

is a description of the alternative algorithm in the form of a sequence.

Consider the collection of all such sequences that compute the complex multiplication. An optimal algorithm is one with the minimum number of multiplications of any algorithm in that collection.

We are now ready to formalize the definition of a computational problem. We shall study only problems of the form

$$s = Hd,$$

where d is an input data vector of length k, and s is an output vector of length n. The elements of the matrix H are linear combinations of the r indeterminates $g_0, \ldots, g_{r-1}$. Typically, r is less than the number of elements of H, and each indeterminate may appear more than once in H. This structure includes all of the convolution problems that we have studied.

We think of the elements of H not as field elements, but as linear combinations of the indeterminates of the form

$$H_{ij} = \sum_{k=0}^{r-1} \alpha_{ijk} g_k,$$

where each α_{ijk} is a scalar. Two such linear forms can be added or multiplied by a scalar, and the set of such linear forms is closed under these operations. Thus the set of such linear forms over the field E is a vector space denoted by $E[g_0, \ldots, g_{r-1}]$. Further, H is a matrix over the set of such linear forms. Many familiar properties of matrices over a field do not hold for H, because H is not a matrix over a field (nor even

5.8 Complexity of convolution algorithms

over a ring), but rather over the set $E[g_0, \ldots, g_{r-1}]$. In particular, the row rank need not equal the column rank. We shall see that the row rank and the column rank each provides a lower bound on the number of multiplications needed to compute $s = Hd$. By interchanging the roles of g and d, each of these two bounds may be applied in two ways.

The row rank is defined in the obvious way as follows; the column rank is defined in a similar way. Each row h_i of H is a vector of length k of elements from the set[1] $E[g_0, \ldots, g_{r-1}]$. A linear combination of rows of H is also a vector of length k of elements of the same set, given by $\sum_{i=1}^{n} \beta_i h_i$ where β_i for $i = 1, \ldots, n$ is an element of the field E. The row rank of H is the size of the largest linearly independent set of rows. This is a set of rows such that no nonzero linear combination is equal to zero.

As an example, the matrix

$$H = \begin{bmatrix} 4g_0 & -g_0 & g_0 \\ 2g_1 & -g_1 & 0 \\ 2g_1 - 2g_0 & g_0 & g_1 \end{bmatrix}$$

has column rank equal to two with respect to the rationals (or the reals), because

$$\frac{1}{2} \begin{bmatrix} 4g_0 \\ 2g_1 \\ 2g_1 - 2g_0 \end{bmatrix} + \begin{bmatrix} -g_0 \\ -g_1 \\ g_0 \end{bmatrix} - \begin{bmatrix} g_0 \\ 0 \\ g_1 \end{bmatrix} = \begin{bmatrix} 0 \\ 0 \\ 0 \end{bmatrix},$$

but no linear combination of only two columns is identically equal to zero.

Theorem 5.8.1 (Row rank theorem) *The number of multiplications used by any algorithm that computes $s = Hd$ is at least as large as the row rank of H.*

Proof Without loss of generality, we can assume that the first ρ rows of H are linearly independent and we shall deal only with these rows. Let M be the matrix formed by these rows, and consider only the partial computation

$$\begin{bmatrix} s_0 \\ \vdots \\ s_{\rho-1} \end{bmatrix} = Md.$$

The proof will identify an appropriate matrix A over the field E for which known properties of matrices over fields can be invoked.

Suppose that, in the algorithm specified by the sequence $f_1, \ldots, f_N$, there are ℓ multiplication steps given by the ℓ terms from the sequence $e_1, \ldots, e_\ell$. Then the first

---

[1] The set is itself a vector space, but it is better to call it a set than to risk confusion by speaking of "vectors of vectors."

182   **Fast algorithms for short convolutions**

ρ components of s must be linear combinations of these product terms and of linear terms and elements of the ground field. That is,

$$\begin{bmatrix} s_0 \\ \vdots \\ s_{\rho-1} \end{bmatrix} = A \begin{bmatrix} e_1 \\ \vdots \\ e_\ell \end{bmatrix} + \begin{bmatrix} b_0 \\ \vdots \\ b_{\rho-1} \end{bmatrix},$$

where the elements of A are from the ground field E and the elements of b are linear combinations of the indeterminates and of elements of the ground field.

Now suppose that ρ is larger than ℓ. Then A has more rows than columns, so the rows of A are linearly dependent, and there exists a vector c over the field E such that $c^T A = 0^T$. Therefore

$$c^T(Md) = c^T(Ae + b),$$

and, because $c^T A = 0^T$, this reduces to

$$(c^T M)d = c^T b.$$

Now the right side has no products of two indeterminates because c contains only elements of E, and b has no products of indeterminates. Therefore the left side can have no products of indeterminates. Then, because every component of d is an indeterminate, $c^T M$ can contain no indeterminates. But $c^T M$ is a vector of linear combinations of indeterminates. Because $c^T M$ contains no indeterminates, it equals zero, and the rows of M are dependent. The contradiction proves that ℓ is larger than ρ, and the number of multiplications is as least as large as the row rank of H.   □

Theorem 5.8.2 (Column rank theorem) *The number of multiplications used by any algorithm that computes $s = Hd$ is at least as large as the column rank of H.*

Proof The proof is by induction. If the column rank is one, then any algorithm must use at least one multiplication. Suppose the theorem is true whenever the column rank is $\ell - 1$. That is, if the column rank of H is $\ell - 1$, then any algorithm that computes $s = Hd$ must use at least $\ell - 1$ multiplications. The induction step will infer the corresponding conclusion when the column rank is ℓ.

Suppose we have an algorithm to compute:

$$s = Hd,$$

where H has column rank ℓ. Without loss of generality, we may assume that the last column of H contains at least one nonzero element (otherwise it could be deleted from H). Hence d_ℓ appears in some product term, say, the last product term.

That is, the sum $\sum_i \alpha_i d_i$, for some set of scalars α_i is a factor in a multiplication with $\alpha_\ell \neq 0$. We are free to suppose that α_ℓ equals one because multiplication by scalars is free. Then $d_\ell + \sum_{i=0}^{\ell-1} \alpha_i d_i$ is a factor in a multiplication.

5.8 Complexity of convolution algorithms

To complete the proof, we devise from any algorithm for the given problem an algorithm for an artificial problem

$$s' = H'd'$$

that has rank $\ell - 1$, and which can be solved with the given algorithm by deleting one multiplication. By the induction assumption, every algorithm for the new problem uses at least $\ell - 1$ multiplications, so the original problem must use at least ℓ multiplications.

To do this, replace d_ℓ in the given problem by $-\sum_{i=0}^{\ell-1} \alpha_i d_i$. Then the last product term that involves the sum $d_\ell + \sum_{i=0}^{\ell-1} \alpha_i d_i$ is a multiplication by zero and so can be deleted from the algorithm. But the algorithm now does solve some problem. Specifically, it solves the problem

$$s' = H'd',$$

where d' is the vector of length $\ell - 1$ obtained from d by deleting the last component, and H' is obtained from H by replacing the jth column h_j with $h_j - \alpha_j h_\ell$. Then

$$H' \begin{bmatrix} d_0 \\ d_1 \\ \vdots \\ d_{\ell-1} \end{bmatrix} = H \begin{bmatrix} d_0 \\ d_1 \\ \vdots \\ d_{\ell-1} \\ -\sum_{i=0}^{\ell-1} \alpha_i d_i \end{bmatrix}.$$

Thus the algorithm computes $H'd'$, and the proof is complete. □

Theorem 5.8.3 *Every algorithm to compute the linear convolution*

$$s(x) = g(x)d(x),$$

where $\deg g(x) = L - 1$ *and* $\deg d(x) = N - 1$, *uses at least* $L + N - 1$ *multiplications.*

Proof The computation can be written as a matrix–vector product

$$s = Hd,$$

where

$$H = \begin{bmatrix} g_0 & 0 & \cdots & 0 & 0 \\ g_1 & g_0 & \cdots & 0 & 0 \\ \vdots & & & & \\ 0 & & \cdots & g_{L-1} & g_{L-2} \\ 0 & & \cdots & 0 & g_{L-1} \end{bmatrix}$$

is an $L + N - 1$ by N matrix. As elements of $E[g_0, \ldots, g_{L-1}]$, the rows of H clearly are independent, so Theorem 5.8.1 implies that there are at least $L + N - 1$ multiplications. □

Theorem 5.8.4 *Let $p(x)$ be a prime polynomial of degree n. Every algorithm to compute the polynomial product*

$$s(x) = g(x)d(x) \pmod{p(x)}$$

uses at least $2n - 1$ multiplications.

Proof Suppose there are t multiplications. Then the output of the computation is a linear combination of these t product terms:

$$s = AS,$$

where A is an n by t matrix over E and S is a vector of length t containing all the product terms.

Clearly, one can always choose $d(x)$ and $g(x)$ to make any component of $s(x)$ nonzero and all other components equal to zero. Therefore the n rows of A must be linearly independent, so A also contains n linearly independent columns. Without loss of generality, we can take the first n columns as linearly independent. Then we can block A as

$$s = [A' \mid A'']S,$$

where A' is an n by n invertible matrix. Multiply by C, the inverse of A',

$$Cs = [I : P]S$$
$$= CHd.$$

Using the fact that $p(x)$ is irreducible, we will show that all columns of any row of H are linearly independent, as are all columns of any linear combination of rows of H. This statement is a consequence of standard results of matrix theory. We will prove it separately as Theorem 5.8.6 at the end of this section. Consequently, all columns of the first row of CH are linearly independent, so by Theorem 5.8.2, it takes at least n multiplications to compute the first element of CHd.

On the other hand, the first row of

$$Cs = [I : P]S$$

uses at most $1 + (t - n)$ multiplications, because S is a vector of length t containing all the products and P has $(t - n)$ columns. Hence

$$1 + (t - n) \geq n$$

5.8 Complexity of convolution algorithms

so that

$$t \geq 2n - 1,$$

which proves the theorem. □

As an application of this theorem consider complex multiplication, which can be viewed as multiplication modulo $(x^2 + 1)$. The theorem says that at least three real multiplications are required to compute one complex multiplication.

Theorem 5.8.4 also can be proved for polynomial multiplication modulo $p(x)^\ell$ as long as $p(x)$ is a prime polynomial, although we shall not do so. We consider, instead, the case where $p(x)$ is not prime but has k prime factors. Then the Chinese remainder theorem can be used to break the problem into k subproblems, each of length n_i, with $\sum_i n_i = n$. These subproblems can be combined without multiplications. Hence using Theorem 5.8.4 for each of the smaller problems says that if the Chinese remainder theorem is used, then at least

$$\sum_{i=1}^{k}(2n_i - 1) = 2n - k$$

multiplications are required. The following theorem says that no algorithm can do better than one that uses the Chinese remainder theorem in this way.

Theorem 5.8.5 *Let $p(x)$, a polynomial of degree n, be a product of k distinct prime polynomials. Any algorithm to compute the polynomial product*

$$s(x) = g(x)d(x) \pmod{p(x)}$$

uses at least $2n - k$ multiplications.

Proof The Chinese remainder theorem is an invertible transformation without multiplications, so it suffices to consider the computation $s = Hd$ in the block-diagonal form

$$\begin{bmatrix} s_1 \\ s_2 \\ \vdots \\ s_k \end{bmatrix} = \begin{bmatrix} H_1 & 0 & \cdots & 0 \\ 0 & H_2 & & \\ \vdots & & & \\ 0 & & & H_k \end{bmatrix} \begin{bmatrix} d_1 \\ d_2 \\ \vdots \\ d_k \end{bmatrix},$$

where

$$s_i = H_i d_i$$

corresponds to the ith subproblem formed with the Chinese remainder theorem.

Now we mimic the proof of Theorem 5.8.4. Suppose there are t multiplications in an algorithm. Then, it must be possible to write

$$s = AS,$$

where S is a vector of length t containing all the product terms and A is an n by t matrix over the field E. Because A must contain n linearly independent rows, it must contain n linearly independent columns, which we can take to be the first n columns. Then

$$s = [A' : A'']S$$

and

$$Cs = [I : P]S,$$

where C is the inverse of A'. Consequently, it takes at most $1 + (t - n)$ multiplications to compute the first element of Cs. But we also have

$$Cs = C \begin{bmatrix} H_1 & & & \\ & H_2 & & \\ & & \ddots & \\ & & & H_k \end{bmatrix} d.$$

Now take any linear combination of the rows of H. Any linear combination of rows of H_i has only linearly independent columns. Hence there are at least $n - (k - 1)$ linearly independent columns in any linear combination of rows of H. By Theorem 5.8.2, it takes at least $n - (k - 1)$ multiplications to compute the first element of CHd. Therefore the upper and lower bounds on the number of multiplications needed to compute the first element of Cs yield the inequality

$$1 + (t - n) \geq n - (k - 1)$$

so that

$$t \geq 2n - k,$$

which proves the theorem. □

Now we must finish a detail that was left hanging in the proof of Theorem 5.8.4.

Theorem 5.8.6 *Let*

$$s = Hd$$

be the matrix representation of the polynomial product

$$s(x) = g(x)d(x) \pmod{p(x)},$$

5.8 Complexity of convolution algorithms

where $p(x)$ is a prime polynomial and $\mathbf{H}$ is a matrix of indeterminates whose elements are coefficients of $g(x)$. Then all columns of any row of $\mathbf{H}$ are linearly independent, as are all columns of any linear combination of rows of $\mathbf{H}$.

Proof

Step 1 Let $\mathbf{C}_p$ be the *companion matrix* of the polynomial $p(x)$, defined as the n by n matrix

$$\mathbf{C}_p = \begin{bmatrix} 0 & 0 & \cdots & 0 & -p_0 \\ 1 & 0 & \cdots & 0 & -p_1 \\ 0 & 1 & \cdots & 0 & -p_2 \\ \vdots & \vdots & & & \vdots \\ 0 & 0 & \cdots & 1 & -p_{n-1} \end{bmatrix}.$$

Then the ith column of $\mathbf{H}$ is equal to $\mathbf{C}_p^i \mathbf{g}$, and $\mathbf{H}$ can be written in terms of its column vectors:

$$\mathbf{H} = [\mathbf{g} \quad \mathbf{C}_p \mathbf{g} \quad \mathbf{C}_p^2 \mathbf{g} \quad \cdots \quad \mathbf{C}_p^{n-1} \mathbf{g}].$$

Let $\mathbf{w}$ be any row vector, and let $\mathbf{w}\mathbf{H}$ be a linear combination of rows of $\mathbf{H}$. We must show that no linear combination of columns of $\mathbf{w}\mathbf{H}$ is zero. Assume that

$$\mathbf{0} = \sum_{i=0}^{n-1} (\mathbf{w} \mathbf{C}_p^i \mathbf{g}) a_i = \left[\mathbf{w} \sum_{i=0}^{n-1} a_i \mathbf{C}_p^i \right] \mathbf{g}.$$

This must hold for any $\mathbf{g}$, so we have

$$\mathbf{w} \cdot \mathbf{a}(\mathbf{C}_p) = \mathbf{0},$$

where

$$\mathbf{a}(\mathbf{C}_p) = \sum_{i=0}^{n-1} a_i \mathbf{C}_p^i$$

is an n by n matrix computed from $\mathbf{C}_p$. Because $\mathbf{w}$ is nonzero, we must have that $\mathbf{a}(\mathbf{C}_p)$ is a singular matrix; otherwise, $\mathbf{w} \cdot \mathbf{a}(\mathbf{C}_p)$ could not be zero. Hence we must show that the only singular matrix computed in this way as a matrix polynomial of degree at most $n-1$ is the all-zero polynomial.

Step 2 It is well-known and easy to verify that any prime polynomial, $p(x)$, is zero when evaluated at its own companion matrix. That is,

$$p(\mathbf{C}_p) = \mathbf{0},$$

where $\mathbf{0}$ is the all-zero n by n matrix. Let $a(x)$ be a polynomial of degree at most $n-1$ such that $\mathbf{a}(\mathbf{C}_p)$ is a singular matrix, and let $\mathbf{v}$ be a nonzero vector in the null space of $a(C_p)$. Then, because $p(x)$ is a prime polynomial of degree n, there exist Bézout polynomials $A(x)$ and $P(x)$ such that

$$A(x)a(x) + P(x)p(x) = 1.$$

Therefore

$$[A(C_p)a(C_p) + P(C_p)p(C_p)]v = Iv.$$

But $p(C_p) = 0$, so we have that

$$A(C_p)a(C_p)v = v,$$

which is a contradiction because $a(C_p)v = 0$. Hence there exists no polynomial $a(x)$ of degree $n - 1$ or less such that $a(C_p)$ is singular, except the all-zero polynomial. This completes the proof of the theorem. $\square$

Problems for Chapter 5

5.1 a The complex multiplication $(e + jf) = (a + jb)(c + jd)$ can be computed with three real multiplications and five real additions by

$$e = (a - b)d + a(c - d),$$
$$f = (a - b)d + b(c + d).$$

Let c and d be constant and represent the multiplication algorithm in the matrix form

$$\begin{bmatrix} e \\ f \end{bmatrix} = BDA \begin{bmatrix} a \\ b \end{bmatrix},$$

where A and B represent the preadditions and the postadditions, and D is a diagonal matrix.

b A two-point cyclic convolution, $s(x) = g(x)d(x) \pmod{x^2 - 1}$, can be computed by the algorithm

$$\begin{bmatrix} s_0 \\ s_1 \end{bmatrix} = \begin{bmatrix} 1 & 1 \\ 1 & -1 \end{bmatrix} \begin{bmatrix} \frac{1}{2}(g_0 + g_1) & 0 \\ 0 & \frac{1}{2}(g_0 - g_1) \end{bmatrix} \begin{bmatrix} 1 & 1 \\ 1 & -1 \end{bmatrix} \begin{bmatrix} d_0 \\ d_1 \end{bmatrix}.$$

Suppose that $d(x)$ and $g(x)$ are complex polynomials. How many real multiplications and real additions are required here if the complex arithmetic is done in the straightforward classical way?

c Now represent the input and output data by real vectors of length four. Give an integrated algorithm that has six real multiplications. How many real additions are there?

Problems

5.2 Given a device that computes the linear convolution of two sequences, each of length n, describe how it can he used to compute the *crosscorrelation function*

$$\sum_{i=0}^{n-1} g_{i+j} d_i$$

of two sequences of length n.

5.3 Use the Cook–Toom algorithm to construct a convolution algorithm for filtering a sequence of four data inputs with a three-tap FIR filter.

5.4 a Starting with the two-point linear convolution algorithm

$$s(x) = g_1 d_1 x^2 + [g_1 d_1 + g_0 d_0 - (g_1 - g_0)(d_1 - d_0)]x + g_0 d_0,$$

construct an algorithm for

$$s(x) = g(x)d(x) \pmod{x^2 - x + 1}.$$

b Repeat, starting with

$$s(x) = g_1 d_1 x^2 + [(g_1 + g_0)(d_1 + d_0) - g_1 d_1 - g_0 d_0]x + g_0 d_0.$$

Which is to be preferred?

5.5 Give two distinct reasons why the Rader prime algorithm combined with the Winograd convolution algorithm leads to good FFT algorithms, while the Bluestein chirp algorithm combined with the Winograd convolution algorithm does not. Can you give a third?

5.6 Show that when the Cook–Toom algorithm with the n points $\beta_i = e^{-(j2\pi/n)i}$ is used to linearly convolve two sequences of length $n/2$, it gives the same algorithm as when a Fourier transform with the convolution theorem is used.

5.7 Suppose that one has a device (a circuit module or a software subroutine) that computes 315-point cyclic convolutions. This device is to be used to pass a vector of 1000 data points through a 100-tap FIR fitter. Describe how to break up the data to feed the convolver and to assemble the convolver outputs to get the desired answer.

5.8 Derive an algorithm for a four-point cyclic convolution of complex sequences that uses four complex multiplications. Compare this with an algorithm for a four-point cyclic convolution of real sequences that is used for convolving complex sequences.

5.9 One way to define the complex field is as an extension of the real field using the prime polynomial $p(x) = x^2 + 1$ over the real field. Complex multiplication becomes

$$e + fx = (a + bx)(c + dx) \pmod{x^2 + 1}.$$

Use this viewpoint to convert a linear convolution algorithm into an algorithm for complex multiplication.

5.10 **a** Use only the factorization

$$x^{16} - 1 = (x-1)(x+1)(x^2+1)(x^4+1)(x^8+1)$$

to get modulus polynomials. How many multiplications are needed to perform a sixteen-point cyclic convolution using the Winograd algorithm?

b Suppose factorization over the "complex rationals" (numbers of the form $(a+jb)$ where a and b are rationals) is allowed. How many complex multiplications are now needed using a set of modulus polynomials that includes 1, $\pm j$ as coefficients? (*Note*: we now have complex numbers as intermediate variables, and we do not count multiplication by j as a multiplication.)

c Does part (b) have an advantage over part (a)? (*Hint*: is the cyclic convolution a real convolution or a complex convolution?)

5.11 Set up all equations for a Winograd four-point by three-point linear convolution algorithm using

$$m(x) = x^2(x+1)(x-1)(x^2+1).$$

5.12 A double-precision representation of a number in a computer may be expressed as a polynomial $d_0 + d_1 x$. Give an algorithm for double-precision multiplication without roundoff that uses three single-precision multiplications and four single-precision additions.

5.13 Use the algorithm

$$\begin{bmatrix} s_0 \\ s_1 \\ s_2 \\ s_3 \\ s_4 \end{bmatrix} = \begin{bmatrix} 1 & 0 & 0 & 0 & 0 & 0 \\ -1 & -1 & 0 & 1 & 0 & 0 \\ -1 & 1 & -1 & 0 & 1 & 0 \\ 0 & -1 & -1 & 0 & 0 & 1 \\ 0 & 0 & 1 & 0 & 0 & 0 \end{bmatrix} \left\{ \begin{bmatrix} 1 & 0 & 0 \\ 0 & 1 & 0 \\ 0 & 0 & 1 \\ 1 & 1 & 0 \\ 1 & 0 & 1 \\ 0 & 1 & 1 \end{bmatrix} \begin{bmatrix} g_0 \\ g_1 \\ g_2 \end{bmatrix} \right\}$$

$$\cdot \left\{ \begin{bmatrix} 1 & 0 & 0 \\ 0 & 1 & 0 \\ 0 & 0 & 1 \\ 1 & 1 & 0 \\ 1 & 0 & 1 \\ 0 & 1 & 1 \end{bmatrix} \begin{bmatrix} d_0 \\ d_1 \\ d_2 \end{bmatrix} \right\}$$

and the transformation principle to construct five new algorithms. Which of these are interesting?

5.14 Prove that, for any even n larger than one, the best algorithm for computing

$$s(x) = g(x)d(x) \pmod{x^n + 1}$$

in the field of real numbers uses more multiplications than the best algorithm for computing

$$s(x) = g(x)d(x) \pmod{x^n - 1}.$$

5.15 Find an n and an n' with n less than n' such that the n-point cyclic convolution requires more multiplications than the n'-point cyclic convolution.

5.16 Using your knowledge of fast convolution algorithms, construct an algorithm for the three-point Fourier transform

$$V_k = \sum_{i=0}^{2} \omega^{ik} v_i, \quad k = 0, 1, 2$$

that uses only two real (nontrivial) multiplications.

5.17 Construct algorithms for the following:
  a $s(x) = g(x)d(x) \pmod{x^3 + x + 1}$;
  b $s(x) = g(x)d(x) \pmod{x^3}$;
  c $s(x) = g(x)d(x) \pmod{x^3 + x^2 + 1}$.

5.18 a How many real multiplications are required by the method of Section 5.7 to compute a complex cyclic convolution modulo $x^8 - 1$ in the complex field?
  b How many complex multiplications are required by the Winograd complexity theorem to compute the cyclic convolution modulo $x^8 - 1$ in the complex field?
  c Explain any apparent inconsistency.

5.19 a Prove that it requires six real multiplications to compute two complex products

$$e + jf = (a + jb)(c + jd),$$
$$e' + jf' = (a' + jb')(c' + jd')$$

with the understanding that all variables are independent real numbers; there are no unstated dependencies.

b How many multiplications are needed to simultaneously compute the following two complex products

$$e + jf = (a + jb)(c + jd),$$
$$e' + jf' = (a - jb)(c + jd),$$

where, in this case, variables are repeated?

c How many multiplications are needed to compute simultaneously the following two complex products

$$e + jf = (a + jb)(c + jd),$$
$$e' + jf' = (a - jb)(c - jd)?$$

5.20 a Prove that the computation

$$\begin{bmatrix} s_0 \\ s_1 \end{bmatrix} = \begin{bmatrix} g_0 & g_1 \\ g_1 & g_0 \end{bmatrix} \begin{bmatrix} d_0 \\ d_1 \end{bmatrix},$$

where all variables are complex, requires six real multiplications.

b Prove that the computation

$$\begin{bmatrix} s'_0 + js''_0 \\ s'_1 + js''_1 \end{bmatrix} = \begin{bmatrix} g'_0 & jg''_1 \\ -jg''_1 & g'_0 \end{bmatrix} \begin{bmatrix} d'_0 + jd''_0 \\ d'_1 + jd''_1 \end{bmatrix},$$

where all indicated variables are real, also requires six real multiplications. Hence setting g''_0 and g'_1 equal to zero does not reduce the number of required multiplications.

c Prove that the computation

$$\begin{bmatrix} s_0 \\ s_1 \end{bmatrix} = \begin{bmatrix} g_0 & g_1 \\ -g_1 & g_0 \end{bmatrix} \begin{bmatrix} d_0 \\ d_1 \end{bmatrix},$$

where all variables are complex, requires six real multiplications.

d Prove that the computation

$$\begin{bmatrix} s'_0 + js''_0 \\ s'_1 + js''_1 \end{bmatrix} = \begin{bmatrix} g'_0 & jg''_1 \\ jg''_1 & g'_0 \end{bmatrix} \begin{bmatrix} d'_0 + jd''_0 \\ d'_1 + jd''_1 \end{bmatrix},$$

where all indicated variables are real, requires four real multiplications. In this case, setting g''_0 and g'_1 equal to zero reduces the number of required multiplications.

5.21 Derive an algorithm for a four-point cyclic convolution over the complex field. Contrast this algorithm with the use of a fast Fourier transform and the convolution theorem.

Notes for Chapter 5

A few fast convolution algorithms of small blocklength were first constructed by Agarwal and Cooley (1977) to go with their multidimensional nesting technique, but without any general theory. Winograd (1978) gave the general method of construction that we have described. His 1980 monographs (Winograd, 1980a, b) gives examples of his methods. Winograd (1977) also proved his important theorems about the optimality of these convolution algorithms in the real field and the complex field.

6 Architecture of filters and transforms

Now that we have a large collection of algorithms for convolutions and for the discrete Fourier transform, it is time to turn to how these algorithms are used in applications of signal processing. Our major purpose in this chapter is to discuss the role of algorithms in constructing digital filters. We shall also study other tasks such as interpolation and decimation. By using the methods of nesting and concatenation, we will build large signal-processing structures out of small pieces. The fast algorithms for short convolutions that were studied in Chapter 5 will be used to construct small filter segments.

The most important device in signal processing is the finite-impulse-response filter. An incoming stream of discrete data samples enters the filter, and a stream of discrete samples leaves. The streams of samples at the input and output are very long; in some instances millions of samples per second pass through the filter. Fast algorithms for filter sections always break the input stream into batches of perhaps a few hundred samples. One batch at a time is processed. The input samples are clocked into an input buffer, then processed one block at a time after that input block has been completed. The resulting block is placed into an output buffer, and the samples are clocked out of the output buffer at the desired rate.

6.1 Convolution by sections

Many algorithms for the discrete Fourier transform were studied in Chapter 3. Some of these algorithms were derived directly, and some were constructed on top of the fast convolution algorithms that will be developed in this chapter. The fast Fourier transform algorithms can be used, in turn, to do cyclic convolutions efficiently, especially when the blocklength is long. This is because the convolution theorem tells us that the cyclic convolution

$$s_i = \sum_{k=0}^{n-1} g_{((i-k))} d_k, \qquad i = 0, \ldots, n-1$$

in the time domain becomes

$$S_k = G_k D_k, \qquad k = 0, \ldots, n-1$$

in the frequency domain. Consequently, one way to compute the cyclic convolution is by using a Fourier transform, followed by a componentwise product in the transform domain, and then an inverse Fourier transform.

A Fourier transform can also be used to do a linear convolution by choosing the parameters of the computation so that the linear convolution is the same as a cyclic convolution. Choose a convenient n at least as long as the blocklength of s, and for which the Fourier transform is easy to compute. Then lengthen the vectors g and d by appending zeros such that those components with index $i = 0, \ldots, n-1$ retain their assigned values, and s is still related to g and d by a cyclic convolution. However, now the length is n'.

This requires that the length of the linear convolution be short enough to be accommodated by the blocklength of a practical Fourier transform. However, most filtering applications involve a linear convolution that is much too long for this, usually of unlimited length. Then other methods must be used.

Many techniques are available for reducing a long linear convolution to a sequence of short cyclic convolutions. It is commonly necessary in applications to pass a sequence that is so long as to appear infinite through an FIR filter with a fixed number of taps. The filtering operation must be computed piecemeal on the fly. It is not feasible to collect all of the input data before the computation of the output is begun because of the delay, because there would be too much data to buffer, and because good algorithms do not use all of the input data at once. Therefore the input data is segmented into blocks. As each block of data becomes available, processing on it begins. When processing of a block is complete, that output block is concatenated with previous output blocks to form the filter output.

First, we shall describe a technique known as the *overlap–save method*. We regard the data sequence d as represented by a polynomial $d(x)$ of arbitrarily large degree, with data sample d_i as the ith coefficient of the polynomial $d(x)$. Suppose we have a device for doing cyclic convolutions of blocklength n, and we want to multiply $g(x)$, a polynomial whose degree A is smaller than n, and $d(x)$, a polynomial whose degree is larger than n. Usually, the degree of $d(x)$ is very large, and we will treat it as unbounded. From $d(x)$, form a set of polynomials $\{d^{(0)}(x), d^{(1)}(x), \ldots\}$, with each polynomial of the set having degree $n-1$ or less. The definition of the (overlapping) coefficients is as follows:

$$d_i^{(0)} = d_i, \qquad i = 0, \ldots, n-1,$$
$$d_i^{(1)} = d_{i+(n-A)}, \qquad i = 0, \ldots, n-1,$$
$$d_i^{(2)} = d_{i+2(n-A)}, \qquad i = 0, \ldots, n-1,$$
$$d_i^{(3)} = d_{i+3(n-A)}, \qquad i = 0, \ldots, n-1,$$
$$\vdots \qquad \vdots \qquad \vdots$$

where enough such polynomials are formed so that all coefficients of $d(x)$ are assigned. This may require that an indefinitely long sequence of such polynomials will be formed, and we regard the sequence as unending. Notice that the coefficients of the polynomials $d^{(\ell)}(x)$ are overlapped by A coefficients. For each ℓ, let

$$s^{(\ell)}(x) = g(x)d^{(\ell)}(x) \pmod{x^n - 1}.$$

But, because

$$s(x) = g(x)d(x),$$

the desired coefficients of $s(x)$, except for the first A coefficients, can be found among the coefficients of the $s^{(\ell)}(x)$. Thus

$$\begin{aligned} s_i &= s_i^{(0)}, & i &= A, \ldots, n-1, \\ s_{i+(n-A)} &= s_i^{(1)}, & i &= A, \ldots, n-1, \\ s_{i+2(n-A)} &= s_i^{(2)}, & i &= A, \ldots, n-1, \end{aligned}$$

and so on. In this way, an unending stream of filter output data is computed. The A low-order coefficients of $s(x)$ are not computed. In many applications of linear convolution, lost data at the start-up of the filter does not matter. However, if the first A output coefficients are needed, simply replace $d(x)$ by the polynomial $x^A d(x)$ in the above discussion. All output coefficients will then be obtained, but with index offset by A.

Each cyclic convolution, except the last one, produces $n - A$ coefficients of the linear convolution and A unwanted coefficients, which are discarded. Because the last input segment $d^{(r)}(x)$ can have degree smaller than $n - 1$, the last convolution may produce more than $n - A$ meaningful coefficients.

The overlap–save method does not actually require that a complete cyclic convolution be computed. Only $n - A$ of the output coefficients need to be computed, not all of them. Consequently, it may be possible to develop an algorithm that is somewhat simpler than a full cyclic convolution algorithm by suppressing the unneeded coefficients. Such an algorithm, whether computed by using a cyclic convolution or computed directly, is called a *filter section*. We shall study algorithms for filter sections in the next section.

A finite-impulse-response filter constructed by using the overlap–save technique is shown in Figure 6.1. Functionally, the FIR filter looks like a tapped delay line, but the actual construction of the filter can look very different. As datawords arrive, they are entered into a cyclic buffer memory that typically is about twice as large as the filter section. As the filter section processes the data in one part of the input buffer, data for the next filter section is filling the next block of memory words in the input buffer and will be ready in time for the next computation. Each new section will be addressed in a manner overlapped with the section that precedes it. The filter section must process a block of length n in the time it takes to collect $n - A$ new data samples. At the same time, data from a preceding block is shifted out of the output buffer to the user.

The input buffer is cyclic, as is the output buffer. This means that, after its last memory location is filled, subsequent inputs are written by starting from the first memory location. In this way, stale data is discarded by being overwritten. It is usually

6.1 Convolution by sections

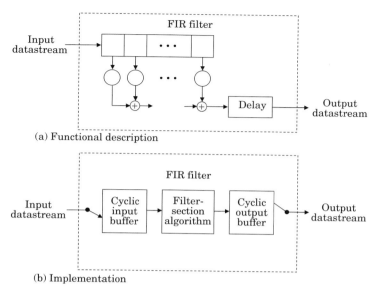

Figure 6.1 Construction of an FIR filter

convenient to make the length of the input buffer a power of two. The number in a binary register designates the address where the next input is to be stored, and this register is incremented after each input. The cyclic property of the buffer is implemented simply by allowing the address register to overflow after it reaches the all-ones value.

The filter section finds its block of data at memory locations with addresses $((b)), \ldots, ((b + n - 1))$, where b is a starting address and the double parentheses denote modulo the buffer memory length. After each section is finished, b is replaced by $((b + A))$, and the next section begins.

The block output of the filter section is placed in an output buffer. This buffer is managed by the same techniques used for the input buffer. Because it takes time to fill and empty buffers as well as to do the computation, the overlap–save method always entails some delay. This is shown in the functional description in Figure 6.1 so as to make the two representations equivalent.

The overlap–save method overlaps the blocks of input data and discards a portion of each output block. It uses a cyclic convolution algorithm with the same blocklength as the data blocks. Another method, known as the *overlap–add method*, does not overlap the input data but uses a linear convolution algorithm with a blocklength larger than the data blocks. It uses all components in the output block, using a small amount of computation to add together the overlapped segment of the output blocks. The two overlap methods are illustrated in Figure 6.2 and Figure 6.3.

To explain the overlap–add method, suppose again that we have a device for doing linear convolutions of output length n. We want to multiply $g(x)$, a polynomial whose degree A is smaller than n, and $d(x)$, a polynomial whose degree is larger than n. The

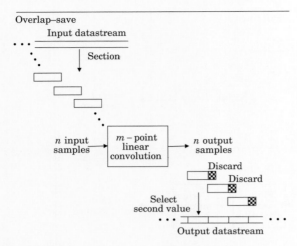

Figure 6.2 Convolution by overlap–save sections

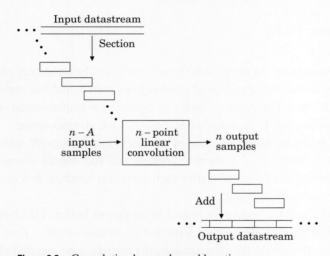

Figure 6.3 Convolution by overlap–add sections

degree of $d(x)$ can be very large, and we shall treat it as indefinitely large. From $d(x)$, form a sequence of polynomials of degree at most $n - A - 1$, defined as follows:

$$d_i^{(1)} = d_i, \qquad i = 0, \ldots, n - A - 1,$$
$$d_i^{(2)} = d_{i+(n-A)}, \qquad i = 0, \ldots, n - A - 1,$$
$$d_i^{(3)} = d_{i+2(n-A)}, \qquad i = 0, \ldots, n - A - 1.$$

Notice that these new polynomials are not overlapped, but each length is smaller than n. Then $d(x)$ can be expressed as

$$d(x) = \sum_{\ell=1}^{\infty} d^{(\ell)}(x) x^{(\ell-1)(n-A)}$$

and

$$s(x) = g(x)d(x) = \sum_{\ell=1}^{\infty} g(x)d^{(\ell)}(x)x^{(\ell-1)(n-A)}.$$

For each ℓ, let

$$s^{(\ell)}(x) = g(x)d^{(\ell)}(x).$$

The coefficients of $s(x)$ are given by

$$\begin{aligned} s_i &= s_i^{(1)}, & i &= n - A - 1, \\ s_{i+(n-A)} &= s_i^{(2)} + s_{i+(n-A)}^{(1)}, & i &= n - A - 1, \\ s_{i+2(n-A)} &= s_i^{(3)} + s_{i+(n-A)}^{(2)}, & i &= n - A - 1, \end{aligned}$$

and so on. This completes the computations of the overlap–add method.

The overlap–add method was described using a linear convolution. However, because $\deg g(x) = A$ and $\deg d^{(\ell)}(x) = n - A - 1$, a cyclic convolution can be used just as well. A cyclic convolution of blocklength n then will actually produce a linear convolution; all output points are correct values for the linear convolution.

Notice that both the overlap–save method and the overlap–add method produce $n - A$ output samples for each n-point convolution computed. The overlap–save method has a slight advantage in that there are no additions after the cyclic convolutions are computed, and so it is usually preferred. On the other hand, the overlap–add method can use a linear convolution with only $n - A$ nonzero values at the input. The linear convolution algorithm can be designed with fewer input additions.

For example, in Section 11.4, we will study ways to iterate algorithms for small linear convolutions into algorithms for large linear convolutions. In particular, a two-point by two-point linear convolution algorithm can be iterated to give a 256-point by 256-point linear convolution algorithm. This can be used with the overlap–add method. The parameters are: n equals 511 output samples; A equals 255, corresponding to a 256-tap filter; and $n - A$ equals 256 input samples. After the overlap and add, 256 output samples are produced for every convolution. To produce each batch of 256 output samples, this construction will use 6561 multiplications and 19 171 additions, including the overlap additions. Thus the 256-tap FIR filter is implemented by using 25.6 multiplications and 74.9 additions per output sample.

6.2 Algorithms for short filter sections

Discrete FIR filters can be constructed by using the overlap methods described in the previous section. To apply these methods, we need good algorithms for the individual filter sections. In this section, we will develop good algorithms for short filter sections directly. In Section 6.3, we will describe how to combine them to obtain longer filter sections.

The overlap–save method can be described as a matrix–vector operation. In this section, we write the matrix equation explicitly, omitting from the computation those output samples of a convolution that would be discarded were they computed. Thus

$$\begin{bmatrix} s_{r-1} \\ s_r \\ s_{r+1} \\ \vdots \\ s_{n+r-2} \end{bmatrix} = \begin{bmatrix} d_{r-1} & \cdots & d_2 & d_1 & d_0 \\ d_r & \cdots & d_3 & d_2 & d_1 \\ \vdots & & & & \vdots \\ d_{n+r-2} & & & \cdots & d_{r-1} \end{bmatrix} \begin{bmatrix} g_0 \\ g_1 \\ \vdots \\ g_{r-1} \end{bmatrix}$$

$$= \begin{bmatrix} g_{r-1} & g_{r-2} & g_{r-3} & \cdots & 0 & 0 \\ 0 & g_{r-1} & g_{r-2} & \cdots & 0 & 0 \\ 0 & 0 & g_{r-1} & \cdots & 0 & 0 \\ \vdots & & & & & \vdots \\ 0 & 0 & 0 & \cdots & & g_{r-1} \end{bmatrix} \begin{bmatrix} d_0 \\ d_1 \\ d_2 \\ \vdots \\ d_{n+r-2} \end{bmatrix}.$$

This equation computes a batch of n outputs of an r-tap FIR filter. An unending sequence of outputs can be computed, one batch at a time, by appropriately advancing the indices after each batch is computed. We call the batch computation an (r, n)-filter section or, when r equals n, an n-tap filter section. Such a problem is a truncated convolution. As we shall prove later, an (r, n)-filter section requires $r + n - 1$ multiplications, and an algorithm that uses $r + n - 1$ multiplications may be regarded as an optimum algorithm. This could be a misleading description, however, because such an algorithm will have too many additions unless r and n are small. These algorithms are valued primarily when r and n are small.

Algorithms for filter sections can be obtained from algorithms for linear convolutions. The following theorem gives a general principle for turning an algorithm for one computation into an algorithm for the other.

Theorem 6.2.1 (Transformation principle) *Given the algorithm*

$$s = Td$$
$$= CGAd,$$

where G is a diagonal matrix and A and C are matrices of small constants, then

$$e = A^T G C^T f$$

is an algorithm for computing $e = T^T f$. If the first algorithm has the minimum number of multiplications for computing s, then the second algorithm has the minimum number of multiplications for computing e.

6.2 Algorithms for short filter sections

Proof The algorithm for computing s is the matrix factorization

$$T = CGA.$$

Therefore, because G is a diagonal matrix, the transpose is

$$T^\mathrm{T} = A^\mathrm{T} G C^\mathrm{T}$$

so that

$$T^\mathrm{T} f = A^\mathrm{T} G C^\mathrm{T} f.$$

The second claim of the theorem is easy to see, because if an algorithm with fewer multiplications exists for computing e, then the first part of the theorem can be used to obtain an algorithm for computing s with that lesser number of multiplications. □

The transformation principle can be used to construct a short filter section. We start with the linear convolution

$$\begin{bmatrix} s_0 \\ s_1 \\ s_2 \end{bmatrix} = \begin{bmatrix} g_0 & 0 \\ g_1 & g_0 \\ 0 & g_1 \end{bmatrix} \begin{bmatrix} d_0 \\ d_1 \end{bmatrix}$$

$$= \begin{bmatrix} 1 & 0 & 0 \\ -1 & 1 & -1 \\ 0 & 0 & 1 \end{bmatrix} \begin{bmatrix} g_0 & & \\ & g_0 + g_1 & \\ & & g_1 \end{bmatrix} \begin{bmatrix} 1 & 0 \\ 1 & 1 \\ 0 & 1 \end{bmatrix} \begin{bmatrix} d_0 \\ d_1 \end{bmatrix}.$$

Then, by Theorem 6.2.1, we can write

$$\begin{bmatrix} e_0 \\ e_1 \end{bmatrix} = \begin{bmatrix} g_0 & g_1 & 0 \\ 0 & g_0 & g_1 \end{bmatrix} \begin{bmatrix} f_0 \\ f_1 \\ f_2 \end{bmatrix}$$

$$= \begin{bmatrix} 1 & 1 & 0 \\ 0 & 1 & 1 \end{bmatrix} \begin{bmatrix} g_0 & & \\ & g_0 + g_1 & \\ & & g_1 \end{bmatrix} \begin{bmatrix} 1 & -1 & 0 \\ 0 & 1 & 0 \\ 0 & -1 & 1 \end{bmatrix} \begin{bmatrix} f_0 \\ f_1 \\ f_2 \end{bmatrix}.$$

Next, replace (e_0, e_1) by (s_2, s_1) and replace (f_0, f_1, f_2) by (d_2, d_1, d_0) to give the algorithm for a two-tap filter section:

$$\begin{bmatrix} s_2 \\ s_1 \end{bmatrix} = \begin{bmatrix} g_0 & g_1 & 0 \\ 0 & g_0 & g_1 \end{bmatrix} \begin{bmatrix} d_2 \\ d_1 \\ d_0 \end{bmatrix}$$

$$= \begin{bmatrix} 1 & 1 & 0 \\ 0 & 1 & 1 \end{bmatrix} \begin{bmatrix} g_0 & & \\ & g_0 + g_1 & \\ & & g_1 \end{bmatrix} \begin{bmatrix} 1 & -1 & 0 \\ 0 & 1 & 0 \\ 0 & -1 & 1 \end{bmatrix} \begin{bmatrix} d_2 \\ d_1 \\ d_0 \end{bmatrix}.$$

Table 6.1 *Performance of some algorithms for short filter sections*

Number of taps	Number of outputs	Number of multiplications	Number of additions
2	2	3	4
2	3	4	8
3	2	4	8
3	3	5	15
3	4	6	20
4	3	6	
4	4	7	
5	3	7	

This is an optimum algorithm for a two-tap filter section. It uses three multiplications and four additions; the addition involving the filter taps $g_0 + g_1$ is not counted.

The algorithm can be used to compute an unending stream of filter outputs, two outputs at a time, by writing

$$\begin{bmatrix} s_{r+2} \\ s_{r+1} \end{bmatrix} = \begin{bmatrix} g_0 & g_1 & 0 \\ 0 & g_0 & g_1 \end{bmatrix} \begin{bmatrix} d_{r+2} \\ d_{r+1} \\ d_r \end{bmatrix},$$

then applying the factorization of the algorithm.

Algorithms for other short filter sections can be derived in the same way. The performance of some of these filter sections is given in Table 6.1.

6.3 Iterated filter sections

By iterating an algorithm for a small filter section, one can obtain an algorithm for a larger filter section that will compute a long block of filter output samples. The performance of some iterated filter sections is tabulated in Table 6.2. An unending stream of output samples can be formed by concatenating any number of these filter sections by the overlap–save method, discussed in Section 6.1.

The small convolution algorithms we have studied do not make use of the fact that multiplication of numbers is commutative. Any algorithm for the linear convolution

$$s_i = \sum_{k=0}^{N-1} g_{i-k} d_k$$

that does not use division and does not use the commutative property of multiplication remains a valid algorithm even if the elements are from an arbitrary ring, possibly a

6.3 Iterated filter sections

Table 6.2 *Performance of some real FIR filter algorithms*

Number of taps	Number of outputs/section	Number of multiplications	Number of additions	Multiplications per output	Additions per output
2	2	3	4	1.5	2
3	3	5	15	1.67	5
9	9	25	120	2.78	13.33
16	16	81	260	5.06	16.25
27	27	125	735	4.63	27.22
32	32	243	844	7.59	26.37
81	81	625	4080	7.72	53.70

noncommutative ring. For example, we may want to compute a convolution of matrices

$$S_i = \sum_{k=0}^{N-1} G_{i-k} D_k,$$

where now D_i is the ith matrix in a list of N matrices, and G_i is the ith matrix in a list of L matrices. The convolution algorithms we have developed can be used here just as before. The additions then become additions of matrices and the multiplications become multiplications of matrices.

To illustrate, we will iterate the algorithm for a two-tap filter section to obtain algorithms for larger filter sections. The small algorithm, developed in Section 6.2, can be written in the form

$$\begin{bmatrix} s_1 \\ s_2 \end{bmatrix} = \begin{bmatrix} 0 & 1 & -1 \\ 1 & 1 & 0 \end{bmatrix} \begin{bmatrix} (d_2 - d_1) & & \\ & d_1 & \\ & & (d_1 - d_0) \end{bmatrix} \begin{bmatrix} 1 & 0 \\ 1 & 1 \\ 0 & 1 \end{bmatrix} \begin{bmatrix} g_0 \\ g_1 \end{bmatrix}.$$

The algorithm requires 1.5 multiplications and two additions per filter output sample. It can be used repeatedly to compute two outputs at a time, always at the cost of 1.5 multiplications and two additions per output sample. However, a two-tap filter is too small for most practical applications. Algorithms for larger filters that compute 2^m outputs from a 2^m-tap filter can be designed by iterating this two-tap algorithm.

To illustrate the idea, we derive an algorithm for a four-tap filter section by computing four outputs of a four-tap filter. This can be written in matrix form as

$$\begin{bmatrix} s_3 \\ s_4 \\ s_5 \\ s_6 \end{bmatrix} = \begin{bmatrix} d_3 & d_2 & d_1 & d_0 \\ d_4 & d_3 & d_2 & d_1 \\ d_5 & d_4 & d_3 & d_2 \\ d_6 & d_5 & d_4 & d_3 \end{bmatrix} \begin{bmatrix} g_0 \\ g_1 \\ g_2 \\ g_3 \end{bmatrix}.$$

If we partition this equation into two by two blocks, we obtain the block matrix equation

$$\begin{bmatrix} S_1 \\ S_2 \end{bmatrix} = \begin{bmatrix} D_1 & D_0 \\ D_2 & D_1 \end{bmatrix} \begin{bmatrix} G_0 \\ G_1 \end{bmatrix},$$

where

$$D_0 = \begin{bmatrix} d_1 & d_0 \\ d_2 & d_1 \end{bmatrix}, \qquad D_1 = \begin{bmatrix} d_3 & d_2 \\ d_4 & d_3 \end{bmatrix}, \qquad D_2 = \begin{bmatrix} d_5 & d_4 \\ d_6 & d_5 \end{bmatrix},$$

$$G_0 = \begin{bmatrix} g_0 \\ g_1 \end{bmatrix}, \qquad G_1 = \begin{bmatrix} g_2 \\ g_3 \end{bmatrix},$$

and

$$S_1 = \begin{bmatrix} s_3 \\ s_4 \end{bmatrix}, \qquad S_2 = \begin{bmatrix} s_5 \\ s_6 \end{bmatrix}.$$

This problem now has exactly the same form as the earlier problem. Scalars have been replaced by matrices, and the algorithm is a valid algebraic identity. We have

$$\begin{bmatrix} S_1 \\ S_2 \end{bmatrix} = \begin{bmatrix} 0 & I & -I \\ I & I & 0 \end{bmatrix} \begin{bmatrix} (D_2 - D_1) & 0 & 0 \\ 0 & D_1 & 0 \\ 0 & 0 & (D_1 - D_0) \end{bmatrix} \begin{bmatrix} I & 0 \\ I & I \\ 0 & I \end{bmatrix} \begin{bmatrix} G_0 \\ G_1 \end{bmatrix}.$$

As before, the additions $G_0 + G_1$ can be precomputed and need not be counted. There is a total of four matrix additions and three matrix multiplications here. First we inspect the matrix additions. Two of them are

$$D_2 - D_1 = \begin{bmatrix} d_5 - d_3 & d_4 - d_2 \\ d_6 - d_4 & d_5 - d_3 \end{bmatrix},$$

$$D_1 - D_0 = \begin{bmatrix} d_3 - d_1 & d_2 - d_0 \\ d_4 - d_2 & d_3 - d_1 \end{bmatrix}.$$

There are five distinct additions of numbers. (Later, we shall see how one of the additions can be reused so that four additions will suffice here.) There are two more additions of two-point vectors after the multiplications are complete. This takes four more real additions.

Each of the three matrix multiplications is itself seen to have exactly the same form as a two-tap filter section. Each matrix multiplication can be computed with three field multiplications and four field additions. Hence there are nine multiplications and 21 additions in this computation of the four-tap filter section.

6.3 Iterated filter sections

If the algorithm is to be used repeatedly to compute four outputs at a time from a four-tap filter, then one of the additions drops away because $d_6 - d_4$ for one batch is the same as $d_2 - d_0$ in the next batch, so this term can be reused.

The iteration process is quite general. We can partition the matrix of the computation of n outputs of an n-tap filter into four $n/2$ by $n/2$ blocks whenever n is even. When we apply the two-tap algorithm, we see that we obtain an n-tap filter section by using an $n/2$-tap filter section three times. The number of multiplications is three times as many as needed by the $n/2$-tap filter section. The number of additional additions needed to compute $D_1 - D_0$ in this case is $n - 1$, and the number of additional additions then needed to compute $D_2 - D_1$ is $n/2$. This is because all the matrices are $n/2$ by $n/2$ Toeplitz matrices. To compute $D_1 - D_0$, only subtractions in the top row and left column need to be done. Some of these terms are repeated in $D_2 - D_1$, so to compute $D_2 - D_1$ only $n/2$ subtractions are needed. If the algorithm is used repeatedly, then, by reusing earlier terms, $D_1 - D_0$ and $D_2 - D_1$ can be computed by using only $n/2$ additions each. The output vector additions use n additional additions. To summarize, if an $n/2$-tap filter section can be done by using M multiplications and A additions, then, from the two-tap algorithm, we have an n-tap filter section using $2n + 3A$ additions and $3M$ multiplications.

If $n = 2^m$, then the procedure can be iterated, building a 2^m-tap filter section out of several 2^{m-1}-tap filter sections, which, in turn, are built out of 2^{m-2}-tap filter sections. The number of multiplications is $3^m = n^{1.585}$. The number of additions satisfies the recursion $A(n) = 2n + 3A(n/2)$ with $A(1) = 0$.

If n is not a power of two, then filter sections of other lengths can be designed to be used as building blocks. Another approach is to pad the filter tap vector with zeros, adding a few more taps equal to zero to bring the number up to a power of two.

For large enough n, the iterated algorithms with $n^{1.585}$ multiplications will become inferior to an FIR filter based on an FFT with $O(n \log n)$ multiplications. The value of n at which this happens can be postponed if the radix-two iterated filter-section algorithm is replaced with a radix-four filter-section algorithm with seven multiplications. Then the number of multiplications will grow as $7^{m/2} = n^{1.404}$.

To further illustrate the richness of the designer's options, we shall construct a sequence of algorithms to compute 16 outputs from a 16-tap filter. The structure of the algorithms is illustrated in Figure 6.4.

Algorithm 1 The outputs of the filter are computed in the obvious way. This algorithm uses $16 \times 16 = 256$ multiplications and $16 \times 15 = 240$ additions.

Algorithm 2 Use the two-tap filter algorithm to obtain an algorithm that uses any eight-tap filter section three times and needs $4 \times 8 = 32$ additional additions. If we compute the eight-tap filter section in the obvious way, we need $8 \times 8 = 64$ multiplications and

	Number of multiplications	Number of additions
Algorithm 1 Execute 16-point filter directly	256	240
Algorithm 2 Call three 8-point filters / 8-point Execute 8-point filter directly	192	200
Algorithm 3 Call three 8-point filters / 8-point Call three 4-point filters / 4-point Execute 4-point filter directly	144	188
Algorithm 4 Call three 8-point filters / 8-point Call three 4-point filters / 4-point Call three 2-point filters / 2-point Execute 2-point filter directly	108	206
Algorithm 5 Call three 8-point filters / 8-point Call three 4-point filters / 4-point Call three 2-point filters / 2-point Use fast 2-point filter algorithm	81	260

Figure 6.4 Some algorithms for 16-point FIR filters

$8 \times 7 = 56$ additions. Altogether, the algorithm uses $3 \times 64 = 192$ multiplications and $32 + 3 \times 56 = 200$ additions.

Algorithm 3 Use the two-tap algorithm to obtain a new algorithm for an eight-tap filter section. This algorithm uses $4 \times 4 = 16$ additions and uses a four-tap filter section three times. Altogether, the algorithm for a 16-tap filter section uses $32 + 3 \times 16 = 80$ additions and nine times the computations of the four-tap filter section. The obvious way of computing a four-tap filter section uses 16 multiplications and 12 additions, so the algorithm uses $80 + 9 \times 12 = 188$ additions and $9 \times 16 = 144$ multiplications.

Algorithm 4 This is a modification of Algorithm 3, obtained by using an algorithm for the four-tap filter section that uses $4 \times 2 = 8$ additions and three times an algorithm for a two-tap filter section. Thus this algorithm for the 16-tap filter section uses $80 + 9 \times 8 = 152$ additions and uses 27 times the computation of the two-tap filter section.

Computing the two-tap filter section in an obvious way uses two additions and four multiplications, so Algorithm 4 uses $152 + 27 \times 2 = 206$ additions and $27 \times 4 = 108$ multiplications.

Algorithm 5 This algorithm is obtained by iterating the two-tap algorithm all the way. Algorithm 5 uses 260 additions and 81 multiplications.

We do not say which of these algorithms is preferable. Only detailed examination of the algorithm in the context of an application and implementation can decide whether, for example, the algorithm with 206 additions and 108 multiplications is preferable to the one with 260 additions and 81 multiplications. All that we intend to do is provide the designer with the various alternatives.

6.4 Symmetric and skew-symmetric filters

The techniques of the previous section can be enhanced when the coefficients of the filter $g(x)$ have special properties. Symmetries in the sequence of coefficients can be exploited in the construction of the algorithms for the small filter sections. This kind of saving is not readily available if Fourier transform techniques are used.

We shall design algorithms for small symmetric and skew-symmetric FIR filters. To obtain longer filters, one will need to expand these small filters in some way. Unfortunately, iteration cannot be used to enlarge a symmetric filter because the filter that results is no longer symmetric. There is another method, however. We shall see that a symmetric polynomial modulo a symmetric polynomial is a polynomial that itself is closely related to a symmetric polynomial. We will pass the savings due to symmetry up through the levels of the Chinese remainder theorem, building algorithms for large symmetric filters out of pieces that are small symmetric filters.

The L-tap filter described by the polynomial $g(x)$ is called a *symmetric filter* if $g(x)$ equals its reciprocal polynomial, that is, if $g(x) = x^{L-1}g(x^{-1})$ or, equivalently, if $g_i = g_{L-1-i}$. The L-tap filter, described by polynomial $g(x)$, is called a *skew-symmetric filter* if $g(x)$ equals the negative of its reciprocal polynomial, that is, if $g(x) = -x^{L-1}g(x^{-1})$ or, equivalently, if $g_i = -g_{L-1-i}$. We will begin with a discussion of symmetric filters. There is an obvious algorithm to produce an L-tap symmetric filter that uses, for each output sample, $L - 1$ additions but only $L/2$ multiplications if n is even, or $(L + 1)/2$ multiplications if n is odd. The obvious algorithm adds together the data point that is to be multiplied by g_i and the data point that is to be multiplied by g_{L-1-i}, before multiplying by the common coefficient g_i. However, one can do much better.

The Winograd convolution algorithm for a symmetric filter is constructed much the same as before. We will construct an algorithm for passing four data points through a three-tap symmetric filter. Let

$$g(x) = g_0 x^2 + g_1 x + g_0,$$
$$d(x) = d_3 x^3 + d_2 x^2 + d_1 x + d_0,$$

and write the polynomial product $g(x)d(x)$ as

$$s(x) = g(x)d(x) \pmod{(x^5 - x)(x - \infty)}.$$

Because $\deg s(x)$ is equal to five, the modulo reduction has no effect on $s(x)$. The modulus polynomial $m(x)$ factors as $m(x) = x(x - 1)(x + 1)(x^2 + 1)(x - \infty)$. It has four prime factors of degree one, and each of these will lead to one multiplication in the final algorithm. There is also the prime factor $x^2 + 1$, which we may expect would lead to three multiplications. However, we shall see that, because of symmetry, it leads to only two multiplications.

Let

$$s^{(0)}(x) = g^{(0)}(x)d^{(0)}(x) \pmod{x^2 + 1},$$

where

$$d^{(0)}(x) = (d_1 - d_3)x + (d_0 - d_2)$$

and

$$g^{(0)}(x) = g_1 x + (g_0 - g_0) = g_1 x.$$

Hence $s^{(0)}(x)$ can be computed with only two multiplications. The rest of the algorithm is developed just as before. The final algorithm is

$$\begin{bmatrix} s_0 \\ s_1 \\ s_2 \\ s_3 \\ s_4 \\ s_5 \end{bmatrix} = \begin{bmatrix} 1 & 0 & 0 & 0 & 0 & 0 \\ 0 & 1 & -1 & 1 & 0 & -1 \\ 0 & 1 & 1 & 0 & -1 & 0 \\ 0 & 1 & -1 & -1 & 0 & 0 \\ -1 & 1 & 1 & 0 & 1 & 0 \\ 0 & 0 & 0 & 0 & 0 & 1 \end{bmatrix} \begin{bmatrix} G_0 \\ G_1 \\ G_2 \\ G_3 \\ G_4 \\ G_5 \end{bmatrix} \begin{bmatrix} 1 & 0 & 0 & 0 \\ 1 & 1 & 1 & 1 \\ 1 & -1 & 1 & -1 \\ 1 & 0 & -1 & 0 \\ 0 & -1 & 0 & 1 \\ 0 & 0 & 0 & 1 \end{bmatrix} \begin{bmatrix} d_0 \\ d_1 \\ d_2 \\ d_3 \end{bmatrix},$$

where

$$\begin{bmatrix} G_0 \\ G_1 \\ G_2 \\ G_3 \\ G_4 \\ G_5 \end{bmatrix} = \begin{bmatrix} 1 & 0 \\ \frac{1}{2} & \frac{1}{4} \\ \frac{1}{2} & -\frac{1}{4} \\ 0 & \frac{1}{2} \\ 0 & \frac{1}{2} \\ 1 & 0 \end{bmatrix} \begin{bmatrix} g_0 \\ g_1 \end{bmatrix}.$$

6.4 Symmetric and skew-symmetric filters

This algorithm has six multiplications and 14 additions. It can be used for an overlap–add implementation of a three-tap symmetric FIR filter with 1.5 multiplications per output sample and 4.5 additions per output sample.

The transposition principle of Theorem 6.2.1 can be used to turn this into a (4, 3) symmetric filter section with 1.5 multiplications per output sample and four additions per output sample. This algorithm is given by

$$\begin{bmatrix} s_5 \\ s_4 \\ s_3 \\ s_2 \end{bmatrix} = \begin{bmatrix} d_5 & d_4 & d_3 \\ d_4 & d_3 & d_2 \\ d_3 & d_2 & d_1 \\ d_2 & d_1 & d_0 \end{bmatrix} \begin{bmatrix} g_0 \\ g_1 \\ g_0 \end{bmatrix}$$

$$= \begin{bmatrix} 1 & 1 & 1 & 1 & 0 & 0 \\ 0 & 1 & -1 & 0 & -1 & 0 \\ 0 & 1 & 1 & -1 & 0 & 0 \\ 0 & 1 & -1 & 0 & 1 & 1 \end{bmatrix} \begin{bmatrix} G_0 \\ G_1 \\ G_2 \\ G_3 \\ G_4 \\ G_5 \end{bmatrix}$$

$$\times \begin{bmatrix} 1 & 0 & 0 & 0 & -1 & 0 \\ 0 & 1 & 1 & 1 & 1 & 0 \\ 0 & -1 & 1 & -1 & 1 & 0 \\ 0 & 1 & 0 & -1 & 0 & 0 \\ 0 & 0 & -1 & 0 & 1 & 0 \\ 0 & -1 & 0 & 0 & 0 & 1 \end{bmatrix} \begin{bmatrix} d_0 \\ d_1 \\ d_2 \\ d_3 \\ d_4 \\ d_5 \\ d_6 \end{bmatrix},$$

where, as before,

$$\begin{bmatrix} G_0 \\ G_1 \\ G_2 \\ G_3 \\ G_4 \\ G_5 \end{bmatrix} = \begin{bmatrix} 1 & 0 \\ \frac{1}{2} & \frac{1}{4} \\ \frac{1}{2} & -\frac{1}{4} \\ 0 & \frac{1}{2} \\ 0 & \frac{1}{2} \\ 1 & 0 \end{bmatrix} \begin{bmatrix} g_0 \\ g_1 \end{bmatrix}.$$

The reason that the algorithm for the symmetric filter turns out to be better than the algorithm for the nonsymmetric filter is because the symmetric polynomial $g(x)$ modulo $x^2 + 1$ has one coefficient equal to zero. This is a simple case of a phenomenon that always happens when a symmetric polynomial is reduced modulo a symmetric polynomial. Let

$$g(x) = g_0 x^4 + g_1 x^3 + g_2 x^2 + g_1 x + g_0$$

and consider the four remainders

$$g^{(1)}(x) = g(x) \pmod{x^2 + 1},$$
$$g^{(2)}(x) = g(x) \pmod{x^2 + x + 1},$$
$$g^{(3)}(x) = g(x) \pmod{x^2 - x + 1},$$
$$g^{(4)}(x) = g(x) \pmod{x^4 + 1}.$$

These are easily found to be

$$g^{(1)}(x) = 2g_0 - g_2,$$
$$g^{(2)}(x) = (g_0 + g_1 - g_2)x + (g_0 + g_1 - g_2)$$
$$= (x + 1)(g_0 + g_1 - g_2),$$
$$g^{(3)}(x) = (-g_0 + g_1 + g_2)x - (-g_0 + g_1 + g_2)$$
$$= (x - 1)(-g_0 + g_1 + g_2),$$
$$g^{(4)}(x) = g_1 x^3 + g_2 x^2 + g_1 x$$
$$= x(g_1 x^2 + g_2 x + g_1).$$

In each case, the number of independent coefficients is two less than the degree of $m^{(i)}(x)$. Because of the symmetry, one of the independent coefficients is missing. In general, an extra polynomial, free of indeterminates, is multiplying a symmetric polynomial whose degree equals $\deg m^{(i)}(x) - 2$. Because we are interested only in computations of the form

$$s^{(i)}(x) = g^{(i)}(x) d^{(i)}(x) \pmod{m^{(i)}(x)},$$

we can bury this extra polynomial factor by including it in a redefined $d^{(i)}(x)$. Let

$$d^{(1)}(x) = d(x) \pmod{x^2 + 1},$$
$$d^{(2)}(x) = (x + 1)d(x) \pmod{x^2 + x + 1},$$
$$d^{(3)}(x) = (x - 1)d(x) \pmod{x^2 - x + 1},$$
$$d^{(4)}(x) = xd(x) \pmod{x^4 + 1},$$

and redefine the $g^{(i)}(x)$ as follows:

$$g^{(1)}(x) = 2g_0 - g_2,$$
$$g^{(2)}(x) = g_0 + g_1 - g_2,$$
$$g^{(3)}(x) = -g_0 + g_1 + g_2,$$
$$g^{(4)}(x) = g_1 x^2 + g_2 x + g_1.$$

Then $s^{(1)}(x)$, $s^{(2)}(x)$, and $s^{(3)}(x)$ are computed with two multiplications each, while $s^{(4)}(x)$ is computed as a linear convolution followed by modulo $x^4 + 1$ reduction. The linear convolution consists of passing four data points through a three-tap symmetric filter. We have already given an algorithm for this problem that uses six multiplications.

6.4 Symmetric and skew-symmetric filters

With these pieces, we can form a linear convolution algorithm for a five-tap symmetric filter. Suppose we want to pass six points through the five-tap symmetric filter $g(x)$. Then choose

$$m(x) = x(x-1)(x+1)(x-\infty)(x^2+1)(x^2-x+1)(x^2+x+1).$$

Each of the first four polynomial factors will lead to one multiplication. As we have seen, each of the last three polynomial factors will lead to two multiplications. The final algorithm will have ten multiplications, which is 1.67 multiplications per filter output sample.

Suppose, instead, that we want to pass 14 points through the filter. Then append the new factor x^4+1 to the $m(x)$ already used. This will lead to six more multiplications. The final algorithm will have 16 multiplications, which is 1.14 multiplications per filter output sample.

In general, we shall want to compute symmetric filters with more than five taps. The following theorem says that we can continue to choose the modulus polynomials so as to break the large symmetric polynomials into pieces that are small symmetric polynomials.

Theorem 6.4.1 *When a symmetric polynomial $g(x)$ of even degree t is divided by a symmetric polynomial $m(x)$ of even degree n with t not less than n, the remainder polynomial $r(x)$ can be written as*

$$r(x) = f(x)r'(x) \pmod{m(x)},$$

where $r'(x)$ is a symmetric polynomial of degree $n-2$ and $f(x)$ is a polynomial that does not depend on the coefficients of $g(x)$.

Proof Without loss of generality, we can suppose that $m(x)$ is monic. Because $m(x)$ is a symmetric polynomial of even degree n,

$$m(x) = x^n m(x^{-1}).$$

Let $m'(x) = m(x)$ if $n = t$. Otherwise, let

$$m'(x) = m(x) + x^{t-n} m(x).$$

This polynomial is a symmetric polynomial of degree t because

$$\begin{aligned}x^t m'(x^{-1}) &= x^t[m(x^{-1}) + x^{-t+n} m(x^{-1})] \\ &= x^{t-n} m(x) + m(x) = m'(x).\end{aligned}$$

Let $g'(x)$ be defined by

$$xg'(x) = g(x) - g_0 m'(x).$$

The polynomial $xg'(x)$ has the same remainder under division by $m(x)$ as does $g(x)$. Moreover, the polynomial $g'(x)$ is symmetric and has an even degree that is less than that of $g(x)$ by at least two. If the degree of $g'(x)$ is not less than n, this process can be repeated. Continue until a polynomial $r'(x)$ of degree $n-2$ is reached. Then $x^{(t-n+2)/2}r'(x)$ has the same remainder as $g(x)$ under division by $m(x)$. To complete the proof, let $f(x) = R_{m(x)}[x^{(t-n+2)/2}]$, which does not depend on $g(x)$. □

Finally, we turn to algorithms for skew-symmetric filters. These we can deal with rather quickly by showing how to change an algorithm for a symmetric filter into an algorithm for a skew-symmetric filter.

Theorem 6.4.2 *If L is even, a skew-symmetric filter with L taps can be computed as a symmetric filter with $L-1$ taps. If L is odd, a skew-symmetric filter with L taps can be computed as a symmetric filter with $L-2$ taps.*

Proof Suppose L is even and $g(x)$ describes a skew-symmetric filter with L taps. Then $g(1)$ equals zero because $g_i = -g_{L-1-i}$ for $i = 0, \ldots, (L/2) - 1$. Therefore $g(x)$ has $x - 1$ as a factor. Define the filter

$$g'(x) = g(x)/(x-1).$$

Then $g'(x)$ is a filter with $L-1$ taps. It must be symmetric because

$$g(x) = -x^{L-1}g(x^{-1})$$

implies

$$(x-1)g'(x) = -x^{L-1}\left(x^{-1} - 1\right)g'\left(x^{-1}\right)$$

or

$$g'(x) = x^{L-2}g'\left(x^{-1}\right).$$

Hence the filter $g(x)$ is equivalent to the filter $g'(x)$, followed (or preceded) by the two-tap filter with taps described by the polynomial $(x-1)$.

Next, suppose that L is odd and $g(x)$ describes a skew-symmetric filter with L taps. Then both one and minus one are zeros of $g(x)$. This is because the center coefficient of $g(x)$ is zero, and we can write

$$g(\pm 1) = \left(\sum_{\text{even } i} g_i + g_{L-1-i}\right) \pm \left(\sum_{\text{odd } i} g_i + g_{L-1-i}\right).$$

Therefore we can define the filter

$$g'(x) = g(x)/(x^2 - 1),$$

where $g'(x)$ is a symmetric filter with $L - 2$ taps. The filter $g(x)$ is equivalent to the filter $g'(x)$, followed (or preceded) by the filter with taps described by the polynomial $x^2 - 1$. □

The filter described by $x - 1$ or by $x^2 - 1$ merely amounts to some input additions or output additions. Hence, by changing the matrix of preadditions or the matrix of postadditions, an algorithm for a symmetric filter becomes an algorithm for a skew-symmetric filter. As always, the algorithm can be left in the form of a linear convolution for use with the overlap–add technique. Alternatively, the transformation principle of Theorem 6.2.1 can be used to turn it into a filter section for use with the overlap–save technique.

Incidentally, the technique used to turn a symmetric filter into a skew-symmetric filter also can be used to turn an $(L + 1)$-point symmetric filter with L odd into an L-point symmetric filter.

Theorem 6.4.3 *An algorithm for an $(L + 1)$-point symmetric filter, with L odd, can be constructed from an algorithm for an L-point symmetric filter with no additional multiplications.*

Proof Let $g(x)$ be a symmetric polynomial of odd degree L describing the $(L + 1)$-point symmetric filter. Then $g(-1) = 0$, so $(x + 1)$ is a factor of $g(x)$. Then

$$g(x) = (x + 1)g'(x),$$

where $g'(x)$ is a symmetric polynomial of degree L. The original filter is the cascade of $g'(x)$ and $(x + 1)$, and the latter filter can be computed without multiplications. □

6.5 Decimating and interpolating filters

A *decimating FIR filter* is one that provides every rth output of the filter, but does not provide the other outputs. Often r is two or three. The intervening outputs of the filter are not wanted, and, were they provided, would only be discarded. One can hope to find a simpler algorithm that does not compute unwanted outputs. An *interpolating FIR filter* does the opposite. It is a filter that produces output samples at a higher rate than the rate at the input to the filter. Decimating and interpolating filters are also called *down-sampling filters* and *up-sampling filters*, respectively.

The design of a decimating or an interpolating FIR filter is similar to the design of a general FIR filter. Their special characteristics must be exploited, however, if a more efficient algorithm is to be obtained. In the case of a decimating filter, omitting the unwanted output samples can lead to a more efficient algorithm. In the case of an interpolating filter, taking note of the omitted input samples, likewise, can lead to a more efficient algorithm.

The transformation principle of Theorem 6.2.1 does not transform a linear convolution algorithm for a decimating filter into a filter-section algorithm for a decimating filter. Instead, the transformation principle transforms a linear convolution algorithm for a decimating filter into a filter-section algorithm for an interpolating filter.

We shall consider a linear convolution of five input points and a two-to-one decimating filter with five taps. The desired output points can be written

$$\begin{bmatrix} s_0 \\ s_2 \\ s_4 \\ s_6 \\ s_8 \end{bmatrix} = \begin{bmatrix} g_0 & 0 & 0 & 0 & 0 \\ g_2 & g_1 & g_0 & 0 & 0 \\ g_4 & g_3 & g_2 & g_1 & g_0 \\ 0 & 0 & g_4 & g_3 & g_2 \\ 0 & 0 & 0 & 0 & g_4 \end{bmatrix} \begin{bmatrix} d_0 \\ d_1 \\ d_2 \\ d_3 \\ d_4 \end{bmatrix}$$

$$= \begin{bmatrix} g_0 & 0 & 0 \\ g_2 & g_0 & 0 \\ g_4 & g_2 & g_0 \\ 0 & g_4 & g_2 \\ 0 & 0 & g_4 \end{bmatrix} \begin{bmatrix} d_0 \\ d_2 \\ d_4 \end{bmatrix} + \begin{bmatrix} 0 & 0 \\ g_1 & 0 \\ g_3 & g_1 \\ 0 & g_3 \\ 0 & 0 \end{bmatrix} \begin{bmatrix} d_1 \\ d_3 \end{bmatrix}.$$

This partition has broken the condition into the combination of a three by three linear convolution and a two by two linear convolution. It takes three extra additions to combine them, so the algorithm for the decimating filter will have eight multiplications and 26 additions for every five input points. If this algorithm is used with the overlap–add technique, there will be four more additions for every five input points to combine the overlap sections. Hence there will be eight multiplications and 30 additions for every five output points, which is 1.6 multiplications and six additions per output.

Next, we will construct a decimating filter section with five taps and three outputs for the same situation. This can be written

$$\begin{bmatrix} s_4 \\ s_6 \\ s_8 \end{bmatrix} = \begin{bmatrix} d_4 & d_3 & d_2 & d_1 & d_0 \\ d_6 & d_5 & d_4 & d_3 & d_2 \\ d_8 & d_7 & d_6 & d_5 & d_4 \end{bmatrix} \begin{bmatrix} g_0 \\ g_1 \\ g_2 \\ g_3 \\ g_4 \end{bmatrix}$$

$$= \begin{bmatrix} d_4 & d_2 & d_0 \\ d_6 & d_4 & d_2 \\ d_8 & d_6 & d_4 \end{bmatrix} \begin{bmatrix} g_0 \\ g_2 \\ g_4 \end{bmatrix} + \begin{bmatrix} d_3 & d_1 \\ d_5 & d_3 \\ d_7 & d_5 \end{bmatrix} \begin{bmatrix} g_1 \\ g_3 \end{bmatrix}.$$

The filter section has broken into the combination of a three-tap filter section and a (3, 2)-filter section. It takes three extra additions to combine them, so the algorithm for the decimating filter section will have nine multiplications and 26 additions.

6.5 Decimating and interpolating filters

We will transpose these two algorithms to see what we get. In both cases, we shall find that we have obtained algorithms for interpolating filters. The transposition of the linear convolution above is

$$\begin{bmatrix} e_0 \\ e_1 \\ e_2 \\ e_3 \\ e_4 \end{bmatrix} = \begin{bmatrix} g_0 & g_2 & g_4 & 0 & 0 \\ 0 & g_1 & g_3 & 0 & 0 \\ 0 & g_0 & g_2 & g_4 & 0 \\ 0 & 0 & g_1 & g_3 & 0 \\ 0 & 0 & g_0 & g_2 & g_4 \end{bmatrix} \begin{bmatrix} f_0 \\ f_1 \\ f_2 \\ f_3 \\ f_4 \end{bmatrix}.$$

Now replace $(e_0, e_1, e_2, e_3, e_4)$ by $(s_8, s_7, s_6, s_5, s_4)$, and replace $(f_0, f_1, f_2, f_3, f_4)$ by $(d_8, d_6, d_4, d_2, d_0)$. Then

$$\begin{bmatrix} s_8 \\ s_7 \\ s_6 \\ s_5 \\ s_4 \end{bmatrix} = \begin{bmatrix} g_0 & g_2 & g_4 & 0 & 0 \\ 0 & g_1 & g_3 & 0 & 0 \\ 0 & g_0 & g_2 & g_4 & 0 \\ 0 & 0 & g_1 & g_3 & 0 \\ 0 & 0 & g_0 & g_2 & g_4 \end{bmatrix} \begin{bmatrix} d_8 \\ d_6 \\ d_4 \\ d_2 \\ d_0 \end{bmatrix}$$

$$= \begin{bmatrix} d_8 & 0 & d_6 & 0 & d_4 \\ 0 & d_6 & 0 & d_4 & 0 \\ d_6 & 0 & d_4 & 0 & d_2 \\ 0 & d_4 & 0 & d_2 & 0 \\ d_4 & 0 & d_2 & 0 & d_0 \end{bmatrix} \begin{bmatrix} g_0 \\ g_1 \\ g_2 \\ g_3 \\ g_4 \end{bmatrix}.$$

This is the equation of a section of an interpolating filter; alternate input samples are zero. Hence the transposition principle gives an algorithm for computing five output samples of a five-tap one to two interpolating filter section from the algorithm for computing five output samples of a five-tap decimating linear convolution. The algorithm has eight multiplications, which is 1.6 multiplications per output sample.

In a similar way, the five-tap decimating filter section, written in the form

$$\begin{bmatrix} s_8 \\ s_6 \\ s_4 \end{bmatrix} = \begin{bmatrix} g_0 & g_1 & g_2 & g_3 & g_4 & 0 & 0 & 0 & 0 \\ 0 & 0 & g_0 & g_1 & g_2 & g_3 & g_4 & 0 & 0 \\ 0 & 0 & 0 & 0 & g_0 & g_1 & g_2 & g_3 & g_4 \end{bmatrix} \begin{bmatrix} d_8 \\ d_7 \\ d_6 \\ d_5 \\ d_4 \\ d_3 \\ d_2 \\ d_1 \\ d_0 \end{bmatrix},$$

is transposed into an algorithm for the computation

$$\begin{bmatrix} s_0 \\ s_1 \\ s_2 \\ s_3 \\ s_4 \\ s_5 \\ s_6 \\ s_7 \\ s_8 \end{bmatrix} = \begin{bmatrix} g_0 & 0 & 0 \\ g_1 & 0 & 0 \\ g_2 & g_0 & 0 \\ g_3 & g_1 & 0 \\ g_4 & g_2 & g_0 \\ 0 & g_3 & g_1 \\ 0 & g_4 & g_2 \\ 0 & 0 & g_3 \\ 0 & 0 & g_4 \end{bmatrix} \begin{bmatrix} d_0 \\ d_2 \\ d_4 \end{bmatrix}.$$

This is a linear convolution with alternate input samples equal to zero. It is an interpolating filter. The transposed algorithm uses nine multiplications.

Next, we consider symmetric decimating filters by way of examples. We describe an algorithm for a 15-tap symmetric filter with two to one decimation and 15 input points. As in the earlier example, this decomposes into an eight by eight symmetric linear convolution and a seven by seven symmetric linear convolution. By Theorem 6.4.3, the first of these can be obtained from an algorithm that passes eight samples through a seven-tap filter.

We can construct these two symmetric linear convolutions by using the methods of the previous section. The polynomial

$$m(x) = x(x-1)(x+1)(x-\infty)(x^2+1)(x^2+x+1)(x^2-x+1)(x^4+1)$$

has degree 14, and so it can be used for constructing an eight by seven linear convolution algorithm. In Section 6.4, we determined the number of multiplications due to each modulus polynomial. There will be a total of 16 multiplications. Similarly, we can drop one of the linear factors to construct a seven by seven linear convolution algorithm that uses 15 multiplications. Hence an algorithm for the original problem with 31 multiplications can be constructed.

6.6 Construction of transform computers

A good FFT algorithm will be successful only if its implementation takes advantage of its special structure. The design of a hardware module or of a software subroutine requires close consideration of the structure of the algorithm.

A radix-two or radix-four Cooley–Tukey FFT has a regular structure, and its implementation is relatively straightforward. The algorithm can be specified with few instructions. Therefore, if the length of the instruction memory in a software implementation is more important than the running time, such a Cooley–Tukey FFT may be a better

6.6 Construction of transform computers

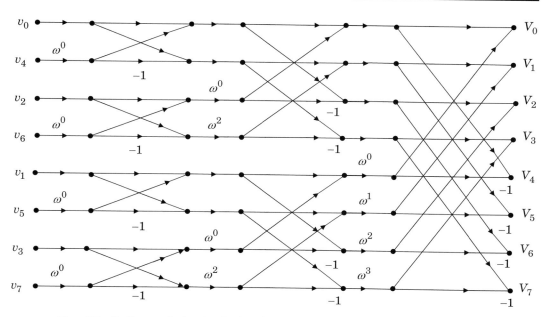

Figure 6.5 Radix-two decimation-in-time Cooley–Tukey FFT

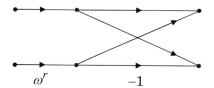

Figure 6.6 A two-point butterfly

design choice than a faster FFT algorithm. If power dissipation or speed is the more important consideration, the Winograd FFT or another fast algorithm may be preferred.

The same considerations may apply to a hardware implementation. The regularity of the architecture of a hardware transform computer may be more important than the total number of components. Then a Cooley–Tukey FFT might be preferred because of its regularity. The regularity of the Cooley–Tukey FFT is made evident by the flow diagram of the decimation-in-time algorithm, shown in Figure 6.5. During each iteration, a set of simple two-point discrete Fourier transforms is computed. The basic computational module consists of a two-point discrete Fourier transform and a phase adjustment, as described by the two-point decimation-in-time butterfly, shown in Figure 6.6. The decimation-in-frequency radix-two Cooley–Tukey FFT, shown in Figure 6.7, has a similar butterfly, this time the two-point decimation-in-frequency butterfly shown in Figure 6.8.

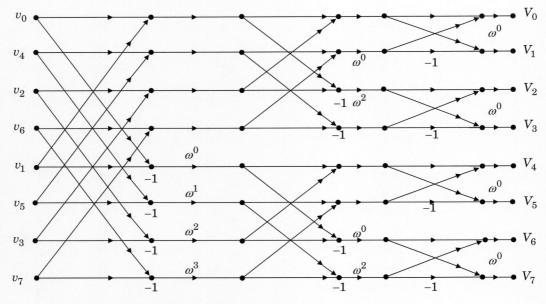

Figure 6.7 Radix-two decimation-in-frequency Cooley–Tukey FFT

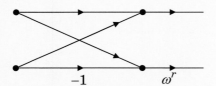

Figure 6.8 A two-point butterfly

On the other hand, in a large recurring problem, power dissipation or speed may be the more important consideration. Then the regularity of the algorithm may not be an important consideration.

Given a two-point Cooley–Tukey butterfly in the form of a software subroutine or a hardware module, a decimation-in-time radix-two FFT can be constructed simply by passing data to the two-point butterfly in the right sequence. As the data is moved through the two-point butterfly, the Fourier transform is developed, but the indices become shuffled. In Figure 6.5 the input data is given in shuffled order so that the data becomes unshuffled during the iterations, and is in the correct order at the output. If desired, the input data in Figure 6.5 could be listed in its natural order, but then lines would have complicated crossovers in the flow diagram.

In writing a computer program for the FFT, the most convenient way to write the program will present the output data in memory in shuffled order. The output data must be unshuffled before it is useful. Alternatively, the program can be written so that the

6.6 Construction of transform computers

unshuffling occurs a little at a time; after each iteration, data is returned to memory in an order such that the addressing of the next iteration can be the same as the past iteration. There are many such ways to handle the shuffling and unshuffling; the details are really part of the structure of a computer program or hardware design. One must decide to what extent data will be reindexed when stored, to what extent it will be reindexed when called, and how much working storage will be allowed beyond the memory locations that hold the input data.

The Winograd large FFT, as described in Section 12.4, is less regular, and so the implementation will not be as neat. If one is careful to study the structure of the algorithm when planning the software or the hardware, however, the implementation can still be quite orderly. We can illustrate some of the ideas by laying out a 63-point Winograd large FFT built out of a seven-point and a nine-point Winograd small FFT, as discussed in Chapter 12. Suppose that the matrix factorizations

$$W_7 = C_7 B_7 A_7$$

and

$$W_9 = C_9 B_9 A_9$$

represent the Winograd small FFT algorithms. The matrix B_7 is a nine by nine diagonal matrix, and the matrix B_9 is an 11 by 11 diagonal matrix.

The Winograd large FFT is then written

$$W_{63} = (C_7 \times C_9)(B_7 \times B_9)(A_7 \times A_9)$$
$$= (C_7 \times C_9) B_{63} (A_7 \times A_9).$$

We shall work with each of the two outside factors left in the form of a Kronecker product, rather than computing new matrices equal to the Kronecker products. The term $B_7 \times B_9$ is replaced by the single matrix B_{63} so that multiplications can be merged off-line.

To implement the FFT, we begin with the data appropriately shuffled into a two-dimensional seven by nine array. First, multiply each of the seven columns by A_9, which will expand each column to length 11. Then multiply each of 11 new rows by A_7, which will expand each row to length nine. Now the data is in the form of a nine by eleven two-dimensional array. Multiply this array, element-by-element, by the nine by eleven array of constants B_{63}, which could be stored in a read-only memory. (To be precise, B_{63} will be described as a 99 by 99 diagonal matrix, but at this point it is easier to think only of the 99 diagonal elements as arranged in a nine by eleven array.)

Next, collapse the array, first multiplying each of the nine columns by C_9, then multiplying each of nine rows by C_7. Hence, all components of the 63-point Fourier

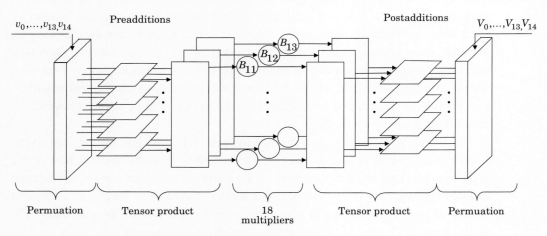

Figure 6.9 Conceptualizing the Winograd large FFT

transform are arranged in a seven by nine array. The 63 components must be unshuffled into a one-dimensional array.

Figure 6.9 illustrates this structure using a smaller Fourier transform, a 15-point Winograd FFT. It should be apparent that simple logic circuits can be used repetitively to implement the indicated functions. Similar structures can be used for other transform blocklengths. If the application requires that a continuing stream of n-point Fourier transforms be computed, then one might choose to build a pipelined circuit. Figure 6.10 shows a possible pipelined architecture for a high-performance 1008-point FFT. Such a structure might be contemplated for a hypothetical radar or sonar problem that needs to compute a never-ending stream of 1008-point Fourier transforms. Speeds approaching a million FFTs per second may be achieved very reasonably in this way. The input data is distributed across a bank of 16-point FFT modules and then is passed to a bank of 63-point FFT modules. The 16-point FFT modules are ready to start on the next batch while the 63-point FFT modules continue to work on the current batch. In this way, computations on two successive FFTs are going on simultaneously. More pipeline stages may be built into the timing to allow time for the transfers across the data bus. Then there could be four stages in the pipeline, two for computations and two for data transfers.

It may be convenient, for example, to choose nine 16-point FFT modules for the first bank and 18 63-point FFT modules for the second bank. Then each 16-point module must compute seven of the 63-point FFTs. Each 16-point module selects from the bus the seven 63-point vectors that it needs by simple techniques of memory management and memory cycle stealing. Similarly, each 16-point module passes data to the center bus according to the fixed schedule. Each 63-point module then selects its respective input data according to its input schedule.

221 6.7 Limited-range Fourier transforms

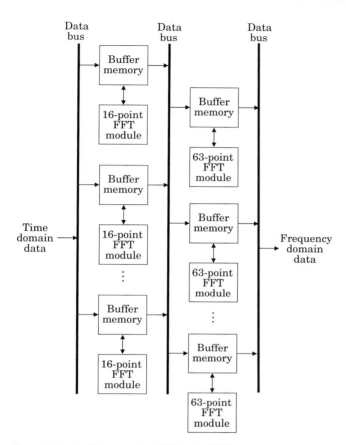

Figure 6.10 Architecture of a 1008-point FFT

6.7 Limited-range Fourier transforms

Suppose that, from n time samples, one wants to compute fewer than n frequency components. We may have an application in which 10,000 time samples are collected. This entitles us to compute 10,000 components of the Fourier spectrum, but we may not need all of them. Perhaps we need only 100 components of the spectrum. One way to get the desired components is to compute all components of the spectrum by using an FFT of blocklength 10,000, and then to discard those that are not needed. A better method is to use decimating filters and the limited-range Fourier transform.

The *limited-range Fourier transform* is a topic that more properly belongs to another branch of the subject of signal processing; it is not a fast algorithm in the sense of this book. We only want to mention it briefly as another kind of technique, one with a much different character than most of the algorithms that we have studied. For the most part, our algorithms have been mathematical identities. Questions of precision enter during

Architecture of filters and transforms

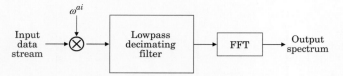

Figure 6.11 Computation of a limited-range Fourier transform

the implementation of the algorithms but not during the design of the algorithms. If the answer is not precise, this is not the fault of the theory of the algorithm. Rather, the lack of precision is because the additions and multiplications were not performed with infinite wordlength.

In the limited-range Fourier transform, a new phenomenon is encountered. It is not a mathematical identity, only an approximation, albeit an arbitrarily good approximation. Figure 6.11 illustrates the general form of a limited-range Fourier transform.

The central theorem for the study of decimation is the sampling theorem. The sampling rate of a discrete-time signal must be at least as high as the Nyquist rate. If one reduces the sampling rate to less than twice the frequency of the highest-frequency nonzero component, then these components will be "folded down" onto smaller frequencies, thereby creating false "images."

The limited-range Fourier transform first passes the input datastream through a lowpass decimating filter that preserves the spectral components of interest and does not pass those spectral components not of interest. Practical lowpass filters, designed by the methods of signal processing, are not perfect lowpass filters. They will change the desired frequency components slightly and will not reject the unwanted frequency components completely; hence the limited-range Fourier transform is not exact.

After decimation takes place, an FFT algorithm of smaller blocklength can be used to obtain the desired spectral components. If the band of desired spectral components does not begin at the zero frequency, then one simply inserts the modulation term ω^{ai}, as shown in Figure 6.11. This translates the spectrum to place the desired frequency components so that they start at the zero frequency.

6.8 Autocorrelation and crosscorrelation

The expression

$$r_i = \sum_{j=0}^{N-1} g_{i+j} d_j, \qquad i = 0, \ldots, L + N - 2,$$

which is called an autocorrelation when g and d are the same, and called a crosscorrelation when g and d are different, is closely related to the convolution. It can be made to look like a convolution simply by reading one of the sequences backwards. In

6.8 Autocorrelation and crosscorrelation

principle, any of the methods that we have studied for computing a linear convolution can be used to compute either correlation.

However, there is one important difference in practice. An FIR filter is usually short in comparison to the sequence of data it filters, but, in a correlation, the "filter" $g(x)$ is actually another data record whose length is about the same as the length of the data record represented by $d(x)$; often the length of both data records is indefinite and quite large – perhaps N equals many thousands. Good algorithms for correlation break both g and d into sections.

We begin with computation of the autocorrelation, given by

$$r_i = \frac{1}{N} \sum_{j=0}^{N-1} d_j d_{j+i}, \qquad i = 0, \ldots, (n/2) - 1,$$

or perhaps this same sum not divided by N. We are interested in problems in which the data blocklength N is much larger than the truncated blocklength $n/2$ of the autocorrelation, so it would be wasteful to choose a Fourier transform with blocklength on the order of N even if such a Fourier transform were practical. The sum can be broken into sections of length n in order to fit the problem to a smaller Fourier transform with blocklength n. Write

$$r_i = \sum_{\ell=0}^{L-1} r_i^{(\ell)},$$

where

$$r_i^{(\ell)} = \frac{1}{N} \sum_{j=0}^{(n/2)-1} d_{j+\ell(n/2)} d_{j+\ell(n/2)+i}, \qquad \begin{aligned} i &= 0, \ldots, (n/2) - 1, \\ \ell &= 0, \ldots, L - 1, \end{aligned}$$

and $L(n/2) = N$. Thus the task is to compute the vector $r^{(\ell)}$ for $\ell = 0, \ldots, L-1$. This we do by computing appropriate cyclic convolutions by using a fast Fourier transform.

We will define two kinds of section of length n. One section is full of data, and one is half-full and padded with zeros. Later, we will show how to avoid using the first of these. Let

$$d_i^{(\ell)} = \begin{cases} d_{i+\ell n/2}, & i = 0, \ldots, (n/2) - 1, \\ 0, & i = n/2, \ldots, n - 1, \end{cases}$$

and

$$g_i^{(\ell)} = d_{i+\ell n/2}, \qquad i = 0, \ldots, n - 1.$$

Then

$$r_i^{(\ell)} = \sum_{k=0}^{n-1} d_k^{(\ell)} g_{k+i}^{(\ell)}, \qquad i = 0, \ldots, (n/2) - 1.$$

Let $\boldsymbol{D}^{(\ell)}$ and $\boldsymbol{G}^{(\ell)}$ denote the Fourier transforms of $\boldsymbol{d}^{(\ell)}$ and $\boldsymbol{g}^{(\ell)}$. Then, as developed in Problem 1.10, the Fourier transform of the cyclic correlation

$$s_i^{(\ell)} = \sum_{j=0}^{n-1} g_{((i+j))} d_j, \qquad i = 0, \ldots, n-1$$

is

$$S_k^{(\ell)} = G_k^{(\ell)} D_k^{(\ell)*},$$

and half of the values of the cyclic convolution are the desired values

$$r_i^{(\ell)} = \frac{1}{N} s_i^{(\ell)}, \qquad i = 0, \ldots, (n/2) - 1.$$

Therefore we can first compute

$$S_k = \sum_{\ell=0}^{L-1} G_k^{(\ell)} D_k^{(\ell)*}, \qquad k = 0, \ldots, n-1,$$

then compute its inverse Fourier transform s, and set

$$r_i = \frac{1}{N} s_i, \qquad i = 0, \ldots, (n/2) - 1.$$

To complete the development, we show how to eliminate some of the work. Instead of using a fast Fourier transform to get $\boldsymbol{G}^{(\ell)}$, use the formula

$$G_k^{(\ell)} = D_k^{(\ell)} + (-1)^k D_k^{(\ell+1)}.$$

This is a direct consequence of the delay property of the Fourier transform; multiplication of the transform by ω^{bk} corresponds to cyclically translating the time function by b positions. When b is equal to $n/2$, this becomes multiplication of $D_k^{(\ell)}$ by $(-1)^k$. In the time domain, this corresponds to moving the nonzero positions of $\boldsymbol{d}^{(\ell+1)}$ into the zero positions of $\boldsymbol{d}^{(\ell)}$; the sum in the time domain is equal to $\boldsymbol{g}^{(\ell)}$, so in the frequency domain, the sum is equal to $\boldsymbol{G}^{(\ell)}$.

The computation then takes the form

$$S_k = \sum_{\ell=0}^{L-1} D_k^{(\ell)} \left[D_k^{(\ell)} + (-1)^k D_k^{(\ell+1)} \right].$$

An inverse Fourier transform completes the computation.

A summary flowchart is shown in Figure 6.12. Notice that there is only one Fourier transform per major loop, except for one more Fourier transform at the start and one at the finish. Even though computing one cyclic convolution would take three Fourier transforms for this problem, we have found an algorithm that averages about one Fourier transform per section.

6.8 Autocorrelation and crosscorrelation

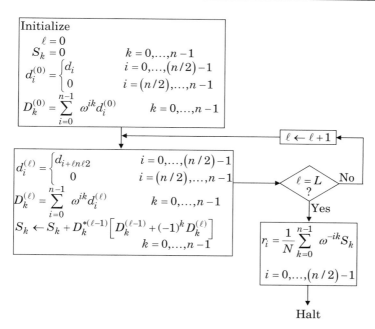

Figure 6.12 Computation of an autocorrelation

The flow is written with the L major loops the same, although the last loop computes the Fourier transform of an all-zero vector padded onto the end of the data sequence. That Fourier transform can be eliminated by modifying the last loop.

Similar methods apply to computing the cross-correlation. Both data records can be processed in sections, with the sections combined in the frequency domain to reduce the number of inverse Fourier transforms. We will show how to bring in other techniques by way of a fairly detailed example.

The cross-correlation

$$s_i = \sum_{j=0}^{r-1} d_{j+i} g_j$$

can be expressed in matrix form as

$$\begin{bmatrix} s_0 \\ s_1 \\ s_2 \\ \vdots \\ s_{n-1} \end{bmatrix} = \begin{bmatrix} d_0 & d_1 & d_2 & \cdots & d_{r-1} \\ d_1 & d_2 & d_3 & \cdots & d_r \\ d_2 & d_3 & d_4 & \cdots & d_{r+1} \\ \vdots & & & & \\ d_{n-1} & d_n & d_{n+1} & \cdots & d_{r+n-2} \end{bmatrix} \begin{bmatrix} g_0 \\ g_1 \\ g_2 \\ \vdots \\ g_{r-1} \end{bmatrix}$$

with r much larger than n.

We shall describe the structure of an algorithm for the specific example in which n equals 120 and r equals 10 000. We arbitrarily decide to break the 120 output points into four batches of thirty output points per batch and to organize the computation around a 60-point Winograd large FFT, which is discussed in Chapter 12.

Block the matrix–vector product as follows:

$$\begin{bmatrix} s_0 \\ \vdots \\ s_{29} \\ s_{30} \\ \vdots \\ s_{59} \\ s_{60} \\ \vdots \\ s_{89} \\ s_{90} \\ \vdots \\ s_{119} \end{bmatrix} = \begin{bmatrix} d_0 & \cdots & d_{29} & d_{30} & \cdots & d_{59} & \cdots & d_{9990} & \cdots & d_{10049} \\ \vdots & & \vdots & \vdots & & \vdots & & & & \\ d_{29} & \cdots & d_{58} & d_{59} & \cdots & d_{88} & \cdots & d_{10019} & \cdots & d_{10048} \\ d_{30} & \cdots & d_{59} & d_{60} & \cdots & d_{89} & & & & \\ \vdots & & \vdots & \vdots & & \vdots & & & & \\ d_{59} & \cdots & d_{88} & d_{89} & \cdots & d_{119} & & & & \\ d_{60} & \cdots & d_{89} & & & & & & & \\ \vdots & & \vdots & & & & & \vdots & & \\ d_{89} & \cdots & d_{118} & & & & & & & \\ d_{90} & \cdots & d_{119} & & & & & & & \\ \vdots & & \vdots & & & & & & & \\ d_{119} & \cdots & d_{148} & \cdots & & & & & & \end{bmatrix} \begin{bmatrix} g_0 \\ \vdots \\ g_{29} \\ g_{30} \\ \vdots \\ g_{59} \\ \vdots \\ g_{9990} \\ \vdots \\ g_{10019} \end{bmatrix}$$

where twenty extra points, each equal to zero, have been appended to $g(x)$ and to $d(x)$ so that the new value of r, 10 020, is divisible by 30. Each of the small blocks, such as

$$\begin{bmatrix} s_0 \\ \vdots \\ s_{29} \end{bmatrix} = \begin{bmatrix} d_0 & \cdots & d_{29} \\ \vdots & & \vdots \\ d_{29} & \cdots & d_{58} \end{bmatrix} \begin{bmatrix} g_0 \\ \vdots \\ g_{29} \end{bmatrix},$$

can be computed as the first 30 points of a 60-point cyclic convolution, which can be computed using a 60-point Winograd FFT.

Now consider the block structure of the computation,

$$\begin{bmatrix} S_0 \\ S_1 \\ S_2 \\ S_3 \end{bmatrix} = \begin{bmatrix} D_0 & D_1 & D_2 & \cdots & D_{333} \\ D_1 & D_2 & D_3 & \cdots & D_{334} \\ D_2 & D_3 & D_4 & \cdots & D_{335} \\ D_3 & D_4 & D_5 & \cdots & D_{336} \end{bmatrix} \begin{bmatrix} G_0 \\ G_1 \\ G_3 \\ \vdots \\ G_{333} \end{bmatrix}$$

$$= \begin{bmatrix} D_0 & D_1 & D_2 & D_3 \\ D_1 & D_2 & D_3 & D_4 \\ D_2 & D_3 & D_4 & D_5 \\ D_3 & D_4 & D_5 & D_6 \end{bmatrix} \begin{bmatrix} G_0 \\ G_1 \\ G_2 \\ G_3 \end{bmatrix} + \begin{bmatrix} D_4 & D_5 & D_6 & D_7 \\ D_5 & D_6 & D_7 & D_8 \\ D_6 & D_7 & D_8 & D_9 \\ D_7 & D_8 & D_9 & D_{10} \end{bmatrix} \begin{bmatrix} G_4 \\ G_5 \\ G_6 \\ G_7 \end{bmatrix}$$

$$+ \cdots + \begin{bmatrix} D_{332} & D_{333} & D_{334} & D_{335} \\ D_{333} & D_{334} & D_{335} & D_{336} \\ D_{334} & D_{335} & D_{336} & D_{337} \\ D_{335} & D_{336} & D_{337} & D_{338} \end{bmatrix} \begin{bmatrix} G_{332} \\ G_{333} \\ G_{334} \\ G_{335} \end{bmatrix},$$

6.8 Autocorrelation and crosscorrelation

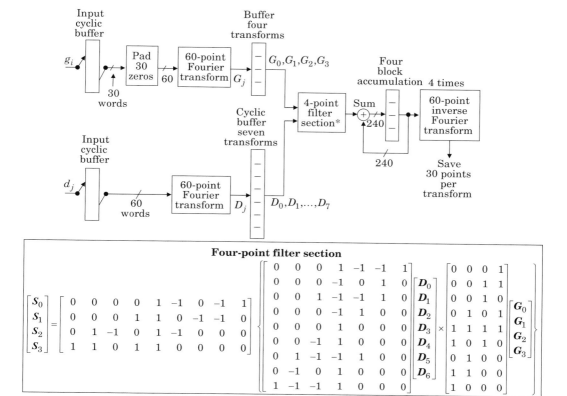

Figure 6.13 A device for correlation

where again zero blocks have been appended to make the number of blocks divisible by four. The computation we need to design is

$$\begin{bmatrix} S_0 \\ S_1 \\ S_2 \\ S_3 \end{bmatrix} = \begin{bmatrix} D_0 & D_1 & D_2 & D_3 \\ D_1 & D_2 & D_3 & D_4 \\ D_2 & D_3 & D_4 & D_5 \\ D_3 & D_4 & D_5 & D_6 \end{bmatrix} \begin{bmatrix} G_0 \\ G_1 \\ G_2 \\ G_3 \end{bmatrix}.$$

This we recognize as a four-point filter section. It does not matter to the algorithm, as designed in Section 6.2, that the blocks are themselves matrices. The algorithm requires nine multiplications and 21 additions. Each of the multiplications is actually a 30-point filter section, which we have already decided to compute with a 60-point Winograd FFT. It must be repeated 84 times, and the outputs added to obtain the desired output.

The correlation algorithm is shown in its final form in Figure 6.13. The symbols D_j and G_j in the figure denote the Fourier transforms of data blocks d_j and g_j. The Fourier transforms of blocks that are used more than once are saved to be reused – not recomputed. Most additions are performed in the frequency domain, although we have described them in the time domain. This reduces the number of inverse FFT computations. One can also reduce the number of additions a bit more if the multiplication by the matrix of postadditions in the filter section is deferred until after all of the filter sections are added together. Then that matrix of additions is needed only once.

We judge the complexity of an implementation, in part, by the number of multiplications. There are 676 applications of the 60-point Winograd FFT, which requires 48 672 multiplications. There are also 84 applications of the filter section; each uses nine vector multiplications, and each vector multiplication is a componentwise product of two 60-point complex vectors, each of which requires three real multiplications. This requires 146 080 multiplications. The total number of multiplications needed to correlate 10 000 samples for 120 lags is 194 752. For problems that need a massive high-speed correlation, such as those that occur in sonar or radar signal processing, the algorithm, shown in Figure 6.13, could be realized as hardwired digital logic.

Problems for Chapter 6

6.1 Prove that an n-tap filter section with n a power of two, constructed by iterating a two-tap filter section, uses

$$M(n) = n^{\log_2 3}$$

multiplications, and that the number of additions satisfies the recursion

$$A(n) = 2n + 3A(n/2)$$

with $A(1) = 0$.

6.2 By iteration, construct a nine-point by nine-point linear convolution algorithm, starting with a three-point by three-point linear convolution algorithm from Chapter 5.

6.3 It is possible to form an overlap technique for linear convolution that, in spirit, is midway between the overlap–add and the overlap–save techniques. Set up the equations for this hybrid technique. What are the advantages and disadvantages?

6.4 Prove that every algorithm that computes three sequential outputs of a symmetric four-tap filter must use at least five multiplications.

6.5 Suppose that one has a device (a circuit module or a software subroutine) that computes 315-point *cyclic* convolutions. It is desired to pass a vector of 1000 data points through a 100-tap FIR filter. Describe how to break up the data to feed the convolver and to assemble the convolver output to get the desired answer.

6.6 **a** Construct an algorithm using seven multiplications for passing four data samples through a four-tap symmetric filter.
b Construct an algorithm using six multiplications for passing four data points through a four-tap skew-symmetric filter.

6.7 When the data sequence d is complex, one usually defines the autocorrelation as

$$r_i = \sum_{k=0}^{N-1} d^*_{i+k} d_k, \qquad i = 0, \ldots, L + N - 2.$$

Show how to revise the flow of Figure 6.12 to handle complex data.

6.8 One can use the overlap–save method with an FFT that is too short by appending correction terms. Lay out a method for computing 25 output values from a 25-tap FIR filter, using a 48-point Winograd FFT. How many multiplications and additions are required?

6.9 Give an algorithm for a three by three filter section in the complex field. How many real additions and real multiplications are used?

6.10 Suppose that $g(x)$ factors as $g(x) = g_1(x)g_2(x)$, where $\deg g_1(x) = \deg g_2(x) = n/2$. Is it more efficient computationally to compute $s(x) = d(x)g(x)$ directly, or to compute $s(x) = d(x)g_1(x)g_2(x)$ as the cascade of two computations? Does this comparison depend on the value of n?

Notes for Chapter 6

The overlap methods for breaking a long linear convolution into pieces are straightforward and have been in use for some time, probably developed independently in many places. Stockham (1966) was the first to point out that the Cooley–Tukey FFT, combined with the convolution theorem, gives a good way to compute cyclic convolutions. Agarwal and Burrus (1974) showed how to change a one-dimensional convolution into a multidimensional convolution by using a reindexing scheme that we have called nesting or iteration. This construction is a form of overlap–save sectioning with the sections arranged into a matrix. The same basic idea, but using a kind of overlap–add sectioning, was proposed by DuBois and Venetsanopoulos (1978) to compute large cyclic convolutions. The method of construction of good algorithms for short filter sections, including symmetric filters and decimating filters, is due to Winograd (1979, 1980a,b). The construction of sections for interpolating filters is a straightforward application of Winograd's methods. The transposition principle was discussed by Hopcroft and Musinski (1973). The method of sectioning for the autocorrelation is due to Rader (1970). There are a great many studies of the organization of FFT computations with regard to the management of memory, such as the paper by Burrus and Eschenbacher (1979).

The use of decimation as an aid in computing the Fourier transform was proposed by Liu and Mintzer (1978). They go far beyond our brief summary and study the design of multistage decimating filters. A survey of methods of interpolation and decimation can be found in the work of Crochiere and Rabiner (1981).

Except for counting multiplications and additions, we have avoided performance comparison of the various FFT algorithms, because, even for a software rather than a hardware implementation, the performance depends on the precise version of the algorithm, the level of the programming language, the architecture of the processor, and the skill of the programmer. Some studies of performance have been given by Kolba and Parks (1977); Silverman (1977); Morris (1978); Nawab and McClellan (1979); Patterson and McClellan (1978); and Panda, Pal, and Chatterjee (1983). It would be rash to say that this diverse work could be summarized by a single conclusion.

7 Fast algorithms for solving Toeplitz systems

A standard method of solving a system of n linear equations in n unknowns is to write the system of equations as a matrix equation $Af = g$, and to solve it either by computing the matrix inverse and writing $f = A^{-1}g$ or, alternatively, by using the method known as *gaussian elimination*. The standard methods of computing a matrix inverse have complexity proportional to n^3. Sometimes, the matrix has a special structure that can be exploited to obtain a faster algorithm.

A *Toeplitz system of equations* is a system of linear equations described by a Toeplitz matrix A. The problem of solving a Toeplitz system of equations arises in a great many applications, including spectral estimation, linear prediction, autoregressive filter design, and error-control codes. Because the Toeplitz system is highly structured, methods of solution are available that are far superior to the general methods of solving systems of linear equations. These methods are the subject of this chapter, and are valid in any algebraic field.

The algorithms of this chapter are somewhat distant from those we have studied for convolution and for Fourier transforms. Convolutions and transforms are essentially problems of matrix multiplication, whereas this chapter deals with the solution of a system of linear equations. The solution of a system of linear equations is closer to the task of matrix inversion. It should be no surprise that we do not build on earlier algorithms directly, though techniques such as doubling may prove useful.

7.1 The Levinson and Durbin algorithms

In certain problems related to deconvolution or spectral estimation, one must solve a Toeplitz system of equations given by the matrix equation

$$Af = g,$$

where A is an n by n Toeplitz matrix, given by

$$A = \begin{bmatrix} a_0 & a_1 & a_2 & \cdots & a_{n-1} \\ a_{-1} & a_0 & a_1 & \cdots & a_{n-2} \\ a_{-2} & a_{-1} & a_0 & \cdots & \\ \vdots & & & & \vdots \\ a_{-n+1} & & & \cdots & a_0 \end{bmatrix}.$$

The computational task is to compute the vector f from the vector g and the Toeplitz matrix A. Of course, one method of solution is to find the inverse of A, but in some applications, n may be greater than 100 or even greater than 1000, so it is important to find a more efficient method of solution.

The *Levinson algorithm* is an efficient iterative algorithm that can be used for the special case in which A is a symmetric Toeplitz matrix. Then we have the system of equations

$$\begin{bmatrix} a_0 & a_1 & a_2 & \cdots & a_{n-2} & a_{n-1} \\ a_1 & a_0 & a_1 & \cdots & a_{n-3} & a_{n-2} \\ a_2 & a_1 & a_0 & \cdots & a_{n-4} & a_{n-3} \\ \vdots & & & & \vdots & \vdots \\ a_{n-2} & a_{n-3} & a_{n-4} & \cdots & a_0 & a_1 \\ a_{n-1} & a_{n-2} & a_{n-3} & \cdots & a_1 & a_0 \end{bmatrix} \begin{bmatrix} f_0 \\ f_1 \\ f_2 \\ \vdots \\ f_{n-2} \\ f_{n-1} \end{bmatrix} = \begin{bmatrix} g_0 \\ g_1 \\ g_2 \\ \vdots \\ g_{n-2} \\ g_{n-1} \end{bmatrix}.$$

The matrix and the vectors have been blocked here to show the repetitive bordering structure upon which the iterative algorithm is based. Each iteration will enlarge A by appending to the A of the previous iteration a half-border consisting of a new row at the bottom and a new column at the right.

By using an exchange matrix J, which satisfies $J^2 = I$, the equation

$$Af = g$$

can be written

$$JAJJf = Jg.$$

Because A is a symmetric Toeplitz matrix,

$$JAJ = A,$$

so we also have

$$AJf = Jg.$$

7.1 The Levinson and Durbin algorithms

Hence we also have the equation

$$\begin{bmatrix} a_0 & a_1 & a_2 & \cdots & a_{n-1} \\ a_1 & a_0 & a_1 & \cdots & a_{n-2} \\ a_2 & a_1 & a_0 & \cdots & a_{n-3} \\ \vdots & & & & \vdots \\ a_{n-1} & a_{n-2} & & \cdots & a_0 \end{bmatrix} \begin{bmatrix} f_{n-1} \\ f_{n-2} \\ \vdots \\ f_1 \\ f_0 \end{bmatrix} = \begin{bmatrix} g_{n-1} \\ g_{n-2} \\ \vdots \\ g_1 \\ g_0 \end{bmatrix}.$$

This form of the equation will be used in the derivation of the Levinson algorithm. The algorithm works with r by r submatrices of A, given by

$$A^{(r)} = \begin{bmatrix} a_0 & a_1 & a_2 & \cdots & a_{r-1} \\ a_1 & a_0 & a_1 & \cdots & a_{r-2} \\ a_2 & a_1 & a_0 & \cdots & a_{r-3} \\ \vdots & & & & \vdots \\ a_{r-1} & & & \cdots & a_0 \end{bmatrix},$$

which are obtained from A by deleting $n - r$ columns from the right and the same number of rows from the bottom.

The Levinson algorithm is an iterative algorithm; we index the iterations by r. At step r, the Levinson algorithm computes the solution to the rth truncated problem:

$$\begin{bmatrix} a_0 & a_1 & a_2 & \cdots & a_{r-1} \\ a_1 & a_0 & a_1 & \cdots & a_{r-2} \\ \vdots & \vdots & \vdots & & \vdots \\ a_{r-1} & a_{r-2} & a_{r-3} & \cdots & a_0 \end{bmatrix} \begin{bmatrix} f_0^{(r)} \\ f_1^{(r)} \\ \vdots \\ f_{r-1}^{(r)} \end{bmatrix} = \begin{bmatrix} g_0 \\ g_1 \\ \vdots \\ g_{r-1} \end{bmatrix},$$

where the superscript r on the vector $f^{(r)}$ denotes that it is the solution of the rth truncated equation. Clearly, even though $f^{(r)}$ solves the truncated equation, it is not a truncated version of the f that solves the original equation. The Levinson algorithm will recursively update $f^{(r)}$, however, in such a way that $f^{(r)}$ is the required solution of the rth truncated equation.

In addition to $f^{(r)}$, the Levinson algorithm also iterates a number of working variables, defined as follows. These working variables are three scalars called α_r, β_r, and γ_r, and a working vector of length r called $t^{(r)}$. The working variables α_r, β_r, γ_r, and $t^{(r)}$ are chosen at each r so that the following side equation is also satisfied:

$$\begin{bmatrix} a_0 & a_1 & a_2 & \cdots & a_{r-1} \\ a_1 & a_0 & a_1 & \cdots & a_{r-2} \\ a_2 & a_1 & a_0 & \cdots & a_{r-3} \\ \vdots & \vdots & \vdots & & \vdots \\ a_{r-1} & a_{r-2} & & \cdots & a_0 \end{bmatrix} \begin{bmatrix} t_0^{(r)} \\ t_1^{(r)} \\ \vdots \\ t_{r-1}^{(r)} \end{bmatrix} = \begin{bmatrix} \alpha_r \\ 0 \\ \vdots \\ 0 \end{bmatrix},$$

where all components, but one, of the vector on the right are equal to zero. The introduction of the working vector $t^{(r)}$ and the side equation is the clever idea that allows the iterations of the Levinson algorithm to continue. We will define iteration $r+1$ so as to perpetuate equations of the same form. The next iteration begins by expanding the equations of the previous iteration as

$$\begin{bmatrix} a_0 & a_1 & a_2 & \cdots & a_{r-1} & a_r \\ a_1 & a_0 & a_1 & \cdots & a_{r-2} & a_{r-1} \\ \vdots & \vdots & & & \vdots & \vdots \\ a_{r-1} & a_{r-2} & & \cdots & a_0 & a_1 \\ a_r & a_{r-1} & & \cdots & a_1 & a_0 \end{bmatrix} \begin{bmatrix} f_0^{(r)} \\ f_1^{(r)} \\ \vdots \\ f_{r-1}^{(r)} \\ 0 \end{bmatrix} = \begin{bmatrix} g_0 \\ g_1 \\ \vdots \\ g_{r-1} \\ \gamma_r \end{bmatrix},$$

which defines γ_r, and

$$\begin{bmatrix} a_0 & a_1 & a_2 & \cdots & a_{r-1} & a_r \\ a_1 & a_0 & a_1 & \cdots & a_{r-2} & a_{r-1} \\ \vdots & \vdots & & & \vdots & \vdots \\ a_{r-1} & a_{r-2} & & \cdots & a_0 & a_1 \\ a_r & a_{r-1} & & \cdots & a_1 & a_0 \end{bmatrix} \begin{bmatrix} t_0^{(r)} \\ t_1^{(r)} \\ \vdots \\ t_{r-1}^{(r)} \\ 0 \end{bmatrix} = \begin{bmatrix} \alpha_r \\ 0 \\ \vdots \\ 0 \\ \beta_r \end{bmatrix},$$

which defines β_r, where the vector on the right side has all zeros in its interior. If $\gamma_r = g_r$ and $\beta_r = 0$, then $f^{(r+1)}$ and $t^{(r+1)}$ are set equal to $f^{(r)}$ and $t^{(r)}$, respectively, but with a zero appended to each to increase the length. In that case, the iteration is complete. Otherwise, $f^{(r)}$ or $t^{(r)}$ must be modified to form $f^{(r+1)}$ and $t^{(r+1)}$.

For the recursive algorithm, we initialize the variables so that the equations hold for $r = 0$, and we suppose that these equations were satisfied at the end of iteration $r-1$. We need only show how to update them so that they are also satisfied at the end of iteration r. But we have already seen that we can also write

$$\begin{bmatrix} a_0 & a_1 & a_2 & \cdots & a_{r-1} & a_r \\ a_1 & a_0 & a_1 & \cdots & a_{r-2} & a_{r-1} \\ \vdots & \vdots & & & \vdots & \vdots \\ a_{r-1} & a_{r-2} & & \cdots & a_0 & a_1 \\ a_r & a_{r-1} & & \cdots & a_1 & a_0 \end{bmatrix} \begin{bmatrix} 0 \\ f_{r-1}^{(r)} \\ \vdots \\ f_1^{(r)} \\ f_0^{(r)} \end{bmatrix} = \begin{bmatrix} \gamma_r \\ g_{r-1} \\ \vdots \\ g_1 \\ g_0 \end{bmatrix}$$

and

$$\begin{bmatrix} a_0 & a_1 & a_2 & \cdots & a_{r-1} & a_r \\ a_1 & a_0 & a_1 & \cdots & a_{r-2} & a_{r-1} \\ \vdots & \vdots & & & \vdots & \vdots \\ a_{r-1} & a_{r-2} & & \cdots & a_0 & a_1 \\ a_r & a_{r-1} & & \cdots & a_1 & a_0 \end{bmatrix} \begin{bmatrix} 0 \\ t_{r-1}^{(r)} \\ \vdots \\ t_1^{(r)} \\ t_0^{(r)} \end{bmatrix} = \begin{bmatrix} \beta_r \\ 0 \\ \vdots \\ 0 \\ \alpha_r \end{bmatrix}.$$

7.1 The Levinson and Durbin algorithms

From these equations, we form the next iteration. Let

$$\begin{bmatrix} t_0^{(r+1)} \\ t_1^{(r+1)} \\ \vdots \\ t_{r-1}^{(r+1)} \\ t_r^{(r+1)} \end{bmatrix} = k_1 \begin{bmatrix} t_0^{(r)} \\ t_1^{(r)} \\ \vdots \\ t_{r-1}^{(r)} \\ 0 \end{bmatrix} + k_2 \begin{bmatrix} 0 \\ t_{r-1}^{(r)} \\ \vdots \\ t_1^{(r)} \\ t_0^{(r)} \end{bmatrix}$$

for constants k_1 and k_2, yet to be chosen. Then write

$$\begin{bmatrix} a_0 & a_1 & \cdots & a_r \\ a_1 & a_0 & \cdots & a_{r-1} \\ \vdots & \vdots & & \vdots \\ a_{r-1} & a_{r-2} & \cdots & a_1 \\ a_r & a_{r-1} & \cdots & a_0 \end{bmatrix} \begin{bmatrix} t_0^{(r+1)} \\ t_1^{(r+1)} \\ \vdots \\ t_{r-1}^{(r+1)} \\ t_r^{(r+1)} \end{bmatrix} = k_1 \begin{bmatrix} \alpha_r \\ 0 \\ \vdots \\ 0 \\ \beta_r \end{bmatrix} + k_2 \begin{bmatrix} \beta_r \\ 0 \\ \vdots \\ 0 \\ \alpha_r \end{bmatrix} = \begin{bmatrix} \alpha_{r+1} \\ 0 \\ \vdots \\ 0 \\ 0 \end{bmatrix},$$

where, to obtain the equality on the right, k_1 and k_2 must be chosen to satisfy the condition

$$0 = k_1 \beta_r + k_2 \alpha_r.$$

We will choose $k_1 = \alpha_r$ and $k_2 = -\beta_r$. Then

$$\alpha_{r+1} = k_1 \alpha_r + k_2 \beta_r = \alpha_r^2 - \beta_r^2.$$

However, one could choose a different k_1, which may give a different numerical accuracy. Finally, let

$$\begin{bmatrix} f_0^{(r+1)} \\ f_1^{(r+1)} \\ \vdots \\ f_{r-1}^{(r+1)} \\ f_r^{(r+1)} \end{bmatrix} = \begin{bmatrix} f_0^{(r)} \\ f_1^{(r)} \\ \vdots \\ f_{r-1}^{(r)} \\ 0 \end{bmatrix} + k_3 \begin{bmatrix} t_r^{(r+1)} \\ t_{r-1}^{(r+1)} \\ \vdots \\ t_1^{(r+1)} \\ t_0^{(r+1)} \end{bmatrix},$$

where the constant k_3 has yet to be chosen; then

$$\begin{bmatrix} a_0 & a_1 & \cdots & a_r \\ a_1 & a_0 & \cdots & a_{r-1} \\ \vdots & \vdots & & \vdots \\ a_{r-1} & & \cdots & a_1 \\ a_r & & \cdots & a_0 \end{bmatrix} \begin{bmatrix} f_0^{(r+1)} \\ f_1^{(r+1)} \\ \vdots \\ f_{r-1}^{(r+1)} \\ f_r^{(r+1)} \end{bmatrix} = \begin{bmatrix} g_0 \\ g_1 \\ \vdots \\ g_{r-1} \\ \gamma_r \end{bmatrix} + k_3 \begin{bmatrix} 0 \\ 0 \\ \vdots \\ 0 \\ \alpha_{r+1} \end{bmatrix} = \begin{bmatrix} g_0 \\ g_1 \\ \vdots \\ g_{r-1} \\ g_r \end{bmatrix},$$

where, to obtain the equality on the right, k_3 is chosen so that

$$\gamma_r + k_3 \alpha_{r+1} = g_r.$$

This completes the iteration.

Fast algorithms for solving Toeplitz systems

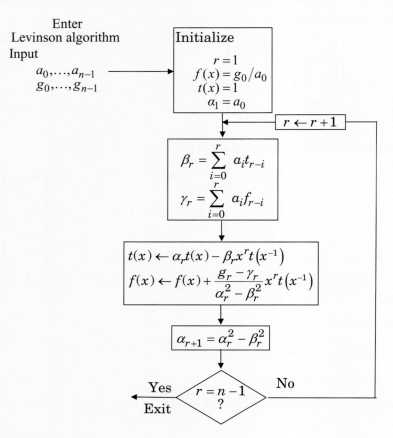

Figure 7.1 Levinson algorithm

The Levinson algorithm is summarized in Figure 7.1. There the vectors $t^{(r)}$ and $f^{(r)}$ are represented by polynomials

$$t(x) = t_r^{(r)} x^r + t_{r-1}^{(r)} x^{r-1} + \cdots + t_1^{(r)} x + t_0^{(r)},$$
$$f(x) = f_r^{(r)} x^r + f_{r-1}^{(r)} x^{r-1} + \cdots + f_1^{(r)} x + f_0^{(r)},$$

and the superscript r is suppressed in the polynomial representation. There is no need to actually form the matrix $A^{(r)}$ in the calculations; it is enough to compute only $f(x)$ and $t(x)$ as iterates.

The rth pass through the loop has a complexity proportional to r, and there are n passes. Therefore the complexity of the Levinson algorithm is proportional to n^2. The recursions of the algorithm fail only if a division by zero arises, and this occurs only if one of the principle submatrices is singular.

The Levinson algorithm holds in any field. In particular, it holds just as stated in the complex field. However, a symmetric Toeplitz matrix in the complex field does not

7.1 The Levinson and Durbin algorithms

often arise in applications. A hermitean Toeplitz matrix is much more common. The Levinson algorithm also holds in this case, provided that one takes complex conjugates at the right points in the computation. It is easy to rework the derivation for this case.

Sometimes, in place of the Levinson algorithm, a better algorithm known as the *Durbin algorithm* can be used. This is possible when the Toeplitz matrix is symmetric and the vector on the right is made up of elements from the Toeplitz matrix in such a way that the system of equations takes the following form:

$$\begin{bmatrix} a_0 & a_1 & a_2 & \cdots & a_{n-2} & a_{n-1} \\ a_1 & a_0 & a_1 & \cdots & a_{n-3} & a_{n-2} \\ a_2 & a_1 & a_0 & \cdots & a_{n-4} & a_{n-3} \\ \vdots & & & & \vdots & \vdots \\ a_{n-2} & & & \cdots & a_0 & a_1 \\ \hline a_{n-1} & & & \cdots & a_1 & a_0 \end{bmatrix} \begin{bmatrix} f_0 \\ f_1 \\ \vdots \\ f_{n-2} \\ \hline f_{n-1} \end{bmatrix} = - \begin{bmatrix} a_1 \\ a_2 \\ a_3 \\ \vdots \\ a_{n-1} \\ \hline a_n \end{bmatrix}.$$

The matrix and the vectors have again been blocked to show the repetitive bordering structure upon which the algorithm is based. Now the vector in the right column is made up of elements of the Toeplitz matrix. This special property allows the equations to be solved with half the work of the Levinson algorithm because only one polynomial needs to be iterated. The Durbin algorithm is important because systems of Toeplitz equations of this form occur frequently in problems of spectral analysis.

At step r, the Durbin algorithm begins with a solution to the truncated problem

$$\begin{bmatrix} a_0 & a_1 & a_2 & \cdots & a_{r-1} \\ a_1 & a_0 & a_1 & \cdots & a_{r-2} \\ \vdots & \vdots & & & \vdots \\ a_{r-1} & a_{r-2} & & \cdots & a_0 \end{bmatrix} \begin{bmatrix} f_0^{(r)} \\ f_1^{(r)} \\ \vdots \\ f_{r-1}^{(r)} \end{bmatrix} = - \begin{bmatrix} a_1 \\ a_2 \\ \vdots \\ a_r \end{bmatrix}.$$

The next iteration begins with

$$\begin{bmatrix} a_0 & a_1 & a_2 & \cdots & a_{r-1} & a_r \\ a_1 & a_0 & a_1 & \cdots & a_{r-2} & a_{r-1} \\ \vdots & \vdots & & & \vdots & \vdots \\ a_{r-1} & a_{r-2} & & \cdots & a_0 & a_1 \\ \hline a_r & a_{r-1} & & \cdots & a_1 & a_0 \end{bmatrix} \begin{bmatrix} f_0^{(r)} \\ f_1^{(r)} \\ \vdots \\ f_{r-1}^{(r)} \\ \hline 0 \end{bmatrix} = - \begin{bmatrix} a_1 \\ a_2 \\ a_3 \\ \vdots \\ a_r \\ \gamma_r \end{bmatrix},$$

which defines γ_r. The iteration must update $f^{(r)}$ to make γ_r become equal to a_{r+1}. Let $f^{(r+1)}$ be given by

$$\begin{bmatrix} f_0^{(r+1)} \\ f_1^{(r+1)} \\ \vdots \\ f_{r-1}^{(r+1)} \\ f_r^{(r+1)} \end{bmatrix} = \begin{bmatrix} f_0^{(r)} \\ f_1^{(r)} \\ \vdots \\ f_{r-1}^{(r)} \\ 0 \end{bmatrix} + k_r \begin{bmatrix} f_{r-1}^{(r)} \\ f_{r-2}^{(r)} \\ \vdots \\ f_0^{(r)} \\ 0 \end{bmatrix} + \beta_r \begin{bmatrix} 0 \\ 0 \\ \vdots \\ 0 \\ 1 \end{bmatrix}.$$

If we can choose k_r and β_r so as to regenerate the desired form, we have a good algorithm. Choose

$$\gamma_r = -\sum_{i=1}^{r} f_{r-i}^{(r)} a_i,$$

$$\gamma_r' = -\sum_{i=1}^{r} f_{r-i}^{(r)} a_i,$$

so that

$$\begin{bmatrix} a_0 & \cdots & a_{r-1} & a_r \\ a_1 & \cdots & a_{r-2} & a_r \\ \vdots & & \vdots & \vdots \\ a_{r-1} & \cdots & a_0 & a_1 \\ \hline a_r & & a_1 & a_0 \end{bmatrix} \begin{bmatrix} f_0^{(r+1)} \\ f_1^{(r+1)} \\ \vdots \\ f_{r-1}^{(r+1)} \\ f_r^{(r+1)} \end{bmatrix} = -\begin{bmatrix} a_1 \\ a_2 \\ \vdots \\ a_r \\ \gamma_r \end{bmatrix} - k_r \begin{bmatrix} a_r \\ a_{r-1} \\ \vdots \\ a_1 \\ \gamma_r' \end{bmatrix} + \beta_r \begin{bmatrix} a_r \\ a_{r-1} \\ \vdots \\ a_1 \\ a_0 \end{bmatrix}$$

$$= -\begin{bmatrix} a_1 \\ \vdots \\ a_r \\ a_{r+1} \end{bmatrix}.$$

To obtain the final equality, we must choose k_r and β_r so that

$$k_r - \beta_r = 0$$

and

$$-\gamma_r - k_r \gamma_r' + \beta_r a_0 = -a_{r+1}.$$

Thus, $k_r = \beta_r$ and

$$\beta_r = -\frac{(a_{r+1} - \gamma_r)}{(a_0 - \gamma_r')}.$$

With these equations, a solution at the rth iteration is propagated into a solution at the $(r+1)$th iteration. Hence the trivial solution at the zeroth iteration can be recursively

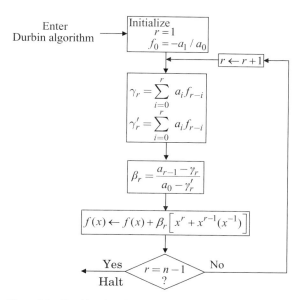

Figure 7.2 Durbin algorithm

propagated into a solution at the nth iteration. This completes the derivation of the Durbin algorithm. It is summarized in the flow diagram of Figure 7.2.

7.2 The Trench algorithm

The Toeplitz system of equations

$$\begin{bmatrix} a_0 & a_1 & a_2 & \cdots & a_{n-1} \\ a_{-1} & a_0 & a_1 & \cdots & a_{n-2} \\ a_{-2} & a_{-1} & a_0 & \cdots & a_{n-3} \\ \vdots & \vdots & \vdots & & \vdots \\ a_{-n+1} & a_{-n+2} & a_{-n+3} & \cdots & a_0 \end{bmatrix} \begin{bmatrix} f_0 \\ f_1 \\ f_2 \\ \vdots \\ f_{n-1} \end{bmatrix} = \begin{bmatrix} g_0 \\ g_1 \\ g_2 \\ \vdots \\ g_{n-1} \end{bmatrix}$$

can be solved by any of a number of different algorithms, depending on which of various side conditions are satisfied. In Section 7.1, we treated the case in which the Toeplitz matrix is symmetric. The Levinson algorithm or the Durbin algorithm then applies. In this section, we shall study the more general case in which the Toeplitz matrix is not symmetric. The *Trench algorithm* solves this more general case. In its fullest form, the Trench algorithm gives the inverse matrix A^{-1} as well as the vector f. We shall only describe the computation of A^{-1}.

In general, a Toeplitz matrix need not be symmetric, but every Toeplitz matrix is persymmetric. A *persymmetric* matrix is one that has symmetry about its antidiagonal.

If an n by n matrix A is persymmetric, then $a_{ij} = a_{n+1-j,n+1-i}$. Another way to state this is that A is a persymmetric matrix if and only if $JAJ = A$, where J is the exchange matrix of the same size as A.

The inverse of a Toeplitz matrix is generally not a Toeplitz matrix, but the inverse does remain persymmetric.

Theorem 7.2.1 *The inverse of a persymmetric matrix A is persymmetric.*

Proof Let J be the exchange matrix of the same size as A, and recall that $J^2 = I$ so that $J^{-1} = J$. But $JAJ = A$. Hence $JA^{-1}J = A^{-1}$ and A^{-1} is persymmetric. $\square$

Another property that is retained by the inverse of a Toeplitz matrix is the property that the matrix is completely specified by its border. Even more, it is completely specified by its half-border, consisting of either its first row and first column, or its last row and last column. This section will proceed by establishing this property, then giving a recursive procedure for computing the inverse matrix from a lower half-border consisting of the last row and the last column. The Trench algorithm computes A^{-1} by computing the half-border of A^{-1}.

Define the column vectors a_+ and a_-, each of length $r - 1$, as

$$a_+ = (a_1, a_2, a_3, \ldots, a_{r-1})^T,$$
$$a_- = (a_{-1}, a_{-2}, a_{-3}, \ldots, a_{-(r-1)})^T.$$

Let $\tilde{a}_+$ and $\tilde{a}_-$ denote these two vectors with their components written in reverse order. That is, $\tilde{a}_+ = Ja_+$ and $\tilde{a}_- = Ja_-$, where J is the $(r-1)$ by $(r-1)$ exchange matrix. The matrix $A^{(r)}$ can be partitioned as

$$A^{(r)} = \left[\begin{array}{c|c} A^{(r-1)} & \tilde{a}_+ \\ \hline \tilde{a}_-^T & a_0 \end{array}\right].$$

Likewise, the inverse matrix $B^{(r)} = (A^{(r)})^{-1}$ will be partitioned as

$$B^{(r)} = \left[\begin{array}{c|c} M^{(r)} & \tilde{b}_+^{(r)} \\ \hline \tilde{b}_-^{(r)T} & b_0^{(r)} \end{array}\right].$$

Our goal is to interpret the blocks of this latter partition.

Theorem 7.2.2 *The blocks of the inverse matrix $B^{(r)}$ satisfy*

$$M^{(r)} = B^{(r-1)} + \frac{1}{b_0^{(r)}} \tilde{b}_+^{(r)} \tilde{b}_-^{(r)T}.$$

7.2 The Trench algorithm

Proof Because $A^{(r)}B^{(r)} = I$, we have the following equation:

$$\left[\begin{array}{c|c} A^{(r-1)}M^r + \tilde{a}_+\tilde{b}_-^{(r)T} & A^{(r-1)}\tilde{b}_+^{(r)} + \tilde{a}_+ b_0^{(r)} \\ \hline \tilde{a}_-^T M^{(r)} + a_0 \tilde{b}_-^{(r)T} & \tilde{a}_-^T \tilde{b}_+^{(r)} + a_0 b_0^{(r)} \end{array}\right] = I.$$

Hence

$$A^{(r-1)}M^{(r)} + \tilde{a}_+\tilde{b}_-^{(r)T} = I,$$
$$A^{(r-1)}\tilde{b}_+^{(r)} + \tilde{a}_+ b_0^{(r)} = 0,$$
$$\tilde{a}_-^T M^{(r)} + a_0 \tilde{b}_-^{(r)T} = 0,$$
$$\tilde{a}_-^T \tilde{b}_+^{(r)} + a_0 b_0^{(r)} = 1.$$

The matrix $M^{(r)}$ can be eliminated by expressing it in terms of the inverse of $A^{(r-1)}$, which is denoted $B^{(r-1)}$. Multiply the first equation by $B^{(r-1)}$ on the left to get

$$M^{(r)} = B^{(r-1)} - B^{(r-1)}\tilde{a}_+\tilde{b}_-^{(r)T}.$$

The theorem then follows from the second equation above. □

Up until this point, we have assumed very little about the matrix $B^{(r)}$, only the existence of $B^{(r-1)}$. Now we will use the property of presymmetry to obtain a recursive procedure for computing $B^{(r)}$ from its half-border. Later, we will give a procedure for computing the half-border of $B^{(r)}$ from the half-border of $B^{(r-1)}$.

Theorem 7.2.3 *Let B be the inverse of a Toeplitz matrix. Then B is completely specified from its upper half-border by the ascending recursion*

$$b_{i+1,j+1} = b_{ij} + \frac{1}{b_0}[b_+ b_-^T - \tilde{b}_+ \tilde{b}_-^T], \quad \begin{array}{l} i = 1, \ldots, n-1, \\ j = 1, \ldots, n-1, \end{array}$$

and from the lower half-border by the descending recursion

$$b_{i-1,j-1} = b_{ij} + \frac{1}{b_0}[\tilde{b}_+ \tilde{b}_-^T - b_+ b_-^T], \quad \begin{array}{l} i = n, \ldots, 2, \\ j = n, \ldots, 2, \end{array}$$

where the first and last rows of the matrix are $(b_0, b_-)^T$ and $(\tilde{b}_-, b_0)^T$, respectively, and the first and last columns are (b_0, b_+) and $(\tilde{b}_+, b_0)$, respectively.

Proof We have already shown in Theorem 7.2.2 that

$$B^{(r)} = \left[\begin{array}{c|c} B^{(r-1)} + \frac{1}{b_0^{(r)}}\tilde{b}_+^{(r)}\tilde{b}_-^{(r)T} & \tilde{b}_+^{(r)} \\ \hline \tilde{b}_-^{(r)T} & b_0^{(r)} \end{array}\right].$$

Because $\boldsymbol{B}^{(r)}$ is persymmetric, we can also write

$$\boldsymbol{B}^{(r)} = \left[\begin{array}{c|c} b_0^{(r)} & \boldsymbol{b}_+^{(r)\mathrm{T}} \\ \hline \boldsymbol{b}_-^{(r)} & \boldsymbol{B}^{(r-1)} + \frac{1}{b_0^{(r)}} \boldsymbol{b}_+^{(r)} \boldsymbol{b}_-^{(r)\mathrm{T}} \end{array}\right].$$

Using these two partitions, we express an element $b_{ij}^{(r)}$ of $\boldsymbol{B}^{(r)}$ in two ways. The first gives

$$b_{ij}^{(r)} = b_{ij}^{(r-1)} + \frac{1}{b_0^{(r)}} (\tilde{\boldsymbol{b}}_+^{(r)} \tilde{\boldsymbol{b}}_-^{(r)\mathrm{T}})_{ij}, \qquad \begin{array}{l} i = 1, \ldots, n-1, \\ j = 1, \ldots, n-1. \end{array}$$

The second gives

$$b_{i+1,j+1}^{(r)} = b_{ij}^{(r-1)} + \frac{1}{b_0^{(r)}} (\boldsymbol{b}_+^{(r)} \boldsymbol{b}_-^{(r)\mathrm{T}})_{ij}, \qquad \begin{array}{l} i = 1, \ldots, n-1, \\ j = 1, \ldots, n-1, \end{array}$$

Eliminating $b_{ij}^{(r-1)}$ gives

$$b_{i+1,j+1}^{(r)} = b_{ij}^{(r)} + \frac{1}{b_0^{(r)}} (\boldsymbol{b}_+^{(r)} \boldsymbol{b}_-^{(r)\mathrm{T}} - \tilde{\boldsymbol{b}}_+^{(r)} \tilde{\boldsymbol{b}}_-^{(r)\mathrm{T}})_{ij}, \qquad \begin{array}{l} i = 1, \ldots, n-1, \\ j = 1, \ldots, n-1, \end{array}$$

which now involves only terms with superscript r. This is the ascending recursion. The descending recursion follows in the same way, and the proof is complete. $\square$

There is one extra piece of information contained in the proof that will be used later. By equating the element in the first column and first row of the two equations, we get the equation

$$b_0^{(r)} = B_{11}^{(r-1)} + \frac{1}{b_0^{(r)}} (\tilde{\boldsymbol{b}}_+^{(r)} \tilde{\boldsymbol{b}}_-^{(r)\mathrm{T}})_{11}.$$

Because $B_{11}^{(r-1)}$ equals $b_0^{(r-1)}$ and $\tilde{\boldsymbol{b}}_+ \tilde{\boldsymbol{b}}_-^{\mathrm{T}} = \boldsymbol{b}_+ \boldsymbol{b}_-^{\mathrm{T}}$, this can be written

$$b_0^{(r)} = b_0^{(r-1)} + \frac{1}{b_0^{(r)}} (\boldsymbol{b}_+^{(r)} \boldsymbol{b}_-^{(r)\mathrm{T}})_{11},$$

which relates $b_0^{(r)}$ and $b_0^{(r-1)}$.

Now we need a recursive algorithm that will compute the half-border of $\boldsymbol{B}^{(n)}$ from the half-border of $\boldsymbol{A}^{(n)}$. But we have already written equations relating the border of $\boldsymbol{B}^{(r)}$ to the border of $\boldsymbol{A}^{(r)}$. We need only to eliminate $\boldsymbol{A}^{(r-1)}$ and $\boldsymbol{M}$ to derive the algorithm. These can be eliminated by pushing back to the half-border at the previous iteration.

7.2 The Trench algorithm

Theorem 7.2.4 *The half-border of $\boldsymbol{B}^{(r)}$ satisfies the following recursions:*

(i) $\boldsymbol{b}_+^{(r)} = \dfrac{b_0^{(r)}}{b_0^{(r-1)}} \left\{ \begin{bmatrix} \boldsymbol{b}_+^{(r-1)} \\ 0 \end{bmatrix} - (\tilde{\boldsymbol{b}}_+^{(r-1)\text{T}} \boldsymbol{a}_+^{(r-1)} + b_0^{(r-1)} a_{r-1}) \begin{bmatrix} \tilde{\boldsymbol{b}}_-^{(r-1)} \\ b_0^{(r-1)} \end{bmatrix} \right\},$

(ii) $\boldsymbol{b}_-^{(r)} = \dfrac{b_0^{(r)}}{b_0^{(r-1)}} \left\{ \begin{bmatrix} \boldsymbol{b}_-^{(r-1)} \\ 0 \end{bmatrix} - (\tilde{\boldsymbol{b}}_-^{(r-1)\text{T}} \boldsymbol{a}_-^{(r-1)} + b_0^{(r-1)} a_{1-r}) \begin{bmatrix} \tilde{\boldsymbol{b}}_+^{(r-1)} \\ b_0^{(r-1)} \end{bmatrix} \right\},$

(iii) $b_0^{(r)} = b_0^{(r-1)} + \dfrac{1}{b_0^{(r)}} (\boldsymbol{b}_+^{(r)} \boldsymbol{b}_-^{(r)\text{T}})_{11}.$

Proof The third expression was already derived as a consequence of the proof of Theorem 7.2.3. By symmetry, it is clear that the second expression will be true if the first is true. It can be derived in the same way. Hence we need only to prove the first expression. We begin with the second of the set of four equations derived earlier in the proof of Theorem 7.2.2:

$$\boldsymbol{A}^{(r-1)} \tilde{\boldsymbol{b}}_+^{(r)} + \tilde{\boldsymbol{a}}_+^{(r)} b_0^{(r)} = \boldsymbol{0}.$$

Because $\boldsymbol{A}^{(r-1)}$ is persymmetric, this can also be written

$$\boldsymbol{A}^{(r-1)\text{T}} \boldsymbol{b}_+^{(r)} + \boldsymbol{a}_+^{(r)} b_0^{(r)} = \boldsymbol{0}$$

and

$$\boldsymbol{b}_+^{(r)} = -b_0^{(r)} \boldsymbol{B}^{(r-1)\text{T}} \boldsymbol{a}_+^{(r)}.$$

Now break this down, using again the same partition for $\boldsymbol{B}^{(r-1)}$ as was used before:

$$\boldsymbol{b}_+^{(r)} = -b_0^{(r)} \left[\begin{array}{c|c} \boldsymbol{B}^{(r-2)} + \dfrac{1}{b_0^{(r-1)}} \tilde{\boldsymbol{b}}_+^{(r-1)} \tilde{\boldsymbol{b}}_-^{(r-1)\text{T}} & \tilde{\boldsymbol{b}}_+^{(r-1)} \\ \hline \tilde{\boldsymbol{b}}_-^{(r-1)\text{T}} & b_0^{(r-1)} \end{array} \right]^{\text{T}} \begin{bmatrix} \boldsymbol{a}_+^{(r-1)} \\ a_{r-1} \end{bmatrix},$$

where $\boldsymbol{a}_+^{(r)}$ has been written as $\boldsymbol{a}_+^{(r-1)}$ with one more component appended. This equation gives the desired recursion, but it can be cleaned up by reusing the equation

$$\boldsymbol{b}_+^{(r-1)} = -b_0^{(r-1)} \boldsymbol{B}^{(r-2)\text{T}} \boldsymbol{a}_+^{(r-1)}.$$

Then

$$\boldsymbol{b}_+^{(r)} = \dfrac{b_0^{(r)}}{b_0^{(r-1)}} \boldsymbol{b}_+^{(r-1)} - b_0^{(r)} \left[\begin{array}{c|c} \dfrac{1}{b_0^{(r-1)}} \tilde{\boldsymbol{b}}_-^{(r-1)} \tilde{\boldsymbol{b}}_+^{(r-1)\text{T}} & \tilde{\boldsymbol{b}}_-^{(r-1)} \\ \hline \tilde{\boldsymbol{b}}_+^{(r-1)\text{T}} & b_0^{(r-1)} \end{array} \right] \begin{bmatrix} \boldsymbol{a}_+^{(r-1)} \\ a_{r-1} \end{bmatrix},$$

which can be rewritten as in the statement of the theorem. □

Fast algorithms for solving Toeplitz systems

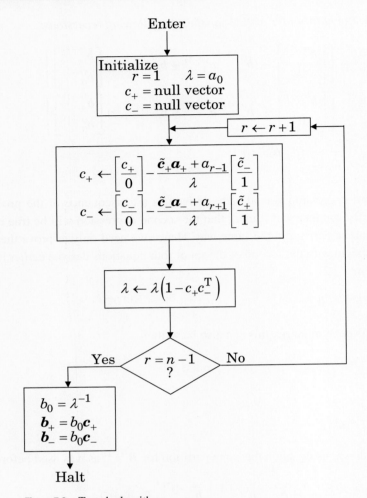

Figure 7.3 Trench algorithm

We now have the basis for the Trench algorithm. All we need to do is to redefine the notation so that all variables with superscript index r are on the left side of the equation, and those with superscript index $r-1$ are on the right side.

The Trench algorithm is summarized in Figure 7.3. The flow is based on Theorem 7.2.4, but the iterates $\boldsymbol{b}_+$, $\boldsymbol{b}_-$, and b_0 have been replaced by the normalized vectors $\boldsymbol{c}_+$, $\boldsymbol{c}_-$, and λ, defined as

$$\boldsymbol{c}_+ = \frac{1}{b_0^{(r)}} \boldsymbol{b}_+^{(r)},$$

$$\boldsymbol{c}_- = \frac{1}{b_0^{(r)}} \boldsymbol{b}_-^{(r)},$$

$$\lambda = \frac{1}{b_0^{(r)}},$$

so that the recursion becomes

$$c_+^{(r)} = \begin{bmatrix} c_+^{(r-1)} \\ 0 \end{bmatrix} - \frac{\tilde{c}_+^{(r-1)T} a_+^{(r-1)} + a_{r-1}}{\lambda^{(r-1)}} \begin{bmatrix} \tilde{c}_-^{(r-1)} \\ 1 \end{bmatrix},$$

$$c_-^{(r)} = \begin{bmatrix} c_-^{(r-1)} \\ 0 \end{bmatrix} - \frac{\tilde{c}_-^{(r-1)T} a_-^{(r-1)} + a_{1-r}}{\lambda^{(r-1)}} \begin{bmatrix} \tilde{c}_+^{(r-1)} \\ 1 \end{bmatrix},$$

$$\lambda^{(r)} = \lambda^{(r-1)}(1 - c_+^{(r)} c_-^{(r)T}).$$

This is the form of the equations used in Figure 7.3.

7.3 Methods based on the euclidean algorithm

Several methods of solving Toeplitz systems of equations make use of the *euclidean algorithm*, as will be described in this section. The conventional formulation of the euclidean algorithm is used in this discussion. However, an accelerated recursive form of the euclidean algorithm, as described in Section 4.7, is also available.

We begin the discussion with the following Toeplitz system of equations:

$$\begin{bmatrix} a_{n-1} & a_{n-2} & a_{n-3} & \cdots & a_0 \\ a_n & a_{n-1} & a_{n-2} & \cdots & a_1 \\ a_{n+1} & a_n & a_{n-1} & \cdots & a_2 \\ \vdots & \vdots & \vdots & & \vdots \\ a_{2n-2} & a_{2n-3} & a_{2n-4} & \cdots & a_{n-1} \end{bmatrix} \begin{bmatrix} f_1 \\ f_2 \\ f_3 \\ \vdots \\ f_n \end{bmatrix} = \begin{bmatrix} -a_n \\ -a_{n+1} \\ -a_{n-2} \\ \vdots \\ -a_{2n-1} \end{bmatrix}.$$

This same Toeplitz system of equations will be inverted in Section 7.4 by using the Berlekamp–Massey algorithm. In this section, it is inverted by using the euclidean algorithm.

Let

$$a(x) = \sum_{i=0}^{2n-1} a_i x^i,$$

$$f(x) = 1 + \sum_{i=1}^{n} f_i x^i$$

and consider the polynomial product

$$g(x) = f(x)a(x).$$

To be in accord with the matrix product above, we see that

$$g_i = 0, \quad i = n, \ldots, 2n-1.$$

However, for i larger than $2n$, g_i may take any value. We shall restate the matrix inverse problem as the task of finding $f(x)$ and $g(x)$ satisfying the conditions $\deg f(x) \leq n$, $\deg g(x) \leq n - 1$, and

$$g(x) = f(x)a(x) \pmod{x^{2n}}.$$

Of course, one way to solve this polynomial problem is to solve the original matrix equations for f and then to compute $g(x)$. Another method is to use the euclidean algorithm for polynomials.

Let us look to the proof of the euclidean algorithm to see how the equation can be solved for $f(x)$ and $g(x)$. From that proof, we see that the euclidean algorithm computes

$$\begin{bmatrix} s^{(r)}(x) \\ t^{(r)}(x) \end{bmatrix} = \begin{bmatrix} A_{11}^{(r)}(x) & A_{12}^{(r)}(x) \\ A_{21}^{(r)}(x) & A_{22}^{(r)}(x) \end{bmatrix} \begin{bmatrix} s(x) \\ t(x) \end{bmatrix},$$

so that

$$t^{(r)}(x) = A_{22}^{(r)}(x)t(x) \pmod{s(x)}.$$

This is the form of the polynomial equation being solved if we take $t(x) = a(x)$ and $s(x) = x^{2n}$. Such a statement holds for each value of r. If we can find an r for which $\deg A_{22}^{(r)}(x) \leq n$ and $\deg t^{(r)}(x) \leq n - 1$, then these polynomials can be given as the solution to the polynomial equation. If such an r exists, then the polynomials $A_{22}^{(r)}(x)$ and $t^{(r)}(x)$ must equal the desired $f(x)$ and $g(x)$. To this end, choose that value of r satisfying

$$\deg t^{(r-1)}(x) \geq n,$$
$$\deg t^{(r)}(x) \leq n - 1.$$

This defines a unique value r because $\deg t^{(0)}(x) = 2n$, and the degree of $t^{(r)}(x)$ is strictly decreasing as r is increasing. By the definition of r, we have satisfied the first requirement

$$\deg t^{(r)}(x) \leq n - 1.$$

As r is increasing, the degree of $A_{22}^{(r)}(x)$ is increasing. We need only to show that

$$\deg A_{22}^{(r)}(x) \leq n.$$

This we prove by working with the inverse of the matrix $A^{(r)}(x)$. First, recall that

$$A^{(r)}(x) = \prod_{\ell=1}^{r} \begin{bmatrix} 0 & 1 \\ 1 & -Q^{(\ell)}(x) \end{bmatrix}.$$

From this equation, it is clear that $\deg A_{22}^{(r)}(x) > \deg A_{12}^{(r)}(x)$. Also recall that $\deg s^{(r)}(x) > \deg t^{(r)}(x)$. From these inequalities and the matrix equation

$$\begin{bmatrix} s(x) \\ t(x) \end{bmatrix} = (-1)^r \begin{bmatrix} A_{22}^{(r)}(x) & -A_{12}^{(r)}(x) \\ -A_{21}^{(r)}(x) & A_{11}^{(r)}(x) \end{bmatrix} \begin{bmatrix} s^{(r)}(x) \\ t^{(r)}(x) \end{bmatrix},$$

it can be concluded that $\deg s(x) = \deg A_{22}^{(r)}(x) + \deg s^{(r)}(x)$, and because $s^{(r)}(x) = t^{(r-1)}(x)$, this becomes

$$\deg A_{22}^{(r)}(x) = \deg s(x) - \deg t^{(r-1)}(x)$$
$$\leq 2n - n = n,$$

where the inequality follows from the definition of r.

We now have developed most of the proof of the following theorem.

Theorem 7.3.1 *Given $s^{(0)}(x) = x^{2n}$ and $t^{(0)}(x) = a(x)$, let*

$$A^{(0)}(x) = \begin{bmatrix} 1 & 0 \\ 0 & 1 \end{bmatrix}.$$

Solve the following recursive equations until $\deg t^{(r)}(x) \leq n - 1$:

$$Q^{(r)}(x) = \left\lfloor \frac{s^{(r-1)}(x)}{t^{(r-1)}(x)} \right\rfloor,$$

$$A^{(r)}(x) = \begin{bmatrix} 0 & 1 \\ 1 & -Q^{(r)}(x) \end{bmatrix} A^{(r-1)}(x),$$

$$\begin{bmatrix} s^{(r)}(x) \\ t^{(r)}(x) \end{bmatrix} = \begin{bmatrix} 0 & 1 \\ 1 & -Q^{(r)}(x) \end{bmatrix} \begin{bmatrix} s^{(r-1)}(x) \\ t^{(r-1)}(x) \end{bmatrix},$$

and let

$$g(x) = \Delta^{-1} t^{(r)}(x),$$
$$f(x) = \Delta^{-1} A_{22}^{(r)}(x),$$

where $\Delta = A_{22}^{(r)}(0)$. Provided that Δ is nonzero, these satisfy the equation

$$g(x) = f(x)a(x) \pmod{x^{2n}}$$

with $\deg f(x) \leq n$, $\deg g(x) \leq n - 1$, and $f_0 = 1$.

Proof The division by Δ ensures that $f_0 = 1$. Otherwise, we have seen, prior to the statement of the theorem, that the final equation and conditions will be satisfied. □

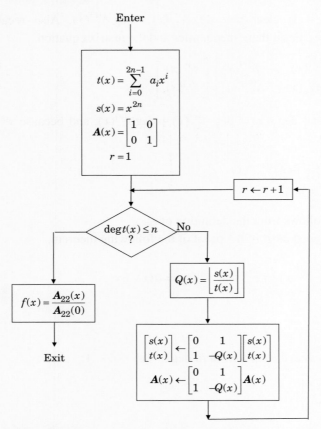

Figure 7.4 Inverting Toeplitz systems with the euclidean algorithm

Hence, provided the Toeplitz system of equations is invertible, we have a way to invert the Toeplitz system of equations, which we will also treat using the Berlekamp–Massey algorithm in the next section. A flow diagram is shown in Figure 7.4. If the Toeplitz system is not invertible, then the use of the euclidean algorithm, as described here, will still produce that $A_{22}^{(r)}(x)$ of least degree satisfying its defining equation, but Δ will be equal to zero, and $f(x)$ is now undefined. When Δ equals zero, the algorithm has produced instead a solution to the equation

$$\begin{bmatrix} a_{n-1} & a_{n-2} & a_{n-3} & \cdots & a_0 \\ a_n & a_{n-1} & a_{n-2} & \cdots & a_1 \\ a_{n+1} & a_n & a_{n-1} & \cdots & a_2 \\ \vdots & \vdots & \vdots & & \vdots \\ a_{2n-2} & a_{2n-3} & a_{2n-4} & \cdots & a_{n-1} \end{bmatrix} \begin{bmatrix} h_1 \\ h_2 \\ h_3 \\ \vdots \\ h_n \end{bmatrix} = \begin{bmatrix} 0 \\ 0 \\ 0 \\ \vdots \\ 0 \end{bmatrix}.$$

7.4 The Berlekamp–Massey algorithm

The *Berlekamp–Massey algorithm* solves a Toeplitz system of equations of the form[1]

$$\begin{bmatrix} a_{n-1} & a_{n-2} & a_{n-3} & \cdots & a_0 \\ a_n & a_{n-1} & a_{n-2} & \cdots & a_1 \\ a_{n+1} & a_n & a_{n-1} & \cdots & a_2 \\ \vdots & \vdots & \vdots & & \vdots \\ a_{2n-2} & a_{2n-3} & a_{2n-4} & \cdots & a_{n-1} \end{bmatrix} \begin{bmatrix} f_1 \\ f_2 \\ \vdots \\ f_n \end{bmatrix} = \begin{bmatrix} -a_n \\ -a_{n+1} \\ \vdots \\ -a_{2n-1} \end{bmatrix},$$

in any field F, for the vector f. In contrast to the formulation for the Levinson algorithm, the Toeplitz matrix is not required to be symmetric. On the other hand, in this problem the vector on the right is not arbitrary – it is made up of elements of the matrix on the left.

The best way to approach the Berlekamp–Massey algorithm is to view the matrix equation as a description of an autoregressive filter. Suppose the vector f is known. Then the first row of the above matrix equation defines the matrix element a_n in terms of $a_0, \ldots, a_{n-1}$. The second row then defines a_{n+1} in terms of $a_1, \ldots, a_n$, and so forth. This sequential process is summarized by the equation

$$a_j = -\sum_{i=1}^{n} f_i a_{j-i}, \qquad j = n, \ldots, 2n-1.$$

For fixed f, this is the equation of an autoregressive filter producing a sequence of a_j starting with the initial terms of the sequence. The autoregressive filter may be implemented as a linear feedback shift register with the values of its taps given by the components of f.

Looked at in this way, the problem of solving the Toeplitz system of equations becomes a problem of finding the autoregressive filter, shown in Figure 7.5, that will produce the specified sequence of filter outputs, as denoted by the a_j. If the Toeplitz matrix is invertible, there will be exactly one autoregressive filter that will produce the sequence of a_j. If the matrix is not invertible, then there may be many solutions or there may be none. In this situation, the Berlekamp–Massey algorithm will always produce an output that is the shortest autoregressive filter that satisfies the equations. If the Toeplitz system has more than one solution, then the Berlekamp–Massey algorithm will find the solution corresponding to the shortest autoregressive filter. If the original Toeplitz system has no solution, then the Berlekamp–Massey algorithm will develop an autoregressive filter having nonzero tap weights f_i with i larger than n.

---

[1] In this section it is more suggestive to index the components of f from 1 to n and to write the elements of A as a_0 to a_{2n-1}.

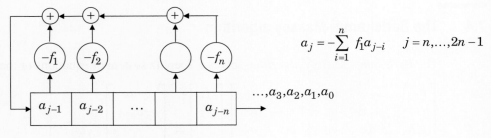

Figure 7.5 An autoregressive filter

Any procedure for finding the autoregressive filter is also a procedure for solving the matrix equation for the vector f. We shall develop such a recursive procedure for finding the autoregressive filter. The procedure does not assume any special properties for the sequence $a_0, a_1, \ldots, a_{2n-1}$.

A linear feedback shift register can be described by giving the length L of the shift register and a feedback polynomial $f(x)$,

$$f(x) = f_n x^n + f_{n-1} x^{n-1} + \cdots + f_1 x + 1.$$

The reason that the length of the shift register must be stated is that the length of the shift register may be larger than the degree of $f(x)$ because one or more of the rightmost stages might not be tapped.

To find the required shift register, we must determine two quantities: the shift-register length L and the feedback connection polynomial $f(x)$, where $\deg f(x) \leq L$. We denote this pair by $(L, f(x))$. We must find the feedback shift register that will produce the sequence $a_0, \ldots, a_{2n-1}$ when properly initialized, and such that it is the shortest feedback shift register with this property.

The design procedure is recursive. For each r, starting with $r = 1$, we will design a feedback shift register for producing the sequence $a_0, \ldots, a_r$. The pair $(L_r, f^{(r)}(x))$ denotes this minimum-length shift register for producing $a_0, \ldots, a_r$. This shift register need not be unique; there may be several such shift registers of length L_r. We only use one of them. At the start of iteration r, we will have constructed a list of shift registers:

$(L_1, f^{(1)}(x))$,
$(L_2, f^{(2)}(x))$,
$\vdots$
$(L_{r-1}, f^{(r-1)}(x))$,

where the lengths are nondecreasing. The Berlekamp–Massey algorithm computes a new shortest-length shift register $(L_r, f^{(r)}(x))$ that generates the sequence $a_0, \ldots, a_{r-1}, a_r$. This will be done by again using the most recent shift register without

7.4 The Berlekamp–Massey algorithm

change, if possible, and otherwise, modifying its length and tap weights as necessary to make it work.

To begin iteration r, first compute the next output of the $(r-1)$th shift register:

$$\widehat{a}_r = -\sum_{j=1}^{L_{r-1}} f_j^{(r-1)} a_{r-j}.$$

Because L_{r-1} may be larger than the degree of $f^{(r-1)}$, some terms in the sum may be equal to zero. The sum could be written as a sum from one to deg $f^{(r-1)}(x)$. However, the chosen notation is less cumbersome.

Let Δ_r be the difference between the desired output a_r and the actual output of the most recent shift register:

$$\Delta_r = a_r - \widehat{a}_r = a_r + \sum_{j=1}^{L_{r-1}} f_j^{(r-1)} a_{r-j}.$$

This can be written compactly as

$$\Delta_r = \sum_{j=0}^{L_{r-1}} f_j^{(r-1)} a_{r-j}.$$

If Δ_r is zero, then set $(L_r, f^{(r)}(x)) = (L_{r-1}, f^{(r-1)}(x))$, and the rth iteration is complete. Otherwise, the shift-register taps are modified by updating the connection polynomial as follows:

$$f^{(r)}(x) = f^{(r-1)}(x) + A x^\ell f^{(m-1)}(x),$$

where A is a field element, ℓ is an integer, and $f^{(m-1)}(x)$ is one of the connection polynomials appearing earlier on the list. Now, with this new polynomial, let

$$\Delta'_r = \sum_{j=0}^{L_{r-1}} f_j^{(r)} a_{r-j}$$

$$= \sum_{j=0}^{L_{r-1}} f_j^{(r-1)} a_{r-j} + A \sum_{j=0}^{L_{r-1}} f_j^{(m-1)} a_{r-j-\ell}.$$

We now choose m, ℓ, and A to make Δ'_r equal to zero. Choose an m smaller than r for which $\Delta_m \neq 0$, choose $\ell = r - m$, and choose $A = -\Delta_m^{-1} \Delta_r$. Then

$$\Delta'_r = \Delta_r - \frac{\Delta_r}{\Delta_m} \Delta_m = 0,$$

which means that the modified shift register will generate the sequence $a_1, \ldots, a_{r-1}, a_r$. We now must ensure that it is the smallest-length such shift register. We have not yet specified which of those m for which $\Delta_m \neq 0$ should be chosen. If we choose m as the most recent iteration at which $L_m > L_{m-1}$, we will get a shortest-length shift register at every iteration, but this last refinement will take some time to develop.

Theorem 7.4.2 asserts that the given procedure does compute a shortest shift register that produces the given sequence. Before giving this theorem, we give a bound on the length of the shift register.

Theorem 7.4.1 *Suppose that* $(L_{r-1}, f^{r-1}(x))$ *is the linear feedback shift register of shortest length that produces* $a_1, \ldots, a_{r-1}$; *and* $(L_r, f^{(r)}(x))$ *is the linear feedback shift register of shortest length that produces* $a_1, \ldots, a_{r-1}, a_r$. *If* $f^{(r)}(x) \neq f^{(r-1)}(x)$, *then* $L_r \geq \max[L_{r-1}, r - L_{r-1}]$.

Proof The inequality to be proved is a combination of two inequalities:

$$L_r \geq L_{r-1}$$

and

$$L_r \geq r - L_{r-1}.$$

The first inequality is obvious, because if a linear feedback shift register produces a sequence, it must also produce any beginning segment of the sequence. The second inequality is obvious if $L_{r-1} \geq r$. Hence assume $L_{r-1} < r$. Suppose the second inequality is not satisfied, and look for a contradiction. Then $L_r \leq r - 1 - L_{r-1}$. Let $c(x) = f^{(r-1)}(x)$, $b(x) = f^{(r)}(x)$, $L = L_{r-1}$, and $L' = L_r$. By assumption we have $r \geq L + L' + 1$ and $L < r$. Next, by the assumptions of the theorem,

$$a_r \neq -\sum_{i=1}^{L} c_i a_{r-i},$$

$$a_j = -\sum_{i=1}^{L} c_i a_{j-i}, \quad j = L+1, \ldots, r-1,$$

and

$$a_j = -\sum_{k=1}^{L'} b_k a_{j-k}, \quad j = L'+1, \ldots, r.$$

Now establish the contradiction. First,

$$a_r = -\sum_{k=1}^{L'} b_k a_{r-k} = \sum_{k=1}^{L'} b_k \sum_{i=1}^{L} c_i a_{r-k-i},$$

where the expansion of a_{r-k} as another sum is valid because $r - k$ runs from $r - 1$ down to $r - L'$, which is in the range $L + 1, \ldots, r - 1$ because of the assumption

7.4 The Berlekamp–Massey algorithm

$r \geq L + L' + 1$. Second,

$$a_r \neq -\sum_{i=1}^{L} c_i a_{r-i} = \sum_{i=1}^{L} c_i \sum_{k=1}^{L'} b_k a_{r-i-k},$$

where the expansion of a_{r-i} as another sum is valid because $r - i$ runs from $r - 1$ down to $r - L$, which is in the range $L' + 1, \ldots, r - 1$, again because of the assumption $r \geq L + L' + 1$. The summations on the right side can be interchanged to agree with the right side of the previous equation. Hence we get the contradiction: $a_r \neq a_r$. The contradiction proves the theorem. □

If we can design a shift register that satisfies the inequality of Theorem 7.4.1 with equality, then it must be of shortest length. The proof of Theorem 7.4.2 gives a construction for this shift register.

Theorem 7.4.2 *Suppose that* $(L_i, f^{(i)}(x))$, $i = 1, \ldots, r$, *is a sequence of minimum-length, linear-feedback shift registers such that* $f^{(i)}(x)$ *produces* $a_1, \ldots, a_i$. *If* $f^{(r)}(x) \neq f^{(r-1)}(x)$, *then*

$$L_r = \max[L_{r-1}, r - L_{r-1}],$$

and any shift register that produces $a_1, \ldots, a_r$ *and has length equal to the right side is a minimum-length shift register.*

Proof By Theorem 7.4.1, L_r cannot be smaller than the right side. We will construct a shift register that produces the required sequence and whose length equals the right side, so it must be a minimum-length shift register. The proof is by induction. We give a construction for a shift register satisfying the conditions of the theorem, assuming that we have already constructed such shift registers for all $k \leq r - 1$. For $k = 1, \ldots, r - 1$, let $(L_k, f^{(k)}(x))$ be the minimum-length shift register that generates $a_1, \ldots, a_k$. For the induction argument, assume that

$$L_k = \max[L_{k-1}, k - L_{k-1}], \quad k = 1, \ldots, n - 1,$$

whenever $f^{(k)}(x) \neq f^{(k-1)}(x)$. This is clearly true for $k = 0$ if a_1 is nonzero because $L_0 = 0$ and $L_1 = 1$. More generally, if a_i is the first nonzero term of the given sequence, then $L_{i-1} = 0$ and $L_i = i$. The induction argument then begins at $k = i$.

Let m denote the value that k had at the most recent iteration step that required a length change. Thus m is the integer such that

$$L_{r-1} = L_m > L_{m-1}.$$

We now have

$$a_j + \sum_{i=1}^{L_{r-1}} f_i^{(r-1)} a_{j-i} = \sum_{i=0}^{L_{r-1}} f_i^{(r-1)} a_{j-i} = \begin{cases} 0, & j = L_{r-1}, \ldots, r-1, \\ \Delta_r, & j = r, \end{cases}$$

If $\Delta_r = 0$, then the shift register $(L_{r-1}, f^{(r-1)}(x))$ also generates the rth term of the sequence, so that $(L_r, f^{(r)}(x)) = (L_{r-1}, f^{(r-1)}(x))$. If $\Delta_r \neq 0$, then a new shift register must be formed. Recall that a change in shift-register length occurred at $k = m$. Hence

$$a_j + \sum_{i=1}^{L_{m-1}} f_i^{(m-1)} a_{j-i} = \begin{cases} 0, & j = L_{m-1}, \ldots, m-1, \\ \Delta_m \neq 0, & j = m, \end{cases}$$

and by the induction hypothesis

$$L_{r-1} = L_m = \max[L_{m-1}, m - L_{m-1}]$$
$$= m - L_{m-1}$$

because $L_m > L_{m-1}$. Now, as before, choose the new polynomial

$$f^{(r)}(x) = f^{(r-1)}(x) - \Delta_r \Delta_m^{-1} x^{r-m} f^{(m-1)}(x)$$

and let $L_r = \deg f^{(r)}(x)$. Because $\deg f^{(r-1)}(x) \leq L_{r-1}$ and $\deg[x^{r-m} f^{(m-1)}(x)] \leq r - m + L_{m-1}$, this gives

$$L_r \leq \max[L_{r-1}, r - m + L_{m-1}]$$
$$\leq \max[L_{r-1}, r - L_{r-1}].$$

Hence, by recalling Theorem 7.4.1, if $f^{(r)}(x)$ produces $a_1, \ldots, a_r$, then $L_r = \max[L_{r-1}, r - L_{r-1}]$. It only remains to prove that the shift register $(L_r, f^{(r)}(x))$ produces the required sequence. This is done by direct computation of the difference between a_j and the shift-register feedback, which by design satisfies

$$a_j - \left(-\sum_{i=1}^{L_r} f_i^{(r)} a_{j-i}\right) = a_j + \sum_{i=1}^{L_{r-1}} f_i^{(r-1)} a_{j-i}$$
$$- \Delta_r \Delta_m^{-1} \left[a_{j-r+m} + \sum_{i=1}^{L_{m-1}} f_i^{(m-1)} a_{j-r+m-i}\right]$$
$$= \begin{cases} 0, & j = L_r, L_r+1, \ldots, r-1, \\ \Delta_r - \Delta_r \Delta_m^{-1} \Delta_m = 0, & j = r. \end{cases}$$

Hence the shift register $(L_r, f^{(r)}(x))$ produces $a_1, \ldots, a_r$. In particular, $(L_n, f^{(n)}(x))$ produces $a_1, \ldots, a_n$, and the theorem is proved. □

Theorem 7.4.3 (Berlekamp–Massey algorithm) *In any field, let the sequence $a_1, \ldots, a_n$ be given. Under the initial conditions $f^{(0)}(x) = 1$, $t^{(0)}(x) = 1$, and $L_0 = 0$,* let the following set of recursive equations be used to compute $f^{(n)}(x)$:

$$\Delta_r = \sum_{j=0}^{n-1} f_j^{(r-1)} a_{r-j},$$

$$L_r = \delta_r(r - L_{r-1}) + (1 - \delta_r)L_{r-1},$$

$$\begin{bmatrix} f^{(r)}(x) \\ t^{(r)}(x) \end{bmatrix} = \begin{bmatrix} 1 & -\Delta_r x \\ \Delta_r^{-1}\delta_r & (1 - \delta_r)x \end{bmatrix} \begin{bmatrix} f^{(r-1)}(x) \\ t^{(r-1)}(x) \end{bmatrix},$$

for $r = 1, \ldots, n$, where $\delta_r = 1$ if both $\Delta_r \neq 0$ and $2L_{r-1} \leq r - 1$ and, otherwise, $\delta_r = 0$. Then $f^{(2t)}(x)$ is the smallest-degree polynomial with the properties that $f_0^{(2t)} = 1$ and

$$a_r + \sum_{j=1}^{n-1} f_j^{(2t)} a_{r-j} = 0, \quad r = L_{2t} + 1, \ldots, 2t.$$

Proof Follows from the proof of Theorem 7.4.2. □

In the theorem, Δ_r may be zero but only when δ_r is zero. The term $\Delta_r^{-1}\delta_r$ is then understood to be zero.

A flow diagram of the Berlekamp–Massey algorithm is shown in Figure 7.6. The algorithm requires a number of multiplications in the rth iteration that is approximately equal to twice the degree of $f^{(r)}(x)$. The degree of $f^{(r)}(x)$ is about $r/2$, and there are $2n$ iterations, so there are about $2n^2 = \Sigma_{r=0}^{2n} r$ multiplications and about the same number of additions. In brief, we say that there are on the order of n^2 multiplications in the Berlekamp–Massey algorithm.

7.5 An accelerated Berlekamp–Massey algorithm

All methods that we have studied so far for inverting a Toeplitz system of equations, including the Berlekamp–Massey algorithm, require a number of multiplications that is proportional to n^2. In this section, we shall use a doubling strategy to reduce the computational complexity of the Berlekamp–Massey algorithm for large n.

The number of multiplications used in the rth iteration of the basic algorithm is approximately equal to twice the degree of $f^{(r)}(x)$. In the early iterations, the degree of $f^{(r)}(x)$ is small. These iterations are easy but as r increases, the complexity grows. The accelerated algorithm will take advantage of the simplicity of the early iterations by doing only a few iterations at a time. After each such batch of iterations, the partial result is used to modify the problem so as to absorb the partial answers computed in that batch. The next batch of iterations starts the Berlekamp–Massey algorithm anew for

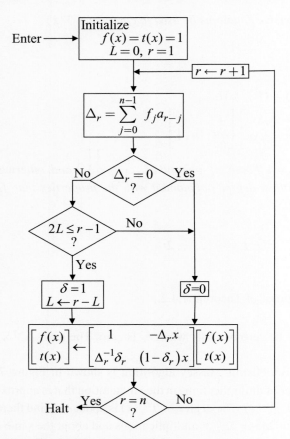

Figure 7.6 Berlekamp–Massey algorithm

the modified problem and with $f(x)$ reinitialized to one. Together with this technique, we will use a doubling structure to reduce the computations.

The development below will begin with a more compact organization of the Berlekamp–Massey algorithm. We replace the polynomials $f^{(r)}(x)$ and $t^{(r)}(x)$ by a two by two matrix of polynomials:

$$\boldsymbol{F}^{(r)}(x) = \begin{bmatrix} F_{11}^{(r)}(x) & F_{12}^{(r)}(x) \\ F_{21}^{(r)}(x) & F_{22}^{(r)}(x) \end{bmatrix}.$$

The element $F_{ij}^{(r)}(x)$ is a polynomial with coefficients denoted by $F_{ij,k}^{(r)}$. The matrix $\boldsymbol{F}^{(r)}(x)$ will be defined in such a way that $f^{(r)}(x)$ and $t^{(r)}(x)$ can be computed by the equations

$$\begin{bmatrix} f^{(r)}(x) \\ t^{(r)}(x) \end{bmatrix} = \begin{bmatrix} F_{11}^{(r)}(x) + F_{12}^{(r)}(x) \\ F_{21}^{(r)}(x) + F_{22}^{(r)}(x) \end{bmatrix} = \boldsymbol{F}^{(r)}(x) \begin{bmatrix} 1 \\ 1 \end{bmatrix}.$$

7.5 An accelerated Berlekamp–Massey algorithm

Recall that the computations of the Berlekamp–Massey algorithm reside primarily in the two equations

$$\Delta_r = \sum_{j=0}^{n-1} f_j^{(r-1)} a_{r-j},$$

$$\begin{bmatrix} f^{(r)}(x) \\ t^{(r)}(x) \end{bmatrix} = \begin{bmatrix} 1 & -\Delta_r x \\ \Delta_r^{-1}\delta_r & (1-\delta_r)x \end{bmatrix} \begin{bmatrix} f^{(r-1)}(x) \\ t^{(r-1)}(x) \end{bmatrix}$$

$$= \begin{bmatrix} 1 & -\Delta_r x \\ \Delta_r^{-1}\delta_r & (1-\delta_r)x \end{bmatrix} \cdots \begin{bmatrix} 1 & -\Delta_1 x \\ \Delta_1^{-1}\delta_1 & (1-\delta_1)x \end{bmatrix} \begin{bmatrix} 1 \\ 1 \end{bmatrix}.$$

Consequently,

$$F^{(r)}(x) = \begin{bmatrix} 1 & -\Delta_r x \\ \Delta_r^{-1}\delta_r & (1-\delta_r)x \end{bmatrix} \cdots \begin{bmatrix} 1 & -\Delta_1 x \\ \Delta_1^{-1}\delta_1 & (1-\delta_1)x \end{bmatrix}.$$

It serves just as well to update $F^{(r)}(x)$ as it does to update $f^{(r)}(x)$ and $t^{(r)}(x)$, although updating $F^{(r)}(x)$ directly can involve about twice as many multiplications because it has four elements rather than two. We accept the greater number of multiplications for now because, in this form, the computation is amenable to a doubling strategy. The penalty can be overcome later by a reorganization of the computations.

The recursive form of the Berlekamp–Massey algorithm is built around the equations

$$\Delta_r = \sum_{j=0}^{n-1} F_{11,j}^{(r-1)} a_{r-j} + \sum_{j=0}^{n-1} F_{12,j}^{(r-1)} a_{r-j},$$

$$F^{(r)}(x) = \begin{bmatrix} 1 & -\Delta_r x \\ \Delta_r^{-1}\delta_r & (1-\delta_r)x \end{bmatrix} F^{(r-1)}(x),$$

which is equivalent to the earlier form.

To split the algorithm into halves, suppose that n is even, and let

$$F^{(n)}(x) = F'^{(n)}(x) F''^{(n/2)}(x),$$

where

$$F'^{(n)}(x) = \prod_{r=n}^{(n/2)+1} \begin{bmatrix} 1 & -\Delta_r x \\ \Delta_r^{-1}\delta_r & (1-\delta_r)x \end{bmatrix},$$

$$F''^{(n/2)}(x) = \prod_{r=n/2}^{1} \begin{bmatrix} 1 & -\Delta_r x \\ \Delta_r^{-1}\delta_r & (1-\delta_r)x \end{bmatrix}.$$

We will compute the two halves separately, and then multiply the two halves together. This will entail less work than the original organization. This revision means that we will also need to reorganize the equations for Δ_r. Think of Δ_r as the rth coefficient of the first component of the two-vector of polynomials

$$\begin{bmatrix} \Delta(x) \\ \Delta'(x) \end{bmatrix} = \begin{bmatrix} F_{11}^{(r-1)}(x) & F_{12}^{(r-1)}(x) \\ F_{21}^{(r-1)}(x) & F_{22}^{(r-1)}(x) \end{bmatrix} \begin{bmatrix} a(x) \\ a(x) \end{bmatrix}.$$

Hence, for r larger than $n/2$,

$$\begin{bmatrix} \Delta(x) \\ \Delta'(x) \end{bmatrix} = \boldsymbol{F}''^{(r-1)}(x) \boldsymbol{F}'''^{(n/2)}(x) \begin{bmatrix} a(x) \\ a(x) \end{bmatrix}$$

$$= \boldsymbol{F}''^{(r-1)}(x) \boldsymbol{a}^{(n/2)}(x),$$

where

$$\boldsymbol{a}^{(n/2)}(x) = \boldsymbol{F}'''^{(n/2)}(x) \begin{bmatrix} a(x) \\ a(x) \end{bmatrix}.$$

This completes the splitting of the Berlekamp–Massey algorithm. The basic algorithm now is written

$$\Delta_r = \sum_{j=0}^{n-1} F_{11,j}^{(r-1)} a_{1,r-j} + \sum_{j=0}^{n-1} F_{12,j}^{(r-1)} a_{2,r-j},$$

$$\boldsymbol{F}^{(r)}(x) = \begin{bmatrix} 1 & -\Delta_r x \\ \Delta_r^{-1} \delta_r & (1 - \delta_r)x \end{bmatrix} \boldsymbol{F}^{(r-1)}(x),$$

where $\boldsymbol{F}^{(r)}(x)$ may represent either $\boldsymbol{F}'^{(r)}(x)$ or $\boldsymbol{F}'''^{(r)}(x)$, and $(a_1(x), a_2(x))$ represents $(a(x), a(x))$ in the first half of the computation and is updated to $\boldsymbol{a}^{(n/2)}(x)$ in the second half of the computation. After both halves are completed, $\boldsymbol{F}^{(n)}(x)$ is obtained by multiplying its two halves.

The Berlekamp–Massey algorithm split into two halves is shown in Figure 7.7. Notice that each of the halves is itself a Berlekamp–Massey algorithm. Hence, if $n/2$ is even, the two halves can, in turn, be split. If n is a power of two, then the splitting can continue until pieces that are only one iteration long are reached. These are executed but are not quite trivial.

Figure 7.8 shows the Berlekamp–Massey algorithm in a fully recursive form. All of the computational work is in the polynomial products that combine the two halves. These are convolutions, and they can be computed by any algorithm for linear convolution.

To estimate the complexity of the recursive Berlekamp–Massey algorithm, notice that the number of multiplications used by iteration m is twice the number of multiplications used at iteration $m - 1$ plus $2n \log n$ multiplications for the polynomial products. The number of multiplications grows more slowly than the recursion,

$$M(n) = 2M(\tfrac{1}{2}n) + 2n \log n,$$

7.5 An accelerated Berlekamp–Massey algorithm

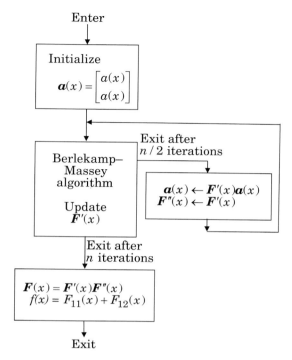

Figure 7.7 Splitting the Berlekamp–Massey algorithm

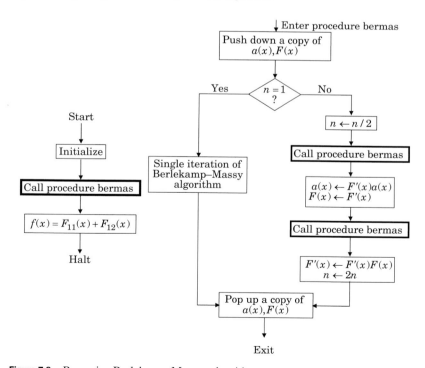

Figure 7.8 Recursive Berlekamp–Massey algorithm

where $n = 2^m$. Let $f(m+1) = M(2^m)$ so that

$$f(m+1) = 2f(m) + 2m2^m.$$

This recursion is approximately satisfied by $f(m) = m2^m \log m$. To show this,

$$\begin{aligned} f(m+1) &= (m+1)2^{m+1}\log(m+1) \\ &\approx (m+1)2^{m+1}\log m \\ &= 2f(m) + 2 \cdot 2^m \log m. \end{aligned}$$

Then, with $n = 2^m$, this gives

$$M(2n) = 2M(n) + 2n,$$

which grows at the rate of our recursion. Thus, because $n = 2^m$, we conclude that the asymptotic complexity of the accelerated Berlekamp–Massey algorithm is on the order of $n \log n \log(\log n)$.

Problems for Chapter 7

7.1 Use the Levinson algorithm to solve the following equation in GF(7):

$$\begin{bmatrix} 1 & 2 & 3 & 4 \\ 2 & 1 & 2 & 3 \\ 3 & 2 & 1 & 2 \\ 4 & 3 & 2 & 1 \end{bmatrix} \begin{bmatrix} f_0 \\ f_1 \\ f_2 \\ f_3 \end{bmatrix} = \begin{bmatrix} 4 \\ 2 \\ 1 \\ 3 \end{bmatrix}.$$

7.2 Rederive the Levinson algorithm in the complex field for the case in which the matrix A is hermitean.

7.3 Suppose that the sequence a that forms the input to the Berlekamp–Massey algorithm is known to be periodic with period N much larger than $2n$.
  a Show how to find all of a, given only its first $2n$ components.
  b Suppose that A, the Fourier transform of a, is given instead of a. Show how the Berlekamp–Massey algorithm can be modified to use A directly without explicitly computing an inverse Fourier transform.

7.4 Apply the Berlekamp–Massey algorithm to the input sequence

$$(a_0, a_1, \ldots, a_7) = (0, 0, 0, 0, 0, 0, 0, 1)$$

to find a solution, f, that has length greater than four. Does the solution depend on the field F? Does the solution hold for the rational field? This is an example of how the Berlekamp–Massey algorithm will produce an autoregressive filter even when the original Toeplitz matrix is singular.

7.5 Use the euclidean algorithm to solve the equation
$$\begin{bmatrix} 3 & 2 & 1 \\ 3 & 3 & 2 \\ 2 & 3 & 3 \end{bmatrix} \begin{bmatrix} f_1 \\ f_2 \\ f_3 \end{bmatrix} = - \begin{bmatrix} 3 \\ 2 \\ 1 \end{bmatrix}.$$

7.6 Prove that the recursions of the Levinson algorithm can be carried out if and only if the nested principal submatrices are all nonsingular.

7.7 Give a condition that avoids any degeneration of the Durbin algorithm.

Notes for Chapter 7

The first fast algorithm for inverting Toeplitz systems was given by Levinson (1947). He was immediately concerned with applications to Wiener filtering problems rather than with the development of fast algorithms in general. Additional fast algorithms for solving other Toeplitz systems were given by Durbin (1960), by Trench (1964), and by Berlekamp (1968). The development of the Trench algorithm was simplified by Zohar (1974). Massey (1969) simplified Berlekamp's algorithm, reformulating it as a method of designing linear feedback shift registers. Welch and Scholtz (1979) presented another view of this algorithm as a continued fraction. The recursive Berlekamp–Massey algorithm is from Blahut (1983b). The method of using the euclidean algorithm to solve a certain Toeplitz system of equations is due to Sugiyama, Kasahara, Hirasawa, and Namekawa (1975). The use of doubling to accelerate the euclidean algorithm when solving Toeplitz systems was studied by Brent, Gustavson, and Yun (1980).

It is not surprising that fast algorithms for solving Toeplitz systems have connections with many other problems. When its steps are written in reverse order, the Levinson algorithm becomes an algorithm known as the Schur–Cohn criteria for testing stability of the autoregressive filter represented by the polynomial $a(x)$. This connection was noticed by Vieira and Kailath (1977).

The length of this chapter would quickly double if we were to also discuss solving generalizations and specializations of Toeplitz matrices such as block-Toeplitz matrices or nearly Toeplitz matrices. A sample of this work can be found in the papers of Wiggins and Robinson (1965); Dickinson (1979); Dickinson, Morf, and Kailath (1974); Friedlander, Morf, Kailath, and Ljung (1979); Morf, Dickinson, Kailath, and Vieira (1977); and Monden and Arimoto (1980).

8 Fast algorithms for trellis search

A finite-state machine that puts out n elements from the field F at each time instant will generate a sequence of elements from the field F. The set of all possible output sequences from the finite-state machine can be represented on a kind of graph known as a *trellis* or, if the number of states is very large, on a kind of graph known as a *tree*. There are many applications in which such an output sequence from a finite-state machine is observed with errors or in noise, and one must estimate either the output sequence itself or the history of the finite-state machine that generated that sequence. This estimation task is a problem in searching a trellis or a tree for the particular path that best fits a given data sequence. Fast path-finding algorithms are available for such problems. This part of the subject of signal processing is quite different from other parts of signal processing. The trellis searching algorithms, which we introduce in this chapter, are quite different in structure from the other algorithms that we have studied.

Among the applications of trellis-searching and tree-searching algorithms are: the decoding of convolutional codes, demodulation of communication signals in the presence of intersymbol interference, demodulation of partial response waveforms or differential phase-shift-keyed waveforms, text character recognition, and voice recognition.

8.1 Trellis and tree searching

A finite-state machine consists of a set of states, a set of transitions between the states, and a set of output symbols from a field F assigned to each transition. A simple finite-state machine is given by the shift register shown in the example of Figure 8.1. The state of the machine in this example is specified by the two bits stored within the shift-register stages. The two bits specify one of four states. The finite-state machine changes state at each discrete instant, referred to as a clock time, as directed by the new input symbols. The input consists of two bits at each clock time, so the input can take one of four values. Thus there are four transitions leaving each state and going to a new state. For this particular example, it is possible to go directly from any state to any other state and there are four transitions entering each new state. The output of

8.1 Trellis and tree searching

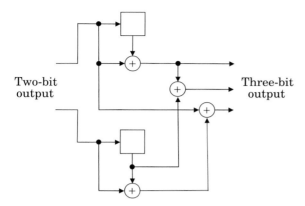

Figure 8.1 One kind of finite-state machine

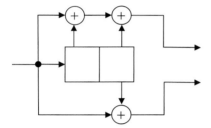

Figure 8.2 Another finite-state machine

the finite-state machine is a function of both the state and the input. In the circuit of Figure 8.1, there are three output bits computed from the two input bits and the two state bits using modulo-two adders (exclusive- or gates). Thus every state transition is associated with a three-bit pattern as the output.

Another example of a finite-state machine is shown in Figure 8.2. At each clock time, there is a single input bit, which specifies one of two possible transitions. For this example, it is not possible to go directly from any state to any other state. Two clock times will sometimes be needed to move from one state to some other state. The circuit of Figure 8.2 has two bits as its output.

A *state-transition diagram* for a finite-state machine is shown in Figure 8.3. The nodes represent the states. The transitions are labeled with vectors of length two. At each time instant, the finite-state machine makes a transition between states across one of the paths of the state-transition diagram and emits the label on the transition path. The transition diagram in Figure 8.3 is the right one for the circuit of Figure 8.2. Then the labels are two-bit patterns.

A graph known as a *trellis* can usefully describe the output sequence of any finite-state machine. A trellis may be regarded as a state-transition diagram embellished with a time axis. A typical trellis with two branches leaving each node is shown in the

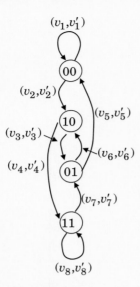

Figure 8.3 State-transition diagram

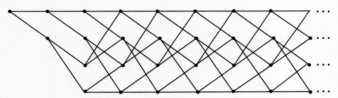

Figure 8.4 A simple trellis

diagram of Figure 8.4. In general, a trellis could have more branches leaving each node, say q branches. The nodes in each column of the trellis represent the set of possible states of the finite-state machine at that time instant. The subsequent column represents the same possible set of states at a subsequent time instant. The set of branches between two columns of nodes is called a *frame*. The branches of a frame represent all of the transitions that are possible during that time interval. In general, a trellis is a graph whose nodes are in a rectangular grid, semi-infinite to the right. The configuration of the branches connecting each column of nodes to the next column of nodes on its right is the same for each column of nodes. Usually, nodes at the beginning of the trellis on the left are not shown if they cannot be reached by starting at the top left node and moving only to the right. The *constraint length* v of the trellis is the number of bits (or symbols) needed to specify the state of the trellis. The trellis of Figure 8.4 has a constraint length of two. The number of nodes in each column is 2^v (or q^v).

In each frame, the finite-state machine changes state. This is represented in the trellis by a branch to the next node. In Figure 8.5, each branch of the trellis is labeled. In

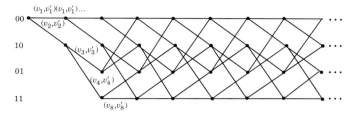

Figure 8.5 A labeled trellis

general, each branch of the trellis is labeled with n numbers from the field F for some fixed n. The set of branches composing a frame will normally have the same labels from frame to frame. (In some instances, the set of labels may vary from frame to frame.) There are many paths along which the finite-state machine can move through the trellis from left to right. Along each such path, it passes along the branches, each branch labeled with n numbers from the field F, and this sequence of numbers labeling the branches of the path becomes the output of the finite-state machine.

Let $\boldsymbol{c} = \{c_i, i = 0, \ldots\}$ be the sequence of symbols of the field F produced by the finite-state machine. Each path through the trellis corresponds to a different path sequence $\boldsymbol{c}$ labeling the branches of that path. Let $\boldsymbol{v} = \{v_i, i = 0, \ldots, \}$ be a sequence of symbols from F. We shall refer to $\boldsymbol{v}$ as the *source sequence*, and each $e_i = v_i - c_i$ as an error term. We may then say that the source sequence $\boldsymbol{v}$ is the *path sequence* $\boldsymbol{c}$ observed in an additive error sequence $\boldsymbol{e}$.

The task of searching the trellis is to find that path through the trellis whose sequence of labels $\boldsymbol{c}$ agrees most closely with the given source sequence $\boldsymbol{v}$. Usually, there is no path whose labels agree exactly with $\boldsymbol{v}$, and one must find the path that agrees most closely. To measure the closeness of agreement between two sequences, we shall define a distance measure.

A *distance function* $d(\alpha, \beta)$ on the set S is a real-valued function on pairs of elements of S that satisfies the following properties: $d(\alpha, \beta) \geq 0$, $d(\alpha, \alpha) = 0$, and $d(\alpha, \beta) = d(\beta, \alpha)$. If the function also satisfies $d(\alpha, \beta) \leq d(\alpha, \gamma) + d(\gamma, \beta)$ for all γ in S, then $d(\alpha, \beta)$ is called a *metric*, which is a stronger term than distance. The *euclidean distance* is a distance on a vector space of n-tuples over the real field. The euclidean distance is given by

$$d(\boldsymbol{v}, \boldsymbol{w}) = \sum_{i=0}^{n-1} (v_i - w_i)^2.$$

The *Hamming distance* is an alternative distance on the space of n-tuples over any set. The Hamming distance between $\boldsymbol{v}$ and $\boldsymbol{w}$ is defined as the number of places i at which $v_i \neq w_i$.

When using the euclidean distance or the Hamming distance as the sequence distance, the Viterbi algorithm can be regarded as a nonparametric algorithm. The

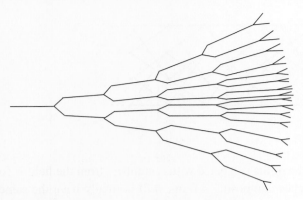

Figure 8.6 A tree with two branches per node

algorithm then does not use a probabilistic model of the data, nor does its response to any given input sequence depend on the probabilistic model that actually generated that sequence. The algorithm examines all possible paths and chooses the path sequence that is closest to the source sequence in the chosen sequence distance.

Consider the first ℓ frames of a trellis. Each path, ℓ frames in length, through the trellis specifies a vector of length ℓn over the field F, where n is the number of symbols labeling each branch. If there are q paths leaving each node, then there are q^ℓ such vectors of length ℓn corresponding to the q^ℓ distinct paths through the first ℓ frames of the trellis. The problem of searching a trellis is the problem of finding which of these vectors, ℓn symbols long, is the closest in the chosen sequence distance to a source sequence v that also is ℓn symbols long.

The method of trellis search that is easiest to understand simply computes the distance from the given source sequence v to each of the q^ℓ sequences produced by the trellis, and then finds the smallest distance in this list of distances. We can imagine simply laying the given source sequence v along every possible path through the trellis and comparing it to that path sequence. This procedure has complexity that is exponential in ℓ, so it is not practical if ℓ is large. Usually, ℓ is so large as to appear infinite; a practical algorithm cannot look at the entire source sequence at the same time. A practical algorithm starts at the beginning of the trellis and works its way along the sequence, making irrevocable decisions as it goes.

In some cases, the number of states of the finite-state machine is large. The number of nodes in each frame of the trellis is large as well, sometimes so large that we do not think about the finiteness of the number of nodes. Then the initial part of the trellis continues to grow, and we ignore the fact that paths do recombine eventually when the finite-state machine revisits a state. This is reasonable because we can look at only tiny pieces of the trellis at one time. Then the appropriate picture to think of is the tree as shown in Figure 8.6. The number of nodes in a tree grows without limit. Sequential algorithms, discussed in later sections, will efficiently search a tree.

8.2 The Viterbi algorithm

If the optimal path through a graph from point A to point C passes through point B, then the segment of the path from point B to point C coincides with the optimal path from point B to point C. This is an evident and well-accepted optimality principle for finding a path in a graph. A trellis is a kind of graph, so this principle applies to finding the best path through a trellis. An iterative procedure for applying this principle to find the best path through a trellis is known as the *Viterbi algorithm*. The key to studying the Viterbi algorithm is to recognize that at each node the Viterbi algorithm does not look forward, asking to which node the preferred path should go to next. Rather, it looks backward, asking at each node how the path might have gotten to that node if it actually did get there. If, for every node in the ℓth frame, the path to that node takes the same branch in the first frame, the Viterbi algorithm concludes that the common branch was, indeed, the branch taken in the first frame.

The Viterbi algorithm for searching a trellis with q branches per node and constraint length ν iteratively updates a set of q^ν path candidates. The complexity of the algorithm is proportional to q^ν. Thus, the algorithm is practical only for small ν. For a constraint length of ten and two branches per node, the Viterbi algorithm iterates a set of 1024 candidate paths. This is practical, but a constraint length of twenty needs to iterate more than one million candidate paths, so it is not practical.

The Viterbi algorithm operates iteratively frame-by-frame tracing through the trellis. At frame time ℓ, the algorithm does not yet know which node the closest path reached, nor does it try to find this node yet. Instead, the algorithm determines the best path from the starting node to each node in the ℓth frame, and also determines the distance between each such best path sequence and the source sequence. This distance is called the *discrepancy* of that path. If all of the q^ν best paths pass through the same node in the first frame, the algorithm has found the first frame of the minimum-distance path, even though it has not yet decided anything about the ℓth frame.

Then the algorithm determines each of the candidate paths to each of the new nodes in frame $\ell + 1$. But, to get to any one of the nodes in frame $\ell + 1$, the path must pass through one of the nodes in frame ℓ. One can get the candidate paths to any new node by extending to this new node each of the old paths that can be so extended. The minimum-distance path is found by adding the incremental discrepancy of each path extension to the discrepancy of the best path to the node in frame ℓ. There are q^k such candidate paths to each of the q^ν new nodes, and the candidate path with the smallest discrepancy is then chosen as the minimum-distance path to the new node. This process is repeated for each of the nodes in frame $\ell + 1$. At the end of the computations for this frame, the algorithm has found a minimum-distance path to each of the nodes in frame $\ell + 1$. Again, if all of these paths pass through the same node in the second frame, the

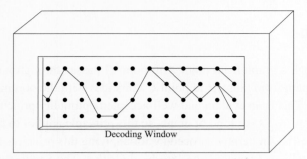

Figure 8.7 Conceptualizing the Viterbi algorithm

Viterbi algorithm has now successfully found the best branch in the second frame of the trellis.

This process continues through successive frames. If there is a tie for the minimum-distance path to a given node, the algorithm can either break the tie by using any arbitrary rule, or simply provide both possible data sequences as its output.

To implement the Viterbi algorithm, one must choose a decoding window width b, usually several times as large as the constraint length v, and validated by simulation. At frame time ℓ, the algorithm examines all surviving paths. If they agree in the first frame, this is the first frame of the estimated path. The first frame can be passed out to the user. Next, the algorithm drops the first frame and takes in a new frame of data for the next iteration. If, again, all surviving paths pass through the same node of the oldest surviving frame, then this frame is now known. The process continues in this way, stepping through the frames indefinitely.

If b is chosen large enough, then a well-defined decision will almost always be made at each iteration. Occasionally, there will be a decision still pending in the frame about to be lost, either because there is a tie for the minimum-distance path or because the decoding window width b is too small to yet make a decision, but this is rare. In a well-designed system, this happens negligibly often.

As the algorithm progresses through many frames, the accumulating discrepancies continue to increase. To avoid overflow problems, they must be reduced occasionally. A simple procedure is to periodically subtract a number, such as the smallest discrepancy, from all of them. This does not affect the choice of the maximum discrepancy.

It may be helpful to think of the Viterbi algorithm as a display in a window through which a portion of the trellis may be viewed, as shown in Figure 8.7. One can see only a finite-length section of the trellis in the window, and on this trellis section are marked the surviving paths, each labeled with a discrepancy. As time increases, the trellis slides to the left within the decoding window, new nodes appear on the right and some old paths are extended to them. Other old paths vanish, and an old column of nodes on the left is shifted out of sight. By the time a column of nodes is lost on the left side, only one of its nodes will have a path through it, except for rare exceptions.

8.2 The Viterbi algorithm

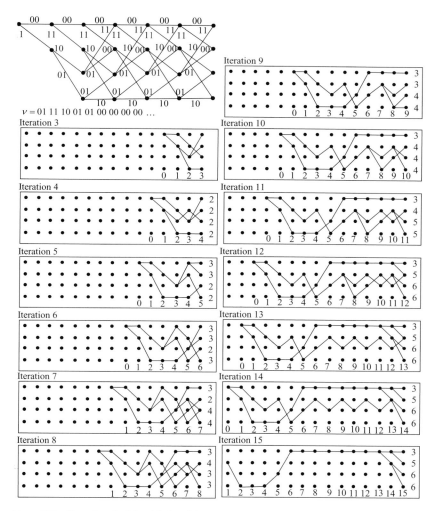

Figure 8.8 Sample of a Viterbi algorithm

An example is shown in Figure 8.8. To keep the example simple, the labels on the trellis are binary, as is the data sequence, and the sequence distance is the Hamming distance. We choose a window width b equal to fifteen. Suppose that the dataword is

$$v = 101000001000000000000000 \ldots$$

We shall find the path through the trellis that most closely agrees with v.

The sequence of states is shown in Figure 8.8. At the third iteration, the algorithm has already identified the shortest path to each node of the third frame. At iteration ℓ, the algorithm finds the shortest path to each node of frame ℓ by extending the paths already formed to each node of frame $\ell - 1$ and keeping the path to each node that has the smallest Hamming distance from the data sequence. In the example, ties are retained until they are eliminated by a better path or until they reach the end of the buffer memory. In the example, the first bit of the sequence closest to the given data sequence is not found until iteration 15. At iteration 15, the initial sequence has been found to be 1101101001 . . . .

The algorithm, shown symbolically in Figure 8.8, might look very different in its actual implementation. For example, the active paths through the trellis could be represented by a table of four 15-bit numbers. At each iteration, each of the 15-bit numbers is shifted left, dropping the leftmost bit and appending a zero or a one at the right. Then, of the resulting list of eight sequences, four sequences are dropped from the table so that the list again consists of one path to each node.

8.3 Sequential algorithms

The Viterbi algorithm can search a small trellis for the best path, but it is an impractical way to search a large trellis. For a trellis with a constraint length of ten, the Viterbi algorithm must store 1024 candidate paths. For a trellis with a constraint length of twenty, the Viterbi algorithm must store more than a million candidate paths, which is clearly unreasonable. For a longer constraint length, say a constraint length of forty, one devises a strategy that searches only the probable paths through the trellis, and ignores the improbable paths. Strategies for searching only those paths through a trellis that appear to be likely are known collectively as *sequential algorithms*. Every sequential algorithm is a backtracking algorithm to visit unexplored paths when the current path no longer appears to be likely. The complexity of a sequential algorithm depends only weakly, or not at all, on the constraint length of the underlying sequences. Thus because a trellis with 2^{40} states is so wide compared to the length of the trellis that is explored, sequential algorithms could just as well be searching a tree as searching a trellis. It is unlikely that two paths that remerge will both be explored.

A sequential algorithm works its way forward in the trellis by exploring only a relatively small number of paths through the trellis and advancing only on the promising paths. Occasionally, a sequential algorithm will decide to backtrack in the trellis and to then explore paths that were previously ignored. It may even backtrack a considerable distance in the trellis, but it can only backtrack as far as symbols of the source sequence have been retained. Old symbols are overwritten when the buffer is full, and at the time of the overwriting the algorithm must have already decided on the corresponding branch. If the computations of the algorithm have still not decided on a branch for this frame, then there is a *buffer overflow* and the algorithm has failed. This failure probability can always be made smaller, but not eliminated, by making the buffer larger.

Consider a trellis with a large constraint length, perhaps a constraint length of forty. This trellis must be searched out through several hundred frames to make a firm decision on the first frame. The number of nodes in each frame is immense. There are 2^{40} such nodes in each frame – approximately one thousand billion. The sequential algorithm must find the path through this trellis whose sequence of labels most closely matches the given sequence; at least it should do this with very high probability, although we shall allow it to occasionally fail to find the correct path because of buffer overflow. This kind of failure is not because the data is inconclusive, but because the processing of the data is incomplete. It is an error mode not found in the Viterbi algorithm.

A sequential algorithm looks at the first frame, chooses the tree branch closest in distance to the data in the first frame, and proceeds to a node at the end of that branch. It then repeats this procedure, chooses a branch leaving the selected node, and proceeds to a node at the end of that branch. At the end of each iteration, it has reached a single node. In the next iteration, based on the next dataframe, it chooses a branch leaving that node, and again proceeds to a single node at the end of the next frame. In this way, it traces a path through the tree. This procedure will work fine if the given data sequence agrees exactly with the labels on some path through the tree. However, if there are disagreements, the sequential algorithm will occasionally choose the wrong branch. As the algorithm proceeds past a wrong choice, it will suddenly perceive that it is finding large discrepancies in every frame and will eventually presume it is on the wrong path. A sequential algorithm will back up through the last few frames to explore alternative paths, perhaps backing up through a few more frames, and even more frames until it finds a likely path on which to move forward. It then proceeds along this alternative path. The rules controlling the search are developed below.

The performance depends on the decoding window width b. When the algorithm has found a path that penetrates b frames into the tree, it makes an irrevocable decision regarding the oldest frame, outputs the estimated data symbols of this frame to the user, and shifts a new dataframe into its window.

There are two classes of sequential algorithm in use: the *Fano algorithm*, which we study in Section 8.4 and the *stack algorithm*, which we study in Section 8.5. These algorithms are quite different in structure. The stack algorithm has less computation than the Fano algorithm. The stack algorithm keeps a list of the paths it has already searched. This is in contrast to the Fano algorithm, which may search its way out to a node, then back up some distance only to repeat in detail the search out to the same node, and recomputing the same quantities as before because they were not saved. The stack algorithm keeps better records so that it need not repeat unnecessary work. On the other hand, the stack algorithm needs considerably more memory and does a considerable amount of list management during each iteration.

A sequential algorithm does not search all paths, so it cannot know for certain that there is not a better path somewhere in the trellis. It must decide that it is on the wrong path only from paths that it has already examined and based on prior information. This means that a sequential algorithm must be given a prior statistical model. If the prior model predicts an average discrepancy that is too large, the algorithm will be insensitive and will not backtrack soon enough. If the prior model predicts an average discrepancy that is too small, the algorithm will be too sensitive and will backtrack excessively often.

The path discrepancy measures the distance between a segment of the source sequence and an initial segment of each path sequence of the same length. There are many different measures of distance that are in use with sequential algorithms. Whenever the source sequence consists of a path sequence disturbed by additive noise that is described by a known probability distribution, then it is appropriate to use this probability distribution to define a distance. The resulting distance measure is known as the *Fano distance*, which we now derive.

The appropriate definition of the discrepancy of a path is the *log-likelihood function* of that path, a function that is defined as follows. The first $N+1$ frames of symbols of the source sequence are written as

$$\boldsymbol{v}^{(N)} = (v_0^1, \ldots, v_0^n, v_1^1, \ldots, v_1^n, \ldots, v_N^1, \ldots, v_N^n)$$

with the subscripts indexing frames and the superscripts indexing symbols within a frame. The first $m+1$ frames of symbols labeling any path sequence can be written as

$$\boldsymbol{c}^{(m)} = (c_0^1, \ldots, c_0^n, c_1^1, \ldots, c_1^n, \ldots, c_m^1, \ldots, c_m^n).$$

The probability that $\boldsymbol{v}^{(N)}$ is the source sequence when $\boldsymbol{c}^{(m)}$ is the path sequence is denoted $\Pr[\boldsymbol{c}^{(m)} | \boldsymbol{v}^{(N)}]$. The sequential algorithms should find the path segment that maximizes $\Pr[\boldsymbol{c}^{(m)} | \boldsymbol{v}^{(N)}]$. It chooses the initial path segment of length $m+1$ frames without looking further out in the tree.

By the Bayes formula,

$$\Pr(\boldsymbol{c}^{(m)} | \boldsymbol{v}^{(N)}) = \frac{\Pr(\boldsymbol{v}^{(N)} | \boldsymbol{c}^{(m)}) \Pr(\boldsymbol{c}^{(m)})}{p(\boldsymbol{v}^{(N)})}$$

because the paths are to be used with equal probability. The term $\Pr(\boldsymbol{c}^{(m)})$ is a constant. It can be deleted without affecting which path achieves the maximum.

The term $\Pr(\boldsymbol{v}^{(N)} | \boldsymbol{c}^{(m)})$ can be broken into two factors:

$$\Pr(\boldsymbol{v}^{(N)} | \boldsymbol{c}^{(m)}) = \left[\prod_{i=0}^{m} \prod_{j=1}^{n} p(v_i^j | c_i^j) \right] \left[\prod_{i=m+1}^{N} \prod_{j=1}^{n} p(v_i^j) \right].$$

The first factor is a product distribution involving the conditional probabilities over the first $m+1$ frames of the true path. The second factor is a product distribution of unconditional probabilities over the frames where the path is not yet specified. The factorization takes this form as a consequence of the definition of a sequential algorithm.

Now maximize

$$\frac{\Pr(\boldsymbol{v}^{(N)} | \boldsymbol{c}^{(m)})}{\Pr(\boldsymbol{v}^{(N)})} = \left[\frac{\prod_{i=0}^{m} \prod_{j=1}^{n} p(v_i^j | c_i^j)}{\prod_{i=0}^{m} \prod_{j=1}^{n} p(v_i^j)} \right]$$

8.3 Sequential algorithms

over path sequences. The *Fano distance* is defined as the log function

$$d_F(\boldsymbol{v}^{(m)}, \boldsymbol{c}^{(m)}) = \log \left[\frac{\Pr(\boldsymbol{v}^{(N)} | \boldsymbol{c}^{(m)})}{\Pr(\boldsymbol{v}^{(N)})} \right]$$

$$= \sum_{i=0}^{m} \left[\sum_{j=1}^{n} \log \frac{p(v_i^j | c_i^j)}{p(v_i^j)} \right],$$

where v_i^j is the jth symbol of the source sequence in the ith frame, and c_i^j is the jth branch symbol in the ith frame of the path sequence $\boldsymbol{c}$. The outer sum is a sum over the m observed frames. The inner sum is the contribution to the Fano distance from the ith frame:

$$d_F(v_i, c_i) = \sum_{j=1}^{n} \log \frac{p(v_i^j | c_i^j)}{p(v_i^j)}.$$

The sum on n runs over the n branch labels of the branch taken in frame i. This term can be either positive or negative. In the ith frame, only the term $d_F(v_i, c_i)$ needs to be computed for each branch considered. The Fano path distance through the first i frames of a given path is obtained by adding the Fano distance increment on the ith branch to the cumulative Fano distance of the path leading to that branch.

The number of computations that a sequential algorithm makes to advance one frame deeper into the source sequence is a random variable because of the wide variability in the amount of backtracking. This is the characteristic behavior of a sequential algorithm. It is the major factor affecting the complexity required to achieve a given level of performance. When there is little noise, the algorithm may proceed along the correct path using only the computations of that frame to advance one node deeper into the tree. However, when the noise is severe, the algorithm may proceed along an incorrect path and may become delayed by exploring a large number of wrong paths before finding the correct path. The variability in the number of computations means that a large memory is required to buffer the incoming data. Any buffer of finite size, no matter how large, when used with a sequential algorithm has a nonzero probability of overflowing. This behavior must be considered in performance calculations. Buffer overflow is a consideration in any backtracking algorithm that performs at least one computation at each node that it visits, and that examines branches sequentially so that at any node the algorithm does not use data on branches deeper in the tree. These conditions lead to the characteristic behavior of sequential algorithms.

In contrast to the Viterbi algorithm, which can be used as a nonparametric algorithm, the sequential algorithms require an adequate model, usually a probabilistic model, describing the source sequence. Therefore, the sequential algorithms can be regarded as parametric algorithms.

8.4 The Fano algorithm

The *Fano algorithm* is a sequential algorithm that uses more computation but less storage than the stack algorithm. This is because it often needs to recompute what it has recently discarded, and perhaps it may even recompute the same thing and discard it many times. The Fano algorithm backtracks when the observed distance between the path it is following and the source sequence is too much larger than expected. This distance is called the *discrepancy*. The Fano algorithm requires that there is enough statistical regularity in the per-frame discrepancy, and requires a good estimate of the average per-frame discrepancy $\overline{d}$ with respect to the correct sequence. As long as the Fano algorithm is following the right path, it expects to see a discrepancy of about $\overline{d}\ell$ through the first ℓ frames. The Fano algorithm will allow a discrepancy a little larger than this, but if it is too much larger, it concludes that it is following the wrong path and so it backtracks to look for another path.

Choose a parameter $\overline{d}'$ larger than $\overline{d}$ and define[1]

$$t(\ell) = \overline{d}'\ell - d(\ell),$$

where $d(\ell)$ is the measured discrepancy between the source sequence and the current candidate path sequence through the tree. For the correct path, by the law of large numbers, $d(\ell)$ is approximately $\overline{d}\ell$, and $t(\ell)$ is positive and increasing. As long as $t(\ell)$ is increasing, the Fano algorithm continues threading its way through the tree. If ever $t(\ell)$ decreases by an unacceptable amount, the algorithm concludes that at some node it might have chosen the wrong branch. It backtracks through the tree and tests other paths. It may find a better path and follow it, or it may return again to the same node, but now with more confidence and continue past it, perhaps permanently, or perhaps it will backtrack again. To decide when $t(\ell)$ shows an unacceptable decrease, the Fano algorithm uses a running threshold T, which is always a multiple of a threshold quantization increment, Δ. As long as the algorithm is moving forward, it keeps the threshold as large as possible, but satisfying the constraints that it is not larger than $t(\ell)$ and it is a multiple of Δ. The quantization of the running threshold T to a multiple of Δ allows $t(\ell)$ to decrease a little without falling below the running threshold.

The Fano algorithm requires that, at each node, the q branches leaving that node be ordered according to some rule. The ordering rule assigns an index j for $j = 0, \ldots, q - 1$ to each branch. This index need not be stored for each branch. It is only necessary that the rule be known so that when the algorithm backs up to a node along the branch with known index j, it can reorder the branches by the rule, find branch j, and thence find branch $j + 1$. An appropriate rule is the minimum-distance rule. The branches are ordered according to their distance from the corresponding frame of the

---

[1] We have chosen the sign convention so that a positive sign occurs when the algorithm is performing properly.

8.4 The Fano algorithm

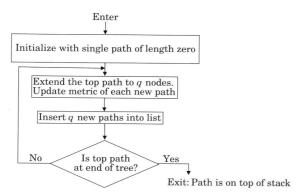

Figure 8.9 Shift-register implementation of Fano algorithm

dataword, and ties are broken by any convenient subrule. However, the algorithm will work if the branches are assigned any fixed order.

To ease the understanding of the structure of the algorithm, it is best to leave the ordering rule unspecified; we only suppose that, at each node, the branch that is closest in distance to the source branch is ordered first. The algorithm will search the branches out of each node according to the chosen order.

A shift-register implementation of the Fano algorithm is outlined in Figure 8.9. The implementation consists of a replica of a finite-state machine for the trellis at hand. The finite-state machine is augmented with auxiliary storage registers. The algorithm attempts to insert symbols into the replica of the finite state machine in order to generate a source sequence that is sufficiently close to the data sequence. At each iteration, it has immediate access to the content of the latest frame entered into the replica finite-state machine. It can increment the symbol in this frame, or it can back up the content to an earlier frame, or it can shift in a new frame. It decides what to do based on a comparison of $t(\ell)$ with the running threshold T.

The Fano algorithm is shown as a flow diagram in simplified form in Figure 8.10. When the data sequence is close to a possible source sequence, the algorithm will circle around the rightmost loop of Figure 8.11 and, in each loop, shift all the registers of Figure 8.10 one frame to the right. As long as $t(\ell)$ remains above the threshold, the algorithm continues to shift right and continues to raise the threshold to keep it tight. If $t(\ell)$ drops below the threshold, the Fano algorithm will test alternative branches in that frame trying to find one above the threshold. If it cannot, it backs up. Once it begins to back up, the logic will force it to continue to back up until it either finds an alternative path that stays above the current threshold or until it finds the node at which the current threshold was set. Then it moves forward again with a lowered threshold, but now the threshold is not raised again until the Fano algorithm reaches new nodes previously unexplored. Each time it passes through the same node while moving forward, it has a smaller threshold. It will never advance to the same node twice with the same threshold.

276   **Fast algorithms for trellis search**

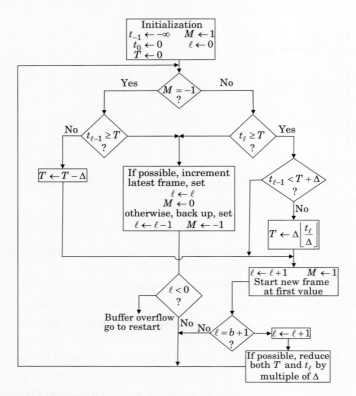

Figure 8.10   Flow diagram for the Fano algorithm

Consequently, it can pass through the same node only a finite number of times. This behavior assures us that the algorithm cannot be trapped in a loop; it must continue to work forward through the data sequence.

Now we must prove two earlier assertions: that if the Fano algorithm cannot find an alternative path, it will move back to the node where the current threshold was set and lower it, and it will not again raise the threshold until reaching a node previously unexplored. The first assertion is obvious because if the Fano algorithm cannot find a new branch on which to move forward, it must eventually back up to the specified node. But, if at any node, the value of $t(\ell - 1)$ at the previous frame is smaller than the current threshold T, then the threshold must have been increased at the ℓth frame. This is just the test contained in Figure 8.11 to find the node at which the current threshold was set, and at this node the threshold is now reduced.

To see the second assertion, notice that after the threshold is lowered by Δ, the Fano algorithm will search the subsequent branches in exactly the same order as before until it finds a place where the threshold test previously failed and is now passed. Until reaching this point, the logic will not allow the threshold T to be changed. This is because once the threshold is lowered by Δ, the quantity $t(\ell)$ will never be smaller

8.4 The Fano algorithm

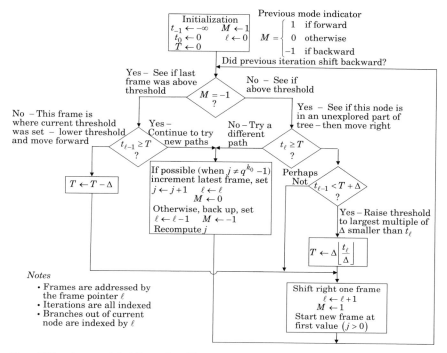

Figure 8.11 An annotated Fano algorithm

than $T + \Delta$ at any node where it previously exceeded the original threshold. When it penetrates into a new part of the tree, it will eventually reach the condition that $t(\ell - 1) < T + \Delta$, while $t(\ell) \geq T$. This is the point at which the threshold is raised. This then is the test to determine if a new node is being visited. There is no need to keep an explicit record of the nodes previously visited. This test appears in Figure 8.11.

The Fano algorithm depends on the two parameters p' and Δ, which can be chosen by means of computer simulation. In practice, because $t(\ell)$ and T are increasing, it will be necessary to reduce them occasionally so that the numbers do not get too large. Subtraction of any multiple of Δ from both will not affect subsequent computations.

In practical implementations, there is also the decoding window width b. Figure 8.11 gives a more complete flow diagram for the Fano algorithm showing the important role of b. Whenever a frame reaches the end of the buffer, as shown by the frame pointer, it is passed out of the window, and the frame pointer is decremented so that it is always a reference to the oldest available frame. The algorithm may occasionally try to back up so far that it tries to look at a frame that has been passed out. This *buffer overflow* occurs when the frame pointer becomes negative. Buffer overflow is the major limitation on the performance of the Fano algorithm. For many applications, the probability of buffer overflow decreases very slowly with buffer size, so that no matter how large one makes b in practice, the problem does not completely go away.

Two ways to handle buffer overflow are available. The most certain way is to periodically force the finite-state machine through a known sequence of transitions of length equal to the constraint length. Upon buffer overflow, the algorithm declares a failure and waits for the start of the next known sequence, where it starts again. All of the intervening data between buffer overflow and the next start-up is lost. Alternatively, if the constraint length is not too large, one can just force the pointer forward. The algorithm will again find the correct path if it can find a correct node. This may be possible if there is a long enough segment of the dataword that agrees exactly with the corresponding segment of the source output sequence.

8.5 The stack algorithm

The *stack*[2] *algorithm* is a sequential algorithm that uses more storage but less computation than the Fano algorithm. The stack algorithm is easy to explain. The algorithm maintains a list of paths of various lengths that have already been examined. The entries in the list are paths in the trellis. The paths are listed in order of increasing discrepancy, which is defined as the distance of the path from the data sequence. At the top of the list is the path with the smallest discrepancy. Each entry in the list is a path that is recorded in two parts: one part is a variable-length sequence of state transitions defining the path; the other part is the path discrepancy. Initially, the list contains only the trivial path of length zero. During each step of the iterative algorithm, the path at the top of the list is removed from the list, then extended to each of its q successor nodes and the discrepancy is updated for each new path by adding the discrepancy of the appropriate branch of the new frame. Each of the q new paths is inserted back into the list at a position determined by its updated discrepancy. In this way, old entries in the list may be pushed down as new entries are inserted in the appropriate place and other entries are pushed out at the bottom of the list.

The stack algorithm is shown in Figure 8.12. It has very little computation. The disadvantage of the stack algorithm is the large buffer memory, as well as the large task of memory management. To insert a new path into the list requires that the discrepancies of the existing entries in the list be examined in a systematic manner. Of course, the entries in memory need not be actually moved within the memory, they can be virtually moved by means of indirect addressing.

For a simple example of the stack algorithm, we refer to the labeled trellis shown in Figure 8.13, searching for the path that is closest in sequence distance to the sequence $v = 01\ 10\ 01\ 10\ 11\ \ldots$. The branch distance that has been chosen for this example is the function $d(0, 0) = d(1, 1) = -1$ and $d(0, 1) = d(1, 0) = 5$. This might occur as a

---

[2] This term is standard in this context, but is at variance with the usual definition of a stack as a list that can be accessed only at the top.

8.5 The stack algorithm

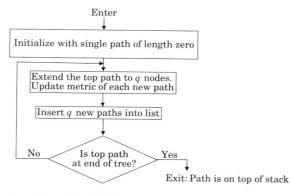

Figure 8.12 Simplified stack algorithm

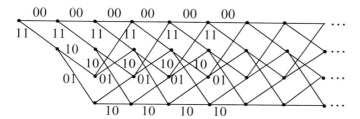

Figure 8.13 Labeled trellis for the example

Step	0	1	2	3	4	5	6	7
List	−, 0	0, 4	1, 4	10, 2	100, 6	100, 6	1010, 4	10100, 2
		1, 4	00, 8	00, 8	101, 6	00, 8	00, 8	00, 8
			01, 8	01, 8	00, 8	01, 8	01, 8	01, 8
				11, 14	01, 8	1000, 10	1000, 10	1000, 9
					11, 14	11, 10	11, 10	11, 10

Figure 8.14 Iterations of stack algorithm

Fano distance for some binary channel, but the Fano distance has been rounded and scaled.

To illustrate the phenomenon of buffer overflow, we have chosen a list of length five for the example. A real problem would use a much longer list. As new items are inserted into the list, some items are pushed down, and the items dropping out the bottom of the list are discarded.

The iterations of the stack algorithm are summarized in Figure 8.14. At each step, the entries contained in the list consist of pairs. The first element of the pair identifies the sequence of branches forming the path through the trellis. This consists of a sequence of bits specifying the upper or lower path out of each node. A zero specifies the upper

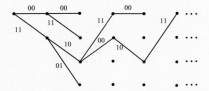

Figure 8.15 Summary of stack algorithm iterations

path, a one specifies the lower path. Some way of indicating the length of the path is also needed. In Figure 8.14, the path length is an explicit part of the notation. The second element of the pair records the path discrepancy. At step 0, the list is initialized with a path of length zero, and a sequence distance equal to zero, as is shown in the first entry of Figure 8.14. After five iterations, the stack algorithm has reached the position shown in Figure 8.15. As the iterations continue past this point, some path of Figure 8.15 will be lengthened. Perhaps the longest path in Figure 8.15 is the path that is lengthened. Occasionally the algorithm will return to another path and lengthen that path for one or more iterations, then possibly returning to the previous path. Possibly the longest path will split into several paths, each growing a little until one of them becomes dominant. Eventually the longest path becomes so long that the tree must be pruned at the root. Normally, when this happens all paths in the list will have the same first branch because all other paths have been pushed out the bottom of the list (or soon will be). This means that pruning at the root simply consists of providing the corresponding information as the current output of the algorithm.

Notice that the list entries in Figure 8.14 are sequences of varying lengths. This is why the choice of the branch distance must correspond to a probabilistic model for the generation of v, such as Fano distance. The use of the Hamming distance would give no extra credit for long runs of correct symbols. With Hamming distance, two paths of vastly different lengths would be of equal discrepancy if the total number of errors were the same.

The stack algorithm attempts to find the path that is closest to the data sequence in the specified branch distance. It attempts this regardless of whether that specific choice of branch distance is appropriate. However, its performance with regard to this goal depends on whether that branch distance is appropriate to this goal. If not, the correct path may fail to find its way to the top of the list. For this reason, the stack algorithm can be regarded as a parametric algorithm.

8.6 The Bahl algorithm

We now turn to another task of searching a trellis for a rather different purpose than the purpose that has occupied this chapter until now. The most common instance of

8.6 The Bahl algorithm

this new task is the computation of posterior probabilities of branch sets when given prior probabilities. We shall use this case as a preferred example. We will describe an algorithm for this computational task known as the *Bahl algorithm*. The Bahl (or BCJR) algorithm is a fast algorithm for calculating all of the distances on branch sets of a trellis.

As does the Viterbi algorithm, the Bahl algorithm searches a trellis, but with a different purpose. The Viterbi algorithm is given the branch labels and infers the best path through the trellis matching those branch labels. The Bahl algorithm is given the path labels and infers the best branch index in each frame based on the path labels. Whimsically, one may regard the Viterbi algorithm as a way of inferring preferred paths when given branch specifications, and correspondingly regard the Bahl algorithm as a way of inferring preferred branch indices when given path specifications.

The goal of the Viterbi algorithm is to find the best path through the trellis when given independent branch distances on successive frames of the path that add to produce a path distance. The goal of the Bahl algorithm is to find the best branch index in each frame when given a certain distance structure on all the paths of the trellis to and from that frame. We will define this task precisely. It may be surprising that the structures of the Viterbi algorithm and the Bahl algorithm have a lot of similarity. However, upon further consideration, this might be expected because the structure of the trellis should be a primary contributor to the structure of any algorithm on that trellis.

The Viterbi algorithm is normally described for an unterminated trellis, one with a beginning but no end. In contrast, the Bahl algorithm must be described for a terminated trellis, one with both a beginning and an end. The Bahl algorithm makes two passes through the trellis, one pass in the forward direction starting at the beginning and one pass in the backward direction starting at the end.

The Bahl algorithm and the Viterbi algorithm both move through the frames of a trellis by an iterative process, so it is not surprising that there are similarities in the logical structure. Both have a similar pattern of walking the trellis. However, they perform very different computations. The fundamental computational unit of the Viterbi algorithm is an "add–compare–select" step. The fundamental computational unit of the Bahl algorithm is a "multiply–add" step, (or a sum-of-products step). The pattern of the data flow is similar in the two algorithms, but the executed arithmetic operations are different.

The Bahl algorithm treats a trellis that has the same number of branches, say q^k branches, leaving each node. At each node, the branches are indexed from 0 to $q^k - 1$. All branches of the same frame with the same index belong to the same branch set. These branches all correspond to the same input to the finite-state machine that the trellis is describing. The goal of the algorithm is to marginalize a global variable on all paths of the trellis to a local variable on the branches with common indices within a frame.

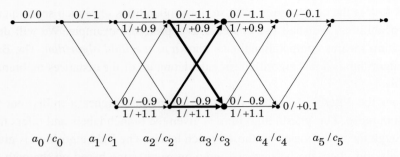

Figure 8.16 Trellis for explaining the Bahl algorithm

A trellis for an example of the Bahl algorithm is shown in Figure 8.16. This example corresponds to a binary communication waveform in which each bit is represented by a ±1 according to its binary value, and each bit has intersymbol interference consisting of one-tenth of the previous value. The waveform has a blocklength of four. The waveform is preceded and followed by zeros in order to begin and end the trellis.

Let each path through the finite trellis corresponds to a vector c denoting the sequence of labels on the branches of that path. Each path has such a vector of labels c. The symbols of c are determined by the sequence of branches forming the path. Thus, we can regard the symbols of c to be labels, taking values in a specified algebraic field, on the branches of the trellis. This is illustrated in Figure 8.16. The path through the trellis is defined by the sequence of indices specifying the branch at each node. In the binary trellis of Figure 8.16, this branch index is a zero or a one. The sequence of indices is denoted $a = (a_0, a_1, \ldots, a_{n-1})$. The individual branch index a_ℓ alone does not determine the individual branch label c_ℓ, nor does c_ℓ alone determine a_ℓ. It is only in the context of the entire trellis path that such an inference can be drawn. Thus we can write $c(a)$, but not $c_\ell(a_\ell)$.

With the trellis structure and the path sequences a fixed, the input to the algorithm is a vector v, of length equal to the length of the path label sequences c. The task of the algorithm is to marginalize a function $d(c(a), v)$ of a and v to the individual symbols of a. In the usual application, the function $d(a, v)$ is a conditional probability distribution, which we write as $p(a|v) = p(a_1, a_2, \ldots, a_n|v)$. The definition of the underlying probability model, however, does not directly give $p(a|v)$. Rather, the model gives a probability distribution conditional on c, giving it as a product distribution $p(v|c) = \prod_{\ell=1} p(v_\ell|c_\ell)$. Then the Bayes formula gives the posterior probability distribution

$$p(a|v) = p(c(a)|v) = \frac{p(v|c(a))p(a)}{\sum_c p(v|c(a)p(a))}.$$

8.6 The Bahl algorithm

ℓ	$\Pr[a_\ell|\mathbf{v}]$	α_ℓ	β_ℓ

Figure 8.17 Partial computations for the Bahl algorithm

The marginalization of $p(\mathbf{a}|\mathbf{v})$ to the single letter a_ℓ is

$$p(a_\ell|\mathbf{v}) = \sum_{a_0}\sum_{a_1}\cdots\sum_{a_{\ell-1}}\sum_{a_{\ell+1}}\cdots\sum_{a_{n-1}} p(\mathbf{a}|\mathbf{v})$$

$$= \sum_{a_0}\sum_{a_1}\cdots\sum_{a_{\ell-1}}\sum_{a_{\ell+1}}\cdots\sum_{a_{n-1}} p(\mathbf{v}|\mathbf{c}(\mathbf{a})).$$

In the special case that a_ℓ can take only the values zero and one, then the task is to compute $p(a_\ell = 0|\mathbf{v})$ by summing $p(\mathbf{a}|\mathbf{v})$ for all paths for which $a_\ell = 0$, and to compute $p(a_\ell = 1|\mathbf{v})$ by summing $p(\mathbf{a}|\mathbf{v})$ over all paths for which $a_\ell = 1$. The marginals $p(a_\ell = 0|\mathbf{v})$ and $p(a_\ell = 1|\mathbf{v})$ must be computed for $\ell = 0, \ldots, n-1$. For each ℓ, $p(a_\ell = 0|\mathbf{v})$ and $p(a_\ell = 1|\mathbf{v})$ sum to one, so it is only necessary to compute a common constant multiple of these two terms. It is convenient to ignore common constants during the computation, and to rescale the result after the computation so that the two probabilities sum to one. Accordingly, the denominator in Bayes formula is not relevant. When one intends to later compute the ratio

$$\Lambda = \frac{p(a_\ell = 0|\mathbf{v})}{p(a_\ell = 1|\mathbf{v})},$$

the step of rescaling may be unnecessary.

The Bahl algorithm starts a computation at each end of the truncated trellis and stores all partial computations as illustrated in Figure 8.17. The nature of the Bahl algorithm is suggested by the structure of the partial computations to be stored as is illustrated in that figure. One partial computation starts at the beginning of the trellis and works its way forward until it reaches the end of the trellis, storing as it goes the results of all partial computations. The other partial computation starts at the end of the trellis and works its way backwards until it reaches the beginning, again storing its results as it goes.

The branches in the third frame of the trellis in Figure 8.16 are highlighted in order to discuss the computation of the posterior probability of the third bit. To marginalize $p(\mathbf{a}|\mathbf{v})$ to a_3, one must sum the probabilities of all paths of the trellis that have $a_3 = 0$; these are the paths that end on the top highlighted node at the end of the frame but may begin at either highlighted node at the beginning of the frame. One must also sum the probabilities of all paths that have $a_3 = 1$; these are the paths that end on the bottom highlighted node.

To compute $p(a_3 = 0|v)$, referring to Figure 8.16, first sum the path probabilities over all paths that start at the beginning of the trellis and reach the top node at the end of the frame of a_3, then sum the path probabilities over all paths from that same top node of the trellis that reach the end of the trellis. Multiply these together. Do the same for the paths that go through the bottom highlighted node at the end of the frame of a_3. Because $p(v|c(a))$ is a product distribution, if one also specifies that $p(a)$ is a product distribution, then the path probabilities, proportional to $p(v|c(a))$ can be computed by multiplying terms one-by-one, either starting from the beginning or the end. This leads to the multiply–add procedure that one-by-one fills the columns labeled α_ℓ and β_ℓ in Figure 8.17.

Problems for Chapter 8

8.1 Give a simple expression for the Fano metric for the case in which the difference between the data sequence and the received sequence is gaussian noise, independent, and identically distributed from sample to sample.

8.2 Repeat the example of Figure 8.8, now with the input sequence

$$v = 10001000001000000000\ldots.$$

8.3 The Viterbi algorithm can be programmed as an "in-place" algorithm, with path discrepancies at the ith iteration stored in the same memory locations as path discrepancies at the $(i - 1)$th iteration. Set up the addressing sequence and computational sequence that will do this.

8.4 The 2^ν surviving paths in the Viterbi algorithm (with $q = 2$) may be represented by 2^ν b-bit binary words. Typically, b may be on the order of forty bits. Devise a memory management scheme for the Viterbi algorithm (using such techniques as bit slices and pointers) that eliminates the necessity to read all 2^ν b-bit words every iteration.

8.5 Develop a flow diagram for the stack algorithm at a level of detail suitable to begin an implementation. Include sufficient definitions of the list structure and the list management.

Notes for Chapter 8

Algorithms for searching a trellis first arose in connection with error-control codes, where the decoding of certain codes, known as convolutional codes, can be viewed as a problem in searching a trellis (or a tree). Much of the literature of the topic of searching a trellis is permeated with the language of error control. However, the trellis-searching algorithms are now seen to have wider interest, and this chapter attempts to present the algorithms disassociated from any application.

Although, in retrospect, the Viterbi algorithm is the best place to begin the study of trellis-searching algorithms, the sequential algorithms were actually developed earlier. Sequential decoding was introduced by Wozencraft (1957) and further described by Wozencraft and Reiffen (1961). Additional developments are due to Fano (1963), Zigangirov (1966), and Jelinek (1969). Further developments by Chevillat and Costello (1978) and by Haccoun and Ferguson (1975), and tutorials by Jelinek (1968) and Forney (1974), as well as pedagogy by Johannesson and Zigangirov (1999), also advanced the subject. Our formulation of the Fano algorithm relies heavily on a version given by Gallager (1968). Viterbi (1967) originally published his trellis-searching algorithm more as a pedagogical device than as a serious algorithm. Its practical applications came later. Rader (1981) discussed procedures for programming the Viterbi algorithm in order to use memory efficiently. Sequential algorithms are tainted by their data-dependent computational complexity. Lower bounds on the distribution of computation were found by Jacobs and Berlekamp (1967), and a complementary upper bound was found by Savage (1966).

The Bahl algorithm is also known as the BCJR algorithm or the forward–backward algorithm. It was described by Bahl, Cocke, Jelinek, and Raviv (1974) and variations of it, including the Baum–Welch algorithm, have been developed independently by many others.

9 Numbers and fields

Number theory has already been seen in earlier chapters of this book. It was used in the design of fast Fourier transform algorithms. We did make use of some ideas that only now will be proved. This chapter, which is a mathematical interlude, will develop the basic facts of number theory – some that were used earlier in the book and some that we may need later.

We also return to the study of fields to develop the topic of an extension field more fully. The structure of algebraic fields will be important to the construction of number theory transforms in Chapter 10 and also to the construction of some multidimensional convolution algorithms in Chapter 11 and for some multidimensional Fourier transform algorithms in Chapter 12.

9.1 Elementary number theory

Within the integer quotient ring $\mathbf{Z}_q$, some of the elements may be coprime to q, and, unless q is a prime, others will divide q. It is important to us to know how many elements there are of each type.

Definition 9.1.1 (Euler) *The totient function, denoted $\phi(q)$, where q is an integer larger than one, is the number of nonzero elements in $\mathbf{Z}_q$ that are coprime to q. For q equal to one, $\phi(q) = 1$.*

When q is a prime p, then all the nonzero elements of $\mathbf{Z}_p$ are coprime to p, and so $\phi(p) = p - 1$ whenever p is a prime. When q is a power of a prime, p^m, then the only elements of $\mathbf{Z}_q$ not coprime to p^m are the p^{m-1} multiples of p. Therefore $\phi(p^m) = p^m - p^{m-1} = p^{m-1}(p-1)$. All other values of the totient function can be obtained by using the following theorem.

Theorem 9.1.2 *If $\mathrm{GCD}(q', q'') = 1$, then*

$$\phi(q'q'') = \phi(q')\phi(q'').$$

Proof Let i index the elements of $\mathbf{Z}_{q'q''}$. Then $i = 0, \ldots, q'q'' - 1$. Imagine the elements of $\mathbf{Z}_{q'q''}$ mapped into a two-dimensional array using new indices, given by

$i' = i \pmod{q'}$,
$i'' = i \pmod{q''}$.

Then i' and i'' are elements of $\mathbf{Z}_{q'}$ and $\mathbf{Z}_{q''}$, respectively. Because $\mathrm{GCD}(q', q'') = 1$, the mapping from i to pairs (i', i'') is one to one with the correspondence between i and (i', i'') given by

$i = q'Q' + i'$,
$i = q''Q'' + i''$

for some Q' and Q''. Every i that is a factor of $q'q''$ is a factor of either q' or q'', and so must be a factor of either i' or i'', and conversely. Thus, all i that are factors of $q'q''$, and only such factors, can be deleted from the array by crossing out all columns for which i' is a factor of q' and all rows for which i'' is a factor of q''. The remaining array has $\phi(q')$ columns and $\phi(q'')$ rows. That is,

$$\phi(q'q'') = \phi(q')\phi(q''),$$

and the theorem is proved. □

Corollary 9.1.3 *If $q = p_1^{c_1} p_2^{c_2} \cdots p_r^{c_r}$ is the prime factorization of q, then*

$$\phi(q) = p_1^{c_1-1} p_2^{c_2-1} \cdots p_r^{c_r-1}(p_1 - 1)(p_2 - 1) \cdots (p_r - 1).$$

Proof Theorem 9.1.2 and the remark prior to the theorem state that

$$\phi(q) = \phi(p_1^{c_1})\phi(p_2^{c_2}) \cdots \phi(p_r^{c_r})$$
$$= p_1^{c_1-1}(p_1 - 1)p_2^{c_2-1}(p_2 - 1) \cdots p_r^{c_r-1}(p_r - 1),$$

as was to be proved. □

There is another important relationship satisfied by the totient function that will prove useful. Suppose that d is a divisor of q, and suppose that $f(q)$ is any function of q. By the expression $\sum_{d|q} f(d)$, we mean the sum of all terms $f(d)$ such that d is a divisor of q. Because $(q/d)d = q$, it is clear that q/d is a factor of q if, and only if, d is a factor of q. Therefore

$$\sum_{d|q} f(d) = \sum_{d|q} f\left(\frac{q}{d}\right),$$

because the two sums have the same set of summands. The next theorem is perhaps more unexpected.

Theorem 9.1.4 *The totient function satisfies*

$$\sum_{d|q} \phi(d) = q.$$

Proof For each d that divides q, including $d = 1$ and $d = q$, consider the following set of elements of $\mathbf{Z}_q$:

$$\mathcal{S}_d = \{i \mid \text{GCD}(i, q) = d\}.$$

Each element of $\mathbf{Z}_q$ will belong to the set $\mathcal{S}_d$ for exactly one value of d. Hence if we sum the number of elements in the sets $\mathcal{S}_d$, we get q. But $i \in \mathcal{S}_d$ if, and only if, $\text{GCD}(i, q/d) = 1$, so $\phi(q/d)$ is equal to the number of elements in $\mathcal{S}_d$. Thus $\sum_{d|q} \phi\left(\frac{q}{d}\right) = q$. Because $\sum_{d|q} \phi(d) = \sum_{d|q} \phi\left(\frac{q}{d}\right)$, the proof is complete. □

The elements of $\mathbf{Z}_q$ always form a group under the operation of addition. We shall see that under the operation of multiplication, the nonzero elements of $\mathbf{Z}_q$ form a group only if q is a prime. However, we can always find a subset of $\mathbf{Z}_q$ that is a group under the operation of multiplication.

Let $\mathbf{Z}_q^*$ be the set of positive integers that are less than q and coprime to q. Let a be an element of $\mathbf{Z}_q^*$. No power of a can be equal to zero, modulo p, because $\text{GCD}(a, p) = 1$. It follows easily that the set $\{a, a^2, a^3, \ldots, a^{m-1}, 1\}$ is a cyclic group generated by a. The *order* of a is the number m of elements in the cyclic group of a. In particular, $aa^{m-1} = 1$ so a^{m-1} is an inverse for a. Because a is an arbitrary element of $\mathbf{Z}_q^*$, every element of $\mathbf{Z}_q^*$ has an inverse under integer multiplication modulo q. Thus $\mathbf{Z}_q^*$ forms a group with $\phi(q)$ elements.

The following theorem gives information about the order of an element a in $\mathbf{Z}_q^*$.

Theorem 9.1.5 (Euler's theorem) *If $a \in \mathbf{Z}_q^*$, then*

$$a^{\phi(q)} = 1 \pmod{q}.$$

Hence the order of any a in $\mathbf{Z}_q^*$ divides $\phi(q)$.

Proof Because $\mathbf{Z}_q$ is a group with $\phi(q)$ elements, the theorem follows from Corollary 2.1.6. □

Corollary 9.1.6 (Fermat's theorem) *If p is a prime, then for every a,*

$$a^{p-1} = 1 \pmod{p}$$

or

$$a^p = a \pmod{p}.$$

Proof This is an immediate consequence of Euler's theorem, because if p is a prime, then $\phi(p) = p - 1$. □

Fermat's theorem is also an immediate consequence of the following theorem.

Theorem 9.1.7 *If p is a prime, then under multiplication modulo p, the nonzero elements of the ring $\mathbf{Z}_p$ form a cyclic group generated by a primitive element π.*

Proof The theorem follows immediately as a restatement of Theorem 9.6.2, which is not stated and proved until the end of the chapter. □

Euler's theorem states that if a is coprime to q, then the order of a divides $\phi(q)$, but the order of a need not equal $\phi(q)$. The next theorem gives more specific information for the case in which q is a power of a prime. It is a rather difficult theorem of number theory, whose proof will occupy the remainder of the section. The proof of Theorem 9.1.8 uses Theorem 9.1.7, the proof of which is pending. However, there is no circular reasoning; Theorem 9.1.7 will be proved without using Theorem 9.1.8.

Before giving the theorem, we will consider two examples. Let $q = 9$. Then $\mathbf{Z}_9^* = \{1, 2, 4, 5, 7, 8\}$. It is simple to verify that the element 2 has order six and so this element can be used to generate $\mathbf{Z}_9^*$ as a cyclic group.

Let $q = 16$. Then $\mathbf{Z}_{16}^* = \{1, 3, 5, 7, 9, 11, 13, 15\}$. By trying all elements, we can verify that the largest order of any element is four. (The element 3 is such an element of order four.) Hence $\mathbf{Z}_{16}^*$ is not cyclic.

In the following theorem, because q is a prime power p^m, we have

$$\phi(p^m) = p^m - p^{m-1}$$
$$= p^{m-1}(p - 1).$$

Theorem 9.1.8 *Let p be a prime; then:*
 (i) *if p is odd, then $\mathbf{Z}_{p^m}^*$ is cyclic and so is isomorphic to $\mathbf{Z}_{p^{m-1}(p-1)}$;*
 (ii) *if p is 2 and p^m is not smaller than 8, then $\mathbf{Z}_{p^m}^*$ is not cyclic and is isomorphic to $\mathbf{Z}_2 \times \mathbf{Z}_2^{m-2}$;*
 (iii) *if $p^m = 4$, then $\mathbf{Z}_{p^m}^*$ is isomorphic to $\mathbf{Z}_2$.*

Proof Part (iii) is trivial because there is only one group with two elements. The remainder of the proof is broken into five steps. Part (i) of the theorem is proved in Steps 1 and 2. Step 1 shows that, for p odd, there is an element of order p^{m-1}, and Step 2 shows that, for p odd, there is an element of order $p - 1$. Because p^m and $p - 1$ are coprime, the product of these two elements has order $p^{m-1}(p - 1)$, which proves part (i) of the theorem.

Part (ii) is proved in Steps 3 and 4. Step 3 shows that for p equal to two, the order of every element divides 2^{m-2}. Step 4 shows there is an element of order 2^{m-2}.

In preparation for these steps, we begin with a useful relationship in Step 0.

Step 0 For all integers a and b and any prime power p^m,
$$(a+bp)^{p^{m-1}} \equiv a^{p^{m-1}} \pmod{p^m}.$$

The proof of Step 0 is by induction. The statement is true by inspection when $m = 1$. Suppose that for some m,
$$(a+bp)^{p^{m-1}} \equiv a^{p^{m-1}} \pmod{p^m}.$$

This can be written
$$(a+bp)^{p^{m-1}} = a^{p^{m-1}} + kp^m$$

for some integer k. By using the general rule
$$(x+y)^p = x^p + px^{p-1}y + \sum_{i=2}^{p} \binom{p}{i} x^{p-i} y^i,$$

raise the expression
$$(a+bp)^{p^{m-1}} = a^{p^{m-1}} + kp^m$$

to the pth power as follows:
$$((a+bp)^{p^{m-1}})^p = (a^{p^{m-1}} + kp^m)^p$$

or
$$(a+bp)^{p^m} = a^{p^m} + pkp^m a^{(p-1)p^{m-1}} + \sum_{i=2}^{p} \binom{p}{i} k^i p^{im} a^{(p-i)p^{m-1}}.$$

The second term on the right side is divisible by p^{m+1}, as are all terms in the final sum. Hence
$$(a+bp)^{p^m} \equiv a^{p^m} \pmod{p^{m+1}},$$

so the statement of Step 0 is also true with m replaced by $m+1$. The statement of Step 0 is clearly true for $m = 1$, so by induction this statement follows for all m.

Step 1 For p an odd prime, the order of $1+p$ modulo p^m is p^{m-1}. To prove this, choose $a = b = 1$ in the equation of Step 0. Then
$$(1+p)^{p^{m-1}} \equiv 1 \pmod{p^m},$$

so $(1+p)$ has order dividing p^{m-1}. We will show that the order of $(1+p)$ equals p^{m-1} when p is an odd prime by showing that the order does not divide p^{m-2}. Specifically,

9.1 Elementary number theory

to complete the proof of Step 1, we shall prove that

$$(1 + p)^{p^{m-2}} \equiv 1 + p^{m-1} \pmod{p^m}$$

for m greater than one, so the order of $(1 + p)$ is not p^{m-2} or a factor of p^{m-2}. The statement is true by inspection when m equals two. Suppose that the congruence is true for the integer $m - 1$. Then the congruence with m replaced by $m - 1$ is equivalent to

$$(1 + p)^{p^{m-3}} = 1 + p^{m-2} + kp^{m-1}$$

for some integer k. Hence, raising this to the power p gives

$$(1 + p)^{p^{m-2}} = 1 + p(p^{m-2} + kp^{m-1})$$
$$+ \sum_{i=2}^{p-1} \binom{p}{i} (p^{m-2} + kp^{m-1})^i + (p^{m-2} + kp^{m-1})^p.$$

For p an odd prime, $\binom{p}{2}, \binom{p}{3}, \ldots, \binom{p}{p-1}$ are each divisible by p. Every term in the sum is divisible by the term $p^{2(m-2)+1}$, and so is divisible by p^m if $m \geq 2$. The last term on the right side is divisible by $p^{p(m-1)}$, and so is divisible by p^m if $p \geq 3$ and $m \geq 2$. (The proof would fail here if $p = 2$.) Therefore by induction, $(1 + p)^{p^{m-2}} = 1 + p^{m-1} \pmod{p^m}$, so the order, modulo p^m, of $1 + p$ does not divide p^{m-2}.

This completes the proof that $1 + p$ has order p^{m-1} in $\mathbf{Z}_{p^m}$.

Step 2 For p an odd prime, there is an element of $\mathbf{Z}^*_{p^m}$ of order $p - 1$. To prove this, note that because $\mathbf{Z}^*_{p^m}$ has $p^{m-1}(p - 1)$ elements, the order of every element divides $p^{m-1}(p - 1)$. To find an element of $\mathbf{Z}^*_{p^m}$ of order $p - 1$, choose an integer π that is a primitive element of the ring $\mathbf{Z}_p$. Then π has order $p - 1$ modulo p. We will show that $\alpha = \pi^{p^{m-1}}$ is the desired element of order $p - 1$ modulo p^m. Because $\alpha^{p-1} = \pi^{(p^{m-1})(p-1)} = 1$, the order of α must divide $p - 1$; we only need to show that the order is not smaller than $p - 1$.

Because π is primitive in $\mathbf{Z}_p$, we know that the $p - 1$ powers π^i for $i = 0, \ldots, p - 2$ can be written as elements of the integer ring $\mathbf{Z}$ as

$$\pi^i = a_i + b_i p,$$

where the a_i are distinct nonzero integers less than p. Hence, by Step 0,

$$(\pi^i)^{p^{m-1}} = (a_i + b_i p)^{p^{m-1}} = a_i^{p^{m-1}} \pmod{p^m}.$$

But, by definition, $\alpha = \pi^{p^{m-1}}$, so

$$\alpha^i = a_i^{p^{m-1}} \pmod{p^m},$$

where the a_i are distinct integers less than p. Hence, if we can show that

$$a^{p^{m-1}} \neq b^{p^{m-1}} \pmod{p^m}$$

for any two distinct integers a and b, both less than p, then we can conclude that the order of α must be larger than $p-2$. But $a^p = a$ for any element of $\mathbf{Z}_p$, and hence $a^{p^{m-1}} = a \pmod{p}$. Therefore, if

$$a^{p^{m-1}} = b^{p^{m-1}} \pmod{p^m},$$

then certainly

$$a^{p^{m-1}} = b^{p^{m-1}} \pmod{p},$$

and so

$$a = b \pmod{p}$$

contrary to the statement that a and b are distinct integers. Hence a has order $p-1$, and part (i) of the theorem is proved.

Step 3 The order of every element of $\mathbf{Z}^*_{2^m}$ divides 2^{m-2}. This is proved by first proving that

$$(a+4b)^{2^{m-2}} \equiv a^{2^{m-2}} \pmod{2^m}$$

for any integers a and b. Because only odd integers are in $\mathbf{Z}^*_{2^m}$, the statement to be proved then follows by choosing $a = \pm 1$ and various values of b.

The congruence follows by inspection for $m=2$. Suppose it is true for some positive integer m. Then

$$(a+4b)^{2^{m-2}} = a^{2^{m-2}} + k2^m$$

for some integer k. Then

$$(a+4b)^{2^{m-1}} = (a^{2^{m-2}} + k2^m)^2$$
$$= a^{2^{m-1}} + ka^{2^{m-2}}2^{m+1} + k^2 2^{2m}.$$

Hence

$$(a+4b)^{2^{m-1}} \equiv a^{2^{m-1}} \pmod{2^{m+1}}.$$

Step 4 It remains only to find an element of order 2^{m-2} when m is at least three. We will show that $3^{2^{m-3}} \neq 1 \pmod{2^m}$, and so the element 3 must have order 2^{m-2}. This claim is immediately evident when $m=3$. For $m \geq 4$, the proof consists of proving that

$$(1+2)^{2^{m-3}} \equiv 1 + 2^{m-1} \pmod{2^m}.$$

This is true when $m=4$. Suppose it is true for some m not smaller than four. Then

$$(1+2)^{2^{m-3}} = 1 + 2^{m-1} + k2^m$$

for some k. Then

$$(1+2)^{2^{m-2}} = (1 + 2^{m-1} + k2^m)^2$$
$$= 1 + 2^m + 2^{2(m-1)} + k2^{m+1} + k2^{2m} + k^2 2^{2m}.$$

Hence, because m is greater than three,

$$(1+2)^{2^{m-2}} \equiv 1 + 2^m \pmod{2^{m+1}}.$$

Therefore, for all m greater than three,

$$(1+2)^{2^{m-3}} \equiv 1 + 2^{m-1} \pmod{2^m}.$$

Therefore the order of the element 3 (mod 2^m) does not divide 2^{m-3}. Hence the order of the element 3 is 2^{m-2}. The proof is now complete. □

Corollary 9.1.9 *The element 3 has order 2^{m-2} modulo 2^m, and the element $2^m - 1$ has order 2 modulo 2^m.*

Proof Step 4 of the proof of Theorem 9.1.8 provides the proof of the first statement. The second statement is immediate because $(2^m - 1)^2 = 2^{2m} - 2 \cdot 2^m + 1 = 1 \pmod{2^m}$. □

9.2 Fields based on the integer ring

There is an important construction by which a new ring, called a *quotient ring*, can be constructed from a given ring. The quotient ring is defined in a technical way that involves the construction of cosets. The construction can be used for an arbitrary ring. In the ring of integers, the construction of a quotient ring is easy. It is the ring of integers modulo q, denoted by $\mathbf{Z}/\langle q \rangle$ or $\mathbf{Z}_q$, that we saw earlier.

Let q be a positive integer. The quotient ring $\mathbf{Z}/\langle q \rangle$ is the set $\{0, \ldots, q-1\}$ with addition and multiplication defined by

$a + b = R_q[a + b],$
$a \cdot b = R_q[ab].$

Elements called $0, \ldots, q-1$ are in both $\mathbf{Z}$ and $\mathbf{Z}/\langle q \rangle$. The elements of $\mathbf{Z}/\langle q \rangle$ have the same names as the first q elements of $\mathbf{Z}$. It is a matter of personal taste whether one thinks of them as the same mathematical objects or as some other objects with the same names.

Two elements a and b of $\mathbf{Z}$ that map into the same element of $\mathbf{Z}/\langle q \rangle$ are congruent modulo q, and $a = b + mq$ for some integer m.

Theorem 9.2.1 *The quotient ring $\mathbf{Z}/\langle q \rangle$ is a ring.*

Proof The proof consists of a straightforward verification of the properties of a ring. □

Theorem 9.2.2 *In $\mathbf{Z}/\langle q \rangle$, a nonzero element s has an inverse under multiplication if and only if s and q are coprime.*

Proof Let s be a nonzero element of the ring such that s and q are coprime. Then $0 < s \leq q - 1$. By Corollary 2.6.5, there exist integers a and b such that

$$1 = aq + bs.$$

Therefore, modulo q, we have

$$1 = R_q[1] = R_q[aq + bs] = R_q\{R_q[aq] + R_q[bs]\}$$
$$= R_q[bs] = R_q\{R_q[b] R_q[s]\}$$
$$= R_q\{R_q[b] \cdot s\}.$$

Hence $R_q[b]$ is a multiplicative inverse for s under modulo q multiplication, so $s^{-1} = R_q[b]$.

Now let s be an element of the ring such that s and q are not coprime. First, we consider the case in which s is a factor of q. Then $q = s \cdot r$. Suppose that s has an inverse s^{-1}. Then

$$r = R_q[r] = R_q[s^{-1} \cdot s \cdot r] = R_q[s^{-1} \cdot q] = 0.$$

But $r \neq 0$ modulo q, so we have a contradiction. Hence s does not have an inverse if it is a factor of q.

Next, consider the case in which s and q are not coprime, but s does not divide q. Let $d = \text{GCD}[s, q]$. Then $s = ds'$ for some s' that is coprime to q, and d is a factor of q. If s has an inverse s^{-1}, then d has an inverse given by $d^{-1} = s^{-1} s'$, contrary to the previous paragraph. Hence s does not have an inverse. The proof of the theorem is complete. □

In general, the nonzero elements of the set $\mathbf{Z}/\langle q \rangle$ do not form a group under multiplication because inverses do not exist for all elements. However, we can find subsets in $\mathbf{Z}/\langle q \rangle$ that are groups. In the ring $\mathbf{Z}/\langle q \rangle$, we can choose any element and form the subset

$$\{\beta, \beta^2, \beta^3, \ldots, \beta^r = 1\}$$

where the construction stops when the identity element one is obtained. We have already seen that the identity element must occur, and that this construction forms a cyclic group. It is a subgroup of $\mathbf{Z}/\langle q\rangle$. The integer r is called the order of the element β in $\mathbf{Z}/\langle q\rangle$.

We can see in the examples of Section 2.3 that the arithmetic of $GF(2)$ and $GF(3)$ can be described as addition and multiplication modulo two or modulo three, respectively, but the arithmetic of $GF(4)$ cannot be so described. That is, in symbols, $GF(2) = \mathbf{Z}/\langle 2\rangle$, $GF(3), = \mathbf{Z}/\langle 3\rangle$, $GF(4) \neq \mathbf{Z}/\langle 4\rangle$. The general fact is given by the following theorem.

Theorem 9.2.3 *The quotient ring $\mathbf{Z}/\langle q\rangle$ is a field if and only if q is a prime.*

Proof Suppose that q is a prime. Then every element of $\mathbf{Z}/\langle q\rangle$ is coprime to q. Hence, by Theorem 9.2.2, every element has a multiplicative inverse, so $\mathbf{Z}/\langle q\rangle$ is a field.

Now suppose that q is not a prime. Then $q = r \cdot s$ and, by Theorem 9.2.2, r and s do not have inverses under multiplication, so $\mathbf{Z}/\langle q\rangle$ is not a field. □

Whenever the quotient ring $\mathbf{Z}/\langle q\rangle$ is a field, it is also called by the name $GF(q)$, which emphasizes that it is a field. Finite fields constructed as quotient rings of integers are not the only finite fields that exist, but all finite fields with a prime number of elements can be so constructed as quotient rings. The finite field $GF(p)$, with p a prime, is also called a *prime field*.

If the field $GF(p)$ does not contain a square root of -1, then $GF(p)$ can be extended to $GF(p^2)$ in the same way that the real field is extended to the complex field. For example, $6 = -1$ in $GF(7)$, and it is easy to check that there is no element in $GF(7)$ whose square is six. Hence we define

$$GF(49) = \{a + jb; a, b \in GF(7)\}$$

with addition and multiplication, defined in the same way as for the complex field:

$$(a + jb) + (c + jd) = (a + c) + j(b + d)$$
$$(a + jb) \cdot (c + jd) = (ac - bd) + j(ab + cd),$$

where operations within the parentheses on the right side are those of $GF(7)$. With these definitions, $GF(49)$ is a field. The integers of $GF(49)$ are the elements of $GF(7)$, so $GF(49)$ has characteristic seven.

Theorem 9.2.4 *Every field contains a unique smallest subfield that is either isomorphic to the field of rationals or has a prime number of elements. Hence the characteristic of every Galois field is either a prime or is zero.*

Proof The field contains the elements zero and one. To define the subfield, consider the subset $G = \{0, 1, 1+1, 1+1+1, \ldots\}$ denoting these $\{0, 1, 2, 3, \ldots\}$. This subset must contain either a finite number, p, of elements, or an infinite number of elements. We will show that if it is finite, p is a prime and $G = GF(p)$. If G is finite, addition is modulo p because G is a cyclic group under addition. Because of the distributive law, multiplication must also be modulo p because

$$\alpha \cdot \beta = (1 + \cdots + 1) \cdot \beta$$
$$= \beta + \cdots + \beta,$$

where there are α copies of β in the sum and the addition is modulo p. Hence multiplication also is modulo p. Each element β has an inverse under multiplication, because the sequence

$$\beta, 2\beta, 3\beta, \ldots$$

is a cyclic subgroup of G. However, $\alpha\beta \neq 0$ for all α, β in the original field, so $\alpha\beta \neq 0$ modulo p for all integers of the field, and so p must be a prime. Thus the subset G is just the prime field given in Theorem 9.2.3.

In the other case, G is infinite. It is isomorphic to the ring of integers. The smallest subfield of F containing G is isomorphic to the smallest field containing the ring of integers, which is the rational field. □

In the Galois field $GF(q)$, we have found the subfield $GF(p)$ with p a prime. In particular, if q were a prime to start with, then we see that it can be interpreted as the field of integers modulo q. Hence, for a given prime, there is really only one field with that number of elements, although of course, it may be represented by many different notations.

9.3 Fields based on polynomial rings

A field can be obtained from a ring of polynomials over a smaller field by using a construction similar to that used to obtain a finite field from the integer ring. The fields that are constructed in this way are either finite or infinite fields, according to whether the polynomial rings are defined over finite or infinite fields. These extension fields prove useful in signal processing in several ways. In Chapter 10, they are used to construct number theory transforms. In Chapter 11, they are used to compute multidimensional convolutions.

Suppose that we have $F[x]$, the ring of polynomials over the field F. Just as we constructed quotient rings in the ring $\mathbf{Z}$, so we can construct quotient rings in $F[x]$. Choosing any polynomial $p(x)$ from $F[x]$, we can define the quotient ring by using

$p(x)$ as a modulus for polynomial arithmetic. We shall restrict the discussion to monic polynomials because this restriction eliminates needless ambiguity in the construction.

Definition 9.3.1 *For any monic polynomial $p(x)$ with nonzero degree over the field F, the ring of polynomials modulo $p(x)$ is the set of all polynomials with degree smaller than that of $p(x)$, together with polynomial addition and polynomial multiplication modulo $p(x)$. This ring is conventionally denoted by $F[x]/\langle p(x) \rangle$.*

Any element $r(x)$ of $F[x]$ can be mapped into $F[x]/\langle p(x) \rangle$ by mapping $r(x)$ to $R_{p(x)}[r(x)]$. Two elements, $a(x)$ and $b(x)$, of $F[x]$ that are congruent modulo $p(x)$ map into the same element of $F[x]/\langle p(x) \rangle$.

Theorem 9.3.2 *The ring of polynomials modulo $p(x)$ is a ring.*

Proof A straightforward verification of the properties of a ring provides the proof.
□

As an example, in the ring of polynomials over $GF(2)$, choose $p(x) = x^3 + 1$. Then the ring of polynomials modulo $p(x)$ is $GF(2)[x]/\langle x^3 + 1 \rangle$. This is the set

$$\{0, 1, x, x+1, x^2, x^2+1, x^2+x, x^2+x+1\},$$

and in this ring, multiplication is as follows:

$$(x^2+1) \cdot (x^2) = R_{x^3+1}[(x^2+1) \cdot x^2]$$
$$= R_{x^3+1}[x(x^3+1) + x^2 + x] = x^2 + x,$$

where we have used the equality $x^4 = x(x^3 + 1) + x$.

As another example, in the ring of polynomials over the rational field Q, choose $p(x) = x^2 + 1$. Then the ring of polynomials modulo $p(x)$ is $Q[x]/\langle x^2 + 1 \rangle$. It contains all polynomials of the form $a + xb$, where a and b are rationals. Multiplication is as follows:

$$(a + xb)(c + xd) = (ac - bd) + x(ad + bc).$$

This is the same structure as complex multiplication. The ring of polynomials over Q modulo $x^2 + 1$ is actually a field. This is a consequence of the following general theorem.

Theorem 9.3.3 *The ring of polynomials over a field F modulo a monic polynomial $p(x)$ is a field if and only if $p(x)$ is a prime polynomial.*[1]

---

[1] Recall that a prime polynomial is both monic and irreducible. It is enough that $p(x)$ be irreducible to get a field, but we insist on the convention of using a polynomial that is monic as well so that later results are less arbitrary.

Proof To prove that the ring is a field if $p(x)$ is a prime polynomial, we must show that every element has a multiplicative inverse. Let $s(x)$ be an element of the ring. Then

$\deg s(x) < \deg p(x)$.

Because $p(x)$ is a prime polynomial, $\text{GCD}[s(x), p(x)] = 1$. By Corollary 2.7.8,

$1 = a(x)p(x) + b(x)s(x)$

for some polynomials $a(x)$ and $b(x)$. Hence

$$\begin{aligned} 1 = R_{p(x)}[1] &= R_{p(x)}[a(x)p(x) + b(x)s(x)] \\ &= R_{p(x)}[R_{p(x)}[b(x)] \cdot R_{p(x)}[s(x)]] \\ &= R_{p(x)}[R_{p(x)}[b(x)] \cdot s(x)]. \end{aligned}$$

Therefore $R_{p(x)}[b(x)]$ is a multiplicative inverse for $s(x)$ in this ring. Because $b(x)$ is an arbitrary nonzero element of the ring, every nonzero element of the ring has an inverse. Therefore $F[x]/\langle p(x) \rangle$ is a field.

Now suppose that $p(x)$ is not a prime polynomial. Then $p(x) = r(x)s(x)$ for some $r(x)$ and $s(x)$. If the ring is a field, then $r(x)$ has an inverse $r^{-1}(x)$. Hence

$$\begin{aligned} s(x) = R_{p(x)}[s(x)] &= R_{p(x)}[r^{-1}(x) \cdot r(x) \cdot s(x)] \\ &= R_{p(x)}[r^{-1}(x)p(x)] = 0. \end{aligned}$$

But $s(x) \neq 0$, so we have a contradiction. Therefore, if $p(x)$ is not a polynomial, the ring $F[x]/\langle p(x) \rangle$ is not a field. □

The theorem gives one way to construct the complex field $\boldsymbol{C}$ as an extension of the real field $\boldsymbol{R}$. Because $x^2 + 1$ is a prime polynomial over $\boldsymbol{R}$, we can form an extension field as the set $\{a + bx\}$, where a and b are elements of $\boldsymbol{R}$. Addition and multiplication are defined modulo $x^2 + 1$. Then

$(a + bx) + (c + dx) = (a + c) + (b + d)x$,
$(a + bx)(c + dx) = (ac - bd) + (ad + bc)x$.

These formulas are more familiar when j is used in place of x, and they are written

$(a + \text{j}b) + (c + \text{j}d) = (a + c) + \text{j}(b + d)$,
$(a + \text{j}b)(c + \text{j}d) = (ac - bd) + \text{j}(ad + bc)$.

The same construction can be used to extend any field in which $x^2 + 1$ is a prime polynomial. The finite field $GF(7)$ is such a field, so it can be extended to $GF(49)$ in the same way that the real field is extended to the complex field. This is the formal background for the construction that was already given prior to the theorem.

On the other hand, $x^2 + 1$ is not a prime polynomial over $GF(5)$, and it cannot be used to extend $GF(5)$. To extend $GF(5)$ to $GF(25)$, we must use a different polynomial, such as $x^2 + x + 1$, which is a prime polynomial over $GF(5)$. With this modulus polynomial, the multiplication rule is

$$(a + xb)(c + xd) = (ac - bd) + x(ad + bc - bd).$$

In the extension field $GF(25)$, multiplication behaves differently than it does in the complex field. This kind of multiplication rule applies whenever $x^2 + x + 1$ is used to extend $GF(p)$ to $GF(p^2)$, and this can be done whenever $x^2 + x + 1$ is a prime polynomial in $GF(p)$. In particular, $GF(2)$ can be extended to $GF(4)$ by using this polynomial.

9.4 Minimal polynomials and conjugates

An element of one field may be the zero of a polynomial over a smaller field. If an element is a zero of a polynomial over a subfield, then associated with that element is a particular polynomial of importance. This relationship leads to the following definition.

Definition 9.4.1 *Let F be a field, and let β be an element of an extension field of F. The minimal polynomial of β over the field F, if there is one, is the nonzero monic polynomial of lowest degree having coefficients in the field F and β as a zero.*

Not all elements of an extension of F have a minimal polynomial over F – elements of the real field that do not have a minimal polynomial over the rationals are called *transcendental* numbers – but if an element has a minimal polynomial, the minimal polynomial is unique. This is because if $f^{(1)}(x)$ and $f^{(2)}(x)$ are both monic of the same degree and with β as a zero, then $f(x) = f^{(1)}(x) - f^{(2)}(x)$ also has β as a zero and has a smaller degree. Consequently, $f(x) = 0$, and $f^{(1)}(x)$ equals $f^{(2)}(x)$.

Theorem 9.4.2 *Every minimal polynomial is a prime polynomial and is unique. Further, if β has a minimal polynomial $f(x)$ and another polynomial $g(x)$ has β as a zero, then $f(x)$ divides $g(x)$.*

Proof We have already seen that the minimal polynomial is unique. By definition, the minimal polynomial $f(x)$ is monic. Suppose $f(x)$ factors as

$$f(x) = a(x)b(x).$$

Then $f(\beta) = a(\beta)b(\beta) = 0$, so either $a(x)$ or $b(x)$ has β as a zero and a degree smaller than $f(x)$. Hence a minimal polynomial is a prime polynomial.

To prove the last part of the theorem, write

$$g(x) = f(x)h(x) + s(x),$$

where $s(x)$ has a smaller degree than $f(x)$ and hence cannot have β as a zero. But

$$0 = g(\beta) = f(\beta)h(\beta) + s(\beta) = s(\beta).$$

Hence $s(x)$ must be zero, and the theorem is proved. □

An element β of the field F itself has a minimal polynomial over F of degree one, given by

$$f(x) = x - \beta.$$

If β is not in F, the minimal polynomial must have a degree of two or greater. Hence the minimal polynomial of β will also have other field elements as zeros. Then $f(x)$ must also be the minimal polynomial for these other field elements.

Definition 9.4.3 *Elements of an extension of the field F that share the same minimal polynomial over F are called conjugates (with respect to F).*

Generally, a single element can have more than one conjugate – in fact, as many conjugates as the degree of its minimal polynomial if the element itself is regarded as its own conjugate. We emphasize that the conjugacy relationship between two elements depends on the base field. Two elements of the complex field might be conjugates with respect to the rationals, but yet not be conjugates with respect to the reals. For example, the complex number $j\sqrt[4]{2}$ has $x^4 - 2$ as its minimal polynomial over $\boldsymbol{Q}$, and $x^2 + \sqrt{2}$ as its minimal polynomial over $\boldsymbol{R}$. The set of $\boldsymbol{Q}$-ary conjugates containing $j\sqrt[4]{2}$ is $\{\sqrt[4]{2}, -\sqrt[4]{2}, j\sqrt[4]{2}, -j\sqrt[4]{2}\}$. The set of $\boldsymbol{R}$-ary conjugates is $\{j\sqrt{2}, -j\sqrt{2}\}$.

9.5 Cyclotomic polynomials

Over any field F, the polynomial $x^n - 1$ can be written in terms of its prime factors:

$$x^n - 1 = p_1(x)p_2(x) \cdots p_K(x).$$

When F is the field of rationals, the prime factors are easy to identify. They are the polynomials named in the following definition.

Definition 9.5.1 *For each n, the polynomial*

$$\Phi_n(x) = \prod_{\text{GCD}(i,n)=1} (x - \omega_n^i),$$

where ω_n is an nth root of unity in the complex field, is called a cyclotomic polynomial.

9.5 Cyclotomic polynomials

Cyclotomic polynomials have simple coefficients. In fact, all the cyclotomic polynomials with n smaller than 105 have coefficients equal only to zero or to ± 1. We shall see later that all coefficients of a cyclotomic polynomial are integers, a fact that is not evident in the definition.

The polynomial $x^n - 1$ always can be factored in the complex field as

$$x^n - 1 = \prod_{i=0}^{n-1}(x - \omega_n^i),$$

where ω_n is an nth root of unity. Hence each of the cyclotomic polynomials can be expressed as a product of some of these first-degree factors, and so the cyclotomic polynomial $\Phi_n(x)$ must divide $x^n - 1$. The definition of a cyclotomic polynomial chooses the minimum number of factors such that the polynomial product has only rational coefficients.

For small n, the cyclotomic polynomials are easy to find by factoring $x^n - 1$. Clearly,

$$\Phi_1(x) = x - 1.$$

To see the general pattern, we shall examine a few more of these polynomials. Let $n = 2$. Then

$$x^2 - 1 = (x - 1)(x + 1) = \Phi_1(x)\Phi_2(x).$$

Let $n = 3$. Then

$$x^3 - 1 = (x - 1)(x^2 + x + 1) = \Phi_1(x)\Phi_3(x).$$

Let $n = 4$. Then

$$x^4 - 1 = (x - 1)(x + 1)(x^2 + 1) = \Phi_1(x)\Phi_2(x)\Phi_4(x).$$

Notice that, at each step, only one new polynomial occurs. Let $n = 5$. Then

$$x^5 - 1 = (x - 1)(x^4 + x^3 + x^2 + x + 1) = \Phi_1(x)\Phi_5(x).$$

Let $n = 6$. Then

$$x^6 - 1 = (x - 1)(x + 1)(x^2 + x + 1)(x^2 - x + 1)$$
$$= \Phi_1(x)\Phi_2(x)\Phi_3(x)\Phi_6(x).$$

Let $n = 7$. Then

$$x^7 - 1 = (x - 1)(x^6 + x^5 + x^4 + x^3 + x^2 + x + 1)$$
$$= \Phi_1(x)\Phi_7(x).$$

Let $n = 8$. Then

$$x^8 - 1 = (x - 1)(x + 1)(x^2 + 1)(x^4 + 1)$$
$$= \Phi_1(x)\Phi_2(x)\Phi_4(x)\Phi_8(x).$$

Let $n = 9$. Then
$$x^9 - 1 = (x - 1)(x^2 + x + 1)(x^6 + x^3 + 1)$$
$$= \Phi_1(x)\Phi_3(x)\Phi_9(x).$$

Some properties of the cyclotomic polynomials are described by the following theorems.

Theorem 9.5.2 *For each n,*

$\deg \Phi_n(x) = \phi(n),$

where $\phi(n)$ is the totient function.

Proof The totient function $\phi(n)$ is defined as the number of integers that do not divide n, and this is the number of linear factors in the definition of $\Phi_n(x)$. □

Theorem 9.5.3 *For each n,*
$$x^n - 1 = \prod_{k|n} \Phi_k(x).$$

Proof This is a straightforward manipulation:
$$x^n - 1 = \prod_{i=1}^{n} (x - \omega_n^i)$$
$$= \prod_{k|n} \prod_{\text{GCD}(i,n)=n/k} (x - \omega_n^i)$$
$$= \prod_{k|n} \Phi_k(x). \quad \square$$

Theorem 9.5.4 *The coefficients of a cyclotomic polynomial are integers.*

Proof The proof is by induction. The polynomial

$\Phi_1(x) = x - 1$

has integer coefficients, and, by Theorem 9.5.3,

$$\Phi_n(x) = \frac{x^n - 1}{\prod_{\substack{k|n \\ k<n}} \Phi_k(x)}.$$

Division of a monic polynomial with integer coefficients by a factor that is a monic polynomial with integer coefficients produces a quotient polynomial that is a monic polynomial with integer coefficients. This is because each iteration of the long-division

9.5 Cyclotomic polynomials

process multiplies the divisor polynomial by an integer times a power of x and subtracts it from the dividend polynomial, thereby producing a new divided polynomial with integer coefficients. Consequently, the theorem is true for any n if it is true for all positive integers less than n. By induction, it is true for all n. □

Theorem 9.5.5 *The cyclotomic polynomials are prime polynomials over the field of rationals, and so they are the minimal polynomials of the roots of unity.*

Proof The cyclotomic polynomials, clearly, are monic. We need only to show that they are irreducible over the rationals. Let $f(x)$ be the minimal polynomial of ω_n over the rationals. Let h be any integer coprime to n, and let $g(x)$ be the minimal polynomial over the rationals of ω_n^h. Suppose that for every such h, $g(x) = f(x)$. This then implies that $f(x)$ is a multiple of $\Phi_n(x)$, and because $\Phi_n(x)$ is monic, has only rational coefficients, and has ω_n as a zero, we conclude that $f(x) = \Phi_n(x)$. Hence we must prove that for every h coprime to n, the minimal polynomial $g(x)$ of ω_n^h is equal to $f(x)$.

Step 1 First, we prove the statement for h equal to a prime p that does not divide n. Let $g(x)$ be the minimal polynomial of ω_n^p, and suppose that $g(x)$ is not equal to $f(x)$. Then, because both $f(x)$ and $g(x)$ divide $x^n - 1$, are irreducible, and have integer coefficients,

$$x^n - 1 = f(x)g(x)t(x)$$

for some $t(x)$ with integer coefficients. Because $g(x^p)$ has a zero at ω_n, it is a multiple of $f(x)$. Then

$$g(x^p) = f(x)k(x),$$

where $k(x)$ is another polynomial with integer coefficients. Now replace both equations by their residue modulo p:

$$x^n - 1 = f(x)g(x)t(x) \pmod{p},$$
$$g(x^p) = f(x)k(x) \pmod{p}.$$

By Theorem 2.2.4, the second equation becomes

$$(g(x))^p = f(x)k(x) \pmod{p}.$$

Now this equation can be regarded as a polynomial equation in the Galois field $GF(p)$ and factored in that field. Every prime factor $p(x)$ of $f(x)$ in $GF(p)$ must also be a prime factor of $(g(x))^p$, and so too of $g(x)$. Therefore $(p(x))^2$ divides $x^n - 1$. Hence the formal derivative nx^{n-1} of $x^n - 1$ is divisible by $p(x)$ in $GF(p)$. But n is not equal to zero modulo p by assumption, and x cannot be a factor of $f(x)$ (mod p) because $f(x)$ divides $x^n - 1$. The contradiction shows that $g(x)$ is not different from $f(x)$.

Step 2 Now we prove the statement for arbitrary h coprime to n. Let h be expressed in terms of its prime factors, each of which is coprime to n. Then

$$h = p_1 p_2 \ldots p_s.$$

By Step 1, if ω_n is a zero of $f(x)$, then so too is $\omega_n^{p_1}$. Then again by Step 1, if $\omega_n^{p_1}$ is a zero of $f(x)$, then so too is $\omega_n^{p_1 p_2}$. Continuing in this way, we conclude that ω_n^h is a zero of $f(x)$. $\square$

9.6 Primitive elements

A primitive element of a Galois field was defined in Section 2.3 as an element α such that every field element except zero can be expressed as a power of α. For example, in the field $GF(7)$, we have

$3^1 = 3,$
$3^2 = 2,$
$3^3 = 6,$
$3^4 = 4,$
$3^5 = 5,$
$3^6 = 1,$

so that 3 is a primitive element of $GF(7)$. Primitive elements are useful for constructing finite fields, because if a primitive element exists, we can construct a multiplication table by multiplying powers of the primitive element. The primitive element then serves the role of a logarithm base in the Galois field. In this section we shall prove that every finite field contains a primitive element.

A finite field forms an abelian group in two ways. The set of field elements forms an abelian group under the operation of addition, and the set of field elements excluding the zero element forms an abelian group under the operation of multiplication. We shall work with the group under the operation of multiplication, a group with $q - 1$ elements. By Theorem 2.1.5, the order of this group is divisible by the order of any of its elements. Thus the order of every element divides $q - 1$.

Theorem 9.6.1 *Let $\beta_1, \beta_2, \ldots, \beta_{q-1}$ denote the nonzero field elements of $GF(q)$. Then*

$$x^{q-1} - 1 = (x - \beta_1)(x - \beta_2) \cdots (x - \beta_{q-1}).$$

9.6 Primitive elements

Proof The set of nonzero elements of the field $GF(q)$ is a finite group under the operation of multiplication. Let β be any nonzero element of $GF(q)$, and let h be its order under the operation of multiplication. Then, by Theorem 2.1.5 (Lagrange's theorem), h divides $q - 1$. Hence

$$\beta^{q-1} = (\beta^h)^{(q-1)/h} = 1^{(q-1)/h} = 1,$$

so β is a zero of $x^{q-1} - 1$. □

Theorem 9.6.2 *The group of nonzero elements of $GF(q)$ under multiplication is a cyclic group.*

Proof The theorem is trivial if $q - 1$ is a prime because then every element except zero and one has order $q - 1$, and so every element is primitive. We need to prove the theorem only for composite $q - 1$. Consider the prime factorization of $q - 1$:

$$q - 1 = \prod_{i=1}^{s} p_i^{v_i}.$$

Because $GF(q)$ is a field, of the $q - 1$ nonzero elements of $GF(q)$, there must be at least one that is not a zero of $x^{(q-1)/p_i} - 1$, because this polynomial has at most $(q-1)/p_i$ zeros. Hence, for each i, a nonzero element a_i of $GF(q)$ can be found for which $a_i^{(q-1)/p_i} \neq 1$. Let $b_i = a_i^{(q-1)/p_i^{v_i}}$, and let $b = \prod_{i=1}^{s} b_i$. We will prove that b has order $q - 1$, and so the group is cyclic.

Step 1 The element b_i has order $p_i^{v_i}$. Proof: Clearly, $b_i^{p_i^{v_i}} = 1$, so the order of b_i divides $p_i^{v_i}$; the order of b_i is of the form $p_i^{n_i}$. If n_i is less than v_i, then $b_i^{p_i^{v_i-1}} = 1$. But $b_i^{p_i^{v_i-1}} \neq 1$. Therefore b_i has order $p_i^{v_i}$.

Step 2 The element b has order $q - 1$. Proof: Suppose $b^n = 1$. We first show that this implies $n \equiv 0 \pmod{p_i^{v_i}}$ for $i = 1, \ldots, s$. For each i, we can write

$$b^{n \prod_{j \neq i} p_j^{v_j}} = 1.$$

Replacing b by $\prod_{i=1}^{s} b_i$ and using that $b_j^{p_j^{v_j}} = 1$, we conclude that

$$b_i^{n \prod_{j \neq i} p_j^{v_j}} = 1.$$

Therefore $n \prod_{j \neq i} p_j^{v_j} \equiv 0$ modulo $p_i^{v_i}$. Because the p_i are distinct primes, it follows that $n \equiv 0 \pmod{p_i^{v_i}}$ for each i. Hence $n \equiv 0 \pmod{q-1}$. The proof is now complete. □

Theorem 9.6.3 *Every Galois field has a primitive element.*

Proof As a cyclic group, the nonzero elements of $GF(q)$ include an element of order $q-1$. This element is a primitive element. □

9.7 Algebraic integers

The conventional way of representing the arithmetic in a Fourier transform is to allow in principle for arbitrary complex numbers in the output vector but, in practice, to reduce the set to a finite number of complex numbers by truncating the binary wordlength; this in a somewhat arbitrary way. The conventional representation manifests itself in wordlength considerations and in considerations of roundoff error. Even when the original time-domain data sequence is a sequence of integers, an exact representation of the transform requires an infinite number of bits. This section provides an alternative representation of the range space of the Fourier transform. It is expressed as a polynomial extension of the field of rationals, or as a polynomial extension of the field of complex rationals.

The ring of integers $\mathbf{Z}$ can be extended to the ring of *gaussian integers*, denoted $\mathbf{Z}[j]$, by introducing the symbol j as an indeterminate. The ring of gaussian integers is defined as the set $\mathbf{Z}[j] = \{a + bj\}$, where a and b are integers, together with the obvious definitions of addition and multiplication using $j^2 = -1$. It is easy to verify that the set of gaussian integers is a ring. The ring of gaussian integers is the simplest instance of the kind of ring known as a ring of algebraic integers, or, more simply, an *algebraic integer ring*. Algebraic integer rings are discussed in this section.

In a similar way, the field of rationals $\mathbf{Q}$ can be extended to the field of *gaussian rationals*, denoted $\mathbf{Q}(j)$, by introducing the symbol j as an indeterminant. The field of gaussian rationals is defined as the set $\mathbf{Q}(j) = \{a + jb\}$, where a and b are rationals, together with the obvious definitions of addition and multiplication using $j^2 = -1$. The set of gaussian rationals is the simplest instance of the kind of field known as a field of algebraic numbers, an algebraic number field, or, more simply, a *number field*. Algebraic number fields are also discussed in this section.

The general construction of a ring of algebraic integers, or of a field of algebraic numbers, is as follows. Choose any irreducible polynomial $p(x)$ of degree m. The cyclotomic polynomial of degree m would be a suitable choice, and our preference. Let $\mathbf{Z}/\langle p(x) \rangle$ be the set of polynomials of degree $m-1$ or less with integer coefficients. A zero of a polynomial $p(x)$ with integer coefficients is called an *algebraic number*. A zero of a *monic* polynomial $p(x)$ with integer coefficients is called an *algebraic integer*. Thus the set of algebraic integers is contained in the set of algebraic numbers, the distinction being whether $p(x)$ is monic or not. Of course, another way of formulating this is to say that $p(x)$ must be monic in both situations – but has integer coefficients in one situation and rational coefficients in the other, the polynomial over

9.7 Algebraic integers

Z being converted to a monic polynomial over **Q** by dividing out the leading integer coefficient.

Each of these two cases, algebraic numbers and algebraic integers, can be viewed in either of two ways. The algebraic integers and the algebraic numbers each can be regarded either as a set of points within the complex plane, or simply as a polynomial extension of **Z** or of **Q**, the difference being whether the zero of $p(x)$ is regarded as the complex number ω or the indeterminant x. These are equivalent mathematically, but the computational interpretations and applications are quite different.

Let $p(x)$ be a polynomial over the rationals, possibly monic, and with integer coefficients. A cyclotomic polynomial would be a good choice for $p(x)$. Then the zeros of $p(x)$ are complex numbers. Let the complex number ω be such a zero. Then the field of algebraic numbers is the smallest extension of **Q** that contains ω. The elements of $\boldsymbol{Q}(\omega)$ are those complex numbers that can be expressed as polynomials in the complex number ω. Because $p(\omega) = 0$, those polynomials may be regarded as elements of the ring $\boldsymbol{Q}[x]/\langle p(x)\rangle$ The algebraic numbers are the elements of this quotient ring evaluated at $x = \omega$.

As a set of points, it is the same as $\boldsymbol{Z}^2$, but embedded in the complex plane. The set of gaussian integers, which can be expressed as $\boldsymbol{Z}[\mathrm{e}^{-\mathrm{j}2\pi/4}]$, is not dense in the complex plane. Surprisingly, the set of algebraic integers $\boldsymbol{Z}[\mathrm{e}^{-\mathrm{j}2\pi/8}]$ is dense in the complex plane. This means that any complex number can be approximated arbitrarily well by a point of $\boldsymbol{Z}[\mathrm{e}^{-\mathrm{j}2\pi/8}]$. In this representation, an arbitrary complex number c is an approximation by a complex number $\tilde{c}$ of the form

$$\tilde{c} = c_7\omega^7 + c_6\omega^6 + \cdots + c_1\omega + c_0,$$

where $\omega = \mathrm{e}^{-\mathrm{j}2\pi/8}$ and the c_j are integers. The approximation can be as exact as desired, though this may require that the integers c_j be large. Even more surprisingly, for most practical criteria of numerical accuracy, the c_j can be restricted to be small integers. In particular, the restriction $c_j \in \{-2, -1, 0, 1, 2\}$ or $c_j \in \{-3, -2, -1, 0, 1, 2, 3\}$ can provide satisfactory approximations to complex numbers in many practical situations. Accordingly, one may execute a Fourier transform of any blocklength by representing all complex numbers including $\mathrm{e}^{-\mathrm{j}2\pi/n}$ by polynomials in ω, performing the corresponding polynomial arithmetic in ω, then substituting $\mathrm{e}^{-\mathrm{j}2\pi/8}$ for ω. For other computational applications, the elements of the number field can also be regarded simply as polynomials in the indeterminant x. Thus

$$\boldsymbol{Q}[x]/\langle p(x)\rangle = \{f(x)|\deg f(x) < n\}$$

with addition and multiplication as polynomials modulo $p(\omega)$. In this way, a vector of rationals can be represented as an element of $\boldsymbol{Q}^m$, and a matrix of elements of $\boldsymbol{Q}$ can be represented as a vector of elements of $\boldsymbol{Q}^m$. This representation is the subject of Section 11.5.

Problems for Chapter 9

9.1 **a** Is the polynomial $p(x) = x^4 + 2x^3 + x^2 - 2$ reducible over the field of rationals? If so, factor it.
b Is the polynomial $p(x) = x^4 + 2x^3 + x^2 - 2$ reducible over the field of reals? If so, factor it.
c Is the polynomial $p(x) = x^4 + 2x^3 + x^2 - 2$ reducible over the complex field? If so, factor it.

9.2 Over the field of rationals, let

$$p_1(x) = x^3 + 1,$$
$$p_2(x) = x^4 + x^3 + x + 1.$$

a Find $\text{GCD}(p_1(x), p_2(x))$.
b Find $\text{LCM}(p_1(x), p_2(x))$.
c Find $A(x)$ and $B(x)$ that satisfy

$$\text{GCD}(p_1(x), p_2(x)) = A(x)p_1(x) + B(x)p_2(x).$$

9.3 Let

$$m_1(x) = x^2 + x + 1,$$
$$m_2(x) = x^2 + 1,$$
$$m_3(x) = x^2 - 1.$$

Given

$$c_1(x) \equiv c(x) \pmod{m_1(x)},$$
$$c_2(x) \equiv c(x) \pmod{m_2(x)},$$
$$c_3(x) \equiv c(x) \pmod{m_3(x)},$$

where $c(x)$ has degree at most five, find all polynomials necessary to compute $c(x)$ from the residues.

9.4 How many cyclotomic polynomials divide $x^{15} - 1$? Find them.

9.5 Let $p(x) = \sum_{i=0}^{59} x^i$. Use the relationship

$$x^{60} - 1 = (x - 1)p(x)$$

to find $R_{p(x)}[x^{120} + x^{70} + x^{20}]$.

9.6 For the formal derivative of polynomials in any field, prove that

$$[r(x)s(x)]' = r'(x)s(x) + r(x)s'(x),$$

and that, if $a(x)^2$ divides $r(x)$, then $a(x)$ divides $r'(x)$.

9.7 Over $\mathbf{Z}/\langle 15 \rangle$, the ring of integers modulo fifteen, show that the polynomial $p(x) = x^2 - 1$ has more than two zeros. Such a polynomial over a field can have only two zeros. Where does the proof of this statement fail for a ring?

9.8 The polynomial $p(x) = x^4 + x^3 + x^2 + x + 1$ is a prime polynomial over $GF(2)$. Therefore the ring of polynomials modulo $p(x)$ is $GF(16)$.

  a Show that the field element represented by x in this construction is not a primitive element.

  b Show that the field element represented by $x + 1$ is primitive.

  c Find the minimal polynomial of the field element $x + 1$.

9.9 Over $GF(16)$:

  a How many distinct second-degree monic polynomials of the form

  $$x^2 + ax + b, \quad b \neq 0$$

  are there?

  b How many distinct polynomials of the form

  $$(x - \beta)(x - \gamma), \quad \gamma, \beta \neq 0$$

  are there?

  c Does this prove that irreducible second-degree polynomials exist? How many second-degree prime polynomials over $GF(16)$ are there?

9.10
  a Extend the real field $\mathbf{R}$ by using the prime polynomial $x^4 + 1$. Explain why this extension field is equivalent to the complex field $\mathbf{C}$, which could also be obtained by using the prime polynomial $x^2 + 1$.

  b Extend the rational field $\mathbf{Q}$ by using the prime polynomial $x^4 + 1$. Explain why this extension field is not equivalent to the extension field obtained by extending $\mathbf{Q}$, using the prime polynomial $x^2 + 1$.

9.11 Construct $GF(7)$ by constructing an addition table and multiplication table.

9.12 Find 3^{200} (mod 7).

9.13 Prove that the quotient ring $\mathbf{Z}_q$ is a ring.

9.14 Find elements of order $(p-1)p^{m-1}$ in each of the fields $GF(9)$, $GF(25)$, $GF(27)$, and $GF(49)$.

9.15 Prove that the polynomial $x^6 + x^3 + 1$ is irreducible over the rationals.

Notes for Chapter 9

This chapter treats topics that are standard in mathematical literature. Good texts on number theory include those of Ore (1948) and Hardy and Wright (1960). The properties of Galois fields and of elementary number theory are developed in any book on abstract algebra, as, for example, the book by Birkhoff and MacLane (1941) or that of Van der Waerden (1949). For the most part, these standard treatments are formal, primarily concerned with abstract properties and little concerned with examples or applications. Berlekamp (1968) concentrates more on the computational aspects of Galois fields. Further discussion of polynomial rings and cyclotomic polynomials can be found in the book by McClellan and Rader (1979) or the book by Blahut (2003). The observation that the cyclotomic polynomials have integer coefficients is originally due to Gauss.

10 Computation in finite fields and rings

Signal-processing computations may arise naturally in a finite field, so it is appropriate to construct fast algorithms in the finite field $GF(q)$. Computations in a finite field might also arise as a surrogate for a computation that originally arises in the real field or the complex field. In this situation, a computational task in one field is embedded into a different field, where that computational task is executed and the answer is passed back to the original field. There are several reasons why one might do this. It may be that the computation is easier to perform in the new field, so one saves work or can use a simpler implementation. It may be that devices that do arithmetic in one field may be readily available and can be used to do computations for a second field, if those computations are suitably reformulated. In other situations, one may want to devise a standard computational module that performs bulk computations and to use that module for a diversity of signal-processing tasks. In seeking standardization, one may want to fit one kind of computational task into a different kind of structure.

Another reason for using a surrogate field is to improve computational precision. Computations in a finite field are exact; there is no roundoff error. If a problem involving real or complex numbers can be embedded into a finite field to perform a calculation, it may be possible to reduce the computational noise in the answer. Most applications of signal processing can be satisfied by fixed-point calculations. If a finite field is big enough to hold sixteen-bit integers, for example, then many computations within it will be correct as integer computations, as long as the results of the integer computation are sixteen-bit integers.

10.1 Convolution in surrogate fields

A convolution in a real field often can be thought of, instead, as a convolution in a suitable Galois field. In practical problems, a real number is represented to a limited number of decimal or binary places; usually, fixed-point numbers are sufficient in signal-processing applications. A binary representation with a limited number of bits is given, perhaps twelve or sixteen bits. For computational purposes, we can suppose that the binary point is to the right – all numbers are integers. It is a simple matter to move the binary point in the output data to accommodate any other case. Consequently, the linear convolution in the real field,

$$s(x) = g(x)d(x),$$

is replaced by a linear convolution in the integer ring. One needs only to regard the data as integers, rather than as real numbers. The equation describing the linear convolution has the same appearance, but the operations are interpreted as ordinary integer arithmetic. The equation in the integer ring then can be embedded into a suitable field such as the prime field, $GF(p)$ provided that the prime p is large enough. If the integers in the convolution are all positive, then the prime field $GF(p)$ must be chosen with p large enough so that the convolution output coefficients s_i are not larger than $p - 1$. Then

$$s(x) = g(x)d(x) \pmod{p}$$

also holds. In this situation, the modulo p condition is superfluous.

If s_i can take on negative values, then the prime field $GF(p)$ must be chosen with p large enough so that s_i is in the range $-(p-1)/2 < s_i \le (p-1)/2$. Then negative integers will be represented as positive integers. The negative integers are folded onto the positive range from $(p+1)/2$ to $p - 1$.

The convolution in the Galois field can be computed by using any appropriate method in that field. Some direct methods are discussed in the next section. One can also use a Fourier transform in the Galois field together with the convolution theorem. Figure 10.1 compares such a calculation in the Galois field with the same calculation in the complex field. Observe that the equations on the two sides of Figure 10.1 are exactly alike, although, of course, the underlying arithmetic is different.

Whenever the cyclic convolution in the Galois field has the same answer as the cyclic convolution in the integer ring, the use of the Galois field Fourier transform will give the right answer. Because the Galois field arithmetic is modulo p, there may be modulo p overflow during the intermediate steps. It does not matter if overflow occurs at intermediate steps; only overflow in the output matters, and overflow in the output cannot happen if p is chosen sufficiently large.

As an example, we take the case of a five-point cyclic convolution in the real field:

$$s(x) = g(x)d(x) \pmod{x^5 - 1}.$$

We choose the prime p equal to 31 and work in $GF(31)$. This is suitable as long as we know that the coefficients of $s(x)$ are not larger than thirty. Of course, for a practical problem, $GF(31)$ is much too small, but it will do for an example.

Because 5 divides $p - 1$, there is a Fourier transform with blocklength five. The element two is a suitable kernel because its order is five. Let

$d_0 = 2, \quad d_1 = 6, \quad d_2 = 4, \quad d_3 = 4, \quad d_4 = 0,$
$g_0 = 3, \quad g_1 = 0, \quad g_2 = 2, \quad g_3 = 1, \quad g_4 = 2.$

10.1 Convolution in surrogate fields

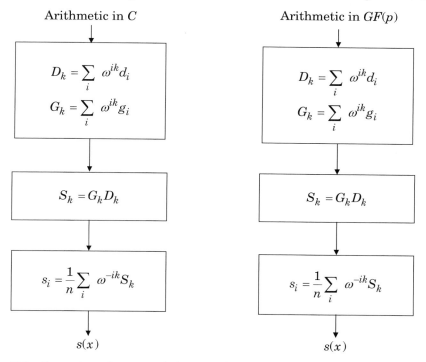

Figure 10.1 Comparison of two convolution procedures

The five-point Fourier transform of d is

$$D_k = \sum_{i=0}^{4} 2^{ik} d_i, \quad k = 0, \ldots, 4$$

with a similar relation for the transform of g. In matrix notation, this Fourier transform becomes

$$\begin{bmatrix} D_0 \\ D_1 \\ D_2 \\ D_3 \\ D_4 \end{bmatrix} = \begin{bmatrix} 1 & 1 & 1 & 1 & 1 \\ 1 & 2 & 4 & 8 & 16 \\ 1 & 4 & 16 & 2 & 8 \\ 1 & 8 & 2 & 16 & 4 \\ 1 & 16 & 8 & 4 & 2 \end{bmatrix} \begin{bmatrix} 2 \\ 6 \\ 4 \\ 4 \\ 0 \end{bmatrix}.$$

Thus the transform vectors are given by

$D_0 = 16, \quad D_1 = 0, \quad D_2 = 5, \quad D_3 = 29, \quad D_4 = 22,$
$G_0 = 8, \quad G_1 = 20, \quad G_2 = 22, \quad G_3 = 0, \quad G_4 = 27.$

Multiplying G_k by D_k gives S_k:

$S_0 = 4, \quad S_1 = 0, \quad S_2 = 17, \quad S_3 = 0, \quad S_4 = 5.$

In $GF(31)$, $5^{-1} = 25$ because $5 \cdot 25 = 1$. Therefore the inverse Fourier transform in $GF(31)$ is

$$\begin{bmatrix} s_0 \\ s_1 \\ s_2 \\ s_3 \\ s_4 \end{bmatrix} = 25 \begin{bmatrix} 1 & 1 & 1 & 1 & 1 \\ 1 & 16 & 8 & 4 & 2 \\ 1 & 8 & 2 & 16 & 4 \\ 1 & 4 & 16 & 2 & 8 \\ 1 & 2 & 4 & 8 & 16 \end{bmatrix} \begin{bmatrix} 4 \\ 0 \\ 17 \\ 0 \\ 5 \end{bmatrix} = \begin{bmatrix} 30 \\ 30 \\ 24 \\ 26 \\ 18 \end{bmatrix},$$

which gives the final result. The direct computation with integer arithmetic would produce the same result. If we had chosen larger input components, some of the output components of the convolution would have been greater than thirty. In that case, computing the convolution in $GF(31)$ would overflow the maximum output and produce the wrong answer. It would give the correct value reduced modulo 31. A larger field should be used; fields with at least 2^{16} elements are especially useful.

If the wordlength is too large for the chosen field size, we can use the Chinese remainder theorem to work with residues of the numbers, thereby breaking the wordlength into a number of shorter wordlengths. This way of using the Chinese remainder theorem is different from the way we have used it previously for shuffling arrays. Now it is used to break the numbers themselves apart. It does not disturb the indexing, so the data is used in its natural order. Of course, the new convolutions may themselves be computed by using the Chinese remainder theorem again, but this time in the way discussed earlier on the new polynomials or on their indices.

Given a set of r coprime integers $m_1, m_2, \ldots, m_r$, let $s_i^{(\ell)}$ denote the residue $s_i \pmod{m_\ell}$. Then the convolution can be broken down as

$$s_i^{(\ell)} = \sum_{k=0}^{N-1} g_{i-k}^{(\ell)} g_k^{(\ell)} \pmod{m_\ell}, \quad \begin{array}{l} i = 0, \ldots, L+N-1, \\ \ell = 1, \ldots, r, \end{array}$$

where $g_i^{(\ell)} = g_i \pmod{m_\ell}$, and $d_i^{(\ell)} = d_i \pmod{m_\ell}$. By using the Chinese remainder theorem, s_i can be recovered from its residues $s_i^{(\ell)}$. Hence we have replaced one convolution involving large integers by a set of r convolutions, each involving small integers. The reduced convolutions can be computed just as written, or they can be computed by using fast algorithms in a prime field.

10.2 Fermat number transforms

The Galois fields in which the algorithms for the computation of a Fourier transform are, perhaps, simplest are those of the form $GF(2^m + 1)$, which is a field whenever

10.2 Fermat number transforms

$p = 2^m + 1$ is a prime. It is known that $2^m + 1$ is not a prime if m is not a binary power. However, the converse is not true, because $2^{32} + 1$ is known to be composite. But primes are found for $m = 2, 4, 8,$ or 16 for which $2^m + 1 = 5, 17, 257,$ or $65{,}537$; a set of integers known as *Fermat primes*.

Whenever q is a Fermat prime, $q - 1$ or any factor of $q - 1$ is a power of two, so every element of $GF(2^m + 1)$ has order equal to a power of two. The Fourier transform

$$V_k = \sum_{i=0}^{n-1} \omega^{ik} v_i, \qquad k = 0, \ldots, n-1$$

exists in $GF(q)$ when q is the Fermat prime $2^m + 1$ for each n that is a divisor of 2^m, and ω is an element of order n. Thus the field $GF(2^{16} + 1)$ has Fourier transforms of sizes $2^{16}, 2^{15}, 2^{14}, \ldots, 2^2, 2$. By using the Cooley–Tukey algorithm, a Fourier transform over $GF(2^m + 1)$ can be broken down into a sequence of radix-two Fourier transforms, which can be implemented rather neatly, using only $(n/2) \log_2 n$ multiplications in $GF(2^{16} + 1)$ and $(n/2) \log_2 n$ additions in $GF(2^{16} + 1)$.

Some Fourier transforms over $GF(2^{16} + 1)$ are especially simple. These are the Fourier transforms over $GF(2^{16} + 1)$ with blocklength 32 or less. This is because the element 2 has order 32. To see this, notice that $2^{16} + 1 = 0$ in $GF(2^{16} + 1)$. Hence $2^{16} = -1$, and so $2^{32} = 1$. The Fourier transform of blocklength 32 is

$$V_k = \sum_{i=0}^{31} 2^{ik} v_i, \qquad k = 0, \ldots, 31,$$

and the multiply operation is actually a shift in a binary arithmetic system because it is a multiplication by a power of two. Because this Fourier transform has no true multiplications, it is easy to compute, but a blocklength of 32 is too short for many applications. To form a larger Fourier transform, a radix-32 Cooley–Tukey FFT is attractive because the inner core is multiply-free.

In general, for any Fermat prime $2^m + 1$, because $2^m = -1$ modulo $2^m + 1$, the element 2 has order $2m$ in $GF(2^m + 1)$, and so it can be used as the kernel of a Fourier transform of blocklength $2m$. To get a larger blocklength, one can use, instead, $\sqrt{2}$ as the kernel to get a Fourier transform of blocklength $4m$. In these fields we can easily verify that $\sqrt{2} = 2^{m/4}(2^{m/2} - 1)$ by calculating the square if we recall that $2^m = -1$. Thus every even power of $\sqrt{2}$ is a power of two, and every odd power of $\sqrt{2}$ is equal to a power of two times $\sqrt{2}$, and so it can be written in the form $2^a \pm 2^b$ modulo $2^m + 1$. Multiplication by the constant $2^a \pm 2^b$ (mod $2^n + 1$) is nearly multiply-free.

For example, the Fourier transform of blocklength 64 in $GF(2^{16} + 1)$ has the form

$$V_k = \sum_{i=0}^{63} (2^{12} - 2^4)^{ik} v_i, \qquad k = 0, \ldots, 63.$$

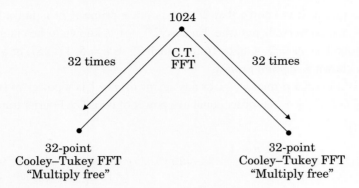

Figure 10.2 Structure of a 1024-point FFT in $GF(2^{16}+1)$.

When this is computed as a radix-two Cooley–Tukey FFT, all multiplications are by a constant of the form 2^a or $2^a \pm 2^b$, which can be implemented as a pair of shifts and a subtraction or addition.

In general, Fourier transforms in $GF(2^m+1)$ with blocklength larger than $2m$ must have some nontrivial multiplications. One can reduce the number of such multiplications by using the Cooley–Tukey FFT to form a multidimensional transform with each dimension, involving at most a $2m$-point Fourier transform.

For example, consider the 1024-point Fourier transform in $GF(2^{16}+1)$:

$$V_k = \sum_{i=0}^{1023} \omega^{ik} v_i, \qquad k = 0, \ldots, 1023,$$

where now ω is an element of order 1024, which will be chosen as π^{64}, where π is a primitive element of the field. The element three is a primitive element of $GF(2^{16}+1)$, but perhaps not the best one to choose. To get the desired form, we want an ω so that $\omega^{32} = 2$; hence we choose π so that $\pi^{64 \cdot 32} = 2$. The Cooley–Tukey FFT puts the transform in the form

$$v_{n''k'+k''} = \sum_{i'=0}^{31} 2^{i'k'} \left[\omega^{i'k''} \sum_{i''=0}^{31} 2^{i''k''} v_{i'+n'i''} \right].$$

The inner sum is a 32-point Fourier transform for each value of i', and the outer sum is a 32-point Fourier transform for each value of k''. The 32-point Fourier transform can itself be broken down by a radix-two Cooley–Tukey FFT that is computed with only shifts and adds. The multiplication by $\omega^{i'k''}$ is a nontrivial multiplication, but there are only 1024 such multiplications and they are integer multiplications. In fact, when i' or k'' equals zero, that multiplication is trivial, so there are only 931 nontrivial multiplications. The structure of this FFT is shown in Figure 10.2. In general, a Fourier transform in $GF(2^{16}+1)$ can be computed in about $n(\lceil \log_{32} n \rceil - 1)$ multiplications in $GF(2^{16}+1)$, $\frac{1}{2}n \log_2 n$ additions in $GF(2^{16}+1)$, and $\frac{1}{2}n \log_2 n$ shifts.

The arithmetic in $GF(2^{16} + 1)$ consists of conventional integer addition and integer multiplication followed by reduction modulo $2^{16} + 1$. But because $2^{16} = -1$, this amounts to setting $2^{16+\ell}$ equal to -2^ℓ. To implement this reduction, high-order bits are downshifted by sixteen bits and subtracted.

10.3 Mersenne number transforms

The Galois fields in which the operation of multiplication is the most straightforward are those of the form $GF(2^m - 1)$, which may be a field if m is a prime but cannot be a field if m is composite, because $2^{ab} - 1$ is divisible by $2^a - 1$. Primes of the form $2^m - 1$ are called *Mersenne primes*. The smallest values of m for which $2^m - 1$ is a prime are 3, 5, 7, 13, 17, 19, and 31; and the corresponding Mersenne primes are 7, 31, 127, 8191, 131,071, 524,287, and 2,147,483,647.

Arithmetic in the Mersenne field $GF(2^m - 1)$ is quite convenient if the integers are represented as m-bit binary numbers. Because $2^m - 1 = 0$ in this field, the overflow 2^m is equal to one. Hence the arithmetic is conventional integer arithmetic, and the overflow bits are added into the low-order bits of the number. This is one's-complement arithmetic.

The field $GF(2^m - 1)$ exists whenever $2^m - 1$ is a prime. In every Galois field $GF(q)$, there is a Fourier transform of blocklength n for every n that divides $q - 1$. Hence in the prime field $GF(2^m - 1)$, there is a Fourier transform of blocklength n for every n that divides $2^m - 2$. These transforms are sometimes called *Mersenne number transforms*.

The Mersenne number transforms cannot be computed by a radix-two Cooley–Tukey FFT because $(2^m - 1) - 1$ is not a power of two. This is considerably different from the Fermat number transforms, for which the radix-two Cooley–Tukey FFT is quite suitable. The Mersenne number transform can be computed by any suitable mixed-radix FFT.

For example, $GF(2^{13} - 1)$ can be used to contain all thirteen-bit one's-complement integers. An integer convolution of two vectors of twelve-bit numbers can be performed as a convolution in $GF(2^{13} - 1)$, as long as there is no overflow in the convolution output. The linear convolution

$s(x) = g(x)d(x),$

where the coefficients of $g(x)$ and $d(x)$ are nonnegative twelve-bit binary integers (or thirteen-bit one's-complement integers), can also be represented by

$s(x) = g(x)d(x) \pmod{2^{13} - 1},$

provided that the coefficients of $s(x)$ are known to be smaller than $2^{13} - 1$. This convolution can be computed by using a Fourier transform in $GF(2^{13} - 1)$.

The Fourier transform blocklengths n that can be used in $GF(2^{13} - 1)$ are those n dividing $2^{13} - 2$. Because $2^{13} - 2 = 2 \cdot 5 \cdot 7 \cdot 9 \cdot 13$, the possible choices for n are readily apparent as the product of any subset of these factors. One possibility is to choose the element -2 for the Fourier transform kernel ω. Because

$$2^{13} - 1 = 0 \pmod{2^{13} - 1},$$

we see that $(-2)^{13} = -1$, so -2 must have order 26. Then we have the Fourier transform

$$V_k = \sum_{i=0}^{25} (-2)^{ik} v_i, \qquad k = 0, \ldots, 25.$$

All multiplications can be executed as multiplications by powers of two, so this Fourier transform can be computed with no true multiplications, only shifts. On the other hand, the factors of 26 are not particularly convenient for constructing a mixed-radix FFT. For this blocklength, one will not do much better than computing the Fourier transform as it is written using 25^2 shifts and $26 \cdot 25$ additions.

To get other blocklengths, one must allow general multiplications. In a prime field, just as in the complex field, a Winograd large FFT algorithm, to be discussed in Chapter 12, could be built out of Winograd small FFT algorithms. A Fourier transform algorithm with a blocklength of 70, for example, can be built out of small FFT algorithms of blocklengths two, five, and seven. These algorithms will be quite similar to the Winograd FFT algorithms in the complex field – except that the constant factors and the structure of the multiplications will be different because ω is in a different field.

Let us construct the five-point Winograd FFT in $GF(2^{13} - 1)$. One can verify by exhaustive calculation that three is primitive in $GF(2^{13} - 1)$; hence, $3^{2 \cdot 7 \cdot 9 \cdot 13}$ (which equals 4794 in this field) has order five. Further, 1904 also has order five because it is equal to 4794^3. We then have the five-point Fourier transform in $GF(2^{13} - 1)$:

$$V_k = \sum_{i=0}^{4} 1904^{ik} v_i, \qquad k = 0, \ldots, 4.$$

We first use the Rader prime algorithm to turn this into a convolution. This construction is the same as for a complex Fourier transform because the Rader algorithm works only on the indices. Then the problem is converted into the cyclic convolution

$$s(x) = g(x)d(x) \pmod{x^4 - 1},$$

where the Rader polynomial is

$$\begin{aligned} g(x) &= (\omega^3 - 1)x^3 + (\omega^4 - 1)x^2 + (\omega^2 - 1)x + (\omega - 1) \\ &= 3001x^3 - 1511x^2 + 4793x + 1903 \end{aligned}$$

10.3 Mersenne number transforms

and

$$d(x) = v_2 x^3 + v_4 x^2 + v_3 x + v_1,$$
$$s(x) = (V_3 - V_0)x^3 + (V_4 - V_0)x^2 + (V_2 - V_0)x + (V_1 - V_0).$$

Next, we need a fast algorithm for the four-point cyclic convolution. Because

$$x^4 - 1 = (x-1)(x+1)(x^2+1)$$

is a prime factorization in $GF(2^{13} - 1)$, an algorithm for this cyclic convolution must have the same form as it did in the real field or the complex field. We use the algorithm shown in Figure 3.14, but executing the arithmetic operations in the field $GF(2^{13} - 1)$. Hence

$$\begin{bmatrix} G_0 \\ G_1 \\ G_2 \\ G_3 \\ G_4 \end{bmatrix} = \frac{1}{4} \begin{bmatrix} 1 & 1 & 1 & 1 \\ 1 & -1 & 1 & -1 \\ 1 & -2 & -2 & 2 \\ 2 & 2 & -2 & -2 \\ 2 & 0 & -2 & 0 \end{bmatrix} \begin{bmatrix} 1903 \\ 4793 \\ -1511 \\ 3001 \end{bmatrix} = \begin{bmatrix} 6142 \\ 2245 \\ 811 \\ 2603 \\ 1707 \end{bmatrix},$$

and the five-point Winograd small FFT can then be expressed as

$$\begin{bmatrix} V_0 \\ V_1 \\ V_2 \\ V_3 \\ V_4 \end{bmatrix} = \begin{bmatrix} 1 & 0 & 0 & 0 & 0 & 0 \\ 1 & 1 & 1 & 0 & -1 & 1 \\ 1 & 1 & -1 & -1 & 0 & 1 \\ 1 & 1 & -1 & 1 & 0 & -1 \\ 1 & 1 & 1 & 0 & 1 & -1 \end{bmatrix} \begin{bmatrix} 1 \\ 6142 \\ 2245 \\ 811 \\ 2603 \\ 1707 \end{bmatrix}$$

$$\times \begin{bmatrix} 1 & 1 & 1 & 1 & 1 \\ 0 & 1 & 1 & 1 & 1 \\ 0 & 1 & -1 & -1 & 1 \\ 0 & 1 & 0 & 0 & -1 \\ 0 & 0 & -1 & 1 & 0 \\ 0 & 1 & -1 & 1 & -1 \end{bmatrix} \begin{bmatrix} v_0 \\ v_1 \\ v_2 \\ v_3 \\ v_4 \end{bmatrix}$$

in the field $GF(2^{13} - 1)$.

For a second example, $GF(2^{17} - 1)$ can be used to contain all seventeen-bit one's-complement integers. A convolution of sixteen-bit numbers can be performed as a convolution in $GF(2^{17} - 1)$ as long as there is no overflow in the output of the convolution. The convolution can be computed using an FFT algorithm in $GF(2^{17} - 1)$ together with the convolution theorem.

A Fourier transform in $GF(2^{17} - 1)$ can only have a blocklength that is a divisor of $2^{17} - 2$. But

$$2^{17} - 2 = 2 \cdot 3 \cdot 5 \cdot 17 \cdot 257.$$

One can choose $n = 510$ and build a 510-point FFT out of a two-point, a three-point, a five-point, and a 17-point FFT. Of these, the two-point, the three-point, and the five-point FFT will be simple modules built as Winograd small FFT algorithms. These Winograd small FFT algorithms are in $GF(2^{17} - 1)$, but they will look the same as the Winograd small FFT algorithms over the complex field.

The 17-point Fourier transform is also a simple module but can be built by another technique as follows. In $GF(2^{17} - 1)$, the element two must have order 17 because $2^{17} = 1$. Hence, the 17-point Fourier transform is

$$V_k = \sum_{i=0}^{16} 2^{ik} v_i, \qquad k = 0, \ldots, 16.$$

It can be computed using only cyclic shifts and additions. There are no multiplications.

If one wants to compute a convolution longer than 510 points, the factor 257 must be used also. The 257-point Fourier transform can be converted to a 256-point cyclic convolution using the Rader prime algorithm, and that cyclic convolution can be computed with a 256-point radix-two Cooley–Tukey FFT and the convolution theorem.

10.4 Arithmetic in a modular integer ring

When arithmetic is specified to be in the ring $\mathbf{Z}_N$, the modulo N reduction is implicit. The modulo N reduction need not be restated, but it often is explicitly stated for clarity and emphasis. When both $\mathbf{Z}$ and $\mathbf{Z}_N$ operations are under discussion at the same time, as in this section, it is best to always restate the modulo N reduction.

Because of the modular reduction, addition in the modular integer ring $\mathbf{Z}_N$ is somewhat more difficult than is addition in the integer ring $\mathbf{Z}$. To compute $x + y$ modulo N, one may compute the integers $x + y$ and $x + y - N$ and keep the smaller of the two, provided it is nonnegative. Even if N is a large integer, this is a straightforward procedure. The task of computing a subtraction in $\mathbf{Z}_N$ can be formulated in a similar way.

The task of modular multiplication is more difficult. This task is to compute $z = xy \pmod{N}$, where $0 \le x < N$, $0 \le y < N$, and $0 \le z < N$. Multiplication in $\mathbf{Z}_N$ is more difficult than multiplication in $\mathbf{Z}$ because of the reduction modulo N, which is an operation defined as a remainder under integer division. Thus the complexity of modular multiplication appears to depend on the complexity of integer division. Instead, we shall see that much of this complexity can be tamed by a method known as the *Montgomery multiplication algorithm*. The Montgomery algorithm can be used for any integer N. Because of its intrinsic computational overhead, the Montgomery algorithm is primarily attractive for long sequences of multiplications, possibly interspersed with additions or exponentiations.

10.4 Arithmetic in a modular integer ring

In the ring Z_N, whenever R has an inverse R^{-1}, we have the obvious identities

$$xy = R^2(R^{-1}x)(R^{-1}y) \pmod{N}$$

and

$$xy = R^{-2}(Rx)(Ry) \pmod{N}.$$

These apparently trivial formulas are actually not trivial computationally when used in the right way.

Select any R coprime to N of the form b^k, where b is the arithmetic base, and such that R is larger than N. For example, if numbers are represented in binary notation, then $b = 2$ and $R = 2^k$, with R larger than and coprime with N, would be the suitable choice. If numbers are represented in decimal notation, then $b = 10$ and $R = 10^k$, with R larger than and coprime with N, would be the suitable choice. The trick of Montgomery multiplication is to exchange integer multiplication modulo N for integer multiplication modulo R at the expense of a few side calculations.

We shall need the two integers $R^{-1} \pmod{N}$ and $N^{-1} \pmod{R}$. As a reminder that these integers do indeed exist, recall that, for any two coprime integers a and b, there exist two integers A and B such that

$$aA + bB = 1.$$

Moreover, because

$$a(A + \ell b) + b(B - \ell a) = 1$$

must also be true for any integer ℓ, it is clear that A and B can be chosen so that $0 < A < b$ or $0 < B < a$. The values of $A \pmod{b}$ and $B \pmod{a}$ are called a^{-1} and b^{-1} because $aa^{-1} = 1 \pmod{b}$ and $bb^{-1} = 1 \pmod{a}$. Thus, for our case, because R and N are coprime, R^{-1} and N^{-1} both exist.

Define the modified multiplicands by $X = Rx \pmod{N}$ and $Y = Ry \pmod{N}$. Because x and y can always be recovered by $x = R^{-1}X \pmod{N}$ and by $y = R^{-1}Y \pmod{N}$, the conversions from x to X and from y to Y must be permutations of $\{0, 1, \ldots, N-1\}$. The quantities X and Y are called the *Montgomery reductions* or the *Montgomery permutations* of x and y. Finally, let $Z = Rxy \pmod{N}$. We shall see that it is easy to recover xy from Z because $xy = R^{-1}Z \pmod{N}$, which, by the choice of R, will prove easy to compute.

Because R is a power of the arithmetic base b, it is easy to compute $Rx \pmod{N}$ and $Ry \pmod{N}$ by an iterative procedure. To compute X and Y from x and y when they are expressed in base-two integer representations and $R = 2^k$, one executes the iteration

$$x^{(i)} = \begin{cases} 2x^{(i-1)} - N & \text{if nonnegative,} \\ 2x^{(i-1)} & \text{otherwise,} \end{cases}$$

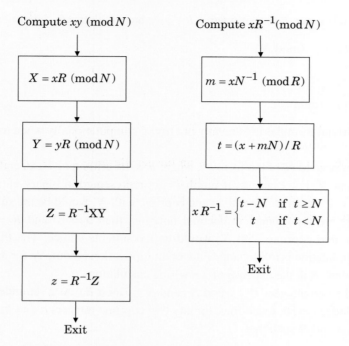

Figure 10.3 Montgomery multiplication

for $i = 1, \ldots, k$, with $x^{(0)} = x$. Then $x^{(k)} = X$. A similar iteration computes Y from y. This is shown in Figure 10.3. Thus, to compute $2a$ for any a, left shift a by one bit and subtract N. Then keep the smaller of $2a$ and $2a - N$ provided it is nonnegative. For any other logarithm base, the computation is similar, with the doubling in the final step replaced by a multiplication by the logarithm base.

Theorem 10.4.1 (Montgomery multiplication) *If $z = xy \pmod{N}$ then*

$$Z = R^{-1}XY \pmod{N}.$$

Proof This is nothing more than the trivial statement

$$Rxy = R^{-1}(Rx)(Ry) \pmod{N},$$

that was mentioned earlier. □

The Montgomery multiplication is to be executed by first multiplying X and Y as integers. The modulo N reduction is deferred until the multiplication by R^{-1} where it becomes easy using the method to be given in Theorem 10.4.2.

It only remains to provide a simple way to multiply by $R^{-1} \pmod{N}$. This is needed both to compute Z from XY and to compute z from Z.

10.4 Arithmetic in a modular integer ring

Let the integer N^{-1} satisfy $N^{-1}N = 1 \pmod{R}$ and define

$$m = xN^{-1} \pmod{R} \quad (0 \leq m < R)$$

and define

$$t = (x + mN)/R.$$

Because R is a power of the base of the number system, the apparent division by R is actually trivial, as we show next.

The following theorem assures us that t is an integer. Because R is the kth power of b, this implies that $x + mN$ must have this power of b as a factor. Thus, in base b representation, $x + mN$ must have k low-order zeros. Division consists merely of removing these k zeros, as by right shifting the base b representation.

Theorem 10.4.2 (Montgomery reduction) *Suppose that* $\mathrm{GCD}(R, N) = 1$, *that* R^{-1} *is an integer that satisfies* $R^{-1}R = 1 \pmod{N}$, *and that* N^{-1} *is an integer that satisfies* $NN^{-1} = 1 \pmod{R}$. *Then*

$$xR^{-1} \pmod{N} = \begin{cases} t + N & \text{if } t \geq N, \\ t & \text{if } t < N, \end{cases}$$

where $t = (x + mN)/R$ *and* $m = xN^{-1} \pmod{R}$.

Proof The integers R^{-1} and N^{-1} exist as is previously asserted. The proof then consists of the following three steps.

Step 1 $mN = xN^{-1}N \pmod{R} = x \pmod{R}$, so R divides $x + mN$, which means that t is an integer.

Step 2 $tR = x - mN = x \pmod{N}$, so $t = xR^{-1} \pmod{N}$

Step 3 $0 \leq x - mN < RN + RN = 2RN$, so $(x - mN)/R < 2N$. □

In some applications, a sequence of many multiplications and additions may be required. Montgomery multiplication can be integrated with these computations in order to obtain greater efficiency. Thus the expressions $xy + uv$ and xyz can be converted to $XY + UV$ and XYZ, then executed as integer arithmetic. The final reduction of the Montgomery algorithm can be deferred until the end and performed only once or perhaps when numbers become too large. Similarly, a sequence of repeated squares $x^2, x^4, x^8, x^{16}, \ldots$ can be executed as $R^{-1}X^2, R^{-2}X^4, R^{-3}X^8, \ldots$ with additional multiplications by R^{-1} to obtain $x^2, x^4, x^8, \ldots$ deferred until later. This is especially important for the task of exponentiation, which can be done efficiently by the squaring and multiply procedure.

10.5 Convolution algorithms in finite fields

Instead of using a Fourier transform and the convolution theorem to compute a convolution, one can also use direct methods to compute a convolution in a Galois field. All of the methods of Chapter 3 apply and can be used to design convolution algorithms directly. Indeed, many of the algorithms that were designed for the real field can be simply modified to provide algorithms for convolution in a Galois field.

To convert a convolution algorithm derived in the real field into a convolution algorithm suitable for a Galois field of characteristic p, start with the convolution algorithm

$$s = C[(Ag) \cdot (Bd)],$$

where A, B, and C are matrices of rational numbers. Multiply through by the smallest integer L that will clear all of the denominators so that the equation can be written

$$Ls = C'[(A'g) \cdot (B'd)],$$

where now L is an integer, and A', B', and C' are matrices of integers. This equation can be viewed as an algorithm for integer convolution. Hence it can be converted to an equation modulo p as follows

$$Ls = C'[(A'g) \cdot (B'd)] \pmod{p}.$$

As long as L is not equal to zero modulo p, this becomes an algorithm for convolution in $GF(p)$ simply by dividing through by L modulo p. Additionally, because it is an algorithm in $GF(p)$, it is also an algorithm in any extension of $GF(p)$.

If L is equal to zero modulo p, then that algorithm for real convolution cannot be moved into the Galois field $GF(p)$. In that case, one must derive convolution algorithms directly in $GF(p)$. Even when L does not equal zero modulo p, one may prefer to derive convolution algorithms directly in the Galois field, because such algorithms may be better than those taken from a different field. All of the techniques of Chapter 5 for constructing convolution algorithms are valid in any field, requiring only that the factorization

$$m(x) = m^{(0)}(x) \cdots m^{(K)}(x)$$

be a factorization in the field of the convolution.

In most of this chapter, we are interested in fields whose characteristic p is large. In these fields, because L will usually be much smaller than p, all of the convolution algorithms can be moved into $GF(p)$ and will look just as they did in the rational field. In the remainder of this section, we shall look at the contrary case – the case in which the field characteristic is small and, even more specifically, equal to two.

10.5 Convolution algorithms in finite fields

Let us find an algorithm for three-point cyclic convolution in fields of characteristic two. In a field of characteristic zero, there is a cyclic convolution algorithm, given by

$$\begin{bmatrix} s_0 \\ s_1 \\ s_2 \end{bmatrix} = \begin{bmatrix} 1 & 1 & 0 & -1 \\ 1 & -1 & -1 & 2 \\ 1 & 0 & 1 & -1 \end{bmatrix} \begin{bmatrix} \frac{1}{3}(g_0 + g_1 + g_2) \\ (g_0 - g_2) \\ (g_1 - g_2) \\ \frac{1}{3}(g_0 + g_1 - 2g_2) \end{bmatrix} \begin{bmatrix} 1 & 1 & 1 \\ 1 & 0 & -1 \\ 0 & 1 & -1 \\ 1 & 1 & -2 \end{bmatrix} \begin{bmatrix} d_0 \\ d_1 \\ d_2 \end{bmatrix}.$$

None of the denominators is a multiple of two. Hence this algorithm can be moved into a field of characteristic two. Integer arithmetic is modulo 2 in a field of characteristic two, so all integers of the field are either zero or one. The new algorithm in fields of characteristic two is

$$\begin{bmatrix} s_0 \\ s_1 \\ s_2 \end{bmatrix} = \begin{bmatrix} 1 & 1 & 0 & 1 \\ 1 & 1 & 1 & 0 \\ 1 & 0 & 1 & 1 \end{bmatrix} \begin{bmatrix} (g_0 + g_1 + g_2) \\ (g_0 + g_2) \\ (g_1 + g_2) \\ (g_0 + g_1) \end{bmatrix} \begin{bmatrix} 1 & 1 & 1 \\ 1 & 0 & 1 \\ 0 & 1 & 1 \\ 1 & 1 & 0 \end{bmatrix} \begin{bmatrix} d_0 \\ d_1 \\ d_2 \end{bmatrix}.$$

On the other hand, the two-point cyclic convolution algorithm

$$\begin{bmatrix} s_0 \\ s_1 \end{bmatrix} = \begin{bmatrix} 1 & 1 \\ 1 & -1 \end{bmatrix} \begin{bmatrix} \frac{1}{2}(g_0 + g_1) \\ \frac{1}{2}(g_0 - g_1) \end{bmatrix} \begin{bmatrix} 1 & 1 \\ 1 & -1 \end{bmatrix} \begin{bmatrix} d_0 \\ d_1 \end{bmatrix}$$

has denominators that are even integers. The algorithm cannot be moved into a field of characteristic two because the denominators would become zero. The best two-point cyclic convolution algorithm for a field of characteristic two is

$$\begin{bmatrix} s_0 \\ s_1 \end{bmatrix} = \begin{bmatrix} 1 & 1 & 0 \\ 1 & 0 & 1 \end{bmatrix} \begin{bmatrix} g_0 + g_1 \\ g_0 \\ g_1 \end{bmatrix} \begin{bmatrix} 1 & 0 \\ 1 & 1 \\ 1 & 1 \end{bmatrix} \begin{bmatrix} d_0 \\ d_1 \end{bmatrix},$$

which has three multiplications. The reason that three multiplications are needed rather than two is that the polynomial $x^2 - 1$ cannot be factored into two coprime polynomials over $GF(2)$. This is because $-1 = 1$ in fields of characteristic two. Hence

$$x^2 + 1 = (x + 1)^2,$$

and the algorithm must be constructed using $(x + 1)^2$ as a modulus polynomial, which has degree two.

Such a situation, in which the number of multiplications in a finite field is larger than the number of multiplications in the rational field, occurs for other cyclic convolution blocklengths. This occurs because several prime factors of $x^n - 1$ are equal in some fields. In compensation, some cyclic convolution algorithms in a field of finite characteristic have fewer multiplications than like convolutions in the rational field. This is because the cyclotomic polynomials can be factored in a finite field, even though they cannot be factored in the rational field.

For example, over the rationals,

$$x^7 - 1 = (x - 1)(x^6 + x^5 + x^4 + x^3 + x^2 + x + 1)$$

is a prime factorization, but over a field of characteristic two, it factors further as

$$x^7 - 1 = (x - 1)(x^3 + x + 1)(x^3 + x^2 + 1).$$

Hence according to the general bounds of Section 5.8, the optimum algorithm (with respect to the number of multiplications) for seven-point cyclic convolution over the rationals must use 12 multiplications, while the optimum algorithm over fields of characteristic two must use 11 multiplications. From a practical point of view, a good algorithm with 16 multiplications is known for the first case, while a good algorithm with 13 multiplications is known for the second. This latter algorithm, which can be derived by using the methods of Chapter 3, is given by

$$\begin{bmatrix} s_0 \\ s_1 \\ s_2 \\ s_3 \\ s_4 \\ s_5 \\ s_6 \end{bmatrix} = \begin{bmatrix} 1 & 1 & 1 & 1 & 1 & 1 \\ 0 & 1 & 0 & 0 & 1 & 1 & 1 \\ 1 & 0 & 0 & 1 & 1 & 1 & 0 \\ 1 & 1 & 0 & 1 & 0 & 0 & 1 \\ 1 & 1 & 1 & 0 & 1 & 0 & 0 \\ 0 & 0 & 1 & 1 & 1 & 0 & 1 \\ 0 & 1 & 1 & 1 & 0 & 1 & 0 \\ 0 & 1 & 1 & 1 & 0 & 0 & 1 \\ 0 & 1 & 0 & 1 & 1 & 1 & 0 \\ 0 & 0 & 1 & 0 & 1 & 1 & 1 \\ 1 & 1 & 0 & 0 & 1 & 0 & 1 \\ 1 & 0 & 0 & 1 & 0 & 1 & 1 \\ 1 & 1 & 1 & 0 & 0 & 1 & 0 \end{bmatrix} \left\{ \begin{bmatrix} 1 & 1 & 1 & 1 & 1 & 1 & 1 \\ 0 & 1 & 1 & 1 & 0 & 0 & 1 \\ 1 & 0 & 1 & 1 & 1 & 0 & 0 \\ 1 & 1 & 0 & 0 & 1 & 0 & 1 \\ 1 & 0 & 0 & 1 & 0 & 0 & 1 \\ 0 & 1 & 0 & 1 & 1 & 1 & 0 \\ 0 & 0 & 1 & 0 & 1 & 1 & 1 \\ 0 & 1 & 1 & 1 & 0 & 1 & 0 \\ 1 & 1 & 0 & 1 & 0 & 0 & 1 \\ 1 & 0 & 1 & 0 & 0 & 1 & 1 \\ 1 & 0 & 0 & 1 & 1 & 1 & 0 \\ 0 & 1 & 0 & 0 & 1 & 1 & 1 \\ 0 & 0 & 1 & 1 & 1 & 0 & 1 \end{bmatrix} \begin{bmatrix} g_0 \\ g_1 \\ g_2 \\ g_3 \\ g_4 \\ g_5 \\ g_6 \end{bmatrix} \times \begin{bmatrix} 1 & 1 & 1 & 1 & 1 & 1 & 1 \\ 1 & 0 & 1 & 1 & 1 & 0 & 0 \\ 0 & 1 & 0 & 1 & 1 & 1 & 0 \\ 0 & 0 & 1 & 0 & 1 & 1 & 1 \\ 1 & 1 & 1 & 0 & 0 & 1 & 0 \\ 1 & 1 & 0 & 0 & 1 & 0 & 1 \\ 1 & 0 & 0 & 1 & 0 & 1 & 1 \\ 0 & 1 & 1 & 1 & 0 & 1 & 0 \\ 1 & 0 & 0 & 1 & 1 & 1 & 0 \\ 0 & 0 & 1 & 1 & 1 & 0 & 1 \\ 1 & 1 & 1 & 0 & 1 & 0 & 0 \\ 1 & 1 & 0 & 1 & 0 & 0 & 1 \\ 1 & 0 & 1 & 0 & 0 & 1 & 1 \end{bmatrix} \begin{bmatrix} d_0 \\ d_1 \\ d_2 \\ d_3 \\ d_4 \\ d_5 \\ d_6 \end{bmatrix} \right\}.$$

10.5 Convolution algorithms in finite fields

One may also find such an improvement in fields of larger characteristic. For example, in $GF(11)$ we have the prime factorization

$$x^7 - 1 = (x - 1)(x^3 + 5x^2 + 4x - 1)(x^3 - 4x^2 - 5x - 1).$$

Consequently, in fields of characteristic 11, there is a seven-point cyclic convolution algorithm that uses 11 multiplications. Such an algorithm is

$$\begin{bmatrix} s_0 \\ s_1 \\ s_2 \\ s_3 \\ s_4 \\ s_5 \\ s_6 \end{bmatrix} = \begin{bmatrix} 1 & 4 & -1 & 0 & -4 & 1 & -5 & 3 & 0 & 2 & 1 \\ 1 & -1 & 2 & 4 & 2 & 4 & -1 & 4 & -5 & -1 & -5 \\ 1 & -5 & 4 & -5 & 0 & -1 & 4 & 2 & 4 & -2 & -1 \\ 1 & 1 & 3 & 0 & -5 & -5 & 1 & -1 & 0 & 5 & 4 \\ 1 & 0 & 1 & 1 & 4 & 1 & 0 & 1 & 1 & 4 & 1 \\ 1 & 0 & 1 & -1 & 2 & 0 & 0 & 1 & -1 & 2 & 0 \\ 1 & 1 & 1 & 1 & 1 & 0 & 1 & 1 & 1 & 1 & 0 \end{bmatrix}$$

$$\times \left\{ \begin{bmatrix} 1 & 1 & 1 & 1 & 1 & 1 & 1 \\ 1 & 0 & 0 & 1 & -5 & -1 & 4 \\ 1 & 1 & 1 & 3 & 4 & 2 & -1 \\ 1 & -1 & 1 & 0 & -5 & 4 & 0 \\ 1 & 2 & 4 & -5 & 0 & 2 & -4 \\ 0 & 0 & 1 & -5 & -1 & 4 & 1 \\ 1 & 0 & 0 & 1 & 4 & -1 & -5 \\ 1 & 1 & 1 & -1 & 2 & 4 & 3 \\ 1 & -1 & 1 & 0 & 4 & -5 & 0 \\ 1 & 2 & 4 & 5 & -2 & -1 & 2 \\ 0 & 0 & 1 & 4 & -1 & -5 & 1 \end{bmatrix} \begin{bmatrix} g_0 \\ g_1 \\ g_2 \\ g_3 \\ g_4 \\ g_5 \\ g_6 \end{bmatrix} \right.$$

$$\left. \times \begin{bmatrix} -3 & -3 & -3 & -3 & -3 & -3 & -3 \\ 5 & 0 & 3 & 1 & 5 & -4 & 1 \\ -4 & -4 & 4 & 3 & -2 & 2 & 1 \\ 3 & -3 & 2 & 5 & -3 & -3 & -1 \\ 0 & 0 & 1 & -5 & -1 & 4 & 1 \\ 0 & -5 & 2 & -1 & 3 & 2 & -1 \\ -5 & 0 & -3 & 5 & 5 & -2 & 0 \\ -4 & -4 & -1 & 5 & 0 & 2 & 2 \\ -3 & 3 & -2 & 4 & -2 & -1 & 1 \\ 0 & 0 & -1 & -4 & 1 & 5 & -1 \\ 0 & 5 & -1 & -1 & -4 & 0 & 1 \end{bmatrix} \begin{bmatrix} d_0 \\ d_1 \\ d_2 \\ d_3 \\ d_4 \\ d_5 \\ d_6 \end{bmatrix} \right\}.$$

Although there are only 11 general multiplications, we are forced to accept many multiplications by the small constant integers ± 2, ± 3, ± 4, and ± 5. Each such multiplication can be replaced by several additions, but then there will be many more additions in total. The extent to which this algorithm is an improvement is a matter of judgment. It depends on the relative cost of addition and multiplication. In a large enough extension field, given by $GF(11^m)$, multiplications cost substantially more than additions, so such an algorithm may then be useful. In the prime field $GF(11)$, there is probably no advantage.

10.6 Fourier transform algorithms in finite fields

A Fourier transform in a finite field has many of the same properties as does a Fourier transform in the complex field. However, a Fourier transform of blocklength n exists in $GF(q)$ only if n divides $q - 1$. This means that the blocklength of a Fourier transform over a finite field is limited by the choice of the field, whereas the blocklength of a Fourier transform over the complex field is arbitrary. Among the fast algorithms for Fourier transforms in a Galois field are the mixed-radix Cooley–Tukey algorithm, the Good–Thomas algorithm, and the Winograd algorithms. These algorithms are appropriate for any field, including the Galois fields.

In this section, we shall discuss semifast algorithms that are specific for Fourier transforms in Galois fields. By a semifast algorithm, we mean an algorithm that significantly reduces the number of multiplications in the field compared with the natural form of the computation, but does not reduce the number of additions. The algorithms that we shall describe for the finite-field Fourier transforms are based on the notion of conjugates in the Galois field, and so resemble the Goertzel algorithm, which is based on the notion of conjugates in the complex field. The algorithms described here can compute a single component of the Fourier transform with about $\log n$ multiplications in the field of the Fourier transform.

Recall that the minimal polynomial, denoted $m_\beta(x)$, of an element β of an extension field is the monic polynomial of smallest degree over the ground field that has β as a zero. The minimal polynomial of β, $m_\beta(x) = \sum_{i=0}^{d} m_i x^i$, must have $(x - \beta)$ as a factor in the extension field, and satisfies $m_\beta(\beta) = \sum_{i=0}^{d} m_i \beta^i = 0$. Then, recalling that $(a + b)^q = a^q + b^q$ in $GF(q)$ because $q \pmod{p} = 0$,

$$[m_\beta(\beta)]^q = \left[\sum_{i=0}^{d} m_i \beta^i\right]^q = \sum_{i=0}^{d} m_i^q \beta^{iq}$$

$$= \sum_{i=0}^{d} m_i \beta^{iq} = m(\beta^q),$$

where $m_i^q = m_i$ for all i because m_i is in $GF(q)$. This means that β^q is also a zero of the minimal polynomial of β. In turn, because β^q is a zero of $m_\beta(x)$, $(\beta^q)^q = \beta^{q^2}$ is a zero as well. Thus, the conjugates of β are $\beta, \beta^q, \beta^{q^2}, \ldots$, where the sequence stops at

10.6 Fourier transform algorithms in finite fields

that $\beta^{q^{r-1}}$ such that $\beta^{q^r} = \beta$. For example, the minimal polynomial over $GF(2)$ of the element β of $GF(8)$ is $m_\beta(x) = (x-\beta)(x-\beta^2)(x-\beta^4)$, which stops here because $\beta^8 = \beta$.

The jth Fourier component $V_j = \sum_{i=0}^{n-1} \omega^{ij} v_i$ can be regarded as the polynomial $v(x) = \sum_{i=0}^{n-1} v_i x^i$ evaluated at ω^j. Let $m_j(x)$ be the minimal polynomial of ω^j, an element of $GF(q^m)$. It has a degree, denoted m_j, that is less than or equal to m. For each $m_j(x)$, we can write

$$v(x) = m_j(x)Q(x) + r(x)$$

for some quotient polynomial $Q(x)$ and remainder polynomial $r(x)$. Then

$$v(\omega^j) = m_j(\omega^j)Q(\omega^j) + r(\omega^j)$$
$$= r(\omega^j)$$

because $m_j(\omega^j) = 0$. Hence to compute V_j, we can divide $v(x)$ by $m_j(x)$ to obtain the remainder $r(x)$, and then evaluate the remainder at ω^j. Because $m_j(x)$ is a polynomial over $GF(q)$ with degree at most m, the division by $m_j(x)$ requires at most $(n-m)m$ multiplications of elements of $GF(q^m)$ by elements of $GF(q)$, and $(n-m)m$ additions in $GF(q^m)$. The remainder polynomial $r(x)$ has a degree equal to at most $m-1$. Therefore computation of $r(\omega^j)$ involves at most $m-1$ multiplications in $GF(q^m)$ and the same number of additions. This means that the component V_j can be computed from $v(x)$ with at most $m-1$ multiplications in $GF(q^m)$, $(n-m)m$ multiplications of elements of $GF(q^m)$ by elements of $GF(q)$, and $(n-m+1)m-1$ additions in $GF(q^m)$.

Because the Fourier transform has n components, this process must be repeated $n = q^m - 1$ times, for a total of at most $m(q^m - 1)$ multiplications in $GF(q^m)$. Two conjugates have the same minimal polynomial, and therefore $r(x)$ is the same for conjugates. Consequently, the remainder polynomial $r(x)$ need be computed only once for each conjugacy class. The number of additions is at most $(n-1)n + n(m-1) = n^2 + nm - 2m$. In particular, if $q = 2$, there are at most $n \log_2(n+1)$ multiplications in $GF(2^m)$. This procedure is valid even if $2^m - 1$ is a prime, in contrast to the Cooley–Tukey FFT and the Good–Thomas FFT. The number of additions, however, is still of order n^2.

If the input vector v is over the small field $GF(q)$, then even though the transform is a vector over $GF(q^m)$, it requires no multiplications in $GF(q^m)$. This is because, in this case, $v(x)$ is a polynomial over $GF(q)$, and the computation of V_j uses only multiplications of elements from $GF(q)$ by elements from $GF(q^m)$. If $q = 2$, this becomes an addition of elements of $GF(q^m)$ as selected by the nonzero coefficients of $r(x)$. This Fourier transform can be computed with no multiplications in $GF(2^m)$, and with at most $n \log n$ additions in $GF(2)$ and at most $\log n$ additions in $GF(2^m)$.

This semifast Fourier transform will be illustrated by forming a seven-point Fourier transform in $GF(8)$ using $\omega = \alpha$, where α is a primitive element of $GF(8)$. The minimal polynomials of $GF(8)$ over $GF(2)$ are

$$m_0(x) = (x - \alpha^0),$$
$$m_1(x) = (x - \alpha^1)(x - x^2)(x - \alpha^4),$$
$$m_3(x) = (x - \alpha^3)(x - \alpha^6)(x - \alpha^5).$$

The first step of the Goertzel algorithm is to obtain each of the remainders $r_j(x)$ by a long division of $v(x)$ by each minimal polynomial $m_j(x)$, as given by

$$v(x) = (x + 1)m_0(x) + r_{00},$$
$$v(x) = (x^3 + x + 1)m_1(x) + r_{21}x^2 + r_{11}x + r_{01},$$
$$v(x) = (x^3 + x^2 + 1)m_3(x) + r_{22}x^2 + r_{12}x + r_{02},$$

where

$$r_{00} = v_0,$$
$$r_{01} = v_0 + v_3 + v_5 + v_6,$$
$$r_{11} = v_1 + v_3 + v_4 + v_5,$$
$$r_{21} = v_2 + v_4 + v_5 + v_6,$$
$$r_{02} = v_0 + v_3 + v_4 + v_5,$$
$$r_{12} = v_1 + v_4 + v_5 + v_6,$$
$$r_{22} = v_2 + v_3 + v_4 + v_6.$$

The second step of the Goertzel algorithm is to evaluate the remainder polynomial $r(x)$ at each element of the finite field. Thus, writing $V_j = r(\alpha^j)$ gives the components of the Fourier transform partitioned into conjugacy classes, as given by

$$\begin{bmatrix} V_1 \\ V_2 \\ V_4 \end{bmatrix} = \begin{bmatrix} 1 & \alpha^1 & \alpha^2 \\ 1 & \alpha^2 & \alpha^4 \\ 1 & \alpha^4 & \alpha^2 \end{bmatrix} \begin{bmatrix} r_{01} \\ r_{11} \\ r_{21} \end{bmatrix}$$

and

$$\begin{bmatrix} V_3 \\ V_6 \\ V_5 \end{bmatrix} = \begin{bmatrix} 1 & \alpha^3 & \alpha^6 \\ 1 & \alpha^6 & \alpha^5 \\ 1 & \alpha^5 & \alpha^3 \end{bmatrix} \begin{bmatrix} r_{02} \\ r_{12} \\ r_{22} \end{bmatrix}.$$

The Fourier transform as a whole is then written

$$\begin{bmatrix} V_0 \\ V_1 \\ V_2 \\ V_4 \\ \hline V_3 \\ V_6 \\ V_5 \end{bmatrix} = \begin{bmatrix} \alpha^0 & & & & & & \\ & \alpha^0 & \alpha^1 & \alpha^2 & & & \\ & \alpha^0 & \alpha^2 & \alpha^4 & & & \\ & \alpha^0 & \alpha^4 & \alpha^1 & & & \\ \hline & & & & \alpha^0 & \alpha^3 & \alpha^6 \\ & & & & \alpha^0 & \alpha^6 & \alpha^5 \\ & & & & \alpha^0 & \alpha^5 & \alpha^6 \end{bmatrix} \begin{bmatrix} 1 & 1 & 1 & 1 & 1 & 1 & 1 \\ 1 & 0 & 0 & 1 & 0 & 1 & 1 \\ 0 & 1 & 0 & 1 & 1 & 1 & 0 \\ 0 & 0 & 1 & 0 & 1 & 1 & 1 \\ 1 & 0 & 0 & 1 & 1 & 1 & 0 \\ 0 & 1 & 0 & 0 & 1 & 1 & 1 \\ 0 & 0 & 1 & 1 & 1 & 0 & 1 \end{bmatrix} \begin{bmatrix} v_0 \\ v_1 \\ v_2 \\ v_3 \\ v_4 \\ v_5 \\ v_6 \end{bmatrix},$$

where the unstated elements are all zeros. Because there are no postadditions, this algorithm can easily reduce the number of multiplications if only a few of the components of V are to be computed. For example, one can write

$$\begin{bmatrix} V_1 \\ V_2 \\ V_4 \end{bmatrix} = \begin{bmatrix} \alpha^0 & \alpha^1 & \alpha^2 \\ \alpha^0 & \alpha^2 & \alpha^4 \\ \alpha^0 & \alpha^4 & \alpha^1 \end{bmatrix} \begin{bmatrix} 1 & 0 & 0 & 1 & 0 & 1 & 1 \\ 0 & 1 & 0 & 1 & 1 & 1 & 0 \\ 0 & 0 & 1 & 0 & 1 & 1 & 1 \end{bmatrix} \begin{bmatrix} v_0 \\ v_1 \\ v_2 \\ v_3 \\ v_4 \\ v_5 \\ v_6 \end{bmatrix}$$

to compute only the three components in the same conjugacy class.

10.7 Complex convolution in surrogate fields

We saw in the previous two sections how to embed a convolution of real numbers into a Galois field, where the convolution may be more convenient to compute. We now do the same thing for a convolution involving complex numbers. There are two distinct situations, which are handled quite differently, depending on whether or not $\sqrt{-1}$ exists in $GF(p)$. For example, if p is a Mersenne prime, then $\sqrt{-1}$ does not exist in $GF(p)$. If p is a Fermat prime, then $\sqrt{-1}$ does exist in $GF(p)$. We will refer to only Mersenne primes and Fermat primes, but any field whose characteristic is odd can be handled by either of the two techniques.

We begin with a field $GF(p)$ for which $p = 2^m - 1$ is a Mersenne prime, in which case, the exponent m must be an odd prime. In this field, $\sqrt{-1}$ does not exist. We will extend $GF(2^m - 1)$ to $GF((2^m - 1)^2)$ in the same way that the real field $\boldsymbol{R}$ is extended to the complex field $\boldsymbol{C}$. The Galois field Fourier transform over $GF((2^m - 1)^2)$ can be used to compute convolutions in the complex field.

In the real field, the polynomial $x^2 + 1$ has no zeros. Hence we extend the real field by inventing an element called j and forming the set $\boldsymbol{C} = \{a + bj\}$, where a and b are real numbers. Addition and multiplication are defined by

$$(a + bj) + (c + dj) = (a + c) + (b + d)j$$
$$(a + bj)(c + dj) = (ac - bd) + (ad + bc)j.$$

One can verify that this procedure forms an extension of the real field. This extension is the complex field.

Similarly, in the Galois field $GF(2^m - 1)$, where $2^m - 1$ is a Mersenne prime, the polynomial $x^2 + 1$ has no zeros. Hence we extend $GF(2^m - 1)$ by inventing an element called j and forming the set $GF((2^m - 1)^2) = \{a + bj\}$, where a and b are elements of $GF(2^m - 1)$. Addition and multiplication are again defined by

$$(a + bj) + (c + dj) = (a + c) + (b + d)j$$
$$(a + bj)(c + dj) = (ac - bd) + (ad + bc)j,$$

where the operations on the right side are operations in the base field $GF(2^m - 1)$. One can verify that this definition forms a field containing $(2^m - 1)^2$ elements.

We summarize the above claims in the form of two theorems.

Theorem 10.7.1 *In $GF(2^m - 1)$, where $2^m - 1$ is a Mersenne prime, the element -1 does not have a square root. Hence $x^2 + 1$ has no zeros.*

Proof The proof is deferred until the end of the section. □

Theorem 10.7.2 *Under the above definitions of addition and multiplication, $GF((2^m - 1)^2)$ forms a field.*

Proof This follows immediately from the statement that $x^2 + 1$ is a prime polynomial. □

Next, we examine the Fourier transform in $GF((2^m - 1)^2)$. The blocklength is $(2^m - 1)^2 - 1$ or a divisor thereof. But we have the obvious factorization

$$(2^m - 1)^2 - 1 = 2^{m+1}(2^{m-1} - 1).$$

Hence, in $GF((2^m - 1)^2)$, we can choose 2^{m+1} as the blocklength of the Fourier transform and compute it with a radix-two Cooley–Tukey FFT. One could also include some factors of $2^{m-1} - 1$ to get other blocklengths.

For example, choose $m = 17$. Then

$$(2^{17} - 1)^2 - 1 = 2^{18} \cdot 3 \cdot 5 \cdot 17 \cdot 257.$$

10.7 Complex convolution in surrogate fields

Table 10.1 *Elements of order n in some complex Mersenne fields*

Blocklength n	$GF((2^{13}-1)^2)$	$GF((2^{17}-1)^2)$
256	$(2^7-2^4-2^2)+j(2^{11}+2^7-2^3+2^0)$	$(2^8+2^5)+j(2^{16}+2^{12}+2^{11}-2^9+2^7-2^4+2^2)$
512	$(2^{11}-2^7-2^5+2^3)+j(2^8+2^5)$	$(2^{12}+2^8)+j(2^{17}-2^{11}+2^8+2^5+2^2)$
1024	$(2^9-2^5+2^0)+j(2^9-2)$	$(2^{11}+2^8-2^2)+j(2^{15}+2^{12}+2^7-2^0)$
2048	$(2^9+2^7+2^4+2)+j(2^{10}+2^5)$	$(2^7+2^3)+j(2^{12}+2^{11}-2^9+2^7+2^5-2^0)$
4096	$(2^{11}+2^7+2^4+2^2)+j(2^9+2)$	$(2^{10}-2^2+2^0)+j(2^{14}-2^6)$

	$GF((2^{19}-1)^2)$	$GF((2^{31}-1)^2)$
256	$(2^{15}+2^{12}-2^7+2^0)$ $+j(2^{15}+2^{13}+2^{11}+2^7+2^4+2^3-2^0)$	$(2^{27}+2^{25}+2^{21}+2^{19}+2^{16}-2^{13}-2^{10}+2^4-2^1)$ $+j(2^{27}+2^{23}-2^{19}-2^{17}+2^{12}+2^{10}+2^8+2^3)$
512	$(2^{17}+2^{15}+2^{12}+2^9-2^4+2^0)$ $+j(2^{15}-2^{10}-2^3)$	$(2^{30}-2^{28}+2^{25}+2^{19}+2^{18}-2^{16}+2^{13}+2^{10}+2^7)$ $+j(2^{30}-2^{28}+2^{22}+2^{20}-2^{14}+2^8+2^2)$
1024	$(2^{12}+2^{10}-2^1)$ $+j(2^{19}-2^{17}+2^{14}+2^{13}-2^7-2^2)$	$(2^{31}-2^{28}+2^{25}-2^{18}-2^{13}+2^{10}+2^5)$ $+j(2^{30}-2^{25}-2^{19}+2^{16}-2^{13}-2^7+2^4-2^2)$
2048	$(2^{19}-2^{17}+2^{12})$ $+j(2^{16}+2^{11}-2^8+2^2+2^0)$	$(2^{28}+2^{24}+2^{18}-2^{16}+2^{14}-2^6+2^4+2^0)$ $+j(2^{27}-2^{24}+2^{17}-2^{14}+2^{11}+2^{10}-2^0)$
4096	$(2^{13}+2^8-2^5+2^2)$ $+j(2^{12}+2^9+2^6+2^3)$	$(2^{25}+2^{22}+2^{21}-2^{16}+2^{13}+2^5)$ $+j(2^{30}+2^{25}-2^{22}+2^{14}+2^{11}+2^6+2^2)$

Any power of two up to 2^{18} can be chosen to be the blocklength of a Fourier transform, and even larger Fourier transforms can be formed by using some of the other factors of $p^2 - 1$.

Let n be a factor of 2^{18}, and let

$$V_k = \sum_{i=0}^{n-1} \omega^{ik} v_i, \qquad k = 0, \ldots, n-1,$$

where ω is an element of $GF((2^{17}-1)^2)$ of order n. (A table of suitable ω, found by computer search, is given in Table 10.1.) Because n is a factor of two, the Fourier transform can be computed by a radix-two Cooley–Tukey FFT.

In $GF((2^{17}-1)^2)$, there are many more choices for the blocklength of a Fourier transform than in $GF(2^{17}-1)$, especially of blocklengths that are a power of two. Therefore one may choose to compute a cyclic convolution in $GF(2^{17}-1)$, which may have originated as a cyclic convolution of integers, by embedding it in $GF((2^{17}-1)^2)$. This is the same technique as computing a real convolution by embedding the real field into the complex field so that suitable Fourier transforms are available. Indeed, a Fourier transform in $GF((2^{17}-1)^2)$ can be used to simultaneously compute the Fourier transforms of two vectors $\boldsymbol{v}'$ and $\boldsymbol{v}''$ in $GF(2^{17}-1)$ by applying it to the sum $\boldsymbol{v}' + \boldsymbol{v}''$ and decomposing the result into V and V' in the standard way.

In the case that p is a Fermat number, $p = 2^m + 1$, it is not possible to form an extension field $GF(p^2)$ with a multiplication rule that behaves like complex multiplication because $\sqrt{-1}$ is an element of $GF(p)$. Specifically, we have $p = 2^m + 1$ with m a power of two, and $\sqrt{-1} = 2^{m/4}(2^{m/2} - 1)$, which can be verified by computing the square of both sides. This shows explicitly that $\sqrt{-1}$ is an element of $GF(2^m + 1)$. Therefore it cannot be used to extend the prime field $GF(p)$ if p is a Fermat prime or is any other p for which $x^2 + 1$ has a zero in $GF(p)$.

The convolution of two sequences $g(x) = g_R(x) + jg_I(x)$ and $d(x) = d_R(x) + jd_I(x)$ of complex numbers, as represented by gaussian integers, can be represented by the polynomial product

$$s(x) = g(x)d(x).$$

In a prime field $GF(p)$ that already has a $\sqrt{-1}$, one can first compute the four convolutions

$$g_R(x)d_R(x), \qquad g_I(x)d_R(x), \qquad g_R(x)d_I(x), \quad \text{and} \quad g_I(x)d_I(x).$$

Then write

$$s_R(x) = g_R(x)d_R(x) - g_I(x)d_I(x)$$
$$s_I(x) = g_R(x)d_I(x) + g_I(x)d_R(x).$$

These satisfy

$$s(x) = s_R(x) + js_I(x)$$
$$ = g(x)d(x)$$

provided the real and imaginary parts of $s(x)$ do not exceed the range of $GF(p)$. This procedure requires four polynomial multiplications.

A better way to use the Fermat field, because it has half the number of polynomial multiplications, is to regard the polynomials to be in the ring $GF(p)[x]$, where p is a Fermat prime, and to define

$$a(x) = \tfrac{1}{2}(g_R(x) - 2^{m/2}g_I(x))(d_R(x) - 2^{m/2}d_I(x)),$$
$$b(x) = \tfrac{1}{2}(g_R(x) + 2^{m/2}g_I(x))(d_R(x) + 2^{m/2}d_I(x)).$$

These equations specify two convolutions to be computed. The output polynomial $s(x)$ is given by

$$s_R(x) = (a(x) + b(x)),$$
$$s_I(x) = 2^{m/2}(a(x) - b(x)),$$

with all computations in the ring $GF(p)[x]/\langle x^n - 1 \rangle$. In this way, a complex convolution over the gaussian integers can be computed with two convolutions in a Fermat field.

10.7 Complex convolution in surrogate fields

To finish the section, we must give a proof of Theorem 10.7.1, which has been left pending. This proof uses the idea of a quadratic residue of $\mathbf{Z}$. In the integer ring $\mathbf{Z}$, those elements that have a square root, modulo p, are called *quadratic residues* modulo p (because they are the squares of their square roots modulo p). When regarded as elements of the prime field $GF(p)$, it is more natural just to call these elements the *squares* of $GF(p)$. Exactly half of the nonzero elements in $GF(p)$, p an odd prime, have square roots. To see this, first note that every even power of a primitive element α has a square root. On the other hand, every element that is a square root can be written as α^i for some i, and so its square is $\alpha^{((2i))}$, because the multiplicative group of the field is cyclic with $p - 1$ elements (where the use of double parentheses denotes modulo $p - 1$ in the exponent). But $p - 1$ is even, so $((2i))$ is even as well. Hence only even powers of α can have square roots.

Theorem 10.7.3 *In $GF(p)$, p odd, r is a square if, and only if, $r^{(p-1)/2} = 1$.*

Proof Suppose that $r^{(p-1)/2} \neq 1$, and recall that $a^{p-1} = 1$ for every element a of $GF(p)$. Then $\sqrt{r}$ cannot exist, because if it did, $(\sqrt{r})^{p-1}$ must equal one, and the premise is that it does not.

Suppose that $r^{(p-1)/2} = 1$, and let α be a primitive element in $GF(p)$. Obviously, all even powers of α are squares, and all odd powers of α are nonsquares. All that we need to prove is that r is an even power of α. Suppose on the contrary that it is odd. Then $r = \alpha^{2i+1}$ and

$$\begin{aligned}
r^{(p-1)/2} &= (\alpha^{2i+1})^{(p-1)/2} \\
&= \alpha^{i(p-1)} \alpha^{(p-1)/2} \\
&= \alpha^{(p-1)/2} \\
&\neq 1,
\end{aligned}$$

because α has order $p - 1$. Thus $r^{(p-1)/2} = 1$ implies that r is an even power of α. Hence r is a square. □

We are now ready to restate and prove Theorem 10.7.1, whose proof has remained pending.

Theorem 10.7.1 *In $GF(2^m - 1)$, where $2^m - 1$ is a Mersenne prime, the element -1 does not have a square root. Hence $x^2 + 1$ has no zeros.*

Proof Suppose that -1 has the square root r. Then $r^2 = -1$, and by Theorem 10.7.3,

$$r^{(p-1)/2} = 1,$$

where $p = 2^m - 1$. Then

$$r^{(2^m-2)/2} = 1$$

or

$$r^{2^{m-1}} r^{-1} = 1.$$

But $r^2 = -1$, by the premise, and $m - 1$ is even. Hence $r^{2^{m-1}}$ is the product of an even number of copies of r^2. Then

$$r^{2^{m-1}} = 1$$

and

$$r^{-1} = 1,$$

so $r = 1$ and r^2 is not equal to -1. This proves that there is no square root of -1 in $GF(p)$ if p is a Mersenne prime. $\square$

10.8 Integer ring transforms

A Fourier transform of blocklength n exists in a field F whenever F contains an element of order n. All of the elementary properties of the Fourier transform are valid whenever it exists.

It is often possible to also define transforms in a ring R, but now the situation is not as straightforward. In this section, we shall study transforms in $\mathbf{Z}/\langle q \rangle$, the ring of integers modulo q. When q is a prime, $\mathbf{Z}/\langle q \rangle$ is a field, and we have already seen that, in a field, the Fourier transform exists together with all its basic properties. It is only necessary here to consider composite q. We shall see that, over $\mathbf{Z}/\langle q \rangle$, meaningful Fourier transforms can be defined even when q is composite. However, the structure of these Fourier transforms is merely a diminished echo of the structure of Fourier transforms based on the prime factors of q. We shall find that not much value is added by working with a composite q.

We want to define an integer ring transform

$$V_k = \sum_{i=0}^{n-1} \omega^{ik} v_i, \qquad k = 0, \ldots, n-1$$

of a vector v over the ring $\mathbf{Z}/\langle q \rangle$ so that the inverse transform

$$v_i = n^{-1} \sum_{k=0}^{n-1} \omega^{-ik} V_k, \qquad i = 0, \ldots, n-1$$

10.8 Integer ring transforms

exists and the convolution theorem holds. To get a satisfactory Fourier transform, we need two things: an element ω of order n, and an inverse for n in $\mathbf{Z}/\langle q \rangle$. We also need an inverse for ω, but this comes without asking because $\omega^{-1} = \omega^{n-1}$ is an immediate consequence of $\omega^n = 1$.

We need to determine the values of n for which an ω of order n exists in $\mathbf{Z}/\langle q \rangle$. We shall begin with the simple case, $q = p^m$ for p an odd prime. This is not a field because in $\mathbf{Z}/\langle p^m \rangle$, multiplication is defined modulo p^m. According to Theorem 9.1.8, this ring contains an element of order $(p-1)p^{m-1}$, and according to Theorem 9.1.5 (Euler's theorem), the order of every element in $\mathbf{Z}/\langle p^m \rangle$ coprime to p^m divides $(p-1)p^{m-1}$. This condition says that we can have an ω of order n for all n that divide $(p-1)p^{m-1}$. However, Theorem 9.2.2 says that the inverse of n exists if and only if n and p^m are coprime. Therefore n cannot have p as a factor. We can choose only an ω whose order n divides $p - 1$. The next theorem shows that for every such ω there is a suitable Fourier transform. However, the blocklengths that can be chosen for a Fourier transform in $\mathbf{Z}/\langle p^m \rangle$ are the same blocklengths as those that can be chosen in $GF(p)$. Only the wordlength is increased from about $\log_2 p$ bits to about $m \log_2 p$ bits. Because the wordlength of a Fourier transform over $GF(p)$ is usually large enough, there is rarely an advantage in going into $\mathbf{Z}/\langle p^m \rangle$.

The situation for an arbitrary q is similar and is proved in the following theorem.

Theorem 10.8.1 *Over the ring $\mathbf{Z}/\langle q \rangle$ there exists an invertible Fourier transform of blocklength n if and only if n divides $p - 1$ for every prime factor p of q.*

Proof The proof will be given first for the case in which q is a prime power. Later, the Chinese remainder theorem will be used to relate the case in which q is a product of prime powers to a set of Fourier transforms.

The converse is the easiest to prove. The blocklength n has an inverse modulo q only if n and q are coprime because

$$nn^{-1} = 1 + Qq.$$

Any factor common to both n and q must then be a factor of one, which is impossible. Further, Theorem 9.1.5 requires that every element ω whose order n is coprime to q has an order that divides $\phi(q) = (p-1)p^{m-1}$. Therefore a Fourier transform of blocklength n does not exist in $\mathbf{Z}/\langle p^m \rangle$ unless n divides $p - 1$.

Now prove the direct part. There is no point in considering p equal to two, because then n equals one and the theorem becomes trivial. Hence p is an odd prime, so Theorem 9.1.8 guarantees the existence of an element π of order $(p-1)p^{m-1}$. We will choose $\omega = \pi^{bp^{m-1}}$ as an element of order $(p-1)/b$ for any b that divides $p - 1$. Hence, for any divisor of $p - 1$, we have an ω of that order. All that remains is to show that the inverse Fourier transform is valid. This we show by examining the inverse

Fourier transform

$$\frac{1}{n}\sum_{k=0}^{n-1}\omega^{-ik}V_k = \frac{1}{n}\sum_{k=0}^{n-1}\omega^{-ik}\sum_{i'=0}^{n-1}\omega^{i'k}v_{i'}$$

$$= \frac{1}{n}\sum_{i'=0}^{n-1}v_{i'}\left[\sum_{k=0}^{n-1}\omega^{-k(i'-i)}\right].$$

The sum on k is equal to n if i' is equal to i, while if i' is not equal to i, then the summation becomes

$$\sum_{k=0}^{n-1}(\omega^{-(i'-i)})^k = \frac{1-\omega^{-(i'-i)k}}{1-\omega^{-(i'-i)}}.$$

The right side is zero because $i' - i \neq 0 \pmod{n}$. This is because i and i' are both less than n. Then

$$\frac{1}{n}\sum_{k=0}^{n-1}\omega^{-ik} = \frac{1}{n}\sum_{i'=0}^{n-1}v_{i'}(n\delta_{ii'}) = v_i,$$

as was to be proven.

Now let $q = p_1^{m_1} p_2^{m_2} \ldots p_r^{m_r}$. The use of the Chinese remainder theorem for integers, as in the Good–Thomas algorithm, provides a mapping from $\mathbf{Z}/\langle q \rangle$ to $\mathbf{Z}/\langle p_1^{m_1} \rangle \times \mathbf{Z}/\langle p_2^{m_2} \rangle \times \cdots \times \mathbf{Z}/\langle p_r^{m_r} \rangle$. The existence of a Fourier transform over $\mathbf{Z}/\langle q \rangle$ can be related to the existence of a Fourier transform in each of the factor groups, and the condition that n divides $p_i - 1$ must hold in each factor group. $\square$

Integer ring transforms, with q not a prime, are so severely constrained by Theorem 10.8.1 that it seems there will be very few applications. Nevertheless, some possibilities remain that might fit the needs of a special application.

For example, we have the factorization

$$2^{40} + 1 = (257)(4278\,255\,361)$$
$$= p_1 p_2.$$

Because 256 divides both $p_1 - 1$ and $p_2 - 1$, there is a 256-point Fourier transform in $\mathbf{Z}/\langle 2^{40} + 1 \rangle$. This transform will have 40 bits of precision in a 41-bit wordlength. A radix-two Cooley–Tukey FFT can be used in this ring. Arithmetic overflow is implemented simply by using $2^{40} = -1$. The element two cannot be used for ω because two has order 80. An element of order 256 is needed as ω. Therefore it will not be possible to turn multiplications into shifts. Multiplications will be general 40-bit by 40-bit multiplications.

In this way, we can construct a procedure for 256-point cyclic convolution with on the order of 40 bits of precision and using about $256 \log_2 256 + 256$ multiplications; each of the multiplications is a 40-bit by 40-bit multiplication. This procedure has both more precision and fewer multiplications than doing the same thing with a complex FFT in a forty-bit wordlength.

10.9 Chevillat number transforms

The *Chevillat numbers* are a loosely defined set of integers for which, in $\mathbf{Z}/\langle q \rangle$, there is a single-radix Fourier transform algorithm of large order. A Chevillat number is usually a prime, but not always. The Chevillat numbers correspond to good Fourier transforms but have no other apparent number-theoretic significance. They were found by computer search.

In $\mathbf{Z}/\langle q \rangle$, the Fourier transform

$$V_k = \sum_{i=0}^{n-1} \omega^{ik} v_i, \qquad k = 0, \ldots, n-1$$

is convenient to compute with a single-radix Cooley–Tukey FFT whenever the order of ω is a power of a small prime. Hence, for each q, we can tabulate the largest order to see when the blocklength n can be made a large power of a small prime. A tabulation of Chevillat numbers that are primes is shown in Table 10.2. The calculation of the residue modulo p, which is necessary in calculations in $\mathbf{Z}/\langle p \rangle$, is not as convenient in the general case as it is when p is a Mersenne prime or a Fermat prime.

10.10 The Preparata–Sarwate algorithm

Just as a computation in the complex field can be embedded into a Galois field, so a computation in a Galois field can be embedded into the complex field. One may perform convolutions in $GF(q)$ by using a complex-valued FFT. Suppose that in the field $GF(q)$,

$$s(x) = g(x)d(x)$$

is to be computed. In general, the elements of $GF(q)$ can be represented as polynomials over some prime field $GF(p)$. We can think of a product of field elements in $GF(q)$ as a convolution of polynomials modulo an irreducible polynomial $p(x)$. The computation of the residue modulo the irreducible polynomial can be held pending until after all

Table 10.2 *Table of Chevillat single-radix Fourier transforms*

Word length	q	Maximum single-radix FFT	Word length	q	Maximum single-radix FFT	Word length	q	Maximum single-radix FFT
8	163	81	12	2917	729	15	17497	2187
	193	64		3329	256		18433	2048
	197	49		3457	128		25601	1024
	241	16		3889	243		28751	625
	251	125		4001	32, 125		28813	2401
9	257	256		4019	49		30871	343
	401	25		4049	16		32077	729
	449	64		4051	81, 25		32251	125
	487	243		5347	243		32257	512
	491	49		7001	125		32401	25
10	751	125		7547	343		32537	49
	769	256		7681	512		32563	243
	883	49		7841	49		32609	32
	919	27		7937	256	16	39367	19683
	929	32		8101	81, 25		40961	8192
	1009	16		8161	32		52489	6561
11	1373	343	14	8263	243		61441	4096
	1409	128		13751	625		62501	15625
	1459	729		14407	2401		64153	729
	1471	49		15361	1024		64513	1024
	1601	64		16001	128, 125		65089	64
	1783	81		16073	49		65101	25
	1951	25		16193	64		65171	343
	1999	27		16301	25		65269	49
	2017	32		16363	81		65449	81
				16369	16		65521	16

other computations of the cyclic convolution are complete. The original convolution will be written as a two-dimensional convolution of integers, which can be embedded into the complex field.

The field elements g_i and d_i in $GF(q)$ are expressed as polynomials over the prime field $GF(p)$ in the form

$$g_i = \sum_{\ell=0}^{m-1} g_{i\ell} z^\ell,$$

$$d_i = \sum_{\ell=0}^{m-1} d_{i\ell} z^\ell,$$

10.10 The Preparata–Sarwate algorithm

where $q = p^m$ and $g_{i\ell}$ and $d_{i\ell}$ are nonnegative integers less than p. The linear convolution of $\mathbf{g}$ and $\mathbf{d}$ is

$$s_i = \sum_{k=0}^{n-1} g_k d_{i-k}$$

$$= \sum_{k=0}^{n-1} \sum_{\ell=0}^{m-1} \sum_{\ell'=0}^{m-1} g_{k\ell} d_{(i-k)\ell'} z^{\ell+\ell'} \pmod{p}\pmod{p(z)},$$

where p is the characteristic of the field and $p(z)$ is a prime polynomial of degree m. Define

$$s_{ii'} = \sum_{k=0}^{n-1} \sum_{k'=0}^{m-1} g_{kk'} d_{(i-k)(i'-k')}, \qquad \begin{aligned} i &= 0, \ldots, n-1, \\ i' &= 0, \ldots, 2m-1, \end{aligned}$$

interpreted as a two-dimensional convolution of integers; each integer in the two-dimensional array is between zero and $p-1$, inclusively. Then

$$s_i = \sum_{i'=0}^{2m-1} s_{ii'} z^{i'} \pmod{p} \pmod{p(z)}.$$

The residue computations are done last; first, the residue of each integer modulo p is found, then the residue of

$$s_i(z) = \sum_{i=0}^{2m-1} s_{ii'} z^{i'}$$

is found modulo $p(x)$. The computations of the two residue computations are simple in comparison with the computations of the two-dimensional convolution of integers. The two-dimensional convolution of integers can be computed in any convenient surrogate field, such as the real field or the complex field.

One may also use as the surrogate field any convenient finite field, even one whose characteristic is different from the original field. For example, to convolve sequences in $GF(2)$, whose blocklength n is smaller than one-half of a Fermat prime $2^m + 1$ for $m = 2, 4, 8$, or 16 (so that $2^m + 1 = 5, 17, 257$, or $65,537$), one can use $GF(2^m + 1)$. The sequences in $GF(2)$ can be regarded as sequences of integers (which happen to only take the values zero or one). The linear convolution of the integer sequences has length smaller than $2^m + 1$, and is nowhere larger than n. Hence the linear convolution in $GF(2)$ can be done as a cyclic convolution in $GF(2^m + 1)$, followed by a modulo 2 reduction. The cyclic convolution in $GF(2^m + 1)$ can be done by any fast convolution method. The field $GF(2^m + 1)$ has a Fourier transform whose size n is any divisor of 2^m, so one scheme is to use a fast Fourier transform with the convolution theorem. The Fourier transform of size n can be computed by the Cooley–Tukey FFT with $(n/2)\log_2 n$ multiplications in $GF(2^{16} + 1)$ and with the same number of additions.

Problems for Chapter 10

10.1 **a** Design logic circuits for addition and multiplication in the Mersenne field $GF(7)$. Is there an advantage in allowing zero to have two representations?
  b Design logic circuits for addition and multiplication in the Fermat field $GF(5)$. Is there an advantage in allowing zero, one, and two each to have two representations?

10.2 **a** Verify that, if $\omega = 2$ in the Fermat field $GF(2^{2^m} + 1)$, then all multiplications are indeed multiplications by powers of two and can be replaced by shifts.
  b In $GF(193)$ the element $\omega = 8$ has order 32, and so it can be used as the kernel of a 32-point Fourier transform in $GF(193)$. In this field, is it meaningful to say that all multiplications are multiplications by powers of two and can be replaced by shifts?

10.3 In the ring $\mathbf{Z}/\langle 15 \rangle$ the element two has order four. Write out four by four matrices in the form of a four-point Fourier transform and inverse Fourier transform. Show that they do not behave in the standard way for Fourier transforms. What goes wrong? Why?

10.4 **a** In the Fermat prime field $GF(17)$, what is the order of the element two?
  b Describe a radix-two, eight-point Cooley–Tukey FFT in $GF(17)$. How many multiplications are there?
  c Describe a radix-two, four-point Cooley–Tukey FFT using the element four as the kernel of the Fourier transform.
  d In $GF(17)$, $\sqrt{2} = 11$. Is eleven a primitive element? Describe a radix-two, sixteen-point Cooley–Tukey FFT in $GF(17)$ with a minimum number of multiplications.

10.5 Let $q = (2^8 + 1)(2^{16} + 1)$. In $\mathbf{Z}/\langle q \rangle$, what is the order of the element two?

10.6 The integer 31 in a Mersenne prime.
  a In the field $GF(31)$, what is the order of the element two?
  b Give a five-point Winograd FFT in $GF(31)$.
  c In $GF(31)$ the element -1 has no square root. What is the order of the element $1 + j$ in the extension field $GF(31^2)$?
  d What is the largest possible blocklength of a Fourier transform in $GF(31^2)$?
  e Describe the structure of an eight-point FFT in $GF(31^2)$.

10.7 In the complex Mersenne field $GF((2^{17} - 1)^2)$, show that the element $1 + j$ has order 136. This means there is a Fourier transform of blocklength 136 in this field. Describe the structure of a fast algorithm for computing this Fourier transform.

10.8 Describe how to simultaneously compute the $GF((2^{17}-1)^2)$-valued Fourier transform of blocklength 4096 of two $GF(2^{17}-1)$-valued vectors using one pass through an FFT algorithm in $GF((2^{17}-1)^2)$.

10.9 Construct a five-point Winograd FFT in $GF(41)$. (**Hint:** What is the order of two in $GF(41)$?)

10.10 There is a 17-point Fourier transform in $GF((2^{17}-1)^2)$.

    a Describe how this can be computed with two 16-point Fourier transforms in the complex field. Portray the computation in the form of a flow diagram.

    b How many real multiplications and real additions are required if the 16-point Fourier transforms in the complex field are computed by using a Winograd small FFT?

    c What wordlength and rounding procedure should be used so that the complex computations do, indeed, give correct answers in $GF((2^{17}-1)^2)$?

10.11 There is a 17-point Fourier transform in the complex field.

    a Describe how this can be computed with two 16-point Fourier transforms in a complex Mersenne field.

    b How many multiplications and additions are required in the surrogate field?

10.12 **a** Show that the polynomial $x^{32}-1$ has 32 distinct zeros in $GF(2^{16}+1)$.

    b How many general multiplications will be used by a 32-point Winograd cyclic convolution in the extension field $GF((2^{16}+1)^m)$?

    c How many general multiplications will be needed to compute a 32-point cyclic convolution in $GF((2^{16}+1)^m)$ by using an FFT and the convolution theorem?

    d Explain the relationship between parts (b) and (c).

10.13 Develop in detail a 16-point Winograd small FFT for $GF(17)$ or an extension of $GF(17)$. How many nontrivial multiplications are needed?.

10.14 Construct a program that computes a 75-point fast Fourier transform in $GF(2^{20})$. Build the FFT out of three-point and five-point Winograd FFTs.

Notes for Chapter 10

Fourier transforms had already been studied in an arbitrary field, but Rader (1972) was the first to point out that a real convolution could be computed by embedding it in a field of integers modulo a Mersenne prime or a Fermat prime. He proposed the application to digital signal processing, noting that the computations could then be simpler. Agarwal and Burrus (1973) also proposed the use of Fermat transforms and developed their structure (1974, 1975). Chevillat (1978) explored the use of other prime fields. Some of the implementation considerations were explored by McClellan (1976) and by Leibowitz (1976). The use of a complex extension of a Galois field was suggested by Reed and Truong (1975) and by Nussbaumer (1976), as a way of doing complex convolutions. Our Table 10.1 is based on the work of Reed and Truong. Applications of the Winograd FFTs in finite fields were studied by Miller, Truong, and Reed (1980).

Convolution algorithms in Galois fields of small characteristics can look different from convolution algorithms in the real field. Rice (1980) studied the construction of convolution algorithms in small Galois fields. We have used some of his examples. The modification of the Goertzel semifast algorithm for finite fields is due to Blahut (1983a). Trifonof and Fedorenko (2003) and Fedorenko (2006) developed alternative semifast algorithms.

The idea of using surrogate fields was inverted by Preparata and Sarwate (1977), who embedded convolutions in a Galois field into the complex field. Games (1985) studied the use of algebraic integers as surrogates to compute real convolutions. Cozzens and Finkelstein (1985) developed a method of computing the Fourier transform in the complex field by using a ring of algebraic integers.

The representation of the data – as opposed to the indices – with a residue number system has been well-studied. One may refer to the textbook by Szabo and Tanaka (1967) or the work by Jenkins and Leon (1977). The Montgomery multiplication algorithm for modular multiplication was introduced by Montgomery (1985). Kaliski (1995), as well as Dussé and Kaliski (1991), extended the methods of Montgomery. The notion of computation with a residue number system has been extended to the use of finite rings in order to obtain large dynamic range by Wigley, Jullien, and Reaume (1994).

11 Fast algorithms and multidimensional convolutions

Just as one can define a one-dimensional convolution, so one can define a multidimensional convolution. Multidimensional linear convolutions and cyclic convolutions can be defined in any field of interest on multidimensional arrays of data and are useful in many ways. We have seen in earlier chapters that multidimensional arrays and multidimensional convolutions can be created artificially as part of an algorithm for processing a one-dimensional data vector. Multidimensional arrays also arise naturally in many signal-processing problems, especially in the processing of image data such as satellite reconnaissance photographs, medical imagery including X-ray images, seismic records, and electron micrographs.

This chapter will begin the study of fast algorithms for multidimensional convolutions by nesting fast algorithms for one-dimensional convolutions. Then we shall study ways to construct a fast algorithm for a one-dimensional cyclic convolution by temporarily mapping it into a multidimensional convolution, a procedure that is known as the *Agarwal–Cooley convolution algorithm*. The Agarwal–Cooley algorithm for one-dimensional cyclic convolution is a powerful adjunct to the convolution methods studied in Chapter 6, which become unwieldy for large blocklength. It gives a way to build algorithms for large one-dimensional cyclic convolutions by combining the small convolution algorithms. Then, in the latter half of the chapter, we shall study methods that are derived specifically to compute two-dimensional convolutions.

11.1 Nested convolution algorithms

A two-dimensional convolution is an operation on a pair of two-dimensional arrays. We can think of the convolution as a filtering operation wherein one two-dimensional array is a two-dimensional filter through which a two-dimensional data array is passed to form a two-dimensional output signal. Such operations are useful for image processing.

Given an N' by N'' array called the data array,

$$d = \{d_{i'i''} \mid i' = 0, \ldots, N' - 1; \quad i'' = 0, \ldots, N'' - 1\},$$

and an L' by L'' array called the filter array,

$$g = \{g_{i'i''} \mid i' = 0, \ldots, L' - 1; \quad i'' = 0, \ldots, L'' - 1\},$$

compute a new array called the *signal array*,

$$s = \{s_{i'i''} \mid i' = 0, \ldots, L' + N' - 2; \quad i'' = 0, \ldots, L'' + N'' - 2\}$$

by the equation

$$s_{i'i''} = \sum_{k'=0}^{N'-1} \sum_{k''=0}^{N''-1} g_{i'-k',i''-k''} d_{k'k''}, \quad \begin{array}{l} i' = 0, \ldots, L' + N' - 2, \\ i'' = 0, \ldots, L'' + N'' - 2. \end{array}$$

The new array s is an $L' + N' - 1$ by $L'' + N'' - 1$ array formed as the two-dimensional convolution of d and g.

One also can form convolutions of higher dimension by a straightforward extension. However, we shall restrict our attention to two-dimensional convolutions because all the ideas can be easily extended to higher dimensions.

We shall express a two-dimensional convolution in the notation of polynomials; the polynomial notation can be introduced along either or both dimensions. Thus the array d can be represented as a vector of polynomials,

$$d_{i'}(y) = \sum_{i''=0}^{N''-1} d_{i'i''} y^{i''}, \quad i' = 0, \ldots, N' - 1,$$

or as a polynomial in two variables,

$$d(x, y) = \sum_{i'=0}^{N'-1} \sum_{i''=0}^{N''-1} d_{i'i''} y^{i''} x^{i'}.$$

Similar representations can be used for g and s:

$$g_{i'}(y) = \sum_{i''=0}^{L''-1} g_{i'i''} y^{i''}, \quad i' = 0, \ldots, N' - 1,$$

$$s_{i'}(y) = \sum_{i''=0}^{L''-N''-2} s_{i'i''} y^{i''}, \quad i' = 0, \ldots, N' - 1,$$

and

$$g(x, y) = \sum_{i'=0}^{L'-1} \sum_{i''=0}^{L''-1} g_{i'i''} y^{i''} x^{i'},$$

$$s(x, y) = \sum_{i'=0}^{L'+N'-2} \sum_{i''=0}^{L''-N''-2} s_{i'i''} y^{i''} x^{i'}.$$

11.1 Nested convolution algorithms

Then the two-dimensional convolution can be written either as a one-dimensional convolution of polynomials,

$$s_{i'}(y) = \sum_{k'=0}^{N'-1} g_{i'-k'}(y) d_{k'}(y),$$

or as a product of polynomials in two variables,

$$s(x, y) = g(x, y) d(x, y).$$

A two-dimensional cyclic convolution also can be written this way in terms of polynomials. Thus

$$s(x, y) = g(x, y) d(x, y) \pmod{x^{n'} - 1} \pmod{y^{n''} - 1},$$

where n' and n'' need not be equal. A two-dimensional cyclic convolution satisfies the convolution theorem. If D and G are the two-dimensional Fourier transforms of d and g, respectively, and $S_{k'k''} = G_{k'k''} D_{k'k''}$, then S is the two-dimensional Fourier transform of s. Hence a two-dimensional cyclic convolution can be computed by using a two-dimensional FFT algorithm, studied in Chapter 12. One can also use direct methods.

There are a great many direct ways in which to work on a two-dimensional linear or cyclic convolution. The simplest thing to try is to compute the two-dimensional convolution as a sequence of one-dimensional convolutions, first along all the rows, then along all the columns. Then, however, in the formula

$$s_{i'i''} = \sum_{k'=0}^{N'-1} \left[\sum_{k''=0}^{N''-1} g_{i'-k', i''-k''} d_{k'k''} \right],$$

the linear convolution on k'', as specified by the brackets, must be computed for each i' and k' of interest.

The two-dimensional convolution may be easier to understand when it is written as a convolution of polynomials. Thus

$$s_{i'}(y) = \sum_{k'=0}^{N'-1} g_{i'-k'}(y) d_{k'}(y), \qquad i' = 0, \ldots, L' + N' - 2,$$

from which it is easy to see that the obvious implementation of the convolution of polynomials uses $L'N'$ polynomial products, and each polynomial product is a convolution and so requires $L''N''$ multiplications. There is a total of $L'L''N'N''$ multiplications, which is usually unacceptable for large problems. This is why fast algorithms are needed for two-dimensional filtering.

The convolution algorithms that we studied in Chapter 3, though derived for convolutions in an arbitrary field, actually are valid in some rings, such as polynomial rings. Hence a convolution of polynomials can be computed by any of these fast algorithms. Additions in the fast algorithm become additions of polynomials, and multiplications become multiplications of polynomials. Those multiplications of polynomials

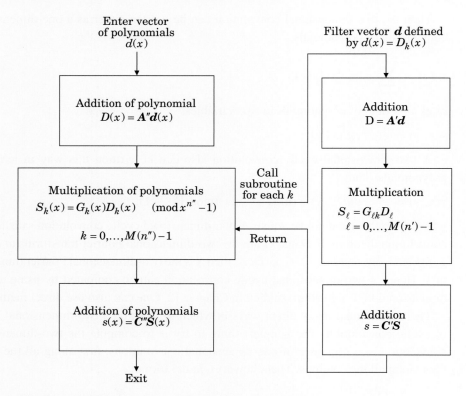

Figure 11.1 An algorithm for two-dimensional cyclic convolution

are themselves convolutions and, in turn, can be computed by any fast convolution algorithm. Thus we have a way to compute a two-dimensional convolution by nesting a fast algorithm for a one-dimensional convolution inside another fast algorithm for a one-dimensional convolution.

Of course, everything that was said for linear convolution applies also to cyclic convolution. Indeed, the methods may be applied to compute a polynomial product modulo any polynomial $m(x)$.

For example, to compute a two-dimensional four by four cyclic convolution, we use the four-point cyclic convolution algorithm in the form

$$s = C[(Bg) \cdot (Ad)],$$

as given in Figure 5.13. This algorithm uses five multiplications and fifteen additions. Hence, when used for a cyclic convolution of polynomials, it requires five polynomial multiplications and 15 polynomial additions. Each polynomial multiplication is itself a cyclic convolution that can be computed with the same algorithm, using five real multiplications and 15 real additions. Each polynomial addition requires four real additions. Hence the four by four cyclic convolution uses 25 real multiplications and 135 real additions. This procedure is shown in Figure 11.1 for an n' by n'' cyclic

11.1 Nested convolution algorithms

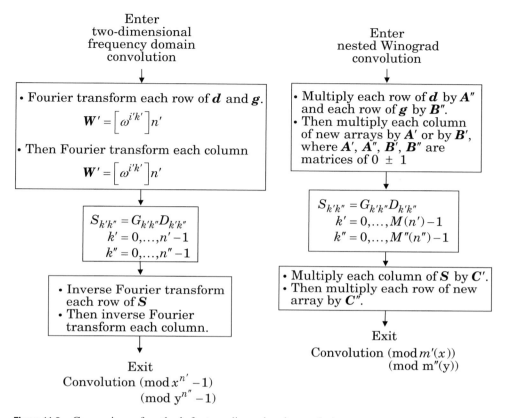

Figure 11.2 Comparison of methods for two-dimensional convolution

convolution. The subroutine, as presented, must know the index k'' and have access to a precomputed array G with elements $G_{k'k''}$.

We can think of this composite algorithm integrated into a single algorithm by using a sequence of operations on the two-dimensional arrays. First, multiply every row of d by the matrix A'' and every row of g by the matrix B''. Then multiply every column of the first new array by A' and every column of the second new array by B'. This produces two $M(n')$ by $M(n'')$ arrays. Multiply their elements componentwise. Then collapse the resulting $M(n')$ by $M(n'')$ array to the n' by n'' output array s by first multiplying every column by C', then multiplying every row by C''.

In Figure 11.2 this procedure is compared to the use of a two-dimensional Fourier transform for cyclic convolution. Clearly, this procedure is a generalization of the use of a two-dimensional Fourier transform. Now the "transforms" use only additions, but the intermediate array is of size $M(n')$ by $M(n'')$, rather than n' by n''.

The nested algorithm is more general in another way; it will compute a polynomial convolution modulo any two polynomials $m'(x)$ and $m''(y)$. The Fourier transform method only computes products modulo $x^{n'} - 1$, $y^{n''} - 1$.

The performance of the nested algorithm is given by the number of multiplications $M(n' \times n'')$ and the number of additions $A(n' \times n'')$ needed for an n' by n'' convolution. The total number of multiplications is easily seen to be given by

$$M(n' \times n'') = M(n')M(n'').$$

Similarly, by looking at Figure 11.1, it is easy to see that

$$A(n' \times n'') = n'A(n'') + M(n'')A(n').$$

The total number of multiplications does not depend on which factor is called n' and which is called n'', but the total number of additions does. Therefore one should check both cases. For example, a seven by nine two-dimensional cyclic convolution can use one-dimensional algorithms with performance given by

$M(7) = 16, \qquad A(7) = 70,$
$M(9) = 19, \qquad A(9) = 74.$

The total number of multiplications is 304. The total number of additions is either 1848 or 1814, depending on which dimension is called n'.

11.2 The Agarwal–Cooley convolution algorithm

The *Agarwal–Cooley convolution algorithm* is a method of breaking an $n'n''$-point one-dimensional cyclic convolution into an n' by n'' two-dimensional cyclic convolution, provided that n' and n'' are coprime. It can be used to combine an n'-point Winograd cyclic convolution algorithm having $M(n')$ multiplications, and an n''-point Winograd cyclic convolution algorithm having $M(n'')$ multiplications to get an $n'n''$-point cyclic convolution algorithm having $M(n')M(n'')$ multiplications.

The Agarwal–Cooley convolution algorithm breaks a one-dimensional cyclic convolution into a multidimensional cyclic convolution by using the Chinese remainder theorem for integers. It is different from the method of using the Chinese remainder theorem for polynomials in the Winograd convolution algorithm. It does not reduce the number of multiplications to as low a level as did the Winograd algorithm. In compensation, it does not have any tendency for the number of additions to get out of hand for large n as did the Winograd algorithm. Further, it is a simpler structure. When n is large, it is more manageable because it can be broken into subroutines. Good convolution algorithms will use the Agarwal–Cooley convolution algorithm to break a long cyclic convolution into short cyclic convolutions, then will use the Winograd convolution algorithm to do the short cyclic convolutions efficiently.

The Agarwal–Cooley algorithm is built from the following three ideas. First, the Chinese remainder theorem for integers maps the one-dimensional cyclic convolution into a multidimensional cyclic convolution. Then, the Winograd cyclic convolution

11.2 The Agarwal–Cooley convolution algorithm

algorithms are used on each component of the multidimensional cyclic convolution. Finally, the Kronecker product theorem is used to bind together the various matrices.

Given d_i and g_i for $i = 0, \ldots, n-1$, we want to compute the cyclic convolution

$$s_i = \sum_{k=0}^{n-1} g_{((i-k))} d_k, \qquad i = 0, \ldots, n-1,$$

where the double parentheses on the indices designate modulo n.

We will turn this one-dimensional convolution into a two-dimensional convolution. By using the Chinese remainder theorem, we will map the one-dimensional data vectors at the input into two-dimensional arrays and also map the two-dimensional array at the output back into a one-dimensional vector. Replace the indices i and k by double indices (i', i'') and (k', k''), given by

$$i' = i \pmod{n'},$$
$$i'' = i \pmod{n''},$$

and

$$k' = k \pmod{n'},$$
$$k'' = k \pmod{n''}.$$

We have already seen in the discussion of the Chinese remainder theorem how the original indices can be recovered from the new indices. The original indices are given by

$$i = N''n''i' + N'n'i'' \pmod{n}, \qquad \begin{array}{l} i' = 0, \ldots, n'-1, \\ i'' = 0, \ldots, n''-1, \end{array}$$

$$k = N''n''k' + N'n'k'' \pmod{n}, \qquad \begin{array}{l} k' = 0, \ldots, n'-1, \\ k'' = 0, \ldots, n''-1, \end{array}$$

where N' and N'' are those integers that satisfy

$$N'n' + N''n'' = 1.$$

This is the same mapping between components of a vector and elements of a two-dimensional array as was used in the Good–Thomas FFT algorithm. The mapping is illustrated in Figure 11.3.

The convolution

$$s_i = \sum_{k=0}^{n-1} g_{((i-k))} d_k$$

can be written

$$s_{N''n''i'+N'n'i''} = \sum_{k'=0}^{n'-1} \sum_{k''=0}^{n''-1} g_{N''n''(i'-k')+N'n'(i''-k'')} d_{N''n''k'+N'n'k''}.$$

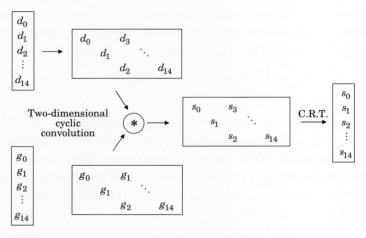

Figure 11.3 Illustration of Agarwal–Cooley algorithm

The double summation on k' and k'' is equivalent to the single summation on k because it picks up the same terms. Now define the two-dimensional variables, also called d, g, and s, given by

$$d_{k'k''} = d_{N''n''k'+N'n'k''}, \quad \begin{array}{l} k' = 0, \ldots, n'-1, \\ k'' = 0, \ldots, n''-1, \end{array}$$

$$g_{k'k''} = g_{N''n''k'+N'n'k''}, \quad \begin{array}{l} k' = 0, \ldots, n'-1, \\ k'' = 0, \ldots, n''-1, \end{array}$$

$$s_{k'k''} = s_{N''n''k'+N'n'k''}, \quad \begin{array}{l} k' = 0, \ldots, n'-1, \\ k'' = 0, \ldots, n''-1, \end{array}$$

so that the convolution now becomes

$$s_{i'i''} = \sum_{k'=0}^{n'-1} \sum_{k''=0}^{n''-1} g_{((i'-k'))((i''-k''))} d_{k'k''},$$

where the first and second indices are interpreted modulo n' and modulo n'', respectively. This is now a two-dimensional cyclic convolution. However, there is not yet any improvement in computational complexity because, for every pair of indices i' and i'', every $d_{k'k''}$ multiplies something. Hence there are still $(n'n'')^2$ multiplications.

For the algorithm to be useful, we must use a fast algorithm for the two-dimensional cyclic convolutions, as studied in the previous section. The Agarwal–Cooley convolution algorithm can be summarized as a map from a one-dimensional array into a multidimensional array, followed by a multidimensional fast convolution algorithm, followed by a map from the multidimensional array back into a one-dimensional array.

The performance of the n-point Agarwal–Cooley algorithm, which is tabulated in Table 11.1, is determined by the equations

$$A(n) = n'A(n'') + M(n'')A(n'),$$
$$M(n) = M(n')M(n'')$$

for the number of additions and multiplications under nesting.

11.2 The Agarwal–Cooley convolution algorithm

Table 11.1 *Performance of the Agarwal–Cooley convolutional algorithm*

Blocklength n	Number of real multiplications $M(n)$	Number of real additions $A(n)$	Real multiplications per point $M(n)/n$	Real additions per point $A(n)/n$
18	38	184	2.11	10.22
20	50	230	2.50	11.50
24	56	272	2.33	11.33
30	80	418	2.67	13.93
36	95	505	2.64	14.03
48	132	900	2.75	18.75
60	200	1120	3.33	18.67
72	266	1450	3.69	20.14
84	320	2100	3.81	25.00
120	560	3096	4.67	25.80
180	950	5470	5.28	30.39
210	1280	7958	6.10	37.90
240	1056	10 176	4.40	42.40
360	2280	14 748	6.33	40.97
420	3200	20 420	7.62	48.62
504	3648	26 304	7.24	52.19
840	7680	52 788	9.14	62.84
1008	10 032	71 265	9.95	70.70
1260	12 160	95 744	9.65	75.99
2520	29 184	241 680	11.58	95.90

One consequence of these equations is that in determining the number of additions in the large algorithm, the number of multiplications in the small algorithms is much more important than the number of additions. For example, suppose we want to form a 504-point cyclic convolution by combining seven-point, nine-point, and eight-point cyclic convolutions. We start with

$M(7) = 16,$ $A(7) = 70,$
$M(9) = 19,$ $A(9) = 74,$
$M(8) = 14,$ $A(8) = 46,$

or

$M(8) = 12,$ $A(8) = 72.$

It seems that the eight-point algorithm with 14 multiplications is better because it trades two more multiplications for 26 fewer additions. However, when these are used to build a 504-point algorithm, the results are surprising:

$M(504) = 4256,$ $A(504) = 28240,$

or

$M(504) = 3648,$ $A(504) = 26304.$

The eight-point algorithm with 12 multiplications leads to a 504-point algorithm with both fewer additions and fewer multiplications.

Recall from Section 5.6 that the Winograd convolution algorithms were written as

$$s = CGAd,$$

where A is an $M(n)$ by n matrix, G is an $M(n)$ by $M(n)$ matrix, and C is an n by $M(n)$ matrix. The computation will proceed as follows. First, each column of the two-dimensional array is multiplied by the matrix A''. Then each column of the resulting array is multiplied componentwise by the vector of constants that are given by the diagonal elements of G''. Then the polynomial products, regarded as convolutions, begin. But this step starts by multiplying each row of the two-dimensional array by A'. Then each row is multiplied componentwise by a vector of constants, given by the diagonal elements of G. Alternatively, the componentwise multiplication of the columns by G'' can be deferred until after the multiplication of the rows by A'. Then the elements of G' and G'' can be multiplied to form an array so that the intermediate array itself is multiplied elementwise by an array of constants. To complete the polynomial multiplication, each column is multiplied by C', then each row is multiplied by C''.

These steps can be collapsed into a more compact form by binding the two cyclic convolution algorithms into one cyclic convolution algorithm. Let the two-dimensional input array be stacked by columns into a one-dimensional input array. Similarly, let the columns of the output two-dimensional array be stacked into a one-dimensional output array. Then the algorithm looks like

$$s = (C'' \times C')(G' \times G'')(A' \times A'')d,$$

using the notation of the Kronecker product. Alternatively, one can form the one-dimensional output array by reading the two-dimensional array by rows. Then the algorithm will be in the more symmetric form:

$$s = (C' \times C'')(G' \times G'')(A' \times A'')d,$$

where this s has the same elements as before but arranged in a different order. In either case, by defining new matrices in the obvious way, this can be written

$$s = CGAd.$$

There is one last detail to take care of: d and s are not in their natural order. They were scrambled by reading them into a two-dimensional array down the extended diagonal and then stacking columns (or rows). To put them back in their natural order, permute the columns of the matrix A appropriately, then permute the rows of the matrix C.

As an example, we will construct a 12-point cyclic convolution algorithm. We use the three-point cyclic convolution algorithm, given by

$$\begin{bmatrix} s_0 \\ s_1 \\ s_2 \end{bmatrix} = \begin{bmatrix} 1 & 1 & 0 & -1 \\ 1 & -1 & -1 & 2 \\ 1 & 0 & 1 & -1 \end{bmatrix} \begin{bmatrix} G_0'' & & & \\ & G_1'' & & \\ & & G_2'' & \\ & & & G_3'' \end{bmatrix} \begin{bmatrix} 1 & 1 & 1 \\ 1 & 0 & -1 \\ 0 & 1 & -1 \\ 1 & 1 & -2 \end{bmatrix} \begin{bmatrix} d_0 \\ d_1 \\ d_2 \end{bmatrix},$$

11.2 The Agarwal–Cooley convolution algorithm

where

$$\begin{bmatrix} G_0'' \\ G_1'' \\ G_2'' \\ G_3'' \end{bmatrix} = \begin{bmatrix} \frac{1}{3} & \frac{1}{3} & \frac{1}{3} \\ 1 & 0 & -1 \\ 0 & 1 & -1 \\ \frac{1}{3} & \frac{1}{3} & -\frac{2}{3} \end{bmatrix} \begin{bmatrix} g_0'' \\ g_1'' \\ g_2'' \end{bmatrix},$$

and the four-point cyclic convolution algorithm, given by

$$\begin{bmatrix} s_0 \\ s_1 \\ s_2 \\ s_3 \end{bmatrix} = \begin{bmatrix} 1 & 1 & 0 & 0 & -1 \\ 1 & -1 & 1 & -1 & 0 \\ 1 & 1 & -1 & 0 & 1 \\ 1 & -1 & -1 & 1 & 0 \end{bmatrix} \begin{bmatrix} G_0' \\ G_1' \\ G_2' \\ G_3' \\ G_4' \end{bmatrix}$$

$$\times \begin{bmatrix} 1 & 1 & 1 & 1 \\ 1 & -1 & 1 & -1 \\ 1 & 1 & -1 & -1 \\ 1 & 0 & -1 & 0 \\ 0 & 1 & 0 & -1 \end{bmatrix} \begin{bmatrix} d_0 \\ d_1 \\ d_2 \\ d_3 \end{bmatrix},$$

where

$$\begin{bmatrix} G_0' \\ G_1' \\ G_2' \\ G_3' \\ G_4' \end{bmatrix} = \begin{bmatrix} \frac{1}{4} & \frac{1}{4} & \frac{1}{4} & \frac{1}{4} \\ \frac{1}{4} & -\frac{1}{4} & \frac{1}{4} & -\frac{1}{4} \\ \frac{1}{2} & 0 & -\frac{1}{2} & 0 \\ \frac{1}{2} & -\frac{1}{2} & -\frac{1}{2} & \frac{1}{2} \\ \frac{1}{2} & \frac{1}{2} & -\frac{1}{2} & -\frac{1}{2} \end{bmatrix} \begin{bmatrix} g_0' \\ g_1' \\ g_2' \\ g_3' \end{bmatrix}.$$

First, write the 12 coefficients of the filter g in a three by four array by writing the 12 coefficients down the extended diagonal, and then stack columns

$$\begin{bmatrix} g_0 \\ g_1 \\ g_2 \\ \vdots \\ g_{11} \end{bmatrix} \rightarrow \begin{bmatrix} g_0 & g_9 & g_6 & g_3 \\ g_4 & g_1 & g_{10} & g_7 \\ g_8 & g_5 & g_2 & g_{11} \end{bmatrix} \rightarrow \begin{bmatrix} g_0 \\ g_4 \\ g_8 \\ g_9 \\ g_1 \\ g_5 \\ g_6 \\ g_{10} \\ g_2 \\ g_3 \\ g_7 \\ g_{11} \end{bmatrix}.$$

Next, express the 20 coefficients of the vector $\mathbf{G}$ as a matrix multiplying the vector $\mathbf{g}$. The matrix is the Kronecker product of two component matrices as follows:

$$\begin{bmatrix} G_0 \\ G_1 \\ G_2 \\ \vdots \\ G_{19} \end{bmatrix} = \left\{ \begin{bmatrix} \frac{1}{4} & \frac{1}{4} & \frac{1}{4} & \frac{1}{4} \\ \frac{1}{4} & -\frac{1}{4} & \frac{1}{4} & -\frac{1}{4} \\ \frac{1}{2} & 0 & -\frac{1}{2} & 0 \\ \frac{1}{2} & -\frac{1}{2} & -\frac{1}{2} & \frac{1}{2} \\ \frac{1}{2} & \frac{1}{2} & -\frac{1}{2} & -\frac{2}{2} \end{bmatrix} \times \begin{bmatrix} \frac{1}{3} & \frac{1}{3} & \frac{1}{3} \\ 1 & 0 & -1 \\ 0 & 1 & -1 \\ \frac{1}{3} & \frac{1}{3} & -\frac{2}{3} \end{bmatrix} \right\} \begin{bmatrix} g_0 \\ g_4 \\ g_8 \\ g_9 \\ \vdots \\ g_{11} \end{bmatrix}.$$

The same construction is used to convert the 12 data coefficients of $\mathbf{d}$ into the 20 components of $\mathbf{D}$. We write the Kronecker product explicitly:

$$\begin{bmatrix} D_0 \\ D_1 \\ D_2 \\ D_3 \\ \vdots \\ D_{19} \end{bmatrix} = \begin{bmatrix} 1 & 1 & 1 & 1 & 1 & 1 & 1 & 1 & 1 & 1 & 1 & 1 \\ 1 & 0 & -1 & 1 & 0 & -1 & 1 & 0 & -1 & 1 & 0 & -1 \\ 0 & 1 & -1 & 0 & 1 & -1 & 0 & 1 & -1 & 0 & 1 & -1 \\ 1 & 1 & -2 & 1 & 1 & -2 & 1 & 1 & -2 & 1 & 1 & -2 \\ 1 & 1 & 1 & -1 & -1 & -1 & 1 & 1 & 1 & -1 & -1 & -1 \\ 1 & 0 & -1 & -1 & 0 & 1 & 1 & 0 & -1 & -1 & 0 & 1 \\ 0 & 1 & -1 & 0 & -1 & 1 & 0 & 1 & -1 & 0 & -1 & 1 \\ 1 & 1 & -2 & -1 & -1 & 2 & 1 & 1 & -2 & -1 & -1 & 2 \\ 1 & 1 & 1 & 1 & 1 & 1 & -1 & -1 & -1 & -1 & -1 & -1 \\ 1 & 0 & -1 & 1 & 0 & -1 & -1 & 0 & 1 & -1 & 0 & 1 \\ 0 & 1 & -1 & 0 & 1 & -1 & 0 & -1 & 1 & 0 & -1 & 1 \\ 1 & 1 & -2 & 1 & 1 & -2 & -1 & -1 & 2 & -1 & -1 & 2 \\ 1 & 1 & 1 & 0 & 0 & 0 & -1 & -1 & -1 & 0 & 0 & 0 \\ 1 & 0 & -1 & 0 & 0 & 0 & -1 & 0 & 1 & 0 & 0 & 0 \\ 0 & 1 & -1 & 0 & 0 & 0 & 0 & -1 & 1 & 0 & 0 & 0 \\ 1 & 1 & -2 & 0 & 0 & 0 & -1 & -1 & 2 & 0 & 0 & 0 \\ 0 & 0 & 0 & 1 & 1 & 1 & 0 & 0 & 0 & -1 & -1 & -1 \\ 0 & 0 & 0 & 1 & 0 & -1 & 0 & 0 & 0 & -1 & 0 & 1 \\ 0 & 0 & 0 & 0 & 1 & -1 & 0 & 0 & 0 & 0 & -1 & 1 \\ 0 & 0 & 0 & 1 & 1 & -2 & 0 & 0 & 0 & -1 & -1 & 2 \end{bmatrix} \begin{bmatrix} d_0 \\ d_4 \\ d_8 \\ d_9 \\ d_1 \\ d_5 \\ d_6 \\ d_{10} \\ d_2 \\ d_3 \\ d_7 \\ d_{11} \end{bmatrix}.$$

Finally, permute columns of the matrix and the components of the vector to rewrite this as

$$\begin{bmatrix} D_0 \\ D_1 \\ D_2 \\ D_3 \\ \vdots \\ D_{18} \\ D_{19} \end{bmatrix} = \begin{bmatrix} 1 & 1 & 1 & 1 & 1 & 1 & 1 & 1 & 1 & 1 & 1 & 1 \\ 1 & 0 & -1 & 1 & 0 & -1 & 1 & 0 & -1 & 1 & 0 & -1 \\ 0 & 1 & -1 & 0 & 1 & -1 & 0 & 1 & -1 & 0 & 1 & -1 \\ 1 & 1 & -2 & 1 & 1 & -2 & 1 & 1 & -2 & 1 & 1 & -2 \\ 1 & -1 & 1 & -1 & 1 & -1 & 1 & -1 & 1 & -1 & 1 & -1 \\ 1 & 0 & -1 & -1 & 0 & 1 & 1 & 0 & -1 & -1 & 0 & 1 \\ 0 & -1 & -1 & 0 & 1 & 1 & 0 & -1 & -1 & 0 & 1 & 1 \\ 1 & -1 & -2 & -1 & 1 & 2 & 1 & -1 & -2 & -1 & 1 & 2 \\ 1 & 1 & -1 & -1 & 1 & 1 & -1 & -1 & 1 & 1 & -1 & -1 \\ 1 & 0 & 1 & -1 & 0 & -1 & -1 & 0 & -1 & 1 & 0 & 1 \\ 0 & 1 & 1 & 0 & 1 & -1 & 0 & -1 & -1 & 0 & -1 & 1 \\ 1 & 1 & 2 & -1 & 1 & -2 & -1 & -1 & -2 & 1 & -1 & 2 \\ 1 & 0 & -1 & 0 & 1 & 0 & -1 & 0 & 1 & 0 & -1 & 0 \\ 1 & 0 & 1 & 0 & 0 & 0 & -1 & 0 & -1 & 0 & 0 & 0 \\ 0 & 0 & 1 & 0 & 1 & 0 & 0 & 0 & -1 & 0 & -1 & 0 \\ 1 & 0 & 2 & 0 & 1 & 0 & -1 & 0 & -2 & 0 & -1 & 0 \\ 0 & 1 & 0 & -1 & 0 & 1 & 0 & -1 & 0 & 1 & 0 & -1 \\ 0 & 0 & 0 & -1 & 0 & -1 & 0 & 0 & 0 & 1 & 0 & 1 \\ 0 & 1 & 0 & 0 & 0 & -1 & 0 & -1 & 0 & 0 & 0 & 1 \\ 0 & 1 & 0 & -1 & 0 & -2 & 0 & -1 & 0 & 1 & 0 & 2 \end{bmatrix} \begin{bmatrix} d_0 \\ d_1 \\ d_2 \\ d_3 \\ d_4 \\ \vdots \\ d_{11} \end{bmatrix}.$$

This defines the computation

D = Ad.

By following through a similar development, we can also obtain the matrix **C**. Hence we have reproduced the standard form

s = CGAd

for a 12-point cyclic convolution. With this form, the scrambling operations have been absorbed into the matrices of preadditions and postadditions.

11.3 Splitting algorithms

We have seen how convolution algorithms can be nested to form a multidimensional convolution algorithm. To compute the two-dimensional cyclic convolution,

$$s(x, y) = g(x, y)d(x, y) \pmod{x^{n'} - 1}\pmod{y^{n''} - 1},$$

we nest algorithms that compute

$$s(x) = g(x)d(x) \pmod{x^{n'} - 1}$$

and

$$s(y) = g(y)d(y) \pmod{y^{n''} - 1}.$$

More generally, to compute

$$s(x, y) = g(x, y)d(x, y) \pmod{p(x)}\pmod{q(y)},$$

where $p(x)$ and $q(y)$ are polynomials of degree n' and n'', respectively, we nest algorithms for the two simpler one-dimensional problems

$$s(x) = g(x)d(x) \pmod{p(x)}$$

and

$$s(y) = g(y)d(y) \pmod{q(y)}.$$

If the number of multiplications used by the component problems are $M(n')$ and $M(n'')$, respectively, then the two-dimensional polynomial product will use $M(n')M(n'')$ multiplications.

The one-dimensional algorithms were designed by using the Chinese remainder theorem on each dimension individually. This suggests one more possibility to explore. One can use the Chinese remainder theorem instead on the level of the two-dimensional problem to break the two-dimensional convolution into smaller two-dimensional pieces. Suppose

$$p(x) = p^{(0)}(x)p^{(1)}(x),$$

where $p^{(0)}(x)$ and $p^{(1)}(x)$ are coprime. The polynomial product

$$s(x, y) = g(x, y)d(x, y) \pmod{p(x)}\pmod{q(y)}$$

can be broken into

$$s^{(0)}(x, y) = g^{(0)}(x, y)d^{(0)}(x, y) \pmod{p^{(0)}(x)}\pmod{q(y)}$$

and

$$s^{(1)}(x, y) = g^{(1)}(x, y)d^{(1)}(x, y) \pmod{p^{(1)}(x)}\pmod{q(y)}.$$

The pieces can be combined into $s(x, y)$ by using the Chinese remainder theorem. Alternatively, if $q(y) = q^{(0)}(y)q^{(1)}(y)$, then this factorization can be used to break down the problem in the y variable.

11.3 Splitting algorithms

Both of these can be used at once to break the problem into four pieces:

$$s^{(0,0)}(x,y) = g^{(0,0)}(x,y)d^{(0,0)}(x,y) \pmod{p^{(0)}(x)}\pmod{q^{(0)}(y)},$$
$$s^{(0,1)}(x,y) = g^{(0,1)}(x,y)d^{(0,1)}(x,y) \pmod{p^{(0)}(x)}\pmod{q^{(1)}(y)},$$
$$s^{(1,0)}(x,y) = g^{(1,0)}(x,y)d^{(1,0)}(x,y) \pmod{p^{(1)}(x)}\pmod{q^{(0)}(y)},$$
$$s^{(1,1)}(x,y) = g^{(1,1)}(x,y)d^{(1,1)}(x,y) \pmod{p^{(1)}(x)}\pmod{q^{(1)}(y)},$$

where

$$g^{(i,j)}(x,y) = g(x,y) \pmod{p^{(i)}(x)}\pmod{q^{(j)}(y)},$$

and so on. The output polynomial $s(x,y)$ can be assembled from the four pieces by using the Chinese remainder theorem for polynomials twice.

Let M_0 and M_1 be the number of multiplications necessary to do a polynomial product modulo $p^{(0)}(x)$ and $p^{(1)}(x)$, respectively, and let M_0' and M_1' be the number of multiplications necessary to do a polynomial product modulo $q^{(0)}(y)$ and $q^{(1)}(y)$, respectively. Then it takes

$$M = (M_0 + M_1)(M_0' + M_1')$$

multiplications to compute

$$s(x,y) = g(x,y)d(x,y) \pmod{p(x)}\pmod{q(y)}$$

directly, while to compute it by using the Chinese remainder theorem on two-dimensional pieces, as above, uses

$$M = M_0 M_0' + M_0 M_1' + M_1 M_0' + M_1 M_1'$$

multiplications, which is the same number, so the number of multiplications is not reduced. Though there is no advantage in the number of multiplications, there may be advantages in the dataflow of the computation and in the number of additions.

There are many ways in which a two-dimensional convolution can be split by using the Chinese remainder theorem. Table 11.2 shows examples of the performance that can be obtained when the Agarwal–Cooley algorithm has been modified by splitting it differently. Table 11.2 should be compared with Table 11.1.

As an example, we will outline the development of a 20-point cyclic convolution algorithm. This one-dimensional cyclic convolution is first mapped into a four by five two-dimensional cyclic convolution

$$s(x,y) = g(x,y)d(x,y) \pmod{x^4 - 1}\pmod{y^5 - 1}.$$

We choose to split this product as

$$s^{(0)}(x,y) = g^{(0)}(x,y)d^{(0)}(x,y) \pmod{x^4 - 1}\pmod{y - 1},$$
$$s^{(1)}(x,y) = g^{(1)}(x,y)d^{(1)}(x,y) \pmod{x^4 - 1}\pmod{y^4 + y^3 + y^2 + y + 1},$$

Table 11.2 *Performance of an enhanced Agarwal–Cooley convolution*

Blocklength n	Number of real multiplications $M(n)$	Number of real additions $A(n)$	Real multiplications per point $M(n)/n$	Real additions per point $A(n)/n$
18	38	184	2.11	10.22
20	50	215	2.50	11.75
24	56	244	2.33	11.17
30	80	392	2.67	13.07
36	95	461	2.64	12.81
48	132	840	2.75	17.50
60	200	964	3.33	16.07
72	266	1186	3.69	16.47
84	320	1784	3.81	21.24
120	560	2468	4.67	20.57
180	950	4382	5.28	24.34
210	1280	6458	6.10	30.75
240	1056	9696	4.40	40.40
360	2280	11840	6.33	32.89
420	3200	15256	7.62	36.32
504	3648	21844	7.24	43.34
840	7680	39884	9.14	47.48
1008	10032	56360	9.95	55.91
1260	12160	72268	9.65	57.36
2520	29184	190148	11.58	75.46

but not to split it any further as a two-dimensional problem. To compute $s^{(0)}(x, y)$, we nest an algorithm for a polynomial product modulo $x^4 - 1$ with an algorithm for a polynomial product modulo $y - 1$. To compute $s^{(1)}(x, y)$, we nest an algorithm for a polynomial product modulo $x^4 - 1$ with an algorithm for a polynomial product modulo $y^4 + y^3 + y^2 + y + 1$. Finally, $s^{(0)}(x, y)$ and $s^{(1)}(x, y)$ are combined into $s(x, y)$ by using

$$s(x, y) = a^{(0)}(y)s^{(0)}(x, y) + a^{(1)}(y)s^{(1)}(x, y) \pmod{y^5 - 1},$$

where $a^{(0)}(y)$ and $a^{(1)}(y)$ are obtained from the Chinese remainder theorem for polynomials. They are the same as would be used to combine $s^{(0)}(y)$ and $s^{(1)}(y)$ if the problem were a one-dimensional polynomial product modulo $y^5 - 1$.

We see that the overall computation is made up of the same subcomputations as before, but they are put together a little differently. Now one application of the Chinese remainder theorem has been pulled out to appear after the nesting is complete, and one polynomial reduction has been pulled out to appear before the nesting begins.

11.3 Splitting algorithms

To determine the number of additions, we must look more closely at the number of additions used by the one-dimensional five-point cyclic convolution algorithm. This is

polynomial multiplication modulo $y - 1$:   0 additions,
polynomial multiplication modulo
$$y^4 + y^3 + y^2 + y + 1:$$ 16 additions,
Chinese remainder theorem: 15 additions.

We denote these by $A^{(0)}(5)$, $A^{(1)}(5)$, and $A_{\text{crt}}(5)$, respectively. The modulo $x^4 - 1$ computation uses 15 additions. The number of additions used by the two-dimensional algorithm is then derived by looking at each of the three subcomputations

$$A(20) = (4 \cdot 0 + 1 \cdot 15) + (4 \cdot 15 + 5 \cdot 16) + (15 \cdot 4)$$
$$= 215,$$

which is fewer than the 230 additions that are needed by the pure form of the Agarwal–Cooley algorithm.

We emphasize that there is no noticeable increase in the organizational complexity of the computation in using the split-nesting algorithms. All that happens is that the computation jumps between steps that perform additions along rows of the two-dimensional array and steps that perform additions along columns. We can think of this simply as a change in the sequence in which subroutines are called.

To reduce the number of multiplications by using the Chinese remainder theorem requires one more innovation, which is going into an extension field to split the problem. We shall allow coprime factorizations of $q(y)$ that use the indeterminate x in the coefficients of the factor polynomials. (Formally stated, we allow factorizations of $q(y)$ in the extension field generated by the indeterminate x.) This means that we can find smaller factors of $q(y)$ than we could find previously when studying one-dimensional problems. This method to find better algorithms is not available in one dimension. In two dimensions, the symbol x becomes mixed up in the y convolutions, so when viewed by itself the y convolution becomes more complicated. When it is combined with the x convolution, however, the symbol x, arising in the factorization of $q(y)$, can be combined algebraically with the symbol x of the x convolution. This means that the two-dimensional algorithm is made simpler.

We will explain the method concretely with an example. Let

$$s(x, y) = g(x, y)d(x, y) \pmod{x^4 - 1}\pmod{y^4 - 1}.$$

This is a two-dimensional cyclic convolution. First, replace it with two polynomial products,

$$s^{(0)}(x, y) = g^{(0)}(x, y)d^{(0)}(x, y) \pmod{x^2 - 1}\pmod{y^4 - 1}$$

and

$$s^{(1)}(x, y) = g^{(1)}(x, y)d^{(1)}(x, y) \pmod{x^2 + 1}\pmod{y^4 - 1}.$$

Solve the first subproblem as already explained. This takes ten multiplications when using a four-point and a two-point cyclic convolution algorithm with five multiplications and two multiplications, respectively.

The second subproblem would require 15 multiplications when using the best possible algorithm for the one-dimensional polynomial product modulo $y^4 - 1$, and the best possible polynomial product for the two-dimensional polynomial product modulo $x^2 + 1$. Then the total number of multiplications is 25, the same as would be obtained by using a four-point cyclic convolution algorithm on both axes.

Instead, choose the factorization

$$y^4 - 1 = (y - 1)(y + 1)(y - x)(y + x) \pmod{x^2 + 1}.$$

This works because modulo $x^2 + 1$, $x^2 = -1$. Now one can build an algorithm modulo $y^4 - 1$ with four multiplications. Each multiplication is a multiplication of first-degree polynomials in x modulo $x^2 + 1$. But the computation requires the multiplication of first-degree polynomials anyway. By using the factorization $y^2 + 1 = (y - x)(y + x)$, the y dependence is partially deflected into x dependence, and the x dependence is absorbed into work that is needed anyway.

In this way, one can compute $s^{(1)}(x, y)$ using 12 multiplications. Hence the original four-point by four-point cyclic convolution is computed with 22 multiplications.

One can press this method even further, introducing even more symbols as zeros of $y^n + 1$. When pressed to the limit, this amounts to computing the Fourier transform in a polynomial representation of an extension field. We shall drop the topic for now, but we shall see it later in a different form, after we study the polynomial representations of an extension field in Section 11.5.

11.4 Iterated algorithms

Construction of convolution algorithms by the method of iteration depends on the fact that the small convolution algorithms over the field F, despite the setting in which they were developed, are actually valid identities in any ring containing F. The final algorithms do not assume that multiplication is commutative, nor do they involve division except for some small scalars. Only addition, subtraction, and multiplication are used for the input variables.

First, consider the one-dimensional four by four linear convolution

$$s(x) = g(x)d(x),$$

11.4 Iterated algorithms

where $g(x)$ and $d(x)$ each has degree three. Write them parenthesized as follows:

$g(x) = (g_3 x + g_2)x^2 + (g_1 x + g_0),$
$d(x) = (d_3 x + d_2)x^2 + (d_1 x + d_0).$

Define the following polynomials in two variables:

$g(y, z) = (g_3 y + g_2)z + (g_1 y + g_0),$
$d(y, z) = (d_3 y + d_2)z + (g_1 y + g_0),$

and

$s(y, z) = g(y, z)d(y, z).$

Then the desired convolution is obtained from $s(y, z)$ by

$s(x) = s(x, x^2).$

We can rewrite the computation in abbreviated form as

$$s_2(y)z^2 + s_1(y)z + s_0(y) = (g_1(y)z + g_0(y))(d_1(y)z + d_0(y)),$$

where now all of the coefficients are actually polynomials in y of degree one.

A two by two linear convolution algorithm is

$$\begin{bmatrix} s_0 \\ s_1 \\ s_2 \end{bmatrix} = \begin{bmatrix} 1 & 0 & 0 \\ -1 & 1 & -1 \\ 0 & 0 & 1 \end{bmatrix} \begin{bmatrix} g_0 \\ g_0 + g_1 \\ g_1 \end{bmatrix} \begin{bmatrix} 1 & 0 \\ 1 & 1 \\ 0 & 1 \end{bmatrix} \begin{bmatrix} d_0 \\ d_1 \end{bmatrix}.$$

This identity holds even if the input variables are from a ring of polynomials. Then

$$\begin{bmatrix} s_0(y) \\ s_1(y) \\ s_2(y) \end{bmatrix} = \begin{bmatrix} 1 & 0 & 0 \\ -1 & 1 & -1 \\ 0 & 0 & 1 \end{bmatrix} \begin{bmatrix} g_0(y) \\ g_0(y) + g_1(y) \\ g_1(y) \end{bmatrix} \begin{bmatrix} 1 & 0 \\ 1 & 1 \\ 0 & 1 \end{bmatrix} \begin{bmatrix} d_0(y) \\ d_1(y) \end{bmatrix}.$$

There are three polynomial multiplications and three polynomial additions, excluding the additions involving the coefficients of $g(x)$. Each polynomial product, in turn, can use the two by two linear convolution algorithm; each requires three multiplications and three additions, so there is a total of nine multiplications in the iterated four by four linear convolution algorithm. This is poorer than the optimum algorithm, which uses seven multiplications. The advantage is in the number of additions. Excluding those additions involving only the coefficients of $g(x)$, the total number of additions used is 15.

The resulting four by four convolution algorithm itself can be iterated to obtain a 16 by 16 convolution algorithm with 81 multiplications and 195 additions, which is 5.06 multiplications and 12.19 additions per output sample. This can be iterated again to obtain a 256 by 256 convolution algorithm with 6561 multiplications and 18 915 additions, which is 25.63 multiplications and 73.89 additions per output sample.

In general, iteration can be used to compute

$$s(x) = g(x)d(x)$$

whenever the number of coefficients in $g(x)$ and the number of coefficients in $d(x)$ have a common factor. Let

$\deg d(x) = MN - 1,$
$\deg g(x) = ML - 1.$

Convert the polynomial $d(x)$ into a two-dimensional polynomial, $d(y, z)$, defining new indices ℓ and k by

$i = M\ell + k, \quad \ell = 0, \ldots, N - 1,$
$ k = 0, \ldots, M - 1,$

and

$$d(y, z) = \sum_{\ell=0}^{N-1} \left(\sum_{k=0}^{M-1} d_{M\ell+k} y^k \right) z^\ell.$$

Similarly, convert the polynomial $g(x)$ into a two-dimensional polynomial $g(y, z)$ by using new indices ℓ and k, satisfying

$i = M\ell + k, \quad \ell = 0, \ldots, L - 1,$
$ k = 0, \ldots, M - 1,$

and

$$g(y, z) = \sum_{\ell=0}^{L-1} \left(\sum_{k=0}^{M-1} g_{M\ell+k} y^k \right) z^\ell.$$

Compute

$$s(y, z) = g(y, z)d(y, z).$$

Then $s(x)$ can be obtained by

$$s(x) = s(x, x^M),$$

which requires only additions. The two-dimensional convolution uses an L by N linear convolution and also an M by M linear convolution. The total number of multiplications is the product of the number of multiplications required by each of the two small algorithms.

Of course, one of the two component algorithms might itself have been constructed by iteration. In this way, the iteration procedure can be repeated any number of times to build large convolution algorithms. Further, in cases in which the number of coefficients in $d(x)$ and $g(x)$ does not have a common factor, it is easy to pad one of them with dummy coefficients, equal to zero, so that iteration then can still be used.

11.4 Iterated algorithms

Iteration can also be used to compute the polynomial product

$$s(x) = g(x)d(x) \pmod{m(x)}.$$

One way is to compute the linear convolution first and then to reduce modulo $m(x)$. One can usually do better than this by trying harder.

We shall consider only the case in which $m(x)$ equals $x^n + 1$ and n is a power of two. This is an important case because

$$x^{2n} - 1 = (x^n - 1)(x^n + 1),$$

so products modulo $x^n + 1$ occur whenever one computes a $2n$-point cyclic convolution using the Chinese remainder theorem.

To compute

$$s(x) = g(x)d(x) \pmod{x^n + 1},$$

replace the one-dimensional polynomial

$$d(x) = \sum_{i=0}^{n-1} d_i x^i$$

by the two-dimensional polynomial

$$d(y, z) = \sum_{i''=0}^{n''-1} \sum_{i'=0}^{n'-1} d_{i'+n'i''} y^{i'} z^{i''},$$

where $n = n'n''$. The original polynomial can be recovered by setting y equal to x and z equal to $x^{n'}$. Similarly, define $g(y, z)$, and let

$$s(y, z) = g(y, z)d(y, z) \pmod{z^{n''} + 1}.$$

We only need to be concerned with computing $s(y, z)$ because then $s(x)$ can be obtained as

$$s(x) = s(x, x^{n'}) \pmod{x^n + 1}$$

with no multiplications.

Consider the bivariate convolution restated as the linear convolution

$$s(y) = g(y)d(y),$$

where the coefficients of the polynomials $g(y)$, $d(y)$, and $s(y)$ are themselves polynomials in z, and products of polynomials in z are understood to be products modulo $z^{n''} + 1$.

It may be that this procedure will be easier to understand if we take the case in which $n'' = 2$ and z is replaced by j. Then we see that we have replaced a product of real polynomials modulo $x^n + 1$ by a product of complex polynomials modulo $x^{n/2} + 1$. The advantage of this form is that the Cook–Toom algorithm can now be used with zeros at $\pm j$. In the general case, we may think of z as representing an n''th root of -1.

We shall use the Cook–Toom algorithm, but allowing "zeros" to occur at powers of z. (The formal statement is that we are choosing zeros in the extension field generated by z.) Choosing powers of z as zeros means that under modulo $y - z^\ell$ reduction, the coefficients of the polynomials $g(y)$ and $d(y)$ will become polynomials in z. The reason this does not hurt here is that they started out as polynomials in z anyway, so the complexity does not grow. If we allow the degrees of these polynomials in z to grow beyond what they started with, however, there will be an increase in complexity. This imposes an inequality relationship between n' and n''. The linear convolution can be computed with the Cook–Toom algorithm by choosing the polynomial with zeros at $\pm z^\ell$ for ℓ smaller than n''. Thus

$$m(y) = y(y - \infty) \prod_{\ell=0}^{n''-1} (y - z^\ell)(y + z^\ell).$$

This is a polynomial of degree $2n'' + 2$, so it can be used if $s(y)$ has degree at most $2n'' + 1$. If $s(y)$ has a smaller degree, some of the factors in $m(y)$ can be discarded. We shall always use a polynomial $m(y)$ satisfying $\deg m(y) = \deg s(y) + 1 = 2n' - 1$.

Because the factors of $m(y)$ are all of first degree, only one multiplication is needed for each coefficient of $s(y)$. Those multiplications, of course, are actually multiplications of polynomials in z modulo $z^{n''} + 1$.

The iterated algorithm requires that $2n' - 1 < 2n'' + 1$. As long as this condition is satisfied, a polynomial product modulo $x'' + 1$ is broken into $2n' - 1$ polynomial products modulo $x^{n''} + 1$. These, in turn, can be broken into smaller pieces by repeating the same procedure, as shown conceptually in the flow diagram of Figure 11.4. The flow diagram omits a criterion for choosing the factorization $n = n'n''$, and a general rule for the preadditions and postadditions. The number of multiplications and the number of additions satisfy the recursions

$$M(n) = (2n' - 1)M(n''),$$
$$A(n) = (2n' - 1)A(n'') + A_1(n) + A_2(n),$$

where $A_1(n)$ is the number of preadditions needed to reduce $d(y)$ modulo the factors of $m(y)$, and $A_2(n)$ is the number of additions needed to recover $s(y)$ from its residues. The actual values of $A_1(n)$ and $A_2(n)$ will depend on the choice of the $2n' - 1$ factors of $m(y)$.

For example, a polynomial product modulo $x^4 + 1$ can be reduced to three polynomial products modulo $x^2 + 1$, and these polynomial products can each be computed with three multiplications. Hence we will obtain an algorithm that uses nine multiplications. Similarly, a polynomial product modulo $x^{16} + 1$ can be reduced to seven

11.4 Iterated algorithms

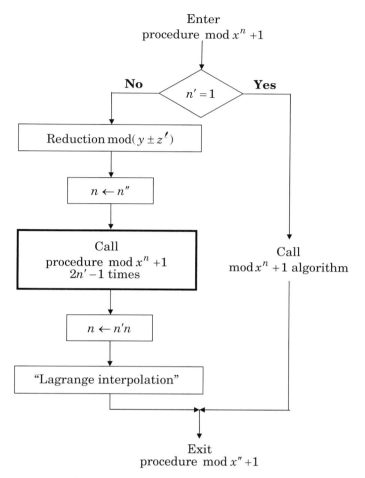

Figure 11.4 Conceptual structure of iterated algorithm

polynomial products modulo $x^4 + 1$, employing a total of 63 multiplications or 49 multiplications, depending on which modulo $x^4 + 1$ algorithm is chosen. The performance of the iterated algorithm for a variety of iteration schemes is tabulated in Table 11.3. The iteration schemes are identified in the last column as n' by n'' iteration schemes, and n'' may be decomposed further.

A slightly different version of the iterated algorithm is obtained if we choose

$$m(x) = \prod_{\ell=0}^{n''-1}(x - z^\ell)(x + z^\ell)$$
$$= \prod_{\ell=0}^{2n''-1}(x - z^\ell) \pmod{z^{n''} + 1}.$$

Table 11.3 *Performance of some iterated algorithms for products modulo $x^n + 1$*

Modulus polynomial	Number of multiplications	Number of additions	Iteration scheme (n' by n'')
$x^2 + 1$	3	3	–
$x^4 + 1$	9	15	2 by 2
	7	41	–
$x^8 + 1$	27	57	2 by (2 by 2)
	21	77	2 by 4
$x^{16} + 1$	63	205	4 by (2 by 2)
	49		4 by 4
$x^{32} + 1$	189	599	4 by (2 by (2 by 2))
	147	739	4 by (2 by 4)
$x^{64} + 1$	405	1599	8 by (2 by (2 by 2))
	315	1899	8 by (2 by 4)
$x^{128} + 1$	945	4563	8 by (4 by (2 by 2))
$x^{256} + 1$	1953	10531	16 by (4 by (2 by 2))
$x^{512} + 1$	5859	26921	16 by (4 by (2 by (2 by 2)))
$x^{1024} + 1$	11907	58889	32 by (4 by (2 by (2 by 2)))
$x^{2048} + 1$	25515	143041	32 by (8 by (2 by (2 by 2)))
$x^{4096} + 1$	51435	304769	64 by (8 by (2 by (2 by 2)))

The polynomial $m(x)$ now has degree $2n''$, although a degree of $2n'' - 1$ would suffice. Hence this choice of $m(x)$ will use one unnecessary polynomial product. It has the advantage that when n' is large, the preadditions and the postadditions can be executed in the form of a radix-two Cooley–Tukey FFT. In fact, we have now pushed the development so far that we are touching on a new topic known as a *polynomial transform*. This is the topic that will be studied in the remainder of the chapter. Thus we shall abandon the line we have been following so that the polynomial transform can be introduced in a more fundamental way.

11.5 Polynomial representation of extension fields

A discrete Fourier transform maps a vector of real numbers into a vector of complex numbers. In practical problems of signal processing, sequences do not consist of arbitrary real numbers. Sequences consist of rational numbers or, perhaps even more simply, of integers. In this section, we shall restrict the domain of the Fourier transform to the set of rationals Q. This restriction really costs nothing from a practical point of view, and yet it leads to radically different computational insights. The Fourier

11.5 Polynomial representation of extension fields

transform of blocklength n then maps vectors of length n of rational numbers into vectors of length n of elements from an extension field. Extension fields of Q were studied in Section 9.7, emphasizing that these fields are subfields of the complex field C. In this section, we shall suppress this fact and regard the elements of the extension field $Q[x]/\langle f(x) \rangle$ simply as polynomials in x over the field Q.

To restate this distinction, recall that the discrete Fourier transform

$$V_k = \sum_{i=0}^{n-1} \omega^{ik} v_i, \quad k = 0, \ldots, n-1$$

with $\omega = e^{-j2\pi/n}$ produces a vector V of complex numbers, where ω is a complex zero of the cyclotomic polynomial of degree n. An arbitrary complex number, however, cannot occur in the transform. Complex numbers can occur only in the subfield known as $Q(\omega)$ or, more simply, as Q^m.

Let $x^n - 1$ be factored into its prime polynomial factors over the rationals:

$$x^n - 1 = p_0(x) p_1(x) \cdots p_S(x).$$

Each of the factors must be a cyclotomic polynomial, $\Phi_n(x)$. When n is small, the cyclotomic polynomial of degree n has coefficients equal only to $-1, 0,$ or $+1$. Because ω is a zero of $x^n - 1$, it is a zero of one of the cyclotomic polynomials, say $\Phi_m(x)$, a polynomial of degree m with the leading coefficient equal to one:

$$p(x) = x^m + p_{m-1} x^{m-1} + \cdots + p_1 x + p_0.$$

Because $p(\omega) = 0$, this gives

$$\omega^m = -p_{m-1} \omega^{m-1} - \cdots - p_1 \omega - p_0.$$

Hence ω can be expressed in terms of lesser powers of ω. For i less than m, ω^i cannot be so expressed because, if it could, ω would be a zero of another polynomial of degree smaller than the degree of $p(x)$.

The field Q^m can be represented as the set of all polynomials in ω with rational coefficients and with degrees at most $m - 1$. Addition is polynomial addition, and multiplication is polynomial multiplication modulo $\Phi_m(x)$. The polynomials are not to be evaluated, but are represented by a list of m coefficients. It takes m words of memory to store one element of Q^m instead of the two words of memory that suffice for the usual representation of complex numbers.

To emphasize that the number representation consists of the polynomials themselves, and not the complex values that the polynomials take on at ω, we may use the variable x in place of ω. Then the numbers are

$$a = a_{m-1} x^{m-1} + a_{m-2} x^{m-2} + \cdots + a_1 x + a_0,$$

as represented by the list of coefficients. Of course, if we wanted to know the "true" complex value of a, we could just substitute ω for x and carry out the indicated calculations. However, our aim is to derive algorithms that use the polynomial representation as intermediate variables, and this form will actually lead to simpler algorithms in some instances.

For example, if n is a power of two, then

$$x^n - 1 = (x^{n/2} + 1)(x^{n/4} + 1) \cdots (x + 1)(x - 1).$$

The cyclotomic polynomial $x^{n/2} + 1$ leads to an extension field whose elements are all the rational-valued polynomials of degree less than $n/2$; addition is polynomial addition; multiplication is polynomial multiplication modulo $x^{n/2} + 1$.

More specifically, the field $\boldsymbol{Q}^8$ consists of all rational-valued polynomials of degree seven or less with polynomial arithmetic modulo $x^8 + 1$. A sample multiplication in the field is

$$\begin{aligned}\left(x^7 - \tfrac{1}{2}x^2 + \tfrac{1}{4}\right)(x^2 - 1) &= x^9 - x^7 - \tfrac{1}{2}x^4 + \tfrac{3}{4}x^2 - \tfrac{1}{4} \\ &= -x^7 - \tfrac{1}{2}x^4 + \tfrac{3}{4}x^2 - x - \tfrac{1}{4}.\end{aligned}$$

The polynomial representation is not limited to extensions of the rationals. One also can so extend the complex rationals. A *complex rational* is a complex number of the form $v = a + jb$, where a and b are rational numbers. The complex rationals form a subfield of the complex numbers, which is sometimes denoted by $\boldsymbol{Q}(j)$. Applications of digital signal processing often must deal with complex-valued vectors. Because of wordlength limitations, these always are vectors of complex rationals (or, even more specifically, of complex integers).

The complex rationals can be extended to contain an nth root of unity. This is the smallest extension field in which the Fourier transform of blocklength n exists. If $\boldsymbol{Q}(\omega)$ contains j (that is, contains a zero of $x^2 + 1$), then the extension field $\boldsymbol{Q}(\omega)$ contains all the complex rationals. It is the extension field we need. This will happen if the cyclotomic polynomial is of the form $x^r + 1$, which occurs only if r is even. Otherwise, the element j must be appended.

Let ω be an nth root of unity, n not a power of two, and let $\Phi_n(x)$ be the cyclotomic polynomial of degree m with ω as a zero. Then the extension field $\boldsymbol{Q}(j, \omega)$ or, more simply, $\boldsymbol{Q}(j)^m$, is the set of polynomials of degree less than m with coefficients in $\boldsymbol{Q}(j)$. Addition is polynomial addition; multiplication is polynomial multiplication modulo $\Phi_n(x)$. Coefficients of the polynomials are added and multiplied as complex numbers.

It requires $2m$ rational numbers to specify one element of $\boldsymbol{Q}(j, \omega)$. With this difference and the more general addition and multiplication of coefficients, everything that we shall discuss in subsequent sections for processing sequences of rationals also holds for processing sequences of complex rationals. Consequently, we shall not again mention the complex rationals.

11.6 Convolution with polynomial transforms

The Fourier transform

$$V_k = \sum_{i=0}^{n-1} \omega^{ik} v_i, \qquad i = 0, \ldots, n-1$$

takes values only in the extension field $Q(\omega)$ if $\boldsymbol{v}$ is a vector over the rationals. More generally, the Fourier transform takes its values only in $Q(\omega)$ if $\boldsymbol{v}$ takes its values only in $Q(\omega)$.

Let m be the degree of the cyclotomic polynomial $\Phi_m(x)$ that has ω as a zero. Then $Q(\omega)$ is represented by the set of polynomials of degree less than m. In the polynomial representation of $Q(\omega)$, the Fourier transform becomes

$$V_k = \sum_{i=0}^{n-1} x^{ik} v_i \pmod{\Phi_m(x)}, \qquad i = 0, \ldots, n-1.$$

This formula is simpler to evaluate than the usual formula because multiplication by x is an indexing operation, and the modulo $\Phi_m(x)$ reduction consists of at most m additions. Any FFT algorithm, such as the Cooley–Tukey FFT, can be used for the computation, but now a multiplication is a multiplication by x, which requires no real multiplications and at most m real additions. Similarly, an addition becomes a polynomial addition, which requires m real additions. When applied to rational-valued inputs, the Fourier transform maps vectors of zero-degree polynomials into vectors of $(m-1)$-degree polynomials. More generally, the Fourier transform maps vectors of $(m-1)$-degree polynomials into vectors of $(m-1)$-degree polynomials.

The validity of the convolution theorem does not depend on the method of representing numbers. It applies to vectors whose components are in the conventional representation, and it applies to vectors whose components are in the polynomial representation. Hence, to perform a cyclic convolution of rational sequences

$$s_i = \sum_{k=0}^{n-1} g_{((i-k))} d_k,$$

take the Fourier transforms of $\boldsymbol{g}$ and $\boldsymbol{d}$ in the field Q^m, multiply in the frequency domain

$$S_k = G_k D_k,$$

and take the inverse Fourier transform. With the polynomial representation of the field Q^m, the transform uses no multiplications, so it is easy to compute. The spectral products, however, are now products of polynomials modulo $p(x)$, and so they require a large number of real multiplications. The computational complexity has been moved from one place to another. The increase in complexity in computing the spectral products more than offsets the savings in computing the Fourier transforms, so there is no net advantage. Later, we shall look at two-dimensional cyclic convolutions. This is where savings will be found.

For example, let $n = 64$; then the cyclotomic polynomial is

$$\Phi_{32}(x) = x^{32} + 1.$$

An element of $\boldsymbol{Q}(\omega)$ is a polynomial of degree 31, described by a list of 32 rational numbers. Each product

$$S_k = G_k D_k$$

is a product of polynomials modulo $x^{32} + 1$. This polynomial is an irreducible polynomial, so the polynomial product requires at least $2 \cdot 32 - 1$ multiplications. In practice, the number of multiplications will be considerably greater than 63. A practical 32-point algorithm, tabulated in Table 11.3, uses 147 multiplications.

There are 64 components in the Fourier transform. Using the practical 32-point algorithm, each component uses 147 real multiplications. This is considerably more multiplications than are needed by a conventional representation using an FFT.

Actually, there are many constraints relating the S_k, because the inverse Fourier transform must be real-valued. The inverse Fourier transform s_i will again consist of 64 polynomials, each represented by 32 rational numbers, but each polynomial must have a degree of zero, so 31 of the polynomial coefficients will equal zero. This constraint on the inverse Fourier transform can be used to set up constraints among the S_k so that all 64 of them need not be computed. If one works through such a procedure, however, it will begin to look like an algorithm based on the Chinese remainder theorem, which can be derived directly. We will not press this procedure further.

There are also shorter Fourier transforms defined in the extension field $\boldsymbol{Q}^m$. Suppose that n, as defined above, is composite, so that $n = n'n''$. Then we also have the n'-point Fourier transform

$$V_k = \sum_{i=0}^{n'-1} (x^{n''})^{ik} v_i, \qquad i = 0, \ldots, n' - 1,$$

where the polynomial arithmetic is understood to be in the field $\boldsymbol{Q}^m$. The kernel of the transform is $x^{n''}$ instead of x. With this definition of the Fourier transform, there is a corresponding inverse Fourier transform, and the convolution theorem holds just as before.

11.7 The Nussbaumer polynomial transforms

The transforms that we have constructed as Fourier transforms in an extension field can be studied in their own regard. We introduced these transforms in an extension of

11.7 The Nussbaumer polynomial transforms

the rational field, so all coefficients of the polynomials are rational numbers. Now that we have the transforms, the inverse transforms, and so on, we have a set of identities involving polynomials. The identities still hold even if we now allow the coefficients to be real numbers (or complex numbers).

In this section, we shall study the polynomial transforms once again; this time we do not think of them as notational variations of the usual Fourier transform, but rather as transforms in their own right.

In the ring of polynomials modulo $p(x)$, a polynomial transform is defined as

$$V_k(x) = \sum_{i=0}^{n-1} \omega(x)^{ik} v_i(x), \qquad k = 0, \ldots, n-1,$$

where $\omega(x)$ is an element of order n in the ring, and, of course, multiplication in the ring is modulo $p(x)$. We will restrict the discussion by considering only the case in which $\omega(x) = x$, and $p(x)$ is a factor of $x^n - 1$, specifically, a cyclotomic polynomial. There seem to be no other cases of real interest anyway.

Definition 11.7.1 *Let $p(x)$ be a polynomial of degree m. Let $v_i(x)$ for $i = 0, \ldots, n-1$ be a vector of polynomials of degree at most $n - 1$ over the field F. The polynomial transform of $v_i(x)$ is the vector of polynomials*

$$V_k(x) = \sum_{i=0}^{n-1} x^{ik} v_i(x) \pmod{p(x)}, \qquad k = 0, \ldots, n-1.$$

The polynomial transform is easy to compute. There are no general multiplications of ring elements. Multiplication by x^{ik} is trivial, and if $p(x)$ is chosen so that it has only 0, 1, and -1 as coefficients, the modulo reduction uses only additions.

The merit of the polynomial transform will be established by proving two theorems: that there is an inverse polynomial transform with the same structure, and that the polynomial transform supports cyclic convolution. Of course, all of this follows immediately if we recognize the polynomial transform as a notational variation of a Fourier transform, as we did in the previous section. Nevertheless, it is instructive to give a more direct proof.

Let n be the smallest integer such that $p(x)$ divides $x^n - 1$. The proof of the theorem will use the fact that, if $a(x)$ divides $x^n - 1$, then for any polynomial $f(x)$,

$$R_{a(x)}[f(x)] = R_{a(x)}[R_{x^n-1}[f(x)]].$$

We begin with the simplest case, which is the case in which n is a prime.

Theorem 11.7.2 *Over a field F, suppose that $p(x)$, a polynomial of degree $m - 1$, divides $x^n - 1$, where n is a prime. A vector $v(x)$ of length n of polynomials of degree*

$m - 1$ and its polynomial transform $V(x)$ are related componentwise by

$$V_k(x) = \sum_{i=0}^{n-1} x^{ik} v_i(x) \pmod{p(x)}, \qquad k = 0, \ldots, n-1,$$

$$v_i(x) = \frac{1}{n} \sum_{k=0}^{n-1} x^{(n-1)ik} V_k(x) \pmod{p(x)}, \qquad i = 0, \ldots, n-1.$$

Proof To avoid the trivial case, we assume that $p(x)$ has a degree of at least two. Evaluate the right side of the second equation:

$$\frac{1}{n} \sum_{k=0}^{n-1} x^{(n-1)ik} V_k(x) = \frac{1}{n} \sum_{k=0}^{n-1} x^{(n-1)ik} \left[\sum_{\ell=0}^{n-1} x^{\ell k} v_\ell(x) \right] \pmod{p(x)}$$

$$= \frac{1}{n} \sum_{\ell=0}^{n-1} \left[v_\ell(x) \sum_{k=0}^{n-1} x^{(\ell+ni-i)k} \right] \pmod{p(x)}.$$

The inner sum must now be evaluated. If i equals ℓ, then

$$\frac{1}{n} \sum_{k=0}^{n-1} x^{(\ell+ni-i)k} = \frac{1}{n} \sum_{k=0}^{n-1} x^{nik}$$

$$= \frac{1}{n} \sum_{k=0}^{n-1} 1 \pmod{x^n - 1}$$

$$= 1 \pmod{p(x)},$$

because $p(x)$ divides $x^n - 1$. If i is not equal to ℓ, consider the ring identity

$$(1 - x^r) \sum_{k=0}^{n-1} x^{rk} = 1 - x^{rn},$$

where $r = \ell + ni - i$ is not a multiple of n. Therefore

$$(1 - x^r) \neq 0 \pmod{x^n - 1}$$

while

$$(1 - x^{rn}) = 0 \pmod{x^n - 1}.$$

Because $p(x)$ divides $x^n - 1$, this implies that

$$\sum_{k=0}^{n-1} x^{rk} = 0 \pmod{p(x)}.$$

Thus we have shown that

$$\frac{1}{n} \sum_{k=0}^{n-1} x^{(\ell+ni-i)k} = \begin{cases} 1 & \text{if } \ell = i \pmod{p(x)}, \\ 0 & \text{if } \ell \neq i \pmod{p(x)}. \end{cases}$$

11.7 The Nussbaumer polynomial transforms

Hence

$$\frac{1}{n}\sum_{k=0}^{n-1} x^{(n-1)ik} V_k(x) = v_i(x) \pmod{p(x)},$$

which completes the proof of the theorem. □

The inverse polynomial transform can be written more simply as

$$v_i(x) = \frac{1}{n}\sum_{k=0}^{n-1} x^{-ik} V_k(x) \pmod{p(x)},$$

because $x^n = 1$ modulo $p(x)$, although technically x^{-ik} is not a polynomial. Just as multiplication by x can be implemented by reindexing the coefficients and reducing x^m modulo $p(x)$, so too multiplication by x^{-1} can be implemented by reindexing coefficients and using $p(x)$ to eliminate x^{-1} by writing

$$p_m x^m + p_{m-1} x^{m-1} + \cdots + p_1 x + p_0 = 0,$$

and so

$$x^{-1} = p_0^{-1}[p_m x^{m-1} + p_{m-1} x^{m-2} + \cdots + p_1].$$

Hence the inverse polynomial transform is as easy to compute as the direct polynomial transform.

Theorem 11.7.3 (Convolution theorem) *In the ring of polynomials modulo $p(x)$, if the vector of polynomials with components $s_i(x)$ is related to the vectors of polynomials $g_i(x)$ and $d_i(x)$ for $i = 0, \ldots, n-1$ by a cyclic convolution of polynomials*

$$s_i(x) = \sum_{\ell=0}^{n-1} g_{i-\ell}(x) d_\ell(x) \pmod{p(x)},$$

then the polynomial spectrum $S_k(x)$ is related to $G_k(x)$ and $D_k(x)$ by a componentwise polynomial product

$$S_k(x) = G_k(x) D_k(x) \pmod{p(x)}, \qquad k = 0, \ldots, n-1.$$

Proof By the inverse transform relationship

$$s_i(x) = \sum_{\ell=0}^{n-1} g_{i-\ell}(x) d_\ell(x)$$

$$= \sum_{\ell=0}^{n-1} \sum_{k=0}^{n-1} x^{(n-1)(i-\ell)k} G_k(x) d_\ell(x)$$

$$= \sum_{k=0}^{n-1} x^{(n-1)ik} G_k(x) \sum_{\ell=0}^{n-1} x^{\ell k} d_\ell(x),$$

where x^{nk} has been set to one because, as before, everything holds modulo $x^n - 1$. Hence

$$s_i(x) = \sum_{k=0}^{n-1} x^{(n-1)k} G_k(x) D_k(x),$$

and so $G_k(x) D_k(x)$ must equal $S_k(x)$. □

11.8 Fast convolution of polynomials

The use of a polynomial representation of an extension field does have an advantage for computing multidimensional convolutions. This is because there is no penalty in turning the spectral products into polynomial products if they started out as polynomial products. The extra work that would otherwise arise can be absorbed into work that must be done anyway.

It is enough to study the two-dimensional cyclic convolution. Write the two-dimensional cyclic convolution as a one-dimensional cyclic convolution of polynomials:

$$s_{i'}(y) = \sum_{k'=0}^{n'-1} g_{((i'-k'))}(y) d_{k'}(y) \pmod{y^{n'} - 1}.$$

Because $y^{n'} - 1$ is a product of cyclotomic polynomials, we can break the problem into a set of computations of the form

$$s_{i'}(y) = \sum_{k'=0}^{n'-1} g_{((i'-k'))}(y) d_{k'}(y) \pmod{p(y)},$$

where $p(y)$, a polynomial of degree m', is one of the cyclotomic polynomials dividing $y^{n'} - 1$. An algorithm for each of these subproblems can be combined to form an algorithm for the original problem.

Consider $g_{i'}(y)$ and $d_{i'}(y)$ for $i' = 0, \ldots, n' - 1$ as elements of $\boldsymbol{Q}(\omega)$. We then have the spectra

$$G_{k'}(y) = \sum_{i'=0}^{n'-1} y^{i'k'} g_{i'}(y) \pmod{p(y)}$$

$$D_{k'}(y) = \sum_{i'=0}^{n'-1} y^{i'k'} d_{i'}(y) \pmod{p(y)},$$

and

$$S_{k'}(y) = G_{k'}(y) D_{k'}(y) \pmod{p(y)}, \qquad k' = 0, \ldots, n' - 1.$$

11.8 Fast convolution of polynomials

Then $s_{i'}(y)$ is given by the inverse Fourier transform

$$s_{i'}(y) = \frac{1}{n'} \sum_{k'=0}^{n'-1} y^{-i'k'} S_{k'}(y) \pmod{p(y)}.$$

The Fourier transforms use no real multiplications.

The multiplications in this procedure all reside in the spectral products $G_{k'}(y)D_{k'}(y)$. These require n' polynomial products modulo $p(y)$; each requires at least $2m' - 1$ multiplications. This is a total of at least $n'(2m' - 1)$ multiplications to process a two-dimensional array containing $n'm'$ numbers. In practice, the number of multiplications needed for polynomial products modulo $p(y)$ will be considerably greater than $2m' - 1$, but still small enough to yield efficient algorithms. Some suitable algorithms for polynomial products are shown in Table 11.3.

Now consider the two-dimensional cyclic convolution, as represented by the polynomial product

$$s(x, y) = g(x, y)d(x, y) \pmod{x^{n'} - 1}\pmod{y^{n''} - 1}.$$

This polynomial product can be broken down by using the Chinese remainder theorem. We shall carry through the details for the special case in which both n' and n'' are powers of two, possibly equal. The method of computation is recursive. It employs a smaller two-dimensional cyclic convolution. Hence, by using the recursion, a radix-two, two-dimensional cyclic convolution can be computed by computing a number of small pieces.

Let $n' = 2^{m'}$ and $n'' = 2^{m''}$ with $m'' \geq m'$. Then we have the factorization

$$\begin{aligned} y^{n''} - 1 &= (y^{n''/2} + 1)(y^{n''/2} - 1) \\ &= (y^{n''/2} + 1)(y^{n''/4} + 1)(y^{n''/4} - 1) \\ &= (y^{n''/2} + 1) \cdots (y^{n'/2} + 1)(y^{n'/2} - 1). \end{aligned}$$

Although the last term could be factored further, we choose to stop here. Let the terms on the right be denoted by $f_{R-1}(y), \ldots, f_0(y)$ and, for $r = 0, \ldots, R-1$,

$$d^{(r)}(x, y) = d(x, y) \pmod{f_r(y)},$$
$$g^{(r)}(x, y) = g(x, y) \pmod{f_r(y)},$$

and

$$s^{(r)}(x, y) = s(x, y) \pmod{f_r(y)},$$

then

$$s^{(r)}(x, y) = g^{(r)}(x, y)d^{(r)}(x, y) \pmod{x^{n'} - 1}\pmod{f_r(x)},$$

and by the Chinese remainder theorem for polynomials,

$$s(x, y) = \sum_{r=0}^{R-1} a^{(r)}(y) s^{(r)}(x, y) \pmod{x^{n''} - 1},$$

where the $a^{(r)}(y)$ for $r = 0, \ldots, R - 1$ form an appropriate set of Bézout polynomials. This last step is straightforward and does not involve any real multiplications.

Most of the computations are in computing $s^{(r)}(x, y)$. There are two types of computation. These are

$$s^{(r)}(x, y) = g^{(r)}(x, y) d^{(r)}(x, y) \pmod{x^{n''} - 1} \pmod{y^{n''} + 1}$$

and

$$s^{(r)}(x, y) = g^{(r)}(x, y) d^{(r)}(x, y) \pmod{x^{n''} - 1} \pmod{y^{n''/2} - 1}.$$

The second kind of computation, if x and y are interchanged, is just a smaller copy of the problem being solved. By choosing a recursive formulation, as shown in Figure 11.5, we can suppose that this smaller problem is solved in the same way that the larger problem is solved.

The first kind of computation, with the superscript r dropped, is

$$s(x, y) = g(x, y) d(x, y) \pmod{x^{n'} - 1} \pmod{y^{m'} + 1}$$

and $m' \geq n'/2$. This computation can be viewed as a one-dimensional convolution in the field $\boldsymbol{Q}^{m'}$. That is,

$$s(x) = g(x) d(x) \pmod{x^{n'} - 1},$$

where the coefficients of the polynomials are elements of $\boldsymbol{Q}^{m'}$. This means that they are represented by polynomials in y of degree less than m'. Because $n' \leq 2m'$, a Fourier transform of blocklength n' exists and can be used to do the convolutions. The Fourier transform uses no multiplications – only additions.

The convolution becomes

$$S_{k'} = G_{k'} D_{k'}$$

in the frequency domain. This requires n' multiplications in $\boldsymbol{Q}^{m'}$. Each multiplication is a polynomial multiplication modulo $x^{m'} + 1$, and so requires $2m' - 1$ real multiplications. Hence, in total, there are $n'(2m' - 1)$ real multiplications to compute

$$s(x, y) = g(x, y) d(x, y) \pmod{x^{n'} - 1} \pmod{y^{m'} + 1}$$

and to compute the original two-dimensional cyclic convolution requires such computations be repeated several times.

For example, a 64 by 64 two-dimensional cyclic convolution requires $64 \cdot 63$ real multiplications plus a 64 by 32 two-dimensional cyclic convolution. In turn, a 32 by

11.8 Fast convolution of polynomials

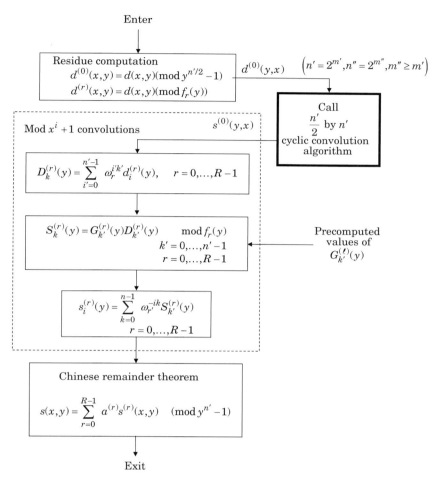

Figure 11.5 A multidimensional cyclic convolution algorithm

64 cyclic convolution requires $32(63) + 32(31)$ real multiplications plus a 32 by 16 cyclic convolution. In turn, a 16 by 32 cyclic convolution requires $16(31) + 16(15)$ multiplications. Continue to break down the problem in this way until the cyclic convolution is trivial. The total number of multiplications, in principle, is 8000. The number of calculations is considerably smaller than the number that are required by using a two-dimensional Cooley–Tukey algorithm with the convolution theorem, which would be 73 728.

In practice, the polynomial products use more than $n(2m - 1)$ real multiplications, as tabulated in Table 11.3, and the total number of multiplications is 17 770 for a practical 64 by 64 two-dimensional cyclic convolution. The performance of a number of two-dimensional cyclic convolution algorithms, constructed in a similar way, is given in Table 11.4.

Fast algorithms and multidimensional convolutions

Table 11.4 *Performance of some two-dimensional cyclic convolutions*

Array size	Number of real multiplications	Number of real additions	Real multiplications per point	Real additions per point
3 by 3	13	70	1.44	7.78
4 by 4	22	122	1.37	7.62
5 by 5	55	369	2.20	14.76
6 by 6	52	424	1.44	11.78
7 by 7	121	1163	2.47	23.73
8 by 8	130	750	2.03	11.72
9 by 9	193	1382	2.38	17.06
10 by 10	220	1876	2.20	18.76
14 by 14	482	5436	2.47	27.73
16 by 16	634	4774	2.48	18.65
18 by 18	772	6576	2.38	20.30
24 by 24	1402	12 954	2.43	22.49
27 by 27	2893	21 266	3.97	29.17
32 by 32	3658	24 854	3.57	24.27
64 by 64	17770	142 902	4.34	34.89
128 by 128	78250	720 502	4.78	43.98

Problems for Chapter 11

11.1 **a** Compute the performance of a 7560-point cyclic convolution algorithm based on the Agarwal–Cooley algorithm.

b Compute the performance of a 504 by 504 two-dimensional convolution algorithm based on nesting.

11.2 An algorithm for a twelve-point cyclic convolution can be constructed by using the Winograd convolution algorithm directly or by combining a three-point cyclic convolution algorithm and a four-point cyclic convolution algorithm using the Agarwal–Cooley algorithm. Compare the number of multiplications used by these two methods.

11.3 **a** Outline a method for designing an algorithm to compute a six-point cyclic convolution

$$s(x) = g(x)d(x) \pmod{x^6 - 1}.$$

How many multiplications will be needed?

b Use the Agarwal–Cooley algorithm to build a six-point cyclic convolution algorithm out of a two-point cyclic convolution algorithm and a three-point cyclic convolution algorithm. How many multiplications are needed?

11.4 Use the Agarwal–Cooley algorithm to set up the matrices for a 15-point cyclic convolution algorithm in the standard form

$$s = CGAd.$$

11.5 A cyclic convolution has blocklength $n = n_1 n_2 n_3 n_4$, where the four factors are coprime. We can obtain an algorithm for this cyclic convolution by combining algorithms for cyclic convolutions of blocklengths n_1, n_2, n_3, and n_4. Two schemes for building the algorithm by pairwise combinations are suggested by the parameterizations $n = ((n_1 n_2)(n_3 n_4))$ and $n = n_1(n_2(n_3(n_4)))$. Prove that each of these schemes uses the same number of multiplications and additions, provided that each pairwise combination is optimal.

11.6 An algorithm for complex convolution modulo $x^n - 1$ was given in Section 5.7.
  a Derive this algorithm by developing an algorithm for two-dimensional polynomial products modulo $x^{2r} + 1$ modulo $y^2 + 1$.
  b One can obtain two algorithms for the same problem by starting with

  $$s(x, y) = g(x, y)d(x, y) \pmod{x^{2r} + 1}\pmod{y^2 + 1}$$

  or

  $$s(x, y) = g(x, y)d(x, y) \pmod{y^2 + 1}\pmod{x^{2r} + 1}.$$

  Which is better?

11.7 Quaternions are defined in Problem 2.16.
  a Write the two by two cyclic convolution

  $$s(x, y) = g(x, y)d(x, y) \pmod{x^2 - 1}\pmod{y^2 - 1}$$

  in the form of a four by four matrix times a four vector. Give an algorithm that uses four multiplications.
  b Write the two by two polynomial product

  $$s(x, y) = g(x, y)d(x, y) \pmod{x^2 + 1}\pmod{y^2 - 1}$$

  in the form of a four by four matrix times a four vector. Give an algorithm that uses six multiplications.
  c Write the two by two polynomial product

  $$s(x, y) = g(x, y)d(x, y) \pmod{x^2 + 1}\pmod{y^2 + 1}$$

  in the form of a four by four matrix times a four vector. Give an algorithm that uses six multiplications.
  d Write the quaternion product

  $$s = gd$$

in the form of a four by four matrix times a four vector. Give an algorithm that uses nine multiplications.

11.8 Determine the number of multiplications and additions used by the Agarwal–Cooley algorithm to compute a 45-point cyclic convolution. How much can this be improved by using a splitting algorithm?

11.9 a Compute the performance of an algorithm for computing a 15 by 15 two-dimensional cyclic convolution that is computed by nesting a three by three cyclic convolution algorithm and a five by five cyclic convolution algorithm. Compare with the performance of an algorithm constructed by nesting a 15-point one-dimensional cyclic convolution with itself.

b Repeat for a 21 by 21 two-dimensional cyclic convolution.

11.10 To compute an n by n cyclic convolution of complex vectors, one can always simply use an algorithm designed for an n by n cyclic convolution of real vectors with each real complex multiplication or addition replaced by a complex multiplication or addition. How much better can you do for a five by five cyclic convolution of complex arrays? How much better for a four by four cyclic convolution?

Notes for Chapter 11

At the simplest level, a two-dimensional Fourier transform consists of two independent Fourier transforms operating sequentially along the two axes of the array; a similar statement applies to a two-dimensional convolution. Hence fast algorithms for a one-dimensional convolution were readily applied to the two-dimensional convolution. Algorithms that were intrinsically two-dimensional came later. Nussbaumer (1977) pioneered this subject with his polynomial transforms. These were defined heuristically by analogy with the Fourier transform because they did what was wanted. Later, it was realized that they were more than just an analogy to a Fourier transform. The interpretation as a Fourier transform using a polynomial representation of extension fields is from Blahut (1983a). Similar ideas are due to Beth, Fumy, and Muhlfeld (1982). Another kind of polynomial transform with a clearer structure but more computation is due to Arambepola and Rayner (1979).

The use of the Good–Thomas type of indexing for computing one-dimensional convolutions is due to Agarwal and Cooley (1977). They turned the one-dimensional convolution into a multidimensional convolution, which they computed by nesting one-dimensional algorithms, a procedure that we described in terms of the Kronecker product. The nesting technique is the same as that proposed by Winograd (1978) for computing multidimensional Fourier transforms.

The method of using the Chinese remainder theorem for polynomials at the level of the multidimensional convolution to reduce the number of additions is due to Nussbaumer (1978). The use of modulus polynomials with coefficients in a polynomial extension field was suggested by Pitas and Strintzis (1982). Many of our tables are based on Nussbaumer's work.

12 Fast algorithms and multidimensional transforms

In earlier chapters, we saw ways in which the Fourier transform can be broken into pieces and ways in which a convolution algorithm can be turned into an algorithm for the Fourier transform. It also is possible to break a multidimensional Fourier transform into pieces and to turn an algorithm for the multidimensional convolution into an algorithm for the multidimensional Fourier transform. The possibilities now are even richer than they were in the one-dimensional case. We shall discuss a variety of methods.

The algorithms for multidimensional Fourier transforms are studied in the easy way by studying only the algorithms for the two-dimensional Fourier transform as representative of the more general case. The discussion and the formulas for the two-dimensional Fourier transforms can be extended directly to a discussion of the multidimensional Fourier transforms.

Multidimensional Fourier transforms arise naturally from problems that are intrinsically multidimensional. They also arise artificially as a way of computing a one-dimensional Fourier transform. This chapter includes such methods for computing a one-dimensional Fourier transform, and so we will give practical methods for building large one-dimensional Fourier transform algorithms from the small one-dimensional Fourier transform algorithms that were studied in Chapter 3.

12.1 Small-radix Cooley–Tukey algorithms

A two-dimensional discrete Fourier transform can be computed by applying the Cooley–Tukey FFT first to each row and then to each column. This can be regarded simply as a parenthesization of the equation of the two-dimensional Fourier transform with either the row sum or the column sum within the inner parentheses.

Let v be an array of elements $v_{i'i''}$, for $i' = 0, \ldots, n' - 1$ and $i'' = 0, \ldots, n'' - 1$, from the field F. The two-dimensional Fourier transform of v is another two-dimensional array V of elements of F, given by

$$V_{k'k''} = \sum_{i'=0}^{n'-1} \sum_{i''=0}^{n''-1} \omega^{i'k'} \mu^{i''k''} v_{i'i''}, \qquad \begin{aligned} k' &= 0, \ldots, n' - 1, \\ k'' &= 0, \ldots, n'' - 1, \end{aligned}$$

12.1 Small-radix Cooley–Tukey algorithms

where ω is an n'th root of unity in the field F and μ is an n''th root of unity in the field F, which ordinarily would be chosen equal to ω when n' equals n''. It follows that

$$\begin{aligned} V_{k'k''} &= \sum_{i'=0}^{n'-1} \omega^{i'k'} \left[\sum_{i''=0}^{n''-1} \mu^{i''k''} v_{i'i''} \right] \\ &= \sum_{i''=0}^{n''-1} \omega^{i''k''} \left[\sum_{i'=0}^{n'-1} \mu^{i'k'} v_{i'i''} \right]. \end{aligned}$$

Hence we see that one can compute the two-dimensional Fourier transform by computing a one-dimensional Fourier transform, either first along every column then along every row, or the other way around. Any FFT algorithm can be used on the rows and any FFT algorithm, possibly a different one, can be used on the columns. Many good FFT algorithms are available, and any one of them can be chosen for the purpose of reducing the number of multiplications and the number of additions.

When the array is large, besides the number of multiplications and additions, one also is concerned with problems of data management. A 1024 by 1024 array of real numbers consists of more than one million numbers, and twice this many if the data values are complex. A processor may store most of the array in bulk memory, and only a portion in local memory. The transfer of data between bulk memory and local memory can be an issue as important as the number of multiplications and the number of additions.

We shall consider a simple model of the memory-transfer mechanism in which the data is stored in bulk memory by rows and is transferred to local memory by rows. Then the processing consists of a Fourier transform along each row, followed by a transpose of the array, then followed by a Fourier transform along each of the new rows. A second transpose operation will be needed if the final result must be stored in bulk memory by its true rows. Fast transposition algorithms were discussed in Section 4.4 of Chapter 4.

To avoid taking the transpose, we shall develop multidimensional algorithms by looking closely at the basic Cooley–Tukey decimation rule. The multidimensional algorithms are formed by decimating the two-dimensional array directly rather than decimating, in turn, the rows, then the columns. A comparison is shown in Figure 12.1. In particular, the two-dimensional radix-two decimation will replace the n by n array with four $n/2$ by $n/2$ arrays, and the two-dimensional radix-four decimation will replace the n by n array with 16 $n/4$ by $n/4$ arrays. The latter algorithm in particular is attractive, not only because the number of data transfers is small, but also because the number of multiplications and additions is reduced.

We want to compute the n by n-point two-dimensional Fourier transform

$$V_{k\ell} = \sum_{i=0}^{n-1} \sum_{j=0}^{n-1} \omega^{ik} \omega^{j\ell} v_{ij},$$

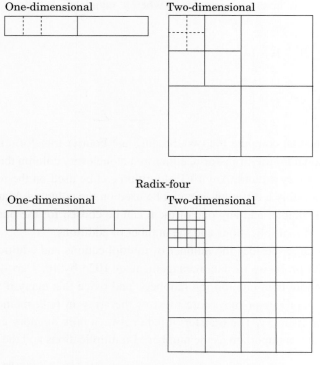

Figure 12.1 Some decimation schemes

where $n = n'n''$. Notice carefully that we have switched notation on indices here to free the primed notation for use in the Cooley–Tukey decimation. Now n' and n'' are used as factors of n, the dimension of the square array.

The Cooley–Tukey decimation formula for a one-dimensional Fourier transform, given in Figure 3.1, is

$$V_{k'k''} = \sum_{i'=0}^{n'-1} \beta^{i'k'} \left[\omega^{i'k''} \sum_{i''=0}^{n''-1} \gamma^{i''k''} v_{i'i''} \right].$$

We use this formula twice for the two-dimensional transform – once on the row index and once on the column index. Then we can arrange the order of the summations to write

$$V_{k'k''\ell'\ell''} = \sum_{i'=0}^{n'-1} \sum_{j'=0}^{n'-1} \beta^{i'k'} \beta^{j'\ell'} \left[\omega^{i'k''} \omega^{j'\ell''} \sum_{i''=0}^{n''-1} \sum_{j''=0}^{n''-1} \gamma^{i''k''} \gamma^{j''\ell''} \right] v_{i'i''j'j''}.$$

This now is in the form of an n'' by n'' two-dimensional Fourier transform for each value of i' and j', followed by an element-by-element multiplication, followed by an n' by n' two-dimensional Fourier transform for each value of k'' and ℓ''.

12.1 Small-radix Cooley–Tukey algorithms

To get a decimation-in-time radix-two two-dimensional Cooley–Tukey FFT, take $n' = 2$, and $n'' = n/2$. Then the equation for the two-dimensional Fourier transform can be put in the matrix form

$$\begin{bmatrix} V_{k,\ell} \\ V_{k+n/2,\ell} \\ V_{k,\ell+n/2} \\ V_{k+(n/2),\ell+n/2} \end{bmatrix} = \begin{bmatrix} 1 & 1 & 1 & 1 \\ 1 & -1 & 1 & -1 \\ 1 & 1 & -1 & -1 \\ 1 & -1 & -1 & 1 \end{bmatrix} \begin{bmatrix} \sum_{i=0}^{(n/2)-1} \sum_{j=0}^{(n/2)-1} \omega^{2ik} \omega^{2j\ell} v_{2i,2j} \\ \omega^k \sum_{i=0}^{(n/2)-1} \sum_{j=0}^{(n/2)-1} \omega^{2ik} \omega^{2j\ell} v_{2i+1,2j} \\ \omega^\ell \sum_{i=0}^{(n/2)-1} \sum_{j=0}^{(n/2)-1} \omega^{2ik} \omega^{2j\ell} v_{2i,2j+1} \\ \omega^k \omega^\ell \sum_{i=0}^{(n/2)-1} \sum_{j=0}^{(n/2)-1} \omega^{2ik} \omega^{2j\ell} v_{2i+1,2j+1} \end{bmatrix}$$

for $k = 0, \ldots, (n/2) - 1$ and $\ell = 0, \ldots, (n/2) - 1$. This FFT breaks the input data array into four arrays according to whether the two indices are each even or odd. The output array is broken into four arrays in a different way, by taking the first $n/2$ rows and the second $n/2$ rows and by taking the first $n/2$ columns and the second $n/2$ columns. The computation now requires four $n/2$ by $n/2$-point Fourier transforms plus $\frac{3}{4}n^2$ multiplications by powers of ω. Here we do not count the $\frac{1}{4}n^2$ trivial multiplications by one that occur in a block and are easily suppressed. The remaining $\frac{3}{4}n^2$ multiplications include a few more trivial multiplications that we do not bother to suppress. Let $M(n \times n)$ be the number of multiplications in the field F needed by this algorithm to compute an n by n-point Fourier transform with n a power of two. It satisfies the recursion

$$M(n \times n) = 4M\left(\frac{n}{2} \times \frac{n}{2}\right) + \frac{3}{4}n^2.$$

This recursion is solved by

$$M(n \times n) = \frac{3}{4}n^2(\log_2 n - C),$$

where C is a constant to be chosen to fit the number of multiplications in the innermost loop. In particular, one can start with either a two by two-point Fourier transform that uses no multiplications so that $M(2 \times 2) = 0$, or with a four by four-point Fourier transform that uses no multiplications, so that $M(4 \times 4) = 0$. Then

$$M(n \times n) = \tfrac{3}{4}n^2(\log_2 n - 1)$$

or

$$M(n \times n) = \tfrac{3}{4}n^2(\log_2 n - 2).$$

Further reductions are possible, but they would complicate the structure of the algorithm.

These formulas should be compared with

$$M(n \times n) = n^2 \log_2 n,$$

which gives the number of required multiplications in the field F if a basic radix-two Cooley–Tukey FFT is used, in turn, on the rows and the columns.

One minor feature of the two-dimensional algorithm is the slight reduction in the number of multiplications. A much more important feature is the sequence in which the data is used because it reduces the number of data transfers between bulk memory and local memory. The entire array is read into local memory, two rows at a time. Local memory needs to be big enough to hold two rows. All of the 2×2 two-dimensional transforms along each pair of rows are computed. This process on the array is repeated $\log_2 n$ times. In each iteration, the pairing of the rows is controlled by the Cooley–Tukey bit-shuffling patterns, and the two by two arrays from within the pair of rows are also selected by the Cooley–Tukey bit-shuffling pattern. In all, because there are n rows in the array and the array is transferred $\log_2 n$ times, there are $n \log_2 n$ rows transferred from bulk memory to local memory and the same number are transferred back.

To get a radix-four two-dimensional Cooley–Tukey FFT, take $n' = 4$, $n'' = n/4$ in the general formula. This will break the two-dimensional array into 16 subarrays. The computations can be expressed in the following matrix equation:

$$\begin{bmatrix} V_{k,\ell} \\ V_{k+n/4,\ell} \\ V_{k+n/2,\ell} \\ V_{k+3n/4,\ell} \\ V_{k,\ell+n/4} \\ V_{k+n/4,\ell+n/4} \\ V_{k+n/2,\ell+n/4} \\ V_{k+3n/4,\ell+n/4} \\ V_{k,\ell+n/2} \\ V_{k+n/4,\ell+n/2} \\ V_{k+n/2,\ell+n/2} \\ V_{k+3n/4,\ell+n/2} \\ V_{k,\ell+3n/4} \\ V_{k+n/4,\ell+3n/4} \\ V_{k+n/2,\ell+3n/4} \\ V_{k+3n/4,\ell+3n/4} \end{bmatrix} = M \begin{bmatrix} \sum_{i=0}^{(n/4)-1} \sum_{j=0}^{(n/4)-1} \omega^{4ik} \omega^{4j\ell} v_{2i,2j} \\ \omega^k \sum_{i=0}^{(n/4)-1} \sum_{j=0}^{(n/4)-1} \omega^{4ik} \omega^{4j\ell} v_{2i+1,2j} \\ \omega^{2k} \sum_{i=0}^{(n/4)-1} \sum_{j=0}^{(n/4)-1} \omega^{4ik} \omega^{4j\ell} v_{2i+2,2j} \\ \omega^{3k} \sum_{i=0}^{(n/4)-1} \sum_{j=0}^{(n/4)-1} \omega^{4ik} \omega^{4j\ell} v_{2i+3,2j} \\ \omega^\ell \sum_{i=0}^{(n/4)-1} \sum_{j=0}^{(n/4)-1} \omega^{4ik} \omega^{4j\ell} v_{2i,2j+1} \\ \omega^{k+\ell} \sum_{i=0}^{(n/4)-1} \sum_{j=0}^{(n/4)-1} \omega^{4ik} \omega^{4j\ell} v_{2i+1,2j+1} \\ \vdots \\ \omega^{3k+3\ell} \sum_{i=0}^{(n/4)-1} \sum_{j=0}^{(n/4)-1} \omega^{4ik} \omega^{4j\ell} v_{2i+3,2j+3} \end{bmatrix},$$

where

$$M = \begin{bmatrix} 1 & 1 & 1 & 1 \\ 1 & -j & -1 & j \\ 1 & -1 & 1 & -1 \\ 1 & j & -1 & -j \end{bmatrix} \times \begin{bmatrix} 1 & 1 & 1 & 1 \\ 1 & -j & -1 & j \\ 1 & -1 & 1 & -1 \\ 1 & j & -1 & -j \end{bmatrix}.$$

In this form, the multiplying 16 by 16 matrix M is written as a Kronecker product of two four by four matrices of elements $\pm 1, \pm j$.

On the right side, all but one of the elements are multiplied by a power of ω. Of these, a few are trivial, but to keep the structure simple, we include them in the count. The number of multiplications satisfies the recursion

$$M(n \times n) = 16 M\left(\frac{n}{4} \times \frac{n}{4}\right) + \frac{15}{16} n^2,$$

which is satisfied by

$$M(n \times n) = \frac{15}{32} n^2 (\log n - C),$$

where C is chosen so that $M(4 \times 4) = 0$. Then

$$M(n \times n) = \frac{15}{32} n^2 (\log n - 2),$$

and we see that the radix-four algorithm is efficient in terms of multiplications as well as having a good partitioning structure.

To use the radix-four two-dimensional decimation, the local memory must be large enough to hold four rows of data at the same time. In all, the array is transferred from bulk memory to local memory $\log_4 n$ times, so there are $\frac{1}{2} n \log_2 n$ transfers of rows from bulk memory to local memory.

12.2 The two-dimensional discrete cosine transform

The two-dimensional discrete cosine transform maps an n by n array of real numbers v into another n by n array of real numbers V, called the *two-dimensional discrete cosine transform* of v. The same term refers to both the mapping and the output of the mapping. Given the n by n array v of real numbers, the two-dimensional discrete cosine transform of v is the array given by

$$V_{k'k''} = \sum_{i'=0}^{n-1} \sum_{i''=0}^{n-1} v_{i'i''} \cos \frac{\pi(2i'+1)k'}{2n} \cos \frac{\pi(2i''+1)k''}{2n}$$

for $k' = 0, \ldots, n-1$ and $k'' = 0, \ldots, n-1$. The two-dimensional discrete cosine transform is an obvious generalization of the one-dimensional discrete cosine transform. Indeed, the two-dimensional discrete cosine transform can be regarded as the one-dimensional discrete cosine transform first applied to every column, then applied to every row, or the same thing but for changing to the other order.

The inverse two-dimensional discrete cosine transform is

$$v_{i'i''} = \sum_{i'=0}^{n-1} \sum_{i''=0}^{n-1} V_{k'k''} \left(1 - \tfrac{1}{2}\delta_{k'}\right)\left(1 - \tfrac{1}{2}\delta_{k''}\right) \cos\frac{\pi(2i'+1)k'}{2n} \cos\frac{\pi(2i''+1)k''}{2n},$$

where $\delta_k = 1$ if $k = 0$ and $\delta_k = 0$ if $k \neq 0$. This expression for the inverse two-dimensional discrete cosine transform follows immediately by applying the inverse one-dimensional discrete cosine transform to every row followed by the inverse one-dimensional discrete cosine transform to every column.

As it is written, the two-dimensional discrete cosine transform must multiply each element of the array v by the product of two cosine terms for every value of i', i'', k', and k''. If the transform were computed as written, then $2n^4$ multiplications would be required. Even if n is as small as eight, a total of 8192 multiplications would be required. Clearly, a better algorithm for the two-dimensional discrete cosine transform is needed.

It is more natural to compute this as a discrete cosine transform of each column followed by a discrete cosine transform of each row. Thus, by writing the equation as

$$V_{k'k''} = \sum_{i'=0}^{n-1} \cos\frac{\pi(2i'+1)k'}{2n} \sum_{i''=0}^{n-1} v_{i'i''} \cos\frac{\pi(2i''+1)k''}{2n},$$

it can be seen as $2n$ one-dimensional discrete cosine transforms. If each one-dimensional transform uses n^2 multiplications, a total of $2n^3$ multiplications are now needed. With this organization, only 1024 multiplications would be needed if n is equal to eight.

An even better procedure is to use a fast algorithm for each one-dimensional discrete cosine transform as described in Section 3.5. Any fast algorithm for computing the one-dimensional discrete cosine transform can be used to compute the two-dimensional discrete cosine transform by applying it first to every column, then to every row. If $M(n)$ multiplications are needed to compute the one-dimensional discrete cosine transform, then $2nM(n)$ multiplications would be needed to compute the two-dimensional discrete cosine transform.

One can go beyond this. Alternative algorithms can be developed based even more directly on the structure of the computation. For example, if n is a power of two, the n by n array v can be partitioned into four subarrays using Theorem 3.5.3 to decimate both rows and columns. Each of the four subarrays could be computed as written by rows, then by columns. Each subarray would require $2\left(\tfrac{n}{2}\right)\left(\tfrac{n}{2}\right)^2$ multiplications. Altogether, there would be n^3 multiplications. Then 512 multiplications are needed if n equals eight. Additional improvements can be made that further reduce the computational complexity by treating each subarray more carefully.

12.3 Nested transform algorithms

Now we turn to another technique, the technique of nesting, to combine one-dimensional fast Fourier transform algorithms, such as the Winograd FFT algorithm, to construct multidimensional FFT algorithms. Recall that as a consequence of the two factorizations

$$V_{k'k''} = \sum_{i'=0}^{n'-1} \omega^{i'k'} \left[\sum_{i''=0}^{n''-1} \mu^{i''k''} v_{i'i''} \right]$$

$$= \sum_{i''=0}^{n''-1} \mu^{i''k''} \left[\sum_{i'=0}^{n'-1} \omega^{i'k'} v_{i'i''} \right],$$

one can compute the two-dimensional Fourier transform by computing a one-dimensional Fourier transform along every column, then along every row, or the other way around. Methods for computing one-dimensional Fourier transforms can be used in any convenient combination to do the row and column Fourier transforms comprising a two-dimensional Fourier transform. We shall look at algorithms that have the structure of the Winograd small FFT, finding ways to combine them efficiently.

Let $M(n')$ and $A(n')$ be the number of multiplications and additions used by some available n'-point one-dimensional Fourier transform algorithm. To do n'' such transforms of blocklength n', it takes $n''M(n')$ multiplications and $n''A(n')$ additions. Similarly, to then do n' transforms of blocklength n'', it takes $n'M(n'')$ more multiplications and $n'A(n'')$ more additions. Thus, when computed as sequential one-dimensional transforms, the computational load of the two-dimensional Fourier transform is

$$M(n' \times n'') = n''M(n') + n'M(n''),$$
$$A(n' \times n'') = n''A(n') + n'A(n'').$$

It does not matter which dimension is processed first. As long as they are processed sequentially in this way, the computational complexity is the same.

A better approach is to nest the algorithms by using a method of Winograd. Because it does not matter whether the rows or the columns of the two-dimensional Fourier transform are transformed first, it seems that it may be possible somehow to do them together. This is what the Winograd nesting does. It binds together the row computations and the column computations in a way that reduces the total number of multiplications. The technique uses the notion of a Kronecker product of matrices. We will still use algorithms that are designed for a one-dimensional Fourier transform on the rows and the columns, but we combine them more efficiently.

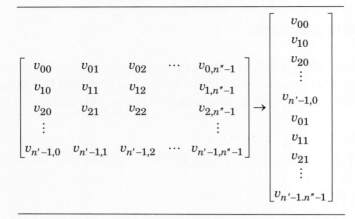

Figure 12.2 Mapping a two-dimensional array into one dimension

Let W' and W'' be matrix representations of Fourier transforms of size n' and n'', respectively. That is,

$$V' = W'v',$$
$$V'' = W''v''$$

are matrix representations of the Fourier transforms

$$V'_k = \sum_{i=0}^{n'-1} \beta^{ik} v'_i,$$

$$V''_k = \sum_{i=0}^{n''-1} \gamma^{ik} v''_i.$$

An n' by n'' two-dimensional Fourier transform of the two-dimensional array $v_{i'i''}$ is obtained by applying W' to each column (a column has n' components) and then applying W'' to each row. An n' by n'' two-dimensional computation can be turned into a one-dimensional computation by stacking columns, as shown in Figure 12.2.

Write the two-dimensional arrays v and V as one-dimensional arrays, also called v and V, by stacking columns. We will write these as the $n'n''$-point one-dimensional input and output vectors, given by

$$v = \begin{bmatrix} v_0 \\ v_1 \\ v_2 \\ \vdots \\ v_{n''-1} \end{bmatrix}, \quad V = \begin{bmatrix} V_0 \\ V_1 \\ V_2 \\ \vdots \\ V_{n''-1} \end{bmatrix},$$

12.3 Nested transform algorithms

where $v_{i''}$ and $V_{i''}$ denote columns of the two-dimensional input and output arrays, respectively, given by

$$v_{i''} = \begin{bmatrix} v_{0i''} \\ v_{1i''} \\ v_{2i''} \\ \vdots \\ v_{n'-1,i''} \end{bmatrix}, \qquad V_{i''} = \begin{bmatrix} V_{0i''} \\ V_{1i''} \\ V_{2i''} \\ \vdots \\ V_{n'-1,i''} \end{bmatrix}.$$

If we think of the arrays v and V rearranged into one-dimensional $n'n''$-point vectors in this way, then the two-dimensional Fourier transform can be written in the form of a Kronecker product. First, write the computation by stacking columns as

$$\begin{bmatrix} V_0 \\ V_1 \\ V_2 \\ \vdots \\ V_{n''-1} \end{bmatrix} = \begin{bmatrix} w''_{00}I & w''_{01}I & \cdots & w''_{0,n''-1}I \\ w''_{10}I & w''_{11}I & \cdots & w''_{1,n''-1}I \\ w''_{20}I & w''_{21}I & \cdots & w''_{2,n''-1}I \\ \vdots & & & \vdots \\ w''_{n''-1,0}I & & \cdots & w''_{n''-1,n''-1}I \end{bmatrix} \begin{bmatrix} W' & 0 & \cdots & 0 \\ 0 & W' & \cdots & 0 \\ 0 & 0 & & \\ \vdots & \vdots & & \vdots \\ 0 & 0 & \cdots & W' \end{bmatrix} \begin{bmatrix} v_0 \\ v_1 \\ v_2 \\ \vdots \\ v_{n''-1} \end{bmatrix}$$

where W' is an n' by n' matrix, 0 is an n' by n' matrix of zeros, and I is an n' by n' identity matrix. But the product of the two matrices is easily recognized as a Kronecker product, so we have

$$V = Wv,$$

where $W = W'' \times W'$ is an $n'n''$ by $n'n''$ matrix.

If we have Winograd small FFT algorithms of blocklength n' and n'', respectively, then we have the matrix factorizations

$$W' = C'B'A',$$
$$W'' = C''B''A'',$$

where A', A'', C', and C'' are matrices of zeros and ones, and B' and B'' are diagonal matrices. The multiplications by the matrix B' or B'' is where the Winograd algorithm collects all of its multiplications. Let $W = W'' \times W'$, and apply Theorem 2.5.5 twice to get

$$W = (C''B''A'') \times (C'B'A')$$
$$= (C'' \times C')(B'' \times B')(A'' \times A')$$
$$= CBA,$$

where the Kronecker products $C = C'' \times C'$ and $A = A'' \times A'$ are matrices of zeros and ones, and the Kronecker product $B = B'' \times B'$ is again a diagonal matrix. Hence we have an algorithm for an $n'n''$-point two-dimensional Fourier transform algorithm in

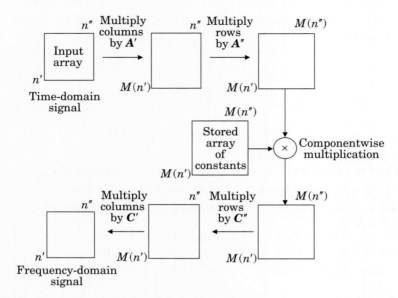

Figure 12.3 Nested computation of a two-dimensional Fourier transform

the same form as the Winograd small FFT. In this way, two-dimensional FFT algorithms can be built up from one-dimensional FFT algorithms.

A good way to organize the computations of a two-dimensional Fourier transform is suggested by the form

$$V = (C'' \times C')(B'' \times B')(A'' \times A')v$$

and is illustrated in Figure 12.3. Think of the data in the original two-dimensional array. To multiply by $A'' \times A'$, first multiply every column by matrix A'; then multiply every row by matrix A''. The first operation uses only additions and expands the array to an $M(n')$ by n'' array; the second operation also uses only additions and expands the array to an $M(n')$ by $M(n'')$ array. Next, one can multiply every column by matrix B'; then multiply every row by matrix B''. This takes $2M(n')M(n'')$ multiplications. A preferred method for this step, however, is to prestore the $M(n')$ by $M(n'')$ array B, given by $B'' \times B'$. In this form, the second step will use only $M(n')M(n'')$ multiplications, but will require more storage of constants. Finally, collapse the $M(n')$ by $M(n'')$ array back to an n' by n'' array by multiplying each column by C', then every row by C''. This last step uses only additions.

The total number of multiplications is

$$M(n' \times n'') = M(n')M(n'').$$

A formula for the total number of additions is a little more difficult to derive and depends on how the number of additions in each small algorithm is distributed between

the preadditions and the postadditions. To simplify the formula and its derivation (and even to reduce the number of additions), we will specify that the order in which matrices C' and C'' are applied is interchanged. The interchange is justified by using the general identity $\sum_{i'} C_{i'k'} \sum_{i''} C_{i''k''} V_{i'i''} = \sum_{i''} C_{i''k''} \sum_{i'} C_{i'k'} V_{i'i''}$. Then
$A(n' \times n'') = n'A(n'') + M(n'')A(n').$

We now have two ways to compute the two-dimensional Fourier transform. One method has

$$M(n' \times n'') = n''M(n') + n'M(n'')$$

multiplications. The other has

$$M(n' \times n'') = M(n')M(n'')$$

multiplications. The second method presumes that the Winograd small FFT is available for the single dimensions. The first method works with any one-dimensional FFT.

For example, a 1008-point by 1008-point FFT with complex input data uses $4 \times 1008 \times 1782$ real multiplications with the first method, and 2×1782^2 real multiplications with the second method. If the input data is real, then only half as many real multiplications are used by the second method, but three-fourths as many are used by the first method because, after the Fourier transforms along rows are computed, the data is complex.

The second method, clearly, is better but it does require a temporary array of size 2 by 1782 by 1782 for complex data (which could initially contain the original 2 by 1008 by 1008 input data array); the first method needs only a one-dimensional temporary array of 3564 words.

12.4 The Winograd large fast Fourier transform

The Winograd large FFT is a method of efficiently computing the one-dimensional discrete Fourier transform when the blocklength n has coprime factors for which one has Winograd small FFT algorithms. It is built from four separate ideas: the Rader prime algorithm, the Winograd small convolution algorithm of Section 3.4, the Good–Thomas prime factor indexing scheme, and the Winograd nesting algorithm. The first two ideas were already combined into the Winograd small FFT in Section 3.8; and the method of nesting was discussed in Section 12.3 for computing two-dimensional Fourier transforms. As measured by the number of multiplications, the Winograd large FFT is better than the Cooley–Tukey FFT, as shown in Table 12.1, but it is more intricate. The price paid for having fewer multiplications is the absence of tight repetitive loops.

The general case of the Winograd large FFT has a blocklength n that is a product of small primes or small prime powers. We shall discuss the case with two factors. Then

Table 12.1 *Comparison of some FFT algorithms*

	Performance of Winograd FFT (complex input data)			(Basic radix-two Cooley–Tukey FFT) (complex input data)	
Blocklength n	Number of real multiplications	Number of real additions	Blocklength n	Number of real multiplications	Number of real additions
30	72	384	32	320	480
48	92	636			
60	144	888	64	768	1152
91	318	2648			
120	288	2076	128	1792	2688
168	432	3492			
240	648	5016	256	4096	6144
420	1296	11 352			
504	1584	14 642	512	9216	13 824
840	2592	24 804			
1 008	3564	34 920	1024	20 480	30 720
2 520	9504	100 188	2048	45 056	67 584
10 920	38 760	320 196	8192	212 992	319 488

$n = n'n''$. The Good–Thomas prime factor algorithm decomposes an n-point Fourier transform into a two-dimensional n' by n''-point Fourier transform. The individual components of this two-dimensional Fourier transform can be computed by an n'-point Winograd small FFT and an n''-point Winograd small FFT, respectively. Because it does not matter which of these small Winograd fast Fourier transforms is computed first, it seems that perhaps they should be done together. Indeed, we shall use Winograd nesting, as in the previous section, to bind the two algorithms together and reduce the amount of computation.

The procedure that we shall develop is illustrated by the example of a twelve-point FFT. Figure 12.4 shows how the one-dimensional data is turned into a two-dimensional array by using the Good–Thomas algorithm to obtain a two-dimensional Fourier transform, and then back into another one-dimensional array by stacking the columns of the two-dimensional array. All of this manipulation can be done with indexing–it is not necessary to physically rearrange the data. A Kronecker product can be found in the two-dimensional matrix, as shown in Figure 12.5, because of the way the two-dimensional data array is mapped into a one-dimensional array by stacking columns.

Let $\mathbf{W}'$ and $\mathbf{W}''$ be matrix representations of Fourier transforms of blocklength n' and n'', respectively, where $n'n'' = n$. In this representation, the Fourier transforms are

$V' = \mathbf{W}'v'$,
$V'' = \mathbf{W}''v''$.

12.4 The Winograd large fast Fourier transform

(a) Input scrambling

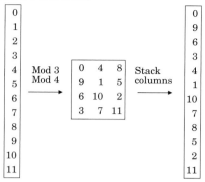

(b) Output unscrambling

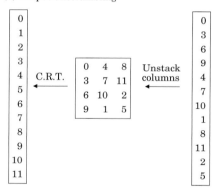

Figure 12.4 Scrambling indices for a 12-point Winograd transform

An n' by n'' two-dimensional Fourier transform of the two-dimensional signal $v_{i'i''}$ is obtained by applying $\boldsymbol{W}'$ to each column, and then applying $\boldsymbol{W}''$ to each row. The n' by n'' two-dimensional signal $v_{i'i''}$, which was previously obtained by rearranging a one-dimensional signal, can be turned back into a one-dimensional signal by reading it by columns.

If we think of $\boldsymbol{v}$ and $\boldsymbol{V}$ as permuted into scrambled one-dimensional n-point vectors in this way, then the transform can be written by using a Kronecker product

$$\boldsymbol{V} = (\boldsymbol{W}'' \times \boldsymbol{W}')\boldsymbol{v}.$$

But

$$\boldsymbol{W}' = \boldsymbol{C}'\boldsymbol{B}'\boldsymbol{A}',$$
$$\boldsymbol{W}'' = \boldsymbol{C}''\boldsymbol{B}''\boldsymbol{A}'',$$

where $\boldsymbol{A}'$, $\boldsymbol{A}''$, $\boldsymbol{C}'$, and $\boldsymbol{C}''$ are matrices of zeros and ones, and $\boldsymbol{B}'$ and $\boldsymbol{B}''$ are diagonal matrices. In the same way as we have seen in Section 12.3, by using Theorem 2.5.5,

398 Fast algorithms and multidimensional transforms

$$\begin{bmatrix} V_0 \\ V_1 \\ V_2 \\ V_3 \\ V_4 \\ V_5 \\ V_6 \\ V_7 \\ V_8 \\ V_9 \\ V_{10} \\ V_{11} \end{bmatrix} = \begin{bmatrix} \omega^0 & \omega^0 & \omega^0 & \omega^0 & \omega^0 & \omega^0 & \omega^0 & \omega^0 & \omega^0 & \omega^0 & \omega^0 & \omega^0 \\ \omega^0 & \omega^1 & \omega^2 & \omega^3 & \omega^4 & \omega^5 & \omega^6 & \omega^7 & \omega^8 & \omega^9 & \omega^{10} & \omega^{11} \\ \omega^0 & \omega^2 & \omega^4 & \omega^6 & \omega^8 & \omega^{10} & \omega^0 & \omega^2 & \omega^4 & \omega^6 & \omega^8 & \omega^{10} \\ \omega^0 & \omega^3 & \omega^6 & \omega^9 & \omega^0 & \omega^3 & \omega^6 & \omega^9 & \omega^0 & \omega^3 & \omega^6 & \omega^9 \\ \omega^0 & \omega^4 & \omega^8 & \omega^0 & \omega^4 & \omega^8 & \omega^0 & \omega^4 & \omega^8 & \omega^0 & \omega^4 & \omega^8 \\ \omega^0 & \omega^5 & \omega^{10} & \omega^3 & \omega^8 & \omega^1 & \omega^6 & \omega^{11} & \omega^4 & \omega^9 & \omega^2 & \omega^7 \\ \omega^0 & \omega^6 & \omega^0 & \omega^6 & \omega^0 & \omega^6 & \omega^0 & \omega^6 & \omega^0 & \omega^6 & \omega^0 & \omega^6 \\ \omega^0 & \omega^7 & \omega^2 & \omega^9 & \omega^4 & \omega^{11} & \omega^6 & \omega^1 & \omega^8 & \omega^3 & \omega^{10} & \omega^5 \\ \omega^0 & \omega^8 & \omega^4 & \omega^0 & \omega^8 & \omega^4 & \omega^0 & \omega^8 & \omega^4 & \omega^0 & \omega^8 & \omega^4 \\ \omega^0 & \omega^9 & \omega^6 & \omega^3 & \omega^0 & \omega^9 & \omega^6 & \omega^3 & \omega^0 & \omega^9 & \omega^6 & \omega^3 \\ \omega^0 & \omega^{10} & \omega^8 & \omega^6 & \omega^4 & \omega^2 & \omega^0 & \omega^{10} & \omega^8 & \omega^6 & \omega^4 & \omega^2 \\ \omega^0 & \omega^{11} & \omega^{10} & \omega^9 & \omega^8 & \omega^7 & \omega^6 & \omega^5 & \omega^4 & \omega^3 & \omega^2 & \omega^1 \end{bmatrix} \begin{bmatrix} v_0 \\ v_1 \\ v_2 \\ v_3 \\ v_4 \\ v_5 \\ v_6 \\ v_7 \\ v_8 \\ v_9 \\ v_{10} \\ v_{11} \end{bmatrix}$$

$$\begin{bmatrix} V_0 \\ V_3 \\ V_6 \\ V_9 \\ V_4 \\ V_7 \\ V_{10} \\ V_1 \\ V_8 \\ V_{11} \\ V_2 \\ V_5 \end{bmatrix} = \begin{bmatrix} \omega^0 & \omega^0 & \omega^0 & \omega^0 & \omega^0 & \omega^0 & \omega^0 & \omega^0 & \omega^0 & \omega^0 & \omega^0 & \omega^0 \\ \omega^0 & \omega^3 & \omega^6 & \omega^9 & \omega^0 & \omega^3 & \omega^6 & \omega^9 & \omega^0 & \omega^3 & \omega^6 & \omega^9 \\ \omega^0 & \omega^6 & \omega^0 & \omega^6 & \omega^0 & \omega^6 & \omega^0 & \omega^6 & \omega^0 & \omega^6 & \omega^0 & \omega^6 \\ \omega^0 & \omega^9 & \omega^6 & \omega^3 & \omega^0 & \omega^9 & \omega^6 & \omega^3 & \omega^0 & \omega^9 & \omega^6 & \omega^3 \\ \omega^0 & \omega^0 & \omega^0 & \omega^0 & \omega^4 & \omega^4 & \omega^4 & \omega^4 & \omega^8 & \omega^8 & \omega^8 & \omega^8 \\ \omega^0 & \omega^3 & \omega^6 & \omega^9 & \omega^4 & \omega^7 & \omega^{10} & \omega^1 & \omega^8 & \omega^{11} & \omega^2 & \omega^5 \\ \omega^0 & \omega^6 & \omega^0 & \omega^6 & \omega^4 & \omega^{10} & \omega^4 & \omega^{10} & \omega^8 & \omega^2 & \omega^8 & \omega^2 \\ \omega^0 & \omega^9 & \omega^6 & \omega^3 & \omega^4 & \omega^1 & \omega^{10} & \omega^7 & \omega^8 & \omega^5 & \omega^2 & \omega^{11} \\ \omega^0 & \omega^0 & \omega^0 & \omega^0 & \omega^8 & \omega^8 & \omega^8 & \omega^8 & \omega^4 & \omega^4 & \omega^4 & \omega^4 \\ \omega^0 & \omega^3 & \omega^6 & \omega^9 & \omega^8 & \omega^{11} & \omega^2 & \omega^5 & \omega^4 & \omega^7 & \omega^{10} & \omega^1 \\ \omega^0 & \omega^6 & \omega^0 & \omega^6 & \omega^8 & \omega^2 & \omega^8 & \omega^2 & \omega^4 & \omega^{10} & \omega^4 & \omega^{10} \\ \omega^0 & \omega^9 & \omega^6 & \omega^3 & \omega^8 & \omega^5 & \omega^2 & \omega^{11} & \omega^4 & \omega^1 & \omega^{10} & \omega^7 \end{bmatrix} \begin{bmatrix} v_0 \\ v_9 \\ v_6 \\ v_3 \\ v_4 \\ v_1 \\ v_{10} \\ v_7 \\ v_8 \\ v_5 \\ v_2 \\ v_{11} \end{bmatrix}$$

$$\begin{bmatrix} V_0 \\ V_3 \\ V_6 \\ V_9 \\ V_4 \\ V_7 \\ V_{10} \\ V_1 \\ V_8 \\ V_{11} \\ V_2 \\ V_5 \end{bmatrix} = \left\{ \begin{bmatrix} \omega^0 & \omega^0 & \omega^0 \\ \omega^0 & \omega^4 & \omega^8 \\ \omega^0 & \omega^8 & \omega^4 \end{bmatrix} \times \begin{bmatrix} \omega^0 & \omega^0 & \omega^0 & \omega^0 \\ \omega^0 & \omega^3 & \omega^6 & \omega^9 \\ \omega^0 & \omega^6 & \omega^0 & \omega^6 \\ \omega^0 & \omega^9 & \omega^6 & \omega^3 \end{bmatrix} \right\} \begin{bmatrix} v_0 \\ v_9 \\ v_6 \\ v_3 \\ v_4 \\ v_1 \\ v_{10} \\ v_7 \\ v_8 \\ v_5 \\ v_2 \\ v_{11} \end{bmatrix}$$

Figure 12.5 Rearranging a 12-point Fourier transform as a Kronecker product

this becomes

$$W = (C''B''A'') \times (C'B'A')$$
$$= (C'' \times C')(B'' \times B')(A'' \times A')$$
$$= CBA,$$

where the Kronecker products $C = C'' \times C'$ and $A = A'' \times A'$ are matrices of zeros and ones, and the Kronecker product $B = B'' \times B'$ is again a diagonal matrix with each element purely real or purely imaginary. Hence we have an $n'n''$ Fourier transform algorithm, again in the form of the Winograd FFT:

$$V = CBAv.$$

The algorithm that we have derived requires that v is in the scrambled order determined by the Good–Thomas algorithm, and it computes V in a scrambled order. However, once matrices A and C are derived, it is trivial to rearrange the columns of A so that v is in its natural order and to rearrange the rows of C so that V is in its natural order. The final form is referred to as the Winograd large FFT.

Let $M(n')$ and $M(n'')$ be the dimensions of matrices B' and B'', respectively. These are the numbers of multiplications required of the n'-point and n''-point Winograd FFT algorithms, respectively, including multiplications by one. Then $M(n) = M(n')M(n'')$ is the number of multiplications needed by the $n'n''$-point Winograd large FFT, including multiplications by one. This is because $B'' \times B'$ is again a diagonal matrix of dimension $M(n')M(n'')$.

Figure 12.6 gives a menu for constructing a Winograd large FFT algorithm from small FFT algorithms, and Table 12.2 shows the performance of the large FFT algorithms. An example of a 1008-point transform is given in Figure 12.7. This FFT algorithm will use 3564 real multiplications, the product of the number of multiplications in the seven, nine, and 16 point Winograd small FFT algorithms, times two for complex data because each element of the diagonal matrix is either purely real or purely imaginary.

If one wants a more regular FFT algorithm with data handled in smaller blocks, one can structure the algorithm as shown in Figure 12.8. In this case, only the nine-point and the seven-point Fourier transforms are merged. The 1008-point transform then appears as a 63-point by 16-point two-dimensional Fourier transform; each component is computed with a Winograd FFT. The number of real multiplications is $4396 = 2(16M(63) + 63M(16))$.

12.5 The Johnson–Burrus fast Fourier transform

We have seen two methods to bind together Winograd small FFT algorithms. These are the Good–Thomas prime factor nesting scheme and the Winograd nesting. The

Table 12.2 *Performance of the Winograd large FFT*

Blocklength n	Number of real multiplications*		Number of real additions*	Nontrivial multiplications per point*	Additions per point*
	Total	Nontrivial			
15	18	17	81	1.13	5.4
21	27	26	150	1.24	7.14
30	36	34	192	1.13	6.40
35	54	53	333	1.51	9.51
48	54	46	318	0.96	6.62
63	99	98	704	1.56	11.17
80	94	86	676	1.07	8.45
120	144	138	1038	1.15	8.65
168	216	210	1746	1.25	10.39
240	324	316	2508	1.32	10.45
420	648	644	5676	1.53	13.51
504	792	786	7270	1.56	14.42
840	1296	1290	12 402	1.54	14.76
1008	1782	1774	17 334	1.76	17.20
2520	4752	4746	49 814	1.88	19.77

* Double for complex input data.

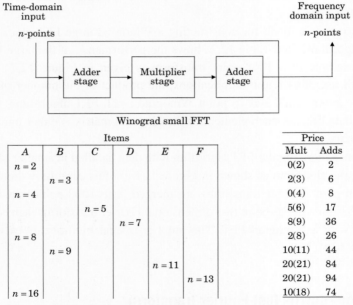

Figure 12.6 A menu of Winograd small transform

12.5 The Johnson–Burrus fast Fourier transform

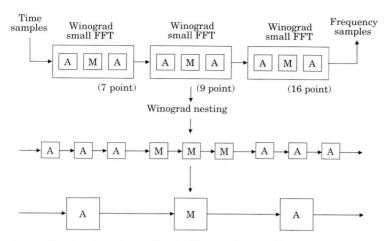

Figure 12.7 Structure of a 1008-point Winograd large FFT

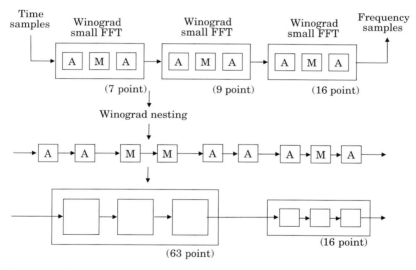

Figure 12.8 Structure of another 1008-point transform

Johnson–Burrus fast Fourier transform algorithms are a whole family of nested algorithms that include the Good–Thomas and Winograd methods as special cases. The idea of the Johnson–Burrus FFT is to moderate the use of the Kronecker product theorem in reordering the computations. In this way, one has the control to reduce the number of multiplications while keeping the number of additions small. One also can obtain FFT architectures that have improved control and data-flow properties.

We shall discuss a Fourier transform whose blocklength n has been broken into two factors n' and n''. One can also apply the same method if there are more than two factors, and the number of design options can become enormous. For example, when n

has four factors, there can be more than 1012 different Johnson–Burrus FFT algorithms built from the same Winograd small FFT algorithms.

Suppose that an n-point input vector has been mapped into a n' by n'' two-dimensional array V along the extended diagonal, provided n' and n'' are coprime. The Good–Thomas algorithm tells us that a one-dimensional Fourier transform of the original data is computed by first taking the n'-point Fourier transform of each column followed by the n''-point Fourier transform of each row. We will represent this by

$$V = W''W'v.$$

This is an unconventional notation because v is a two-dimensional array. By $W'v$, we mean to multiply every column of v, one column at a time, by the matrix W'. By $W''v$, we mean to multiply every row of v, one row at a time, by W''. It would be more precise to append further notation to distinguish matrices that operate on columns from matrices that operate on rows, but we will rely on the prime and double prime to convey this information.

Because it does not matter whether the Fourier transforms along rows or the Fourier transforms along columns are computed first, we can also write

$$V = W'W''v.$$

That is, operations on rows commute with operations on columns. This is nothing more than a straightforward interchange in the order of summation.

Now suppose we have the Winograd small FFT algorithms

$$W' = C'B'A',$$
$$W'' = C''B''A''.$$

Then

$$V = (C''B''A'')(C'B'A')v$$
$$= (C'B'A')(C''B''A'')v.$$

Two matrices with the same number of primes do not commute. However, the Winograd nesting tells us that we can commute matrices having primes with matrices having double primes. We have already appealed to the Kronecker product theorem to prove this. Again, this amounts to nothing more than a straightforward interchange in the order of summation. Thus we can write

$$V = C'C''B'B''A'A''v,$$

which is the form of a Winograd large FFT. We also have other options such as

$$V = C'B'C''B''A'A''v.$$

To get the Johnson–Burrus FFT, we need one more trick. Each of the addition matrices can be factored further before reordering them. We write

$$W' = (G'H')B'(E'F'),$$
$$W'' = (G''H'')B''(E''F''),$$

anticipating that C' can be factored as $G'H'$, and so forth. Now we have

$$V = G'H'B'E'F'G''H''B''E''F''v,$$

which can be rendered in many ways, such as

$$V = (G'G''H'H'')(B'B'')(E'E''F'F'')v.$$

In this instance, the two diagonal matrices B' and B'' have been brought together, and the parentheses are intended only to highlight this.

A 35-point Fourier transform illustrates the ideas nicely. In Figure 12.9, the five-point and seven-point Winograd FFTs are given, but each is factored into five stages. There are many ways to combine these to obtain a 35-point FFT. Figure 12.9 gives three that are the most interesting. Of these, one method uses Good–Thomas nesting and one uses Winograd nesting. The Good–Thomas method results in the fewest additions, while the Winograd method results in the fewest multiplications. There is no need to decide between these two alternatives, however. There is a Johnson–Burrus nesting method that has the same number of multiplications as the Winograd method and nearly the same number of additions as the Good–Thomas method, and so will usually be preferred.

There is no penalty in using the Johnson–Burrus FFT. All that it entails is breaking the additions into several packages that can form subroutines and then calling these subroutines in the right sequence. The computations jump back and forth between additions along rows and additions along columns.

12.6 Splitting algorithms

This chapter opened with a discussion of how fast algorithms for a two-dimensional Fourier transform can be constructed by combining fast algorithms for a one-dimensional Fourier transform. Then we discussed fast algorithms for a one-dimensional Fourier transform by turning it into a two-dimensional Fourier transform. Now we return to the case of two-dimensional Fourier transforms, this time working with the two-dimensional Fourier transform directly. The Winograd small FFT algorithm, developed in Chapter 3, uses the Rader prime algorithm to turn a one-dimensional Fourier transform into a one-dimensional cyclic convolution. In this chapter we shall apply the same procedure for a multidimensional Fourier transform.

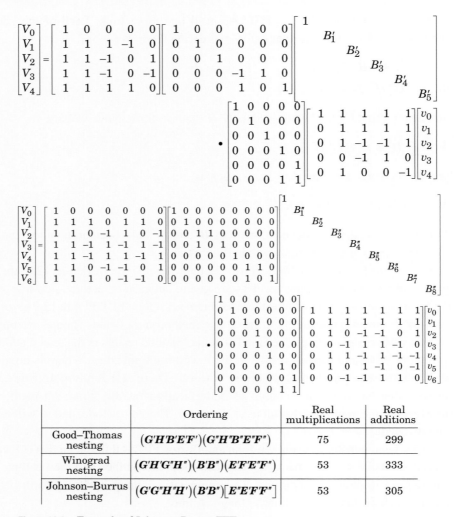

Figure 12.9 Example of Johnson–Burrus FFT

Generally, the transform will split into several one-dimensional and two-dimensional cyclic convolutions. Each dimension of the multidimensional Fourier transform must have a blocklength that is a prime or a prime power, but the blocklength need not be the same on each axis. First, the Rader algorithm or its generalization is applied on each axis to change the multidimensional Fourier transform into a multidimensional convolution. Then a fast multidimensional convolution algorithm is used. In those cases in which the blocklength is the same on every axis, the method to be described in Section 12.7 gives better performance and so should usually be preferred. Therefore the splitting algorithms are of interest primarily for the case in which the blocklength is not the same on every axis.

12.6 Splitting algorithms

The idea is a simple generalization of the method used in the one-dimensional case. We begin with the n' by n'' two-dimensional Fourier transform

$$V_{k'k''} = \sum_{i'=0}^{n'-1} \sum_{i''=0}^{n''-1} \omega^{i'k'} \mu^{i''k''} v_{i'i''}, \qquad \begin{array}{l} k' = 0, \ldots, n'-1, \\ k'' = 0, \ldots, n''-1, \end{array}$$

where n' and n'' are each a prime, possibly different primes. To apply the Rader algorithm, we must restructure the equations to exclude zero as an exponent of ω or μ. Therefore break the equation into four cases by analogy with the Rader algorithm

$$V_{00} = \sum_{i'=0}^{n'-1} \sum_{i''=0}^{n''-1} v_{i'i''},$$

$$V_{0k''} - V_{00} = \sum_{i''=1}^{n''-1} (\mu^{i''k''} - 1) \sum_{i'=0}^{n'-1} v_{i'i''}, \qquad k'' = 1, \ldots, n''-1,$$

$$V_{k'0} - V_{00} = \sum_{i'=1}^{n'-1} (\omega^{i'k'} - 1) \sum_{i''=0}^{n''-1} v_{i'i''}, \qquad k' = 1, \ldots, n'-1,$$

$$V_{k'k''} - V_{k'0} - V_{0k''} + V_{00} = \sum_{i'=1}^{n'-1} \sum_{i''=1}^{n''-1} (\omega^{i'k'} - 1)(\mu^{i''k''} - 1) v_{i'i''}, \qquad \begin{array}{l} k' = 1, \ldots, n'-1, \\ k'' = 1, \ldots, n''-1. \end{array}$$

With the problem so broken into four equations, and the permutations of the Radar algorithm applied to the last three equations, the computation of the Fourier transform is replaced with a simple sum, an $(n'-1)$-point convolution, an $(n''-1)$-point convolution, and an $(n'-1)$ by $(n''-1)$-point two-dimensional convolution. The Winograd convolution algorithms, studied in Chapter 3, can be used for the two one-dimensional convolutions. Theorem 3.8.1 and Theorem 3.8.2 tell us that each multiplication in the algorithm is a multiplication by a purely real or a purely imaginary constant, and so each takes only one real multiplication if the input array is real.

Any good method may be used for the two-dimensional convolution. One such method is the use of a polynomial transform, which was studied in Chapter 11. An $(n'-1)$ by $(n''-1)$-point two-dimensional convolution algorithm can be obtained by nesting any suitable small two-dimensional cyclic convolution algorithms or even small one-dimensional cyclic convolution algorithms. For example, a four by 12 two-dimensional convolution can be converted to a four by four by three three-dimensional cyclic convolution algorithm by using the Agarwal–Cooley algorithm on the second axis. This then can be computed by nesting a four by four two-dimensional cyclic convolution algorithm with a three-point cyclic convolution algorithm.

When a two-dimensional convolution algorithm is used to form a two-dimensional Fourier transform, the multiplications always turn out to have one factor as a purely real or a purely imaginary number. A special case of this is given in the next theorem as an analog of Theorem 3.8.1.

Table 12.3 *Performance of some splitting algorithms*

Array size n by n	Number of real multiplications*		Number of real additions*	Nontrivial multiplications* per output point	Additions* per output point
	Total	Nontrivial			
5 by 5	33	32	230	1.28	9.20
7 by 7	69	68	650	1.39	13.26
9 by 9	109	108	908	1.33	11.21
7 by 9	87	86	712	1.36	11.30
5 by 13	114	113	917	1.74	14.10

* Double for complex input data.

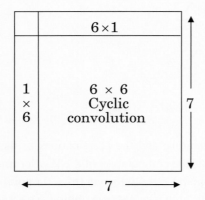

Figure 12.10 Breaking a seven by seven Fourier transform

Theorem 12.6.1 *Let $g(x, y)$ be a two-dimensional $(p-1)$ by $(p-1)$ Rader polynomial, where p is an odd prime. Let $\Phi(x)$ and $\Phi'(x)$ be two cyclotomic polynomials dividing $x^p - 1$. Then $g(x, y)$ (mod $\Phi(x)$) and (mod $\Phi'(y)$) has coefficients that are either purely real or purely imaginary.*

Proof Similar to the proof of Theorem 3.8.1. □

A more general theorem can be proved for the case in which the two blocklengths are primes or prime powers, not necessarily equal.

A tabulation of the performance for some two-dimensional Fourier transform algorithms is shown in Table 12.3. The performance of the algorithms should be compared with the performance of the algorithms described later in Section 12.7 to see that the p by p FFT algorithms described here are not quite as good as those FFT algorithms.

As an example, consider the seven by seven two-dimensional Fourier transform. This transform decomposes into two six-point cyclic convolutions and one six by six two-dimensional cyclic convolution, as shown in Figure 12.10. Each six-point convolution uses eight real multiplications. The six by six two-dimensional convolution can be

12.6 Splitting algorithms

formed by nesting the two by two cyclic convolution algorithm and the three by three cyclic convolution algorithm given in Table 11.4. This takes 52 multiplications, so there is a total of 68 multiplications. An additional trivial multiplication (by 1) is needed to bring the computation of V_{00} into the same format. This becomes essential if the seven by seven FFT is to be nested with another two-dimensional FFT to get a larger two-dimensional FFT.

To count additions is a little more cumbersome, and the count can be changed by making finer adjustments in the organization of the equations. First, we look at the equations as we have already structured them. The computation consists of some preadditions, followed by calling a six-point cyclic convolution algorithm twice and a six by six cyclic convolution algorithm once, followed by some postadditions. There will also be some preadditions and postadditions within the convolution algorithms.

The additions can be counted as follows:

Preadditions	84
Two six-point cyclic convolutions	68
Six by six cyclic convolution	424
Postadditions	84
	660

However, we know that if all of the preadditions are gathered together, including those in the convolutions, it may be possible to rearrange them to reduce their number. Similarly, it may be possible to reduce the number of postadditions by gathering them together. This we will not do in full generality because we have no general theory for minimizing the number of additions, and because it would result in an algorithm that is not neatly broken into small subroutines as we now have. However, we can still do something. The six-point cyclic convolutions can be combined with some of the preadditions and postadditions to become six-point Fourier transforms. This does not change the number of multiplications. Instead, we can rearrange the computation into the form of some preadditions, followed by one seven-point Fourier transform, followed by one six-point cyclic convolution, followed by a six by six cyclic convolution, followed by some postadditions. The number of additions will reduce as follows.

Preadditions	78
Seven-point FFT	36
Six-point cyclic convolution	34
Six by six cyclic convolution	424
Postadditions	78
	650

which is a small improvement.

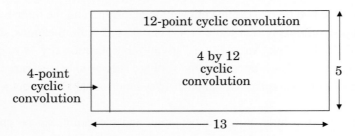

Figure 12.11 Breaking a five by 13 Fourier transform

The second example is a five by 13 two-dimensional Fourier transform. This will break into a four by 12 two-dimensional cyclic convolution, a four-point one-dimensional cyclic convolution, and a 12-point one-dimensional cyclic convolution, as shown in Figure 12.11. The four by 12 two-dimensional cyclic convolution can be computed by changing it to a four by four by three three-dimensional cyclic convolution, then nesting a four by four cyclic convolution algorithm and a three-point cyclic convolution algorithm to get an algorithm with 88 real multiplications and 608 real additions. Appending a four-point cyclic convolution and a 12-point cyclic convolution algorithm (with 20 real multiplications and 100 real additions) and one trivial multiplication for the term V_{00} will yield a five by 13 FFT with 114 real multiplications and 939 real additions.

A slightly better five by 13 FFT can be formed if a subroutine is available for the 13-point FFT that uses 20 nontrivial real multiplications and 94 real additions. Using this in place of the 12-point cyclic convolution will yield a five by 13 FFT with 114 real multiplications and 917 real additions.

The method also works when the blocklengths are powers of a prime. Now the index sets must be broken into more complicated subsets. For an example, consider a seven by nine two-dimensional FFT. Because nine is not itself a prime, this construction requires a more general method than we have discussed above. When generalizing the Rader prime algorithm to nine points in Chapter 3, we had to pull out all indices not coprime to nine to get a six-point cyclic convolution and two two-point cyclic convolutions. We do the same thing here to find a six by six two-dimensional cyclic convolution and two copies of a two by six two-dimensional cyclic convolution. Figure 12.12 shows that the seven by nine FFT can be built from one six by six cyclic convolution, two six by two cyclic convolutions, one nine-point FFT, and one six-point cyclic convolution. This appears to take a total of 102 multiplications ($52 + 2 \cdot 16 + 10 + 8$), but we can do better if we build the structure carefully. We have not yet used the fact that the six-point convolution that comes from the nine-point Fourier transform by the Rader algorithm leads to several multiplications by zero. These multiplications, identified by Theorem 3.8.2, need not be counted.

The six by six convolution arises from a computation of the two-dimensional version of the generalized Rader algorithm. The two-dimensional Rader prime algorithm for

12.6 Splitting algorithms

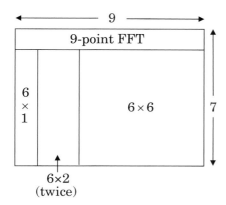

Figure 12.12 Breaking a seven by nine Fourier transform

	x^2+x+1	x^2-1	$x+1$	x^2-x+1	
	6	0 (3)	0 (3)	6	y^2+y+1
	3	0 (1)	0 (1)	3	$y-1$
	3	0 (1)	0 (1)	3	$y+1$
	6	0 (3)	0 (3)	6	y^2-y+1

Figure 12.13 Partition of a six by six convolution

the seven by nine Fourier transform is

$$g(x, y) = \left[\sum_{j=0}^{5}(\omega^{3j} - 1)x^j\right]\left[\sum_{k=0}^{5}(\mu^{2k} - 1)y^k\right],$$

where ω is a sixth root of unity and μ is a ninth root of unity.

To construct the six by six cyclic convolution algorithm, we write

$$x^6 - 1 = (x^2 + x + 1)(x - 1)(x + 1)(x^2 - x + 1),$$
$$y^6 - 1 = (y^2 + y + 1)(y - 1)(y + 1)(y^2 - y + 1)$$

and compute the residues of $g(x, y)$ modulo these factors. Figure 12.13 shows how the computation will break into 16 pieces. But by Theorem 3.8.2, we know that the pieces modulo $(y - 1)$ and $(y + 1)$ are zero and can be dropped.

In Figure 12.13, each rectangle shows the number of multiplications needed for that piece and, in parentheses, the number of additional multiplications needed if the residues in those blocks are not equal to zero. Hence a total of 36 multiplications is needed for the six by six cyclic convolution, and so 86 nontrivial multiplications are needed for the seven by nine-point FFT.

12.7 An improved Winograd fast Fourier transform

The performance of the methods developed in the previous section for computing a two-dimensional Fourier transform motivates us to return to the computation of the one-dimensional Fourier transform and to apply these same techniques to that problem. Whenever a one-dimensional Fourier transform can be converted into a multidimensional Fourier transform by using the Good–Thomas algorithm, then by using the methods of Section 12.6, it can be computed by turning it into a multidimensional convolution. A one-dimensional Fourier transform algorithm formed in this way is like a Winograd large FFT, but with an extra twist. Recall that the Winograd large FFT combines two or more small FFT algorithms but does not change the structure of the underlying small algorithms. However, each small FFT algorithm was constructed from a convolution algorithm, so we may expect to find a multidimensional convolution in the large problem. The Winograd large FFT does not go deep enough to find this multidimensional convolution, instead it nests two one-dimensional FFT algorithms that are each formed from a one-dimensional convolution algorithm. If the multidimensional convolution is large enough, there may be an improvement in performance if a stronger algorithm is used. The performance of an $n'n''$-point improved Winograd FFT is better than the $n'n''$-point Winograd large FFT whenever $\phi(n')$ and $\phi(n'')$ have a common factor at least as large as four. The performance of some of these improved FFT algorithms is tabulated in Table 12.4. This table should be compared with Table 12.2.

The improved algorithm is a combination of three ideas: the use of the Good–Thomas algorithm to convert a one-dimensional Fourier transform into a multidimensional Fourier transform, a multidimensional version of the Rader algorithm, and the use of a polynomial transform to compute the multidimensional Rader convolution. The first of these ideas is an indexing operation. The second and third ideas are already combined in the splitting algorithm of the previous section. To obtain the improved FFT, we need only to modify one of these two-dimensional algorithms by permuting the columns of the matrix of preadditions and permuting the rows of the matrix of postadditions. First consider a fifteen-point Fourier transform. This can be converted into a three by five two-dimensional Fourier transform. By using the Rader algorithm on each axis, we extract a two by four two-dimensional cyclic convolution, which we express as

$$s(x, y) = g(x, y)d(x, y) \pmod{x^4 - 1}\pmod{y^2 - 1}.$$

12.8 The Nussbaumer–Quandalle permutation algorithm

Table 12.4 *Performance of improved fast Fourier transform algorithms*

Blocklength n	Number of real multiplications*		Number of real additions*	Nontrivial multiplications per point*	Additions* per point
	Total	Nontrivial			
15**	18	17	103	1.13	6.87
21**	27	26	184	1.24	8.76
35**	54	53	411	1.51	11.74
63	86	85	712	1.37	11.30
65	114	113	917	1.74	14.10
91	159	158	1560	1.75	17.14

* Double for complex input data.
** Same number of multiplications and Winograd large FFT.

These convolutions are so short that the two-dimensional convolution is best computed by nesting two one-dimensional convolution algorithms. When n equals 15, the improved FFT is not better than the Winograd large FFT. In fact, it is not as good because there are more additions.

Our last example is a 63-point Fourier transform. The Winograd nesting algorithm gives an FFT with 98 nontrivial real multiplications and 704 real additions. The improved algorithm is derived by starting with the 63-point Fourier transform expressed in the form of a seven by nine two-dimensional Fourier transform and using the algorithm with the structure that was shown in Figure 12.12. The improved 63-point algorithm uses 85 nontrivial real multiplications and 712 real additions.

12.8 The Nussbaumer–Quandalle permutation algorithm

The last topic of this chapter deals again with the computation of the two-dimensional Fourier transform, now using yet another method. Recall that in Chapter 11, dealing with multidimensional convolution, we posed the multidimensional convolution as a convolution of polynomials, using the polynomial representation of extension fields to construct algorithms for multidimensional convolutions. Now we shall use the polynomial representation of extension fields to construct algorithms for the multidimensional Fourier transform. Of course, one of the applications of the multidimensional Fourier transform is to compute multidimensional convolutions, so indirectly we will have another way of computing multidimensional convolutions.

Once again, we are able to obtain good algorithms for one class of problems by twisting that class into the form of a class of problems already solved. This instance might appear especially tantalizing because we had introduced the polynomial transform as

Table 12.5 *Performance of the Nussbaumer–Quandalle FFT*

Array size n by n	Number of real multiplications*		Number of real additions*	Nontrivial multiplications* per output point	Additions* per output point
	Total	Nontrivial			
2 by 2	4	0	8	0	2
3 by 3	9	8	36	0.89	4
4 by 4	16	0	64	0	4
5 by 5	31	30	221	1.20	8.84
7 by 7	65	64	635	1.31	12.96
8 by 8	64	24	408	0.375	6.37
9 by 9	105	104	785	1.28	9.69
16 by 16	304	216	2264	0.84	8.85

* Double for complex input.

a notational variation of the Fourier transform, but then suppressed that viewpoint, treating the polynomial transform as its own case. Now we are shifting ground again and using the polynomial transform to compute a Fourier transform, albeit not the same one that originally gave rise to the polynomial transform.

The *Nussbaumer–Quandalle permutation algorithm* is a multidimensional fast Fourier transform algorithm. It is derived by changing a multidimensional Fourier transform into a number of one-dimensional Fourier transforms. This is different from the splitting algorithms of Section 11.3, which change a multidimensional Fourier transform into several multidimensional convolutions. The Nussbaumer–Quandalle fast Fourier transform has superior performance, but it can be used only when the blocklength is the same in all dimensions or has a common factor in all dimensions that can be extracted by using the Chinese remainder theorem. We shall study the Nussbaumer–Quandalle FFT in this section. The performance of the Nussbaumer–Quandalle FFT is shown in Table 12.5 for two dimensions and in Table 12.6 for three dimensions. Table 12.6 should be compared with Table 12.3.

The Nussbaumer–Quandalle algorithm can be used to compute a multidimensional Fourier transform with the same number of points n in each dimension, whenever n is a prime or a power of a prime. Three subcases must be treated separately: n is a prime p; n is a power of an odd prime p^m, and n is a power of two 2^m. These can be nested to form a larger two-dimensional Fourier transform algorithms. The nesting of multidimensional Fourier transform algorithms can be done in the same way as was done for a one-dimensional transform. For example, a 63 by 63 Fourier transform is first mapped into a (seven by nine) by (seven by nine) transform. This four-dimensional Fourier transform is viewed as a (seven by seven) by (nine by nine) transform, which is computed by nesting a seven by seven algorithm and a nine by nine algorithm.

12.8 The Nussbaumer–Quandalle permutation algorithm

Table 12.6 *Performance of the three-dimensional Nussbaumer–Quandalle*

Array size n by n by n	Number of real multiplications* Total	Number of real multiplications* Nontrivial	Number of real additions*	Nontrivial multiplications* per output point	Additions* per output point
2 by 2 by 2	8	0	24	0.00	3.0
3 by 3 by 3	27	26	162	0.96	6.0
4 by 4 by 4	64	0	384	0.00	6.0
5 by 5 by 5	156	155	1686	1.24	13.5
7 by 7 by 7	457	456	6767	1.33	19.7
8 by 8 by 8	512	224	4832	0.44	9.4
9 by 9 by 9	963	962	10 383	1.32	14.2
16 by 16 by 16	4992	3808	52 960	0.93	12.9

* Double for complex input.

This will entail 49 applications of a nine by nine algorithm to 49 subarrays of the data set, followed by 81 applications of a seven by seven algorithm to 81 subarrays of the data set, provided that we view each seven by seven subarray as a 49 vector and each nine by nine subarray, in turn, as an 81 vector. The algorithms can be put in the standard form of a matrix of postadditions, followed by a diagonal matrix, followed by a matrix of postadditions. The required manipulation of the 63 by 63 array is done by indexing. No arithmetic is required, nor is it necessary to physically move the data to view it differently. Then, just as for the one-dimensional algorithms, the components can be merged by using the Kronecker product theorem. This amounts to moving all preadditions before all multiplications and all multiplications before all postadditions. The total number of multiplications needed for an $n'n''$ by $n'n''$ Fourier transform when built in this way by nesting is

$$M(n'n'' \times n'n'') = M(n' \times n')M(n'' \times n''),$$

and the number of trivial multiplications satisfies an equation of the same form. The number of additions is

$$A(n'n'' \times n'n'') = (n')^2 A(n'' \times n'') + M(n'' \times n'')A(n' \times n').$$

These formulas have the same explanation as the corresponding formulas for nesting one-dimensional transform algorithms. The performance of nested Nussbaumer–Quandalle FFT algorithms is shown in Table 12.7.

The algorithms we study here will replace a p by p two-dimensional Fourier transform with $p+1$ one-dimensional Fourier transforms whenever p is a prime. To illustrate the structure that we are working toward, we begin with the three by three

Fast algorithms and multidimensional transforms

Table 12.7 *Performance of some nested FFT algorithms*

Array size n by n	Number of real multiplications* Total	Number of real multiplications* Nontrivial	Number of real additions*	Nontrivial multiplications* per output point	Additions* per output point
12 by 12	144	128	1 152	0.89	8.00
15 by 15	279	278	2 889	1.24	12.84
21 by 21	585	584	7 499	1.32	16.96
30 by 30	1 116	1 112	13 356	1.24	14.84
35 by 35	2 015	2 014	30 240	1.64	24.90
48 by 48	2 736	2 648	29 592	1.15	12.84
63 by 63	6 825	6 824	102 460	1.72	25.81
80 by 80	9 424	9 336	123 784	1.46	19.34
120 by 120	17 856	17 816	276 696	1.24	19.21
168 by 168	37 440	37 400	658 584	1.32	23.33
240 by 240	84 816	84 728	1 344 456	1.47	23.34
420 by 420	290 160	290 144	5 765 760	1.65	32.69
504 by 504	436 800	436 760	8 176 792	1.72	32.19
840 by 840	1 160 640	1 160 600	24 738 840	1.64	35.06
1008 by 1008	2 074 800	2 074 712	40 133 656	2.04	39.50
2520 by 2520	13 540 800	13 540 760	298 481 560	2.13	47.00

* Double for complex input.

Fourier transform as an example. If the three by three input array and the three by three output array are represented as column vectors by stacking columns, then the three by three Fourier transform can be expressed as

$$\begin{bmatrix} V_{00} \\ V_{10} \\ V_{20} \\ V_{01} \\ V_{11} \\ V_{21} \\ V_{02} \\ V_{12} \\ V_{22} \end{bmatrix} = \begin{bmatrix} 1 & 1 & 1 & 1 & 1 & 1 & 1 & 1 & 1 \\ 1 & \omega & \omega^2 & 1 & \omega & \omega^2 & 1 & \omega & \omega^2 \\ 1 & \omega^2 & \omega & 1 & \omega^2 & \omega & 1 & \omega^2 & \omega \\ 1 & 1 & 1 & \omega & \omega & \omega & \omega^2 & \omega^2 & \omega^2 \\ 1 & \omega & \omega^2 & \omega & \omega^2 & 1 & \omega^2 & 1 & \omega \\ 1 & \omega^2 & \omega & \omega & 1 & \omega^2 & \omega^2 & \omega & 1 \\ 1 & 1 & 1 & \omega^2 & \omega^2 & \omega^2 & \omega & \omega & \omega \\ 1 & \omega & \omega^2 & \omega^2 & 1 & \omega & \omega & \omega^2 & 1 \\ 1 & \omega^2 & \omega & \omega^2 & \omega & 1 & \omega & 1 & \omega^2 \end{bmatrix} \begin{bmatrix} v_{00} \\ v_{10} \\ v_{20} \\ v_{01} \\ v_{11} \\ v_{21} \\ v_{02} \\ v_{12} \\ v_{22} \end{bmatrix}$$

with ω an element of order three. The matrix has been blocked to show its origin as a Kronecker product. The equation can be abbreviated as

$$V = Wv,$$

12.8 The Nussbaumer–Quandalle permutation algorithm

where V and v denote the above output vector and input vector, respectively, and W is the matrix of powers of ω. Our goal in this section is to obtain the factorization

$$W = CBA,$$

where A and C are matrices of preadditions and postadditions, respectively, and B is a matrix representing a collection of one-dimensional Fourier transforms. The matrix B has the form

$$B = \begin{bmatrix} 1 & 1 & 1 & & & & & & & & & \\ 1 & \omega & \omega^2 & & & & & & & & & \\ 1 & \omega^2 & \omega & & & & & & & & & \\ & & & 1 & 1 & 1 & & & & & & \\ & & & 1 & \omega & \omega^2 & & & & & & \\ & & & 1 & \omega^2 & \omega & & & & & & \\ & & & & & & 1 & 1 & 1 & & & \\ & & & & & & 1 & \omega & \omega^2 & & & \\ & & & & & & 1 & \omega^2 & \omega & & & \\ & & & & & & & & & 1 & 1 & 1 \\ & & & & & & & & & 1 & \omega & \omega^2 \\ & & & & & & & & & 1 & \omega^2 & \omega \end{bmatrix},$$

where the unfilled blocks of B are all three by three zero matrices.

The factorization $W = CBA$ has a striking resemblance to the form of the Winograd small FFT. Now, however, we are working at a different level; the block diagonal matrix in the center represents a batch of one-dimensional Fourier transforms instead of a batch of multiplications. Once we find this factorization, the three by three Fourier transform is computed by executing the preadditions specified by the A matrix, followed by calling the three-point one-dimensional FFT four times, followed by executing the postadditions specified by the C matrix. We shall see later that we do not always need to compute all of the terms for some of the one-dimensional Fourier transforms. Routines called punctured FFT algorithms (and discussed later) could be used here to attain a slight further reduction in the amount of computation.

The one-dimensional n-point Fourier transform can be thought of as the process of evaluating a polynomial $v(x)$ at ω^k, where ω is an element of order n. Thus

$$V_k = v(\omega^k), \qquad k = 0, \ldots, n-1,$$

where

$$v(x) = \sum_{i=0}^{n-1} v_i x^i.$$

The same idea can be used for a two-dimensional Fourier transform. It turns out that it is more fruitful to treat only one dimension of the two-dimensional n by n array in this way. Toward this purpose, the n by n array is represented by a vector of polynomials of blocklength n:

$$v_{i''}(x) = \sum_{i'=0}^{n-1} v_{i'i''} x^{i'}, \qquad i'' = 0, \ldots, n-1.$$

Define

$$V_{k''}(x) = \sum_{i''=0}^{n-1} \omega^{i''k''} v_{i''}(x), \qquad k'' = 0, \ldots, n-1.$$

This is a one-dimensional Fourier transform of polynomials. Then the original two-dimensional Fourier transform, written as a polynomial remainder, is

$$V_{k'k''} = R_{x-\omega^{k'}}[V_{k''}(x)] = V_{k''}(\omega^{k'}), \qquad \begin{array}{l} k' = 0, \ldots, n-1, \\ k'' = 0, \ldots, n-1. \end{array}$$

With this description, the two axes of the Fourier transform are described differently and will play different roles.

Now we are ready to begin to change this into the form of a polynomial transform by a clever permutation technique. Let r be any positive integer coprime to n. Let k'' be expressed as $k'' = kr$ modulo n for some k in the set $\{0, \ldots, n-1\}$. This modulo n representation is a valid, though indirect, way to rewrite k'' because r and n are coprime, and so as k runs from 0 to $n-1$, k'' will take on all values from 0 to $n-1$. Then, because ω has order n, $\omega^{k''} = \omega^{kr}$, and we can write the previous two equations as

$$V_{((kr))}(x) = \sum_{i''=0}^{n-1} \omega^{i''kr} v_{i''}(x),$$

$$V_{k'((kr))} = R_{x-\omega^{k'}}[V_{((kr))}(x)], \qquad \begin{array}{l} k' = 0, \ldots, n-1, \\ k = 0, \ldots, n-1 \end{array}$$

for any r coprime to n, where the double parentheses denote modulo n. Indeed, when k' itself is coprime to n, we can even choose r equal to k' so that the second equation becomes

$$V_{k'((kk'))} = R_{x-\omega^{k'}}[V_{((kk'))}(x)], \qquad \begin{array}{l} k = 0, \ldots, n-1, \\ \mathrm{GCD}[k', n] = 1. \end{array}$$

Every component of the output array with first index coprime to n can be written in this way.

With this much as preliminaries, we are ready to bring in the polynomial transform. We will jump directly to the final result in the following theorem. When reading the theorem, notice that all k' that share the same cyclotomic polynomial will share the same computations, that the same index k' is used both as an output index and to form the permutation, and that division by k' modulo n is defined because k' is coprime to n.

12.8 The Nussbaumer–Quandalle permutation algorithm

Theorem 12.8.1 *Let k' be a fixed integer coprime to n. Let $\Phi(x)$ be the cyclotomic polynomial having $\omega^{k'}$ as a zero. Then $V_{k'k''}$ for $k'' = 0, \ldots, n-1$ and k' coprime to n can be evaluated by the following three steps: first the polynomial transform*

$$V'_k(x) = \sum_{i''=0}^{n-1} v_{i''}(x) x^{i''k} \pmod{\Phi(x)}, \qquad k = 0, \ldots, n-1,$$

followed by the set of n Fourier transforms

$$V''_{k'k} = \sum_{i'=0}^{n-2} \omega^{i'k'} V'_{i'k}, \qquad \begin{array}{l} k' : \mathrm{GCD}(k', n) = 1, \\ k = 0, \ldots, n-1, \end{array}$$

followed by the permutation

$$V_{k'k''} = V''_{k'((k''/k'))}, \qquad \begin{array}{l} k' : \mathrm{GCD}(k', n) = 1, \\ k'' = 0, \ldots, n-1. \end{array}$$

Proof We shall work with the equations

$$V_{((kk'))}(x) = \sum_{i''=0}^{n-1} \omega^{i''kk'} v_{i''}(x),$$

$$V_{k'((kk'))} = R_{x-\omega^{k'}}[V_{((kk'))}(x)].$$

Because $x - \omega^{k'}$ divides the cyclotomic polynomial $\Phi(x)$, the validity of the second equation is not changed by taking the first equation modulo $\Phi(x)$. Then the equations can be rewritten as

$$V_{((kk'))}(x) = \sum_{i''=0}^{n-1} v_{i''}(x) \omega^{k'i''k} \pmod{\Phi(x)},$$

$$V_{k'((kk'))} = R_{x-\omega^{k'}}[V_{((kk'))}(x)].$$

But now the second equation makes $\omega^{k'}$ congruent to x. Therefore we can replace $\omega^{k'}$ by x in the first equation without changing the final result. Then rewrite the two equations once more as

$$V'_{((kk'))}(x) = \sum_{i''=0}^{n-1} v_{i''}(x) x^{i''k} \pmod{\Phi(x)},$$

$$V_{k'((kk'))} = R_{x-\omega^{k'}}[V_{((kk'))}(x)] = \sum_{i'=0}^{n-2} \omega^{i'k'} V'_{i'((kk'))}.$$

We can now clean up the notation and rewrite the equations in final form to complete the proof of the theorem. □

Table 12.8 *Subroutines used by Nussbaumer–Quandalle FFT*

Blocklength n	Polynomial transforms	Fourier transforms	Extra additions
prime p	p-point transform modulo $(x^p - 1)/(x - 1)$	$(p + 1)$ transforms* of blocklength p	$p^3 + p^2 - 5p + 4$
prime power p^2	p^2-point transform modulo $(x^{p^2} - 1)/(x^p - 1)$	$(p^2 + p)$ transforms** of blocklength p^2	$2p^5 + p^4 - 5p^3 + p^2 + 6$
	p-point transform modulo $(x^{p^2} - 1)/(x^p - 1)$	$(p + 1)$ transforms* of blocklength p	
	p-point transform modulo $(x^p - 1)/(x - 1)$		
power of two 2^m	2^m-point transform modulo $x^{2^{m-1}} + 1$	$\frac{3}{2} 2^m$ transforms** of blocklength 2^m	$(3m + 5)2^{2t-2}$
	2^{m-1}-point transform modulo $x^{2^{m-1}} + 1$	1 two-dimensional transform of size $2^{m-1} \times 2^{m-1}$	

* of which p are punctured.
** of which all are punctured.

There are three cases to which the theorem can be applied: blocklength a prime, blocklength a power of an odd prime, and blocklength a power of two. We shall discuss each of these cases, in turn. Table 12.8 summarizes the conclusions that will follow.

Blocklength a prime For p a prime, we can use the theorem to change a two-dimensional p by p Fourier transform into $p + 1$ distinct one-dimensional Fourier transforms. In a more general form, the method will change an N-dimensional p-point Fourier transform into $(p^N - 1)/(p - 1)$ distinct one-dimensional p-point Fourier transforms. In contrast, the straightforward method, one dimension at a time, uses Np^{N-1} distinct one-dimensional p-point Fourier transforms.

For p prime we have the factorization

$$x^p - 1 = (x - 1)(x^{p-1} + x^{p-2} + \cdots + x + 1).$$

Every nonzero k'' is a zero of the cyclotomic polynomial at the right. Theorem 12.8.1 tells how to compute $V_{k'k''}$ for $k' = 0, \ldots, p - 1$ and $k'' = 1, \ldots, p - 1$, using one polynomial transform and p Fourier transforms of blocklength p. To complete the computation of $V_{k'k''}$, we need those elements with k'' equal to zero. But these elements can be obtained immediately with one more Fourier transform:

$$V_{0k''} = \sum_{i''=0}^{p-1} \omega^{i''k''} \sum_{i'=0}^{p-1} v_{i'i''}, \qquad k'' = 0, \ldots, p - 1.$$

12.8 The Nussbaumer–Quandalle permutation algorithm

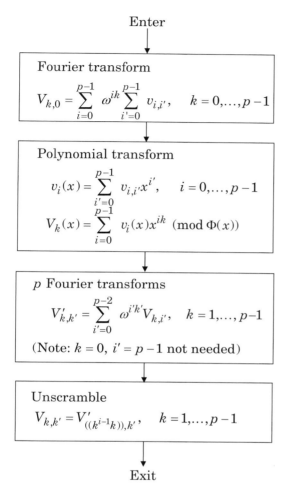

Figure 12.14 The Nussbaumer–Quandalle permutation algorithm

Thus there is a total of $p+1$ one-dimensional Fourier transforms of blocklength p. Figure 12.14 summarizes the algorithm for a prime blocklength. The last step unscrambles the data by using the inverse of k'' modulo p.

There is one more simplification. Most of the Fourier transforms have some missing coefficients, so their algorithm can be simplified by puncturing, that is, by dropping those input terms that are known to be zero and those output terms that are not wanted. In Figure 12.14 the major step computes p-point Fourier transforms of vectors only of length $p-1$ because the high-order coefficient of the vector is always zero. Further, only $p-1$ output values are needed; the low-order coefficient of the transform is discarded. These one-dimensional Fourier transforms can be computed by the punctured FFT algorithms, given in Table 12.9.

Table 12.9 *Performance of some punctured transform algorithms*

Punctured Fourier transform

$$V_k = \sum_{i=0}^{p^{m-1}(p-1)-1} \omega^{ik} v_i \quad k \neq 0 \pmod{p}$$

where $\omega^n = 1, n = p^m$

Blocklength n	Normal Winograd FFT			Punctured Winograd FFT		
	Number* of multiplications		Number of additions	Number* of multiplications		Number* of additions
	Total	Nontrivial		Total	Nontrivial	
2	2	0	2			
3	3	2	6	2	2	4
4	4	0	8	2	0	2
5	6	5	17	5	5	15
7	9	8	36	8	8	34
8	8	2	26	4	2	10
9	11	10	44	8	8	28
16	18	10	74	10	8	32

* Double for complex input.

As an example of the Nussbaumer–Quandalle FFT, we construct a three by three two-dimensional FFT algorithm. The first piece is the easiest to write down:

$$\begin{bmatrix} V_{00} \\ V_{01} \\ V_{11} \end{bmatrix} = \begin{bmatrix} 1 & 1 & 1 \\ 1 & \omega & \omega^2 \\ 1 & \omega^2 & \omega \end{bmatrix} \begin{bmatrix} v_{00} + v_{10} + v_{20} \\ v_{01} + v_{11} + v_{21} \\ v_{02} + v_{12} + v_{22} \end{bmatrix}.$$

Next, we note that the appropriate cyclotomic polynomial is

$$\Phi(x) = x^2 + x + 1 = (x^3 - 1)/(x - 1)$$

and

$$v_0(x) = v_{20}x^2 + v_{10}x + v_{00},$$
$$v_1(x) = v_{21}x^2 + v_{11}x + v_{01},$$
$$v_2(x) = v_{22}x^2 + v_{12}x + v_{02}.$$

The polynomial transform is computed modulo $x^2 + x + 1$ on polynomials of degree one. We replace these polynomials by their residues modulo $\Phi(x)$:

$$v_0(x) = (v_{10} - v_{20})x + (v_{00} - v_{20}),$$
$$v_1(x) = (v_{11} - v_{21})x + (v_{01} - v_{21}),$$
$$v_2(x) = (v_{12} - v_{22})x + (v_{02} - v_{22}),$$

12.8 The Nussbaumer–Quandalle permutation algorithm

although it would do no harm to defer the modulo $\Phi(x)$ reduction. The polynomial transforms are

$$\begin{bmatrix} V'_0(x) \\ V'_1(x) \\ V'_2(x) \end{bmatrix} = \begin{bmatrix} 1 & 1 & 1 \\ 1 & x & x^2 \\ 1 & x^2 & x \end{bmatrix} \begin{bmatrix} v_0(x) \\ v_1(x) \\ v_2(x) \end{bmatrix} \quad (\mod \Phi(x)),$$

which can be written out as follows:

$$\begin{bmatrix} V'_{00} \\ V'_{10} \\ V'_{01} \\ V'_{11} \\ V'_{02} \\ V'_{12} \end{bmatrix} = \begin{bmatrix} 1 & 0 & 1 & 0 & 1 & 0 \\ 0 & 1 & 0 & 1 & 0 & 1 \\ 1 & 0 & 0 & -1 & -1 & 1 \\ 0 & 1 & 1 & -1 & -1 & 0 \\ 1 & 0 & -1 & 1 & 0 & -1 \\ 0 & 1 & -1 & 0 & 1 & -1 \end{bmatrix} \begin{bmatrix} v_{00} - v_{20} \\ v_{10} - v_{20} \\ v_{01} - v_{21} \\ v_{11} - v_{21} \\ v_{02} - v_{22} \\ v_{12} - v_{22} \end{bmatrix}.$$

The next step consists of the punctured FFT algorithms. Rather than the complete Fourier transforms of the form

$$\begin{bmatrix} V''_{0k} \\ V''_{1k} \\ V''_{2k} \end{bmatrix} = \begin{bmatrix} 1 & 1 & 1 \\ 1 & \omega & \omega^2 \\ 1 & \omega^2 & \omega \end{bmatrix} \begin{bmatrix} V'_{0k} \\ V'_{1k} \\ 0 \end{bmatrix},$$

we need to compute only the punctured Fourier transforms:

$$\begin{bmatrix} V''_{1k} \\ V''_{2k} \end{bmatrix} = \begin{bmatrix} 1 & \omega \\ 1 & \omega^2 \end{bmatrix} \begin{bmatrix} V'_{0k} \\ V'_{1k} \end{bmatrix}$$

for $k = 0, 1, 2$. Put together, these become

$$\begin{bmatrix} V''_{10} \\ V''_{20} \\ V''_{11} \\ V''_{21} \\ V''_{12} \\ V''_{22} \end{bmatrix} = \begin{bmatrix} 1 & \omega & 0 & 0 & 0 & 0 \\ 1 & \omega^2 & 0 & 0 & 0 & 0 \\ 0 & 0 & 1 & \omega & 0 & 0 \\ 0 & 0 & 1 & \omega^2 & 0 & 0 \\ 0 & 0 & 0 & 0 & 1 & \omega \\ 0 & 0 & 0 & 0 & 1 & \omega^2 \end{bmatrix} \begin{bmatrix} V'_{00} \\ V'_{10} \\ V'_{01} \\ V'_{11} \\ V'_{02} \\ V'_{12} \end{bmatrix}.$$

Next, we have the unscrambling:

$$\begin{bmatrix} V_{10} \\ V_{20} \\ V_{11} \\ V_{21} \\ V_{12} \\ V_{22} \end{bmatrix} = \begin{bmatrix} 1 & 0 & 0 & 0 & 0 & 0 \\ 0 & 1 & 0 & 0 & 0 & 0 \\ 0 & 0 & 1 & 0 & 0 & 0 \\ 0 & 0 & 0 & 0 & 0 & 1 \\ 0 & 0 & 0 & 0 & 1 & 0 \\ 0 & 0 & 0 & 1 & 0 & 0 \end{bmatrix} \begin{bmatrix} v''_{10} \\ v''_{20} \\ v''_{11} \\ v''_{21} \\ v''_{12} \\ v''_{22} \end{bmatrix}.$$

Finally, all of the pieces are put together. The final algorithm is shown in Figure 12.15.

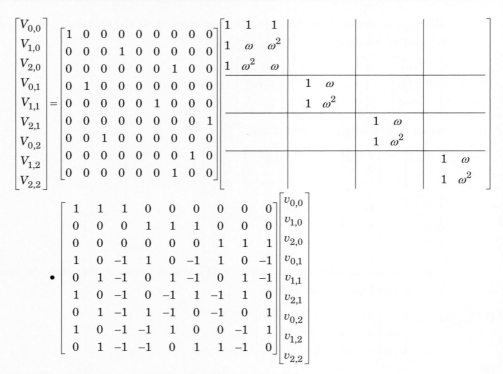

Figure 12.15 A three by three Nussbaumer–Quandalle FFT

Blocklength a power of an odd prime To illustrate the technique, it is enough to treat only the case in which the blocklength is the square of a prime. The construction is essentially the same as before, but it is more complicated because first we must pull away all troublesome indices for special handling. The polynomial notation is a helpful aid in doing this. For example, the polynomial factorization

$$x^{p^2} - 1 = (x-1)(x^{p-1} + x^{p-2} + \cdots + x + 1)$$
$$(x^{p(p-1)} + x^{p(p-2)} + \cdots + x^p + 1)$$
$$= \Phi_1(x)\Phi_p(x)\Phi_{p^2}(x)$$

must be a factorization into cyclotomic polynomials by Theorem 9.5.3 because the only factors of p^2 are 1, p, and p^2. The indices k are divided into three sets, corresponding to the three sets of conjugates into which the roots of unity ω^k are divided by their cyclotomic polynomials. Theorem 12.8.1 can be applied for the indices of the set of conjugates that are zeros of $\Phi_{p^2}(x)$. This requires one polynomial transform of p^2 terms modulo $\Phi_{p^2}(x)$ and p^2 punctured Fourier transforms of length p^2.

To complete the computation of $V_{k'k''}$, we need the elements with k'' equal to a multiple of p. These elements can be written

$$V_{k',\ell p} = \sum_{i'=0}^{p^2-1} \omega^{i'k'} \sum_{i''=0}^{p^2-1} \omega^{i''\ell p} v_{i'i''}, \qquad \begin{matrix} k' = 0,\ldots, p^2-1, \\ \ell = 0,\ldots, p-1, \end{matrix}$$

where ω is an element of order p^2. The inner sum can be collapsed to a p-point Fourier transform because the exponent in that term is a multiple of p. Let $\gamma = \omega^p$. Then γ is an element of order p. Let

$$v'_{i'r} = \sum_{\ell=0}^{p-1} v_{i',r+\ell p}, \quad \begin{array}{l} i' = 0, \ldots, p^2 - 1, \\ r = 0, \ldots, p - 1, \end{array}$$

and let $V'_{k'\ell} = V_{k',\ell p}$. Then

$$V'_{k\ell} = \sum_{i'=0}^{p^2-1} \omega^{i'k'} \sum_{r=0}^{p-1} \gamma^{r\ell} v'_{i'r}, \quad \begin{array}{l} k' = 0, \ldots, p^2 - 1, \\ \ell = 0, \ldots, p - 1. \end{array}$$

Now interchange the order of summation and apply Theorem 12.8.1 again. This requires another polynomial transform modulo $\Phi_{p^2}(x)$ of p terms and p more punctured Fourier transforms. There will still remain a p by p two-dimensional Fourier transform, which can be computed as before.

There is a total of $p^2 + p$ one-dimensional Fourier transforms of blocklength p^2 and $p + 1$ one-dimensional Fourier transforms of blocklength p. Except for a single Fourier transform of length p, all of these are punctured transforms.

Blocklength a power of two The construction for the case with blocklength n equal to 2^m is essentially the same as before, yet different enough to require individual attention. The odd integers less than 2^m are coprime to 2^m, so Theorem 12.8.1 applies. We can compute $V_{k'k''}$ for k' odd and $k'' = 0, \ldots, n - 1$ with a polynomial transform modulo $x^{n/2} + 1$ of blocklength n, and n Fourier transforms of blocklength $n = 2^m$. The polynomial transform can be organized in the form of a radix-two Cooley–Tukey FFT because the blocklength is a power of two. The number of additions will be proportional to $n \log_2 n$. The n Fourier transforms are all punctured Fourier transforms that have input components with even indices equal to zero.

We now have described the computation for half of the components of $V_{k'k''}$, those components with k' odd and $k'' = 0, \ldots, n - 1$. We next interchange the roles of k' and k'' to compute the components with k'' odd and k' even. Only points with k' even need to be computed because we have already computed the others. To compute these points requires another polynomial transform and $n/2$ more Fourier transforms.

Finally, we need to compute those components of $V_{k'k''}$ with k' and k'' both even. But this is just a 2^{m-1} by 2^{m-1} Fourier transform and can be treated the same way as the 2^m by 2^m Fourier transform.

Problems for Chapter 12

12.1 Prepare a table of the number of multiplications and additions used by a two-dimensional Fourier transform, computed by using various one-dimensional Fourier transforms along the rows, then along the columns.

12.2 Determine the number of additions used by the radix-two and the radix-four two-dimensional Cooley–Tukey FFT algorithms discussed in the text.

12.3 Let $A(n)$ and $M(n)$ denote the number of additions and multiplications used by selected Winograd FFT algorithms of blocklength n.

a Prove that to nest two such algorithms, the blocklength designations n' and n'' should be assigned so that

$$[M(n') - n']/A(n') \geq [M(n'') - n'']/A(n'')$$

in order to minimize the number of additions. This ratio of "excess multiplications" to additions specifies the order in which algorithms should be nested.

b Give a similar analysis for preadditions and for postadditions treated separately.

12.4 a Given the two-point and five-point Winograd small FFT algorithms

$$\begin{bmatrix} V_0 \\ V_1 \end{bmatrix} = \begin{bmatrix} 1 & 0 \\ 0 & 1 \end{bmatrix} \begin{bmatrix} 1 & 0 \\ 0 & 1 \end{bmatrix} \begin{bmatrix} 1 & 1 \\ 1 & -1 \end{bmatrix} \begin{bmatrix} v_0 \\ v_1 \end{bmatrix},$$

$$\begin{bmatrix} V_0 \\ V_1 \\ V_2 \\ V_3 \\ V_4 \end{bmatrix} = \begin{bmatrix} 1 & 0 & 0 & 0 & 0 & 0 \\ 1 & 1 & 1 & 1 & -1 & 0 \\ 1 & 1 & -1 & 1 & 0 & 1 \\ 1 & 1 & -1 & -1 & 0 & -1 \\ 1 & 1 & 1 & -1 & 1 & 0 \end{bmatrix} \begin{bmatrix} 1 \\ -1.25 \\ .559 \\ j.951 \\ j.1538 \\ -j.363 \end{bmatrix}$$

$$\bullet \begin{bmatrix} 1 & 1 & 1 & 1 & 1 \\ 0 & 1 & 1 & 1 & 1 \\ 0 & 1 & -1 & -1 & 1 \\ 0 & 1 & -1 & 1 & -1 \\ 0 & 0 & -1 & 1 & 0 \\ 0 & 1 & 0 & 0 & -1 \end{bmatrix} \begin{bmatrix} v_0 \\ v_1 \\ v_2 \\ v_3 \\ v_4 \end{bmatrix},$$

construct a ten-point Winograd large FFT algorithm.

b How many real multiplications are there?

c How many real multiplications are there if the two-point and five-point transforms are combined by using the Good–Thomas algorithm?

d Suppose the ten-point FFT algorithm is used with the Cooley–Tukey algorithm to construct a 1000-point FFT. How many real multiplications are there? How does this compare with a 1024-point radix-two Cooley–Tukey FFT?

e Suppose that we use instead the two-point algorithm

$$\begin{bmatrix} V_0 \\ V_1 \end{bmatrix} = \begin{bmatrix} 1 & 1 \\ 1 & -1 \end{bmatrix} \begin{bmatrix} 1 & 0 \\ 0 & 1 \end{bmatrix} \begin{bmatrix} 1 & 0 \\ 0 & 1 \end{bmatrix} \begin{bmatrix} v_0 \\ v_1 \end{bmatrix}.$$

Is there an advantage or disadvantage?

12.5 The Winograd large FFT algorithm can be used to compute Fourier transforms of blocklengths 72, 315, and 5040. Find the number of additions and multiplications used for each of these cases.

12.6 Combine the Cooley–Tukey FFT with the Nussbaumer–Quandalle permutation algorithm to give a 256 by 256 two-dimensional FFT. How many multiplications and additions are used?

12.7 Sketch the derivation of a 1040-point FFT based on a three-dimensional cyclic convolution. How many multiplications are required?

12.8 A 63 by 63 two-dimensional FFT can be constructed in many ways. Find the number of multiplications used by each of the following procedures:

a By nesting a 63-point FFT with a 63-point FFT. (What are some of the options here?)

b By nesting a seven by seven FFT with a nine by nine FFT.

c By using the Nussbaumer–Quandalle splitting algorithm once on the entire problem to construct a seven by seven by nine by nine four-dimensional FFT.

12.9 For which values of n does the Nussbaumer–Quandalle permutation algorithm for two-dimensional transforms offer no advantage over nesting of one-dimensional algorithms? Does the conclusion change for three-dimensional transforms (if the blocklength is a prime)?

12.10 Give a splitting structure for a five by five Fourier transform that uses 230 additions and for one that uses 236 additions.

12.11 **a** Prove that the two-dimensional Rader polynomial $g(x, y)$ is a product of a polynomial in x and a polynomial in y.

b State and prove a version of Theorem 3.8.2 for two-dimensional Fourier transforms.

12.12 Construct the five by five Nussbaumer–Quandalle FFT.

12.13 Starting with the fast algorithms for a four-point Fourier transform and for a three by three two-dimensional Fourier transform given in this book, construct a twelve by three two-dimensional FFT.

Notes for Chapter 12

At the simplest level, a two-dimensional Fourier transform consists of two independent Fourier transforms operating sequentially along the two axes of the array. Hence, early on, fast algorithms for the one-dimensional case were readily applied to the two-dimensional case. Algorithms that were intrinsically two-dimensional came later. The two-dimensional Cooley–Tukey decimation was first proposed by Rivard (1977), and expanded by Harris, McClellan, Chan, and Schuessler (1977). Mersereau and Speake (1981) gave a slick unification of the various two-dimensional Cooley–Tukey FFT algorithms.

Winograd (1976) proposed a way to bind his small one-dimensional algorithms together by his nesting technique to make a more efficient package. His interest was in two-dimensional problems formed artificially as a way to compute one-dimensional Fourier transforms, but the same technique also applies to two-dimensional Fourier transforms that arise naturally. The use of Winograd's nesting for computing one-dimensional Fourier transforms was further discussed by Kolba and Parks (1977) and by Silverman (1977). Methods for developing a sequence of FFT algorithms intermediate between the Good–Thomas FFT and the Winograd FFT were discussed by Johnson and Burrus (1983). Nussbaumer and Quandalle (1979) gave a method of using Nussbaumer's polynomial transforms to compute Fourier transforms. Another development of a two-dimensional transform algorithm using permutations – in this case using the cyclic structure in a Galois extension field – is due to Auslander, Feig, and Winograd (1983). Their algorithm has nearly the same performance as the Nussbaumer–Quandalle FFT.

Appendix A
A collection of cyclic convolution algorithms

Cyclic convolution algorithms over the field of reals are given for $n = 2, 3, 4, 5, 7, 8,$ and 9. The algorithms are in the form

$s = CGAd.$

The matrix G is a diagonal matrix, and the diagonal elements are given as the elements of the vector $G = Bd$. The matrices A and C are given in full. Using the notation

$D = Ad,$
$S = GD,$
$s = CS,$

a sequence of additions that will multiply by matrix A or C is given as well.

Two-point cyclic convolution; two real multiplications, four real additions:

$$A = \begin{bmatrix} 1 & 1 \\ 1 & -1 \end{bmatrix}, \quad C = \begin{bmatrix} 1 & 1 \\ 1 & -1 \end{bmatrix},$$

$$\begin{bmatrix} G_0 \\ G_1 \end{bmatrix} = \frac{1}{2} \begin{bmatrix} 1 & 1 \\ 1 & -1 \end{bmatrix} \begin{bmatrix} g_0 \\ g_1 \end{bmatrix},$$

$D_0 = d_0 + d_1, \quad s_0 = S_0 + S_1,$
$D_1 = d_0 - d_1, \quad s_1 = S_0 - S_1.$

Three-point cyclic convolution; four real multiplications, 11 real additions:

$$A = \begin{bmatrix} 1 & 1 & 1 \\ 1 & 0 & -1 \\ 0 & 1 & -1 \\ 1 & 1 & -2 \end{bmatrix}, \quad C = \begin{bmatrix} 1 & 1 & 0 & -1 \\ 1 & -1 & -1 & 2 \\ 1 & 0 & 1 & -1 \end{bmatrix},$$

$$\begin{bmatrix} G_0 \\ G_1 \\ G_2 \\ G_3 \end{bmatrix} = \frac{1}{3} \begin{bmatrix} 1 & 1 & 1 \\ 3 & 0 & -3 \\ 0 & 3 & -3 \\ 1 & 1 & -2 \end{bmatrix} \begin{bmatrix} g_0 \\ g_1 \\ g_2 \end{bmatrix},$$

$$t_0 = d_0 + d_1, \quad T_0 = S_1 - S_3,$$
$$D_0 = t_0 + d_2, \quad T_1 = S_2 - S_3,$$
$$D_1 = d_0 - d_2, \quad s_0 = S_0 + T_0,$$
$$D_2 = d_1 - d_2, \quad s_1 = S_0 - T_0 - T_1,$$
$$D_3 = d_1 + d_2, \quad s_2 = S_0 + T_1.$$

Four-point cyclic convolution; five real multiplications, five real additions:

$$A = \begin{bmatrix} 1 & 1 & 1 & 1 \\ 1 & -1 & 1 & -1 \\ 1 & 1 & -1 & -1 \\ 1 & 0 & -1 & 0 \\ 0 & 1 & 0 & -1 \end{bmatrix}, \quad C = \begin{bmatrix} 1 & 1 & 1 & 0 & -1 \\ 1 & -1 & 1 & 1 & 0 \\ 1 & 1 & -1 & 0 & 1 \\ 1 & -1 & -1 & -1 & 0 \end{bmatrix},$$

$$\begin{bmatrix} G_0 \\ G_1 \\ G_2 \\ G_3 \\ G_4 \end{bmatrix} = \frac{1}{4} \begin{bmatrix} 1 & 1 & 1 & 1 \\ 1 & -1 & 1 & -1 \\ 2 & 0 & -2 & 0 \\ -2 & 2 & 2 & -2 \\ 2 & 2 & -2 & -2 \end{bmatrix} \begin{bmatrix} g_0 \\ g_1 \\ g_2 \\ g_3 \end{bmatrix},$$

$$t_0 = d_0 + d_2, \quad T_0 = S_0 + S_1,$$
$$t_1 = d_1 + d_3, \quad T_1 = S_0 - S_1,$$
$$D_0 = t_0 + t_1, \quad T_2 = S_2 - S_4,$$
$$D_1 = t_0 - t_1, \quad T_3 = S_2 + S_3,$$
$$D_3 = d_0 - d_2, \quad s_0 = T_0 + T_2,$$
$$D_4 = d_1 - d_3, \quad s_1 = T_1 + T_3,$$
$$D_2 = D_3 + D_4, \quad s_2 = T_0 - T_2,$$
$$s_3 = T_1 - T_3.$$

Five-point cyclic convolution; ten real multiplications, 31 real additions:

$$A = \begin{bmatrix} 1 & 0 & 0 & 0 & -1 \\ 0 & 1 & 0 & 0 & -1 \\ 1 & 1 & 0 & 0 & -2 \\ 0 & 0 & 1 & 0 & -1 \\ 0 & 0 & 0 & 1 & -1 \\ 0 & 0 & 1 & 1 & -2 \\ 1 & 0 & -1 & 0 & 0 \\ 0 & 1 & 0 & -1 & 0 \\ 1 & 1 & -1 & -1 & 0 \\ 1 & 1 & 1 & 1 & 1 \end{bmatrix}, \quad C = \begin{bmatrix} 1 & 0 & 1 & 0 & 0 & 0 & -1 & 0 & -1 & 1 \\ -1 & -1 & -2 & -1 & -1 & -2 & 0 & 0 & 0 & 1 \\ 0 & 0 & 0 & 0 & 1 & 1 & 0 & 1 & 1 & 1 \\ 0 & 0 & 0 & 1 & 0 & 1 & 1 & 0 & 1 & 1 \\ 0 & 1 & 1 & 0 & 0 & 0 & 0 & -1 & -1 & 1 \end{bmatrix},$$

A collection of cyclic convolution algorithms

$$\begin{bmatrix} G_0 \\ G_1 \\ G_2 \\ G_3 \\ G_4 \\ G_5 \\ G_6 \\ G_7 \\ G_8 \\ G_9 \end{bmatrix} = \frac{1}{5} \begin{bmatrix} 5 & 0 & -5 & 5 & -5 \\ 0 & 5 & -5 & 5 & -5 \\ -2 & -2 & 3 & -2 & 3 \\ -5 & 5 & -5 & 5 & 0 \\ -5 & 5 & -5 & 0 & 5 \\ 3 & -2 & 3 & -2 & -2 \\ 0 & 0 & -5 & 5 & 0 \\ 0 & 5 & -5 & 0 & 0 \\ -1 & -1 & 4 & -1 & -1 \\ 1 & 1 & 1 & 1 & 1 \end{bmatrix} \begin{bmatrix} g_0 \\ g_1 \\ g_2 \\ g_3 \\ g_4 \end{bmatrix},$$

$D_0 = d_0 - d_4,$
$D_1 = d_1 - d_4,$
$D_2 = D_0 + D_1,$
$D_3 = d_2 - d_4,$
$D_4 = d_3 - d_4,$
$D_5 = D_3 + D_4,$
$D_6 = D_0 - D_3,$
$D_7 = D_1 - D_4,$
$D_8 = D_2 - D_5,$
$D_9 = d_0 + d_1 + d_2 + d_3 + d_4,$

$T_0 = S_0 + S_2,$
$T_1 = S_1 + S_2,$
$T_2 = S_3 + S_5,$
$T_3 = S_4 + S_5,$
$T_4 = S_6 + S_8,$
$T_5 = S_7 + S_8,$
$s_0 = T_0 - T_4 + S_9,$
$s_1 = -T_0 - T_1 - T_2 - T_3 + S_9,$
$s_2 = T_3 + T_5 + S_9,$
$s_3 = T_2 + T_4 + S_9,$
$s_4 = T_1 - T_5 + S_9.$

Seven-point cyclic convolution; 16 real multiplications, 70 real additions:

$$A = \begin{bmatrix} 1 & 0 & 0 & 0 & 0 & 0 & -1 \\ 1 & 1 & 1 & 0 & 0 & 0 & -3 \\ 1 & -1 & 1 & 0 & 0 & 0 & -1 \\ 1 & 2 & 4 & 0 & 0 & 0 & -7 \\ 0 & 0 & 1 & 0 & 0 & 0 & -1 \\ 0 & 0 & 0 & 1 & 0 & 0 & -1 \\ 0 & 0 & 0 & 1 & 1 & 1 & -3 \\ 0 & 0 & 0 & 1 & -1 & 1 & -1 \\ 0 & 0 & 0 & 1 & 2 & 4 & -7 \\ 0 & 0 & 0 & 0 & 0 & 1 & -1 \\ -1 & 0 & 0 & 1 & 0 & 0 & 0 \\ -1 & -1 & -1 & 1 & 1 & 1 & 0 \\ -1 & 1 & -1 & 1 & -1 & 1 & 0 \\ -1 & -2 & -4 & 1 & 2 & 4 & 0 \\ 0 & 0 & -1 & 0 & 0 & 1 & 0 \\ 1 & 1 & 1 & 1 & 1 & 1 & 1 \end{bmatrix},$$

$$C = \begin{bmatrix} 1 & 1 & 1 & 1 & 0 & 0 & 0 & 0 & 0 & 0 & 1 & 1 & 1 & 1 & 0 & 1 \\ -1 & -3 & -1 & -7 & -1 & -1 & -3 & 1 & -7 & -1 & 0 & 0 & 0 & 0 & 0 & 1 \\ 0 & 0 & 0 & 0 & 0 & 0 & 1 & 1 & 4 & 1 & 0 & -1 & -1 & -4 & -1 & 1 \\ 0 & 0 & 0 & 0 & 0 & 0 & 1 & -1 & 2 & 0 & 0 & -1 & 1 & -2 & 0 & 1 \\ 0 & 0 & 0 & 0 & 0 & 1 & 1 & 1 & 1 & 0 & -1 & -1 & -1 & -1 & 0 & 1 \\ 0 & 1 & 1 & 4 & 1 & 0 & 0 & 0 & 0 & 0 & 0 & 1 & 1 & 4 & 1 & 1 \\ 0 & 1 & -1 & 2 & 0 & 0 & 0 & 0 & 0 & 0 & 1 & -1 & 2 & 0 & 1 \end{bmatrix},$$

$$\begin{bmatrix} 2G_0 \\ 14G_1 \\ 6G_2 \\ 6G_3 \\ G_4 \\ G_5 \\ 14G_6 \\ 6G_7 \\ 6G_8 \\ G_9 \\ 2G_{10} \\ 14G_{11} \\ 6G_{12} \\ 6G_{13} \\ G_{14} \\ 7G_{15} \end{bmatrix} = \begin{bmatrix} 2 & 1 & -2 & -1 & 3 & -2 & -1 \\ -4 & -11 & 3 & 10 & -11 & 3 & 10 \\ 0 & -1 & 3 & -2 & -1 & 3 & -2 \\ 0 & 1 & 0 & -1 & 1 & 0 & -1 \\ 1 & -2 & -1 & 3 & -1 & -1 & 2 \\ -1 & 3 & -2 & -1 & 2 & 1 & -2 \\ 10 & -11 & 3 & 10 & -4 & -11 & 3 \\ -2 & -1 & 3 & -2 & 0 & -1 & 3 \\ -1 & 1 & 0 & -1 & 0 & 1 & 0 \\ 3 & -2 & -1 & 2 & 1 & -2 & -1 \\ 0 & 1 & -2 & -1 & 2 & 0 & 0 \\ -2 & -9 & 5 & 12 & -2 & -2 & -2 \\ 0 & -1 & 3 & -2 & 0 & 0 & 0 \\ 0 & 1 & 0 & -1 & 0 & 0 & 0 \\ 1 & -2 & -1 & 2 & 0 & 0 & 0 \\ 1 & 1 & 1 & 1 & 1 & 1 & 1 \end{bmatrix} \begin{bmatrix} g_0 \\ g_1 \\ g_2 \\ g_3 \\ g_4 \\ g_5 \\ g_6 \end{bmatrix},$$

$t_0 = d_1 - d_6,$
$t_1 = d_2 - d_6,$
$t_2 = d_4 - d_6,$
$t_3 = d_5 - d_6,$
$t_4 = t_0 + t_1,$
$t_5 = -t_0 + t_1,$
$t_6 = t_2 + t_3,$
$t_7 = -d_4 + d_5,$
$D_0 = d_0 - d_6,$
$D_1 = D_1 + t_4,$
$D_2 = D_0 + t_5,$
$D_3 = D_1 + t_4 + t_4 + t_5,$
$D_4 = t_1,$
$D_5 = d_3 - d_6,$
$D_6 = D_5 + t_6,$

$D_7 = D_5 + t_7,$
$D_8 = D_6 + t_6 + t_6 + t_7,$
$D_9 = t_3,$
$D_{10} = D_5 - D_0,$
$D_{11} = D_6 - D_1,$
$D_{12} = D_7 - D_2,$
$D_{13} = D_8 - D_3,$
$D_{14} = D_9 - D_4,$
$D_{15} = D_{11} + 2(d_2 + d_1 + d_0) + d_6,$
$T_0 = S_0 + S_{10},$
$T_1 = S_1 + S_{11},$
$T_2 = S_2 + S_{12},$
$T_3 = S_3 + S_{13},$
$T_4 = S_4 + S_{14},$
$T_5 = S_5 - S_{10},$

$$T_6 = S_6 - S_{11},$$
$$T_7 = S_7 - S_{12},$$
$$T_8 = S_8 - S_{13},$$
$$T_9 = S_9 - S_{14},$$
$$T_{10} = T_1 + T_3,$$
$$T_{11} = T_{10} + T_2,$$
$$T_{12} = T_0 + T_{11},$$
$$T_{13} = T_{10} + T_3,$$
$$T_{14} = T_{13} - T_2,$$
$$T_{15} = (T_{13} + T_3 + T_3 + T_4) + T_2,$$
$$T_{16} = -T_{12} - T_{13} - (T_{13} + T_3 + T_3 + T_4),$$
$$T_{17} = T_6 + T_8,$$
$$T_{18} = T_{17} + T_7,$$

$$T_{19} = T_5 + T_{18},$$
$$T_{20} = T_{17} + T_8,$$
$$T_{21} = T_{20} - T_7,$$
$$T_{22} = (T_{20} + T_8 + T_8 + T_9) + T_7,$$
$$T_{23} = -T_{19} - T_{20} - (T_{20} + T_8 + T_8 + T_9),$$
$$s_0 = T_{12} + S_{15},$$
$$s_1 = T_{16} + T_{23} + S_{15},$$
$$s_2 = T_{22} + S_{15},$$
$$s_3 = T_{21} + S_{15},$$
$$s_4 = T_{19} + S_{15},$$
$$s_5 = T_{15} + S_{15},$$
$$s_6 = T_{14} + S_{15}.$$

Eight-point cyclic convolution; 14 real multiplications, 46 real additions:

$$A = \begin{bmatrix} 1 & 0 & 0 & 0 & -1 & 0 & 0 & 0 \\ 0 & 1 & 0 & 0 & 0 & -1 & 0 & 0 \\ 0 & 0 & 1 & 0 & 0 & 0 & -1 & 0 \\ 0 & 0 & 0 & 1 & 0 & 0 & 0 & -1 \\ 1 & 0 & -1 & 0 & 1 & 0 & -1 & 0 \\ 0 & 1 & 0 & -1 & 0 & 1 & 0 & -1 \\ 1 & 1 & 1 & 1 & 0 & 0 & 0 & 0 \\ 1 & -1 & 1 & -1 & 0 & 0 & 0 & 0 \\ 0 & 1 & 0 & 1 & 0 & -1 & 0 & -1 \\ 1 & 0 & 1 & 0 & -1 & 0 & -1 & 0 \\ 1 & -1 & 1 & -1 & -1 & 1 & -1 & 1 \\ 0 & 0 & 1 & -1 & 0 & 0 & -1 & 1 \\ 1 & -1 & 0 & 0 & -1 & 1 & 0 & 0 \\ 1 & 1 & -1 & -1 & 1 & 1 & -1 & -1 \end{bmatrix},$$

$$C = \begin{bmatrix} 0 & 0 & 0 & 1 & 0 & -1 & 1 & 1 & 1 & 0 & 1 & -1 & 0 & 1 \\ 0 & 0 & -1 & 0 & 1 & 0 & 1 & -1 & 0 & 1 & -1 & 1 & 0 & 1 \\ 0 & 1 & 0 & 0 & 0 & 1 & 1 & 1 & 1 & 0 & 1 & 0 & 1 & -1 \\ 1 & 0 & 0 & 0 & -1 & 0 & 1 & -1 & 0 & 1 & -1 & 0 & -1 & -1 \\ 0 & 0 & 0 & 1 & 0 & -1 & 1 & 1 & -1 & 0 & -1 & 1 & 0 & 1 \\ 0 & 0 & 1 & 0 & 1 & 0 & 1 & -1 & 0 & -1 & 1 & -1 & 0 & 1 \\ 0 & -1 & 0 & 0 & 0 & 1 & 1 & 1 & -1 & 0 & -1 & 0 & -1 & -1 \\ -1 & 0 & 0 & 0 & -1 & 0 & 1 & -1 & 0 & -1 & 1 & 0 & 1 & -1 \end{bmatrix},$$

$$\begin{bmatrix} 2G_0 \\ 2G_1 \\ 2G_2 \\ 2G_3 \\ 4G_4 \\ 4G_5 \\ 8G_6 \\ 8G_7 \\ 2G_8 \\ 2G_9 \\ 2G_{10} \\ 2G_{11} \\ 2G_{12} \\ 2G_{13} \end{bmatrix} = \begin{bmatrix} -1 & -1 & 1 & 1 & 1 & 1 & -1 & -1 \\ -1 & 1 & 1 & 1 & 1 & -1 & -1 & -1 \\ 1 & 1 & 1 & 1 & -1 & -1 & -1 & -1 \\ 1 & 1 & 1 & -1 & -1 & -1 & -1 & 1 \\ -1 & 1 & 1 & -1 & -1 & 1 & 1 & -1 \\ 1 & 1 & -1 & -1 & 1 & 1 & -1 & -1 \\ 1 & 1 & 1 & 1 & 1 & 1 & 1 & 1 \\ 1 & -1 & 1 & -1 & 1 & -1 & 1 & -1 \\ 1 & 0 & 0 & -1 & -1 & 0 & 0 & 1 \\ 1 & 1 & 0 & 0 & -1 & -1 & 0 & 0 \\ 1 & 0 & 0 & 0 & -1 & 0 & 0 & 0 \\ 1 & 0 & 1 & 0 & -1 & 0 & -1 & 0 \\ -1 & 0 & 1 & 0 & 1 & 0 & -1 & 0 \\ 1 & 0 & -1 & 0 & 1 & 0 & -1 & 0 \end{bmatrix} \begin{bmatrix} g_0 \\ g_1 \\ g_2 \\ g_3 \\ g_4 \\ g_5 \\ g_6 \\ g_7 \end{bmatrix},$$

$t_0 = d_0 + d_4,$ $T_3 = S_{11} - S_2,$
$t_1 = d_1 + d_5,$ $T_4 = S_1 + S_{12},$
$t_2 = d_2 + d_6,$ $T_5 = S_0 - S_{12},$
$t_3 = d_3 + d_7,$ $T_6 = S_{13} - S_5,$
$t_4 = t_0 + t_2,$ $T_7 = S_{13} + S_4,$
$t_5 = t_1 + t_3,$ $T_8 = S_7 + S_6,$
$D_0 = d_0 - d_4,$ $T_9 = S_6 - S_7,$
$D_1 = d_1 - d_5,$ $T_{10} = T_0 - T_2,$
$D_2 = d_2 - d_6,$ $T_{11} = T_6 + T_8,$
$D_3 = d_3 - d_7,$ $T_{12} = T_1 + T_3,$
$D_4 = t_0 - t_2,$ $T_{13} = T_7 + T_9,$
$D_5 = t_1 - t_3,$ $T_{14} = T_0 + T_4,$
$D_6 = t_4 + t_5,$ $T_{15} = -T_6 + T_8,$
$D_7 = t_4 - t_5,$ $T_{16} = T_1 + T_5,$
$D_8 = D_1 + D_3,$ $T_{17} = -T_7 + T_9,$
$D_9 = D_0 + D_2,$ $s_0 = T_{10} + T_{11},$
$D_{10} = D_9 - D_8,$ $s_1 = T_{12} + T_{13},$
$D_{11} = D_2 - D_3,$ $s_2 = T_{14} + T_{15},$
$D_{12} = D_0 - D_1,$ $s_3 = T_{16} + T_{17},$
$D_{13} = D_4 + D_5,$ $s_4 = -T_{10} + T_{11},$
$T_0 = S_8 + S_{10},$ $s_5 = -T_{12} + T_{13},$
$T_1 = S_9 - S_{10},$ $s_6 = -T_{14} + T_{15},$
$T_2 = S_3 + S_{11},$ $s_7 = -T_{16} + T_{17}.$

Nine-point cyclic convolution; 19 real multiplications, 74 real additions:

$$A = \begin{bmatrix} 1 & 1 & 1 & 1 & 1 & 1 & 1 & 1 & 1 \\ 0 & 0 & 0 & 1 & 1 & 1 & -1 & -1 & -1 \\ 1 & 1 & 1 & 0 & 0 & 0 & -1 & -1 & -1 \\ 1 & 1 & 1 & -1 & -1 & -1 & 0 & 0 & 0 \\ 0 & 0 & 0 & 1 & -1 & 1 & -1 & 1 & -1 \\ 1 & -1 & 1 & 0 & 0 & 0 & -1 & 1 & -1 \\ 1 & -1 & 1 & -1 & 1 & -1 & 0 & 0 & 0 \\ 0 & 0 & 0 & 1 & 0 & 0 & -1 & 0 & 0 \\ 1 & 0 & 0 & -1 & 0 & 0 & 0 & 0 & 0 \\ 1 & 0 & 0 & 0 & 0 & 0 & -1 & 0 & 0 \\ 0 & 0 & 0 & 0 & 0 & 1 & 0 & 0 & -1 \\ 0 & 0 & 1 & 0 & 0 & -1 & 0 & 0 & 0 \\ 0 & 0 & 1 & 0 & 0 & 0 & 0 & 0 & -1 \\ 1 & 0 & -1 & 0 & -1 & 1 & -1 & 1 & 0 \\ 1 & -1 & 0 & -1 & 0 & 1 & 0 & 1 & -1 \\ 0 & 1 & -1 & 1 & -1 & 0 & -1 & 0 & 1 \\ 1 & 0 & -1 & 1 & 0 & -1 & 1 & 0 & -1 \\ 0 & 1 & -1 & 0 & 1 & -1 & 0 & 1 & -1 \\ 1 & 1 & -2 & 1 & 1 & -2 & 1 & 1 & -2 \end{bmatrix},$$

$$C = \begin{bmatrix} 1 & 1 & 0 & 1 & 1 & 0 & 1 & 0 & -1 & 1 & 0 & 0 & 0 & 1 & -1 & 0 & 1 & 0 & -1 \\ 1 & 0 & 1 & -1 & 0 & 1 & -1 & 0 & 0 & 0 & 0 & -1 & 1 & 0 & 1 & 1 & -1 & -1 & 2 \\ 1 & 0 & 1 & -1 & 0 & -1 & 1 & 0 & 0 & 0 & 0 & 0 & 0 & 1 & -1 & 0 & 0 & 1 & -1 \\ 1 & 0 & 1 & -1 & 0 & 1 & -1 & -1 & 0 & -1 & 0 & 0 & 0 & -1 & 0 & -1 & 1 & 0 & -1 \\ 1 & -1 & -1 & 0 & -1 & -1 & 0 & 0 & 0 & 0 & -1 & 0 & -1 & 1 & -1 & 0 & -1 & -1 & 2 \\ 1 & -1 & -1 & 0 & 1 & 1 & 0 & 0 & 0 & 0 & 0 & 0 & -1 & 0 & -1 & 0 & 1 & -1 \\ 1 & -1 & -1 & 0 & -1 & -1 & 0 & 1 & 1 & 0 & 0 & 0 & 0 & 1 & 1 & 1 & 0 & -1 \\ 1 & 1 & 0 & 1 & 1 & 0 & 1 & 0 & 0 & 0 & 1 & 1 & 0 & -1 & 0 & -1 & -1 & -1 & 2 \\ 1 & 1 & 0 & 1 & -1 & 0 & -1 & 0 & 0 & 0 & 0 & 0 & 0 & 1 & 1 & 0 & 1 & -1 \end{bmatrix},$$

$$\begin{bmatrix} 9G_0 \\ 6G_1 \\ 6G_2 \\ 6G_3 \\ 6G_4 \\ 6G_5 \\ 6G_6 \\ 3G_7 \\ 3G_8 \\ 3G_9 \\ 3G_{10} \\ 3G_{11} \\ 3G_{12} \\ 3G_{13} \\ 3G_{14} \\ 3G_{15} \\ 3G_{16} \\ 3G_{17} \\ 3G_{18} \end{bmatrix} = \begin{bmatrix} 1 & 1 & 1 & 1 & 1 & 1 & 1 & 1 & 1 \\ -1 & -2 & -1 & 1 & 1 & 0 & 0 & 1 & 1 \\ 0 & 1 & 1 & -1 & -2 & -1 & 1 & 1 & 0 \\ -1 & -1 & 0 & 0 & -1 & -1 & 1 & 2 & 1 \\ 1 & -2 & 1 & 1 & 1 & -2 & -2 & 1 & 1 \\ -2 & 1 & 1 & 1 & -2 & 1 & 1 & 1 & -2 \\ -1 & -1 & 2 & 2 & -1 & -1 & -1 & 2 & -1 \\ -4 & -1 & 2 & 2 & -1 & -1 & 2 & 2 & -1 \\ -2 & 1 & 1 & -2 & -2 & 1 & 4 & 1 & -2 \\ 2 & 2 & -1 & -4 & -1 & 2 & 2 & -1 & -1 \\ -1 & 2 & 2 & -1 & -1 & 2 & 2 & -1 & -4 \\ 1 & 1 & -2 & -2 & 1 & 4 & 1 & -2 & -2 \\ 2 & -1 & -4 & -1 & 2 & 2 & -1 & -1 & 2 \\ -1 & 0 & 1 & 1 & 0 & -1 & 0 & 0 & 0 \\ 0 & 0 & 0 & 1 & 0 & -1 & -1 & 0 & 1 \\ 1 & 0 & -1 & 0 & 0 & 0 & -1 & 0 & 1 \\ 1 & 0 & -1 & 1 & 0 & -1 & 1 & 0 & -1 \\ 0 & 1 & -1 & 0 & 1 & -1 & 0 & 1 & -1 \\ 1 & 1 & -2 & 1 & 1 & -2 & 1 & 1 & -2 \end{bmatrix} \begin{bmatrix} g_0 \\ g_1 \\ g_2 \\ g_3 \\ g_4 \\ g_5 \\ g_6 \\ g_7 \\ g_8 \end{bmatrix},$$

$t_0 = d_0 - d_6,$
$t_1 = d_1 - d_7,$
$t_2 = d_2 - d_8,$
$t_3 = d_3 - d_6,$
$t_4 = d_4 - d_7,$
$t_5 = d_5 - d_8,$
$t_6 = d_0 + d_3 + d_6,$
$t_7 = d_1 + d_4 + d_7,$
$t_8 = d_2 + d_5 + d_8,$
$t_9 = t_0 + t_2,$
$t_{10} = t_3 + t_5,$
$D_0 = t_6 + t_7 + t_8,$
$D_1 = t_{10} + t_4,$
$D_2 = t_9 + t_1,$
$D_3 = D_2 - D_1,$
$D_4 = t_{10} - t_4,$
$D_5 = t_9 - t_1,$
$D_6 = D_5 - D_4,$
$D_7 = t_3,$
$D_8 = t_0 - t_3,$
$D_9 = t_0,$
$D_{10} = t_5,$
$D_{11} = t_2 - t_5,$
$D_{12} = t_2,$
$D_{13} = -D_{11} + t_0 - t_4,$
$D_{14} = D_8 + t_5 - t_1,$
$D_{15} = -D_{14} + D_{13},$
$D_{16} = t_6 - t_8,$
$D_{17} = t_7 - t_8,$
$D_{18} = D_{16} + D_{17},$

$T_0 = S_1 + S_2,$
$T_1 = S_4 + S_5,$
$T_2 = S_{14} + S_{15},$
$T_3 = T_0 + T_1,$
$T_4 = S_1 + S_3,$
$T_5 = S_4 + S_6,$
$T_6 = S_{13} + S_{15},$
$T_7 = -T_3 + S_7,$
$T_8 = T_4 + T_5,$
$T_9 = S_{10} - T_6,$
$T_{10} = S_8 + T_2 + T_7,$
$T_{11} = T_8 + S_{11} + T_9,$
$T_{12} = T_4 - T_5 + T_2,$
$T_{13} = T_7 + T_8 + S_9 + T_6,$
$T_{14} + T_3 + S_{12} + T_9 + T_2,$
$T_{15} = T_0 - T_1 + T_6,$
$T_{16} = S_{16} - S_{18},$
$T_{17} = S_{17} - S_{18},$
$T_{18} = S_0 + T_{16},$
$T_{19} = S_0 - T_{16} - T_{17},$
$T_{20} = S_0 + T_{17},$
$s_0 = T_{13} - T_{10} + T_{18},$
$s_1 = T_{14} - T_{11} + T_{19},$
$s_2 = T_{15} - T_{12} + T_{20},$
$s_3 = -T_{13} + T_{18},$
$s_4 = -T_{14} + T_{19},$
$s_5 = -T_{15} + T_{20},$
$s_6 = T_{10} + T_{18},$
$s_7 = T_{11} + T_{19},$
$s_8 = T_{12} + T_{20}.$

Appendix B
A collection of Winograd small FFT algorithms

Winograd small FFT algorithms are given for $n = 2, 3, 4, 5, 7, 8, 9$, and 16. The algorithms are in the form

$$V = CBAv.$$

The matrix B is a diagonal matrix, and only the diagonal elements are given. The matrices A and C are given in full. In addition, a sequence of additions that will multiply by matrix A or C is given, using the notation

$$a = Av,$$
$$b = Ba,$$
$$V = Cb.$$

The algorithms have been presented with all appearances of j moved from the diagonal matrix and absorbed into the matrix of postadditions C. Trivial additions (additions of a purely real number to a purely imaginary number) are labeled as such but are included in the count of additions. They are not trivial if the input data is complex. The number of multiplications is stated as the number of nontrivial real multiplications followed, in parentheses, by the total number of real multiplications.

Two-point Fourier transform; 0(2) real multiplications, two real additions

$$A = \begin{bmatrix} 1 & 1 \\ 1 & -1 \end{bmatrix}, \qquad C = \begin{bmatrix} 1 & 0 \\ 0 & 1 \end{bmatrix},$$

$$a_0 = v_0 + v_1, \qquad B_0 = 1, \qquad V_0 = b_0,$$
$$a_1 = v_0 - v_1, \qquad B_1 = 1, \qquad V_1 = b_1.$$

A collection of Winograd small FFT algorithms

Three-point Fourier transform; two(three) real multiplications, six real additions:

$$A = \begin{bmatrix} 1 & 1 & 1 \\ 0 & 1 & 1 \\ 0 & 1 & -1 \end{bmatrix}, \quad C = \begin{bmatrix} 1 & 0 & 0 \\ 1 & 1 & -j \\ 1 & 1 & j \end{bmatrix},$$

$a_2 = v_1 + v_2,$
$a_1 = v_1 - v_2,$
$a_0 = v_0 + a_2,$

$\theta = 2\pi/3,$
$B_0 = 1,$
$B_1 = \cos\theta - 1,$
$B_2 = \sin\theta,$

$V_0 = b_0,$
$T_0 = b_0 + b_1,$
$V_1 = T_0 - jb_2,$
$V_2 = T_0 + jb_2.$

Four-point Fourier transform; zero(four) real multiplications, eight real additions:

$$A = \begin{bmatrix} 1 & 1 & 1 & 1 \\ 1 & -1 & 1 & -1 \\ 1 & 0 & -1 & 0 \\ 0 & 1 & 0 & -1 \end{bmatrix}, \quad C = \begin{bmatrix} 1 & 0 & 0 & 0 \\ 0 & 0 & 1 & -j \\ 0 & 1 & 0 & 0 \\ 0 & 0 & 1 & j \end{bmatrix},$$

$a_2 = v_0 - v_2,$
$a_3 = v_1 - v_3,$
$t_0 = v_0 + v_2,$
$t_1 = v_1 + v_3,$
$a_0 = t_0 + t_1,$
$a_1 = t_0 - t_1,$

$B_0 = 1,$
$B_1 = 1,$
$B_2 = 1,$
$B_3 = 1,$

$V_0 = b_0,$
$V_1 = b_2 - jb_3,$
$V_2 = b_1,$
$V_3 = b_2 + jb_3.$

Five-point Fourier transform; five(six) real multiplications, 17 real additions:

$$A = \begin{bmatrix} 1 & 1 & 1 & 1 & 1 \\ 0 & 1 & 1 & 1 & 1 \\ 0 & 1 & -1 & -1 & 1 \\ 0 & 1 & -1 & 1 & -1 \\ 0 & 0 & -1 & 1 & 0 \\ 0 & 1 & 0 & 0 & -1 \end{bmatrix}, \quad C = \begin{bmatrix} 1 & 0 & 0 & 0 & 0 & 0 \\ 1 & 1 & 1 & -j & j & 0 \\ 1 & 1 & -1 & -j & 0 & -j \\ 1 & 1 & -1 & j & 0 & j \\ 1 & 1 & 1 & j & -j & 0 \end{bmatrix},$$

$t_0 = v_1 + v_4,$
$t_1 = v_2 + v_3,$
$a_4 = v_3 - v_2,$
$a_5 = v_1 - v_4,$
$a_1 = t_0 + t_1,$
$a_2 = t_0 - t_1,$
$a_3 = a_4 + a_5,$
$a_0 = v_0 + a_1,$

$\theta = 2\pi/5,$
$B_0 = 1,$
$B_1 = \frac{1}{2}(\cos\theta + \cos 2\theta) - 1,$
$B_2 = \frac{1}{2}(\cos\theta - \cos 2\theta),$
$B_3 = \sin\theta,$
$B_4 = \sin\theta + \sin 2\theta,$
$B_5 = \sin 2\theta - \sin\theta,$

$V_0 = b_0,$
$T_0 = b_0 + b_1,$
$T_1 = b_3 - b_4,$
$T_2 = b_3 + b_5,$
$T_3 = T_0 + b_2,$
$T_4 = T_0 - b_2,$
$V_1 = T_3 - jT_1,$
$V_2 = T_4 - jT_2,$
$V_3 = T_3 + jT_1,$
$V_4 = T_4 + jT_2.$

A collection of Winograd small FFT algorithms

Seven-point Fourier transform; eight(nine) real multiplications, 36 real additions:

$$A = \begin{bmatrix} 1 & 1 & 1 & 1 & 1 & 1 & 1 \\ 0 & 1 & 1 & 1 & 1 & 1 & 1 \\ 0 & 1 & 0 & -1 & -1 & 0 & 1 \\ 0 & 0 & -1 & 1 & 1 & -1 & 0 \\ 0 & -1 & 1 & 0 & 0 & 1 & -1 \\ 0 & 1 & 1 & -1 & 1 & -1 & -1 \\ 0 & 1 & 0 & 1 & -1 & 0 & -1 \\ 0 & 0 & -1 & -1 & 1 & 1 & 0 \\ 0 & -1 & 1 & 0 & 0 & -1 & 1 \end{bmatrix},$$

$$C = \begin{bmatrix} 1 & 0 & 0 & 0 & 0 & 0 & 0 & 0 & 0 \\ 1 & 1 & 1 & 1 & 0 & -j & -j & -j & 0 \\ 1 & 1 & -1 & 0 & -1 & -j & j & 0 & j \\ 1 & 1 & 0 & -1 & 1 & j & 0 & -j & j \\ 1 & 1 & 0 & -1 & 1 & -j & 0 & j & -j \\ 1 & 1 & -1 & 0 & -1 & j & -j & 0 & -j \\ 1 & 1 & 1 & 1 & 0 & j & j & j & 0 \end{bmatrix},$$

$t_0 = v_1 + v_6,$
$t_1 = v_1 - v_6,$
$t_2 = v_2 + v_5,$
$t_3 = v_2 - v_5,$
$t_4 = v_4 + v_3,$
$t_5 = v_4 - v_3,$
$t_6 = t_2 + t_0,$
$a_4 = t_2 - t_0,$
$a_2 = t_0 - t_4,$
$a_3 = t_4 - t_2,$
$t_7 = t_5 + t_3,$
$a_7 = t_5 - t_3,$
$a_6 = t_1 - t_5,$
$a_8 = t_3 - t_1,$
$a_1 = t_6 + t_4,$
$a_5 = t_7 + t_1,$
$a_0 = v_0 + a_1,$

$\theta = 2\pi/7,$
$B_0 = 1,$
$B_1 = \frac{1}{3}(\cos\theta + \cos 2\theta + \cos 3\theta) - 1,$
$B_2 = \frac{1}{3}(2\cos\theta - \cos 2\theta - \cos 3\theta),$
$B_3 = \frac{1}{3}(\cos\theta - 2\cos 2\theta + \cos 3\theta),$
$B_4 = \frac{1}{3}(\cos\theta + \cos 2\theta - 2\cos 3\theta),$
$B_5 = \frac{1}{3}(\sin\theta + \sin 2\theta - \sin 3\theta),$
$B_6 = \frac{1}{3}(2\sin\theta - \sin 2\theta + \sin 3\theta),$
$B_7 = \frac{1}{3}(\sin\theta - 2\sin 2\theta - \sin 3\theta),$
$B_8 = \frac{1}{3}(\sin\theta + \sin 2\theta + 2\sin 3\theta),$

$T_0 = b_0 + b_1,$
$T_1 = b_2 + b_3,$
$T_2 = b_4 - b_3,$
$T_3 = -b_2 - b_4,$
$T_4 = b_6 + b_7,$
$T_5 = b_8 - b_7,$
$T_6 = -b_8 - b_6,$
$T_7 = T_0 + T_1,$
$T_8 = T_0 + T_2,$
$T_9 = T_0 + T_3,$
$T_{10} = T_4 + b_5,$
$T_{11} = T_5 + b_5,$
$T_{12} = T_6 + b_5,$
$V_0 = b_0,$
$V_1 = T_7 - jT_{10},$
$V_2 = T_9 - jT_{12},$
$V_3 = T_8 + jT_{11},$
$V_4 = T_8 - jT_{11},$
$V_5 = T_9 + jT_{12},$
$V_6 = T_7 + jT_{10}.$

Eight-point Fourier transform; two(eight) real multiplications, 26 real additions:

$$A = \begin{bmatrix} 1 & 1 & 1 & 1 & 1 & 1 & 1 & 1 \\ 1 & -1 & 1 & -1 & 1 & -1 & 1 & -1 \\ 1 & 0 & -1 & 0 & 1 & 0 & -1 & 0 \\ 1 & 0 & 0 & 0 & -1 & 0 & 0 & 0 \\ 0 & 1 & 0 & -1 & 0 & -1 & 0 & 1 \\ 0 & 1 & 0 & -1 & 0 & 1 & 0 & -1 \\ 0 & 0 & 1 & 0 & 0 & 0 & -1 & 0 \\ 0 & 1 & 0 & 1 & 0 & -1 & 0 & -1 \end{bmatrix},$$

$$C = \begin{bmatrix} 1 & 0 & 0 & 0 & 0 & 0 & 0 & 0 \\ 0 & 0 & 0 & 1 & 1 & 0 & -j & -j \\ 0 & 0 & 1 & 0 & 0 & -j & 0 & 0 \\ 0 & 0 & 0 & 1 & -1 & 0 & j & -j \\ 0 & 1 & 0 & 0 & 0 & 0 & 0 & 0 \\ 0 & 0 & 0 & 1 & -1 & 0 & -j & j \\ 0 & 0 & 1 & 0 & 0 & j & 0 & 0 \\ 0 & 0 & 0 & 1 & 1 & 0 & j & j \end{bmatrix},$$

$t_0 = v_0 + v_4$,
$a_3 = v_0 - v_4$,
$t_1 = v_1 + v_5$,
$t_2 = v_1 - v_5$,
$t_3 = v_2 + v_6$,
$a_6 = v_2 - v_6$,
$t_4 = v_3 + v_7$,
$t_5 = v_3 - v_7$,
$t_6 = t_0 + t_3$,
$a_2 = t_0 - t_3$,
$t_7 = t_1 + t_4$,
$a_5 = t_1 - t_4$,
$a_4 = t_2 - t_5$,
$a_7 = t_2 + t_5$,
$a_0 = t_6 + t_7$,
$a_1 = t_6 - t_7$,

$\theta = 2\pi/8$,
$B_0 = 1$,
$B_1 = 1$,
$B_2 = 1$,
$B_3 = 1$,
$B_4 = \cos\theta$,
$B_5 = 1$,
$B_6 = 1$,
$B_7 = \sin\theta$,

$V_0 = b_0$,
$V_4 = b_1$,
$V_2 = b_2 - jb_5$,
$V_6 = b_2 + jb_5$,
$T_0 = b_3 + b_4$,
$T_1 = b_3 - b_4$,
$T_2 = b_6 + b_7$,
$T_3 = b_6 - b_7$,
$V_1 = T_0 - jT_2$,
$V_5 = T_0 + jT_2$,
$V_7 = T_1 - jT_3$,
$V_3 = T_1 + jT_3$.

A collection of Winograd small FFT algorithms

Nine-point Fourier transform; 10(11) real multiplications, 44 real additions:

$$A = \begin{bmatrix} 1 & 1 & 1 & 1 & 1 & 1 & 1 & 1 & 1 \\ 0 & 0 & 0 & 1 & 0 & 0 & 1 & 0 & 0 \\ 0 & 1 & 1 & 0 & 1 & 1 & 0 & 1 & 1 \\ 0 & 1 & -1 & 0 & 0 & 0 & 0 & -1 & 1 \\ 0 & 0 & 1 & 0 & -1 & -1 & 0 & 1 & 0 \\ 0 & 0 & 0 & 0 & 1 & 1 & 0 & 0 & 1 \\ 0 & 0 & -1 & 0 & 2 & 0 & 0 & 1 & 0 \\ 0 & 0 & 0 & 0 & 1 & -1 & 0 & 0 & 0 \\ 0 & -1 & -1 & 0 & 0 & 0 & 0 & 1 & 1 \\ 0 & 0 & -1 & 0 & -1 & 1 & 0 & 1 & 0 \\ 0 & 1 & 0 & 0 & -1 & 1 & 0 & 0 & -1 \end{bmatrix},$$

$$C = \begin{bmatrix} 1 & 0 & 0 & 0 & 0 & 0 & 0 & 0 & 0 & 0 & 0 \\ 1 & 0 & -1 & 1 & 1 & 1 & 0 & 0 & j & j & j \\ 1 & 0 & -1 & 1 & 0 & -1 & j & 0 & -j & -j & 0 \\ 1 & -1 & 0 & 0 & 0 & 0 & 0 & j & 0 & 0 & 0 \\ 1 & 0 & -1 & 1 & -1 & 0 & -j & 0 & j & j & -j \\ 1 & 0 & -1 & 1 & -1 & 0 & -j & 0 & -j & -j & j \\ 1 & -1 & 0 & 0 & 0 & 0 & 0 & -j & 0 & 0 & 0 \\ 1 & 0 & -1 & 1 & 0 & -1 & j & 0 & j & j & 0 \\ 1 & 0 & -1 & 1 & 1 & 1 & 0 & 0 & -j & -j & -j \end{bmatrix},$$

$t_0 = v_1 + v_8,$
$t_1 = v_2 + v_7,$
$a_1 = v_3 + v_6,$
$t_2 = v_4 + v_5,$
$a_2 = t_0 + t_1 + t_2,$
$t_3 = v_1 - v_8,$
$t_4 = v_7 - v_2,$
$t_5 = v_3 - v_6,$
$a_7 = v_4 - v_5,$
$a_6 = t_3 + t_4 + a_7,$
$a_3 = t_0 - t_1,$
$a_4 = t_1 - t_2,$
$a_8 = t_4 - t_3,$
$a_9 = t_4 - a_7,$
$a_0 = v_0 + a_1 + a_2,$
$a_5 = -a_4 - a_3,$
$a_{10} = -a_8 + a_9,$

$\theta = 2\pi/9,$
$B_0 = 1,$
$B_1 = \frac{3}{2},$
$B_2 = -\frac{1}{2},$
$B_3 = \frac{1}{3}(2\cos\theta - \cos 2\theta - \cos 4\theta),$
$B_4 = \frac{1}{3}(\cos\theta + \cos 2\theta - 2\cos 4\theta),$
$B_5 = \frac{1}{3}(\cos\theta - 2\cos 2\theta + \cos 4\theta),$
$B_6 = \sin 3\theta,$
$B_7 = \sin 3\theta,$
$B_8 = -\sin\theta,$
$B_9 = -\sin 4\theta,$
$B_{10} = -\sin 2\theta,$

$T_0 = -b_3 - b_4,$
$T_1 = b_5 - b_4,$
$T_2 = -b_8 - b_9,$
$T_3 = b_9 - b_{10},$
$T_4 = b_0 + b_2 + b_2,$
$T_5 = T_4 - b_1,$
$T_6 = T_4 + b_2,$
$T_7 = T_5 - T_0,$
$T_8 = T_1 + T_5,$
$T_9 = T_0 - T_1 + T_5,$
$T_{10} = b_7 - T_2,$
$T_{11} = b_7 - T_3,$
$T_{12} = b_7 + T_2 + T_3,$
$V_0 = b_0,$
$V_1 = T_7 + jT_{10},$
$V_2 = T_8 - jT_{11},$
$V_3 = T_6 + jb_6,$
$V_4 = T_9 + jT_{12},$
$V_5 = T_9 - jT_{12},$
$V_6 = T_6 - jb_6,$
$V_7 = T_8 + jT_{11},$
$V_8 = T_7 - jT_{10}.$

Sixteen-point Fourier transform; 10(18) real multiplications, 74 real additions:

$$A = \begin{bmatrix}
1 & 1 & 1 & 1 & 1 & 1 & 1 & 1 & 1 & 1 & 1 & 1 & 1 & 1 & 1 & 1 \\
1 & -1 & 1 & -1 & 1 & -1 & 1 & -1 & 1 & -1 & 1 & -1 & 1 & -1 & 1 & -1 \\
1 & 0 & -1 & 0 & 1 & 0 & -1 & 0 & 1 & 0 & -1 & 0 & 1 & 0 & -1 & 0 \\
1 & 0 & 0 & 0 & -1 & 0 & 0 & 0 & 1 & 0 & 0 & 0 & -1 & 0 & 0 & 0 \\
1 & 0 & 0 & 0 & 0 & 0 & 0 & 0 & -1 & 0 & 0 & 0 & 0 & 0 & 0 & 0 \\
0 & 1 & 0 & -1 & 0 & -1 & 0 & 1 & 0 & 1 & 0 & -1 & 0 & -1 & 0 & 1 \\
0 & 0 & 1 & 0 & 0 & 0 & -1 & 0 & 0 & 0 & -1 & 0 & 0 & 0 & 1 & 0 \\
0 & 1 & 0 & -1 & 0 & 1 & 0 & -1 & 0 & -1 & 0 & 1 & 0 & -1 & 0 & 1 \\
0 & 1 & 0 & 0 & 0 & 0 & 0 & -1 & 0 & -1 & 0 & 0 & 0 & 0 & 0 & 1 \\
0 & 0 & 0 & -1 & 0 & 1 & 0 & 0 & 0 & 0 & 0 & 1 & 0 & -1 & 0 & 0 \\
0 & -1 & 0 & 1 & 0 & -1 & 0 & 1 & 0 & -1 & 0 & 1 & 0 & -1 & 0 & 1 \\
0 & 0 & -1 & 0 & 0 & 0 & 1 & 0 & 0 & 0 & -1 & 0 & 0 & 0 & 1 & 0 \\
0 & 0 & 0 & 0 & 1 & 0 & 0 & 0 & 0 & 0 & 0 & 0 & -1 & 0 & 0 & 0 \\
0 & 1 & 0 & 1 & 0 & -1 & 0 & -1 & 0 & 1 & 0 & 1 & 0 & -1 & 0 & -1 \\
0 & 0 & 1 & 0 & 0 & 0 & 1 & 0 & 0 & 0 & -1 & 0 & 0 & 0 & -1 & 0 \\
0 & 1 & 0 & 1 & 0 & 1 & 0 & 1 & 0 & -1 & 0 & -1 & 0 & -1 & 0 & -1 \\
0 & 1 & 0 & 0 & 0 & 0 & 0 & 1 & 0 & -1 & 0 & 0 & 0 & 0 & 0 & -1 \\
0 & 0 & 0 & 1 & 0 & 1 & 0 & 0 & 0 & 0 & 0 & -1 & 0 & -1 & 0 & 0
\end{bmatrix},$$

$$C = \begin{bmatrix}
1 & 0 & 0 & 0 & 0 & 0 & 0 & 0 & 0 & 0 & 0 & 0 & 0 & 0 & 0 & 0 & 0 & 0 \\
0 & 0 & 0 & 0 & 1 & 0 & 1 & -1 & 1 & 0 & 0 & 0 & j & 0 & j & j & j & 0 \\
0 & 0 & 0 & 1 & 0 & 1 & 0 & 0 & 0 & 0 & 0 & j & 0 & j & 0 & 0 & 0 & 0 \\
0 & 0 & 0 & 0 & 1 & 0 & -1 & 1 & 0 & -1 & 0 & 0 & -j & 0 & j & j & 0 & -j \\
0 & 0 & 1 & 0 & 0 & 0 & 0 & 0 & 0 & 0 & j & 0 & 0 & 0 & 0 & 0 & 0 & 0 \\
0 & 0 & 0 & 0 & 1 & 0 & -1 & -1 & 0 & 1 & 0 & 0 & j & 0 & -j & j & 0 & -j \\
0 & 0 & 0 & 1 & 0 & -1 & 0 & 0 & 0 & 0 & 0 & -j & 0 & j & 0 & 0 & 0 & 0 \\
0 & 0 & 0 & 0 & 1 & 0 & 1 & 1 & -1 & 0 & 0 & 0 & -j & 0 & -j & j & j & 0 \\
0 & 1 & 0 & 0 & 0 & 0 & 0 & 0 & 0 & 0 & 0 & 0 & 0 & 0 & 0 & 0 & 0 & 0 \\
0 & 0 & 0 & 0 & 1 & 0 & 1 & 1 & -1 & 0 & 0 & 0 & j & 0 & j & -j & -j & 0 \\
0 & 0 & 0 & 1 & 0 & -1 & 0 & 0 & 0 & 0 & 0 & j & 0 & -j & 0 & 0 & 0 & 0 \\
0 & 0 & 0 & 0 & 1 & 0 & -1 & -1 & 0 & 1 & 0 & 0 & -j & 0 & j & -j & 0 & j \\
0 & 0 & 1 & 0 & 0 & 0 & 0 & 0 & 0 & 0 & -j & 0 & 0 & 0 & 0 & 0 & 0 & 0 \\
0 & 0 & 0 & 0 & 1 & 0 & -1 & 1 & 0 & -1 & 0 & 0 & j & 0 & -j & -j & 0 & j \\
0 & 0 & 0 & 1 & 0 & 1 & 0 & 0 & 0 & 0 & 0 & -j & 0 & -j & 0 & 0 & 0 & 0 \\
0 & 0 & 0 & 0 & 1 & 0 & 1 & -1 & 1 & 0 & 0 & 0 & -j & 0 & -j & -j & -j & 0
\end{bmatrix},$$

$t_0 = v_0 + v_8,$
$t_1 = v_4 + v_{12},$
$t_2 = v_2 + v_{10},$
$t_3 = v_2 - v_{10},$
$t_4 = v_6 + v_{14},$
$t_5 = v_6 - v_{14},$
$t_6 = v_1 + v_9,$
$t_7 = v_1 - v_9,$
$t_8 = v_3 + v_{11},$
$t_9 = v_3 - v_{11},$
$t_{10} = v_5 + v_{13},$
$t_{11} = v_5 - v_{13},$
$t_{12} = v_7 + v_{15},$
$t_{13} = v_7 - v_{15},$
$t_{14} = t_0 + t_1,$
$t_{15} = t_2 + t_4,$
$t_{16} = t_{14} + t_{15},$
$t_{17} = t_6 + t_{10},$
$t_{18} = t_6 - t_{10},$
$t_{19} = t_8 + t_{12},$
$t_{20} = t_8 - t_{12},$
$t_{21} = t_{17} + t_{19},$
$a_{16} = t_7 + t_{13},$
$a_8 = t_7 - t_{13},$
$a_{17} = t_{11} + t_9,$
$a_9 = t_{11} - t_9,$
$a_0 = t_{16} + t_{21},$
$a_1 = t_{16} - t_{21},$
$a_2 = t_{14} - t_{15},$
$a_3 = t_0 - t_1,$
$a_4 = v_0 - v_8,$
$a_5 = t_{18} - t_{20},$
$a_6 = t_3 - t_5,$
$a_7 = a_8 + a_9,$
$a_{10} = t_{19} - t_{17},$
$a_{11} = t_4 - t_2,$
$a_{12} = v_{12} - v_4,$
$a_{13} = t_{18} + t_{20},$
$a_{14} = t_3 + t_5,$
$a_{15} = a_{16} + a_{17},$

$\theta = 2\pi/16,$
$B_0 = 1,$
$B_1 = 1,$
$B_2 = 1,$
$B_3 = 1,$
$B_4 = 1,$
$B_5 = \cos 2\theta,$
$B_6 = \cos 2\theta,$
$B_7 = \cos 3\theta,$
$B_8 = \cos \theta + \cos 3\theta,$
$B_9 = -\cos \theta + \cos 3\theta,$
$B_{10} = 1,$
$B_{11} = 1,$
$B_{12} = 1,$
$B_{13} = -\sin 2\theta,$
$B_{14} = -\sin 2\theta,$
$B_{15} = -\sin 3\theta,$
$B_{16} = -\sin \theta + \sin 3\theta,$
$B_{17} = -\sin \theta - \sin 3\theta,$

$T_0 = b_3 + b_5,$
$T_1 = b_3 - b_5,$
$T_2 = b_{11} + b_{13},$
$T_3 = b_{13} - b_{11},$
$T_4 = b_4 + b_6,$
$T_5 = b_4 - b_6,$
$T_6 = b_8 - b_7,$
$T_7 = b_9 - b_7,$
$T_8 = T_4 + T_6,$
$T_9 = T_4 - T_6,$
$T_{10} = T_5 + T_7,$
$T_{11} = T_5 - T_7,$
$T_{12} = b_{12} + b_{14},$
$T_{13} = b_{12} - b_{14},$
$T_{14} = b_{15} + b_{16},$
$T_{15} = b_{15} - b_{17},$
$T_{16} = T_{12} + T_{14},$
$T_{17} = T_{12} - T_{14},$
$T_{18} = T_{13} + T_{15},$
$T_{19} = T_{13} - T_{15},$
$V_0 = b_0,$
$V_1 = T_8 + jT_{16},$
$V_2 = T_0 + jT_2,$
$V_3 = T_{11} - jT_{19},$
$V_4 = b_2 + jb_{10},$
$V_5 = T_{10} + jT_{18},$
$V_6 = T_1 + jT_3,$
$V_7 = T_9 - jT_{17},$
$V_8 = b_1,$
$V_9 = T_9 + jT_{17},$
$V_{10} = T_1 - jT_3,$
$V_{11} = T_{10} - jT_{18},$
$V_{12} = b_2 - jb_{10},$
$V_{13} = T_{11} + jT_{19},$
$V_{14} = T_0 - jT_2,$
$V_{15} = T_8 - jT_{16}.$

Bibliography

R. C. Agarwal and C. S. Burrus, Fast Digital Convolution Using Fermat Transforms, *Southwest IEEE Conference Record, Houston*, 538–543, 1973.

R. C. Agarwal and C. S. Burrus, Fast Convolution Using Fermat Number Transforms with Applications to Digital Filtering, *IEEE Transaction on Acoustics, Speech, and Signal Processing*, **ASSP-22**, 87–97, 1974a.

R. C. Agarwal and C. S. Burrus, Fast One-Dimensional Digital Convolution by Multidimensional Techniques, *IEEE Transactions on Acoustics, Speech, and Signal Processing*, **ASSP-22**, 1–10, 1974b.

R. C. Agarwal and C. S. Burrus, Number Theoretic Transforms to Implement Fast Digital Convolution, *Proceedings of the IEEE*, **63**, 550–560, 1975.

R. C. Agarwal and J. W. Cooley, New Algorithms for Digital Convolution, *IEEE Transactions on Acoustics, Speech, and Signal Processing*, **ASSP-25**, 392–410, 1977.

N. Ahmed, T. Natarajan, and K. R. Rao, Discrete Cosine Transform, *IEEE Transactions on Computers*, **C-23**, 90–93, 1974.

A. V. Aho, J. E. Hopcroft, and J. D. Ullman, *The Design and Analysis of Computer Algorithms*, Reading, MA, Addison-Wesley, 1974.

B. Arambepola and P. J. W. Rayner, Efficient Transforms for Multidimensional Convolutions, *Electronic Letters*, **15**, 189–190, 1979.

L. Auslander, E. Feig, and S. Winograd, New Algorithms for the Multidimensional Discrete Fourier Transform, *IEEE Transactions on Acoustics, Speech, and Signal Processing*, **ASSP-31**, 388–403, 1983.

L. Bahl, J. Cocke, F. Jelinek, and J. Raviv, Optimal Decoding of Linear Codes for Minimizing Symbol Error Rate, *IEEE Transactions on Information Theory*, **IT-20**, 284–287, 1974.

B. Baumslag and B. Chandler, *Theory and Problems of Group Theory*, Schaum's Outline Series, New York, McGraw-Hill, 1968.

E. R. Berlekamp, *Algebraic Coding Theory*, New York, McGraw-Hill, 1968.

R. Bernardini, G. M. Cortelazzo, and G. A. Mian, A New Technique for Twiddle-Factor Elimination in Multidimensional FFTs, *IEEE Transactions on Signal Processing*, **SP-42**, 2176–2178, 1994.

T. Beth, W. Fumy, and R. Muhlfeld, On Algebraic Discrete Fourier Transforms, *Abstracts on the 1982 IEEE International Symposium on Information Theory*, Les Arcs, France, 1982.

G. Birkhoff and S. MacLane, *A Survey of Modern Algebra*, New York, MacMillan, 1941; rev. ed., 1953.

R. E. Blahut, Fast Convolution of Rational Sequences, *Abstracts on the 1983 IEEE International Symposium on Information Theory*, St. Jovite, Quebec, Canada, 1983a.

R. E. Blahut, *Theory and Practice of Error Control Codes*, Reading, MA, Addison-Wesley, 1983b.

R. E. Blahut, *Algebraic Methods for Signal Processing and Communications Coding*, New York, Springer-Verlag, 1992.

R. E. Blahut, *Algebraic Codes for Data Transmission*, Cambridge, Cambridge University Press, 2003.

R. E. Blahut and D. E. Waldecker, Half-Angle Sine-Cosine Generator, *IBM Technical Disclosure Bulletin*, **13**, no. 1, 222–223, 1970.

L. I. Bluestein, A Linear Filtering Approach to the Computation of the Discrete Fourier Transform, *IEEE Transactions on Audio Electroacoustics*, **AU-18**, 451–455, 1970.

R. P. Brent, F. O. Gustavson, and D. Y. Y. Yun, Fast Solution of Toeplitz Systems of Equations and Computation of Pade Approximants, *Journal on Algorithms*, **1**, 259–295, 1980.

C. S. Burrus and P. W. Eschenbacher, An In-Place, In-Order Prime Factor FFT Algorithm, *IEEE Transactions on Acoustics, Speech, and Signal Processing*, **ASSP-29**, 806–817, 1979.

J. Butler and R. Lowe, Beam-forming Matrix Simplifies Design of Electronically Scanned Antennas, *Electronic Design*, **9**, 170–173, 1961.

W.-H. Chen, C. H. Smith, and S. C. Fralick, A Fast Computational Algorithm for the Discrete Cosine Transform, *IEEE Transactions on Communications*, **COM-25**, 1004–1009, 1977.

P. R. Chevillat, Transform-Domain Filtering with Number Theoretic Transforms, and Limited Word Length, *IEEE Transactions on Acoustics, Speech, and Signal Processing*, **ASSP-26**, 284–290, 1978.

P. R. Chevillat and D. J. Costello, Jr., An Analysis of Sequential Decoding for Specific Time-Invariant Convolutional Codes, *IEEE Transactions on Information Theory*, **IT-24**, 443–451, 1978.

J. W. Cooley, P. A. W. Lewis, and P. D. Welch, Historical Notes on the Fourier Transform, *IEEE Transactions on Audio Electroacoustics*, **AU-15**, 76–79, 1967.

J. W. Cooley and J. W. Tukey, An Algorithm for the Machine Computation of Complex Fourier Series, *Mathematics of Computation*, **19**, 297–301, 1965.

D. Coppersmith and S. Winograd, Matrix Multiplication via Arithmetic Progressions, *Journal of Symbolic Computation*, **9**, 251–280, 1990.

J. H. Cozzens and L. A. Finkelstein, Computing the Discrete Fourier Transform Using Residue Number Systems in a Ring of Algebraic Integers, *IEEE Transactions on Information Theory*, **IT-31**, 580–588, 1985.

R. E. Crochiere and L. R. Rabiner, Interpolation and Decimation of Digital Signals – A Tutorial Review, *Proceedings of the IEEE*, **69**, 330–331, 1981.

B. W. Dickinson, Efficient Solution of Linear Equations with Banded Toeplitz Matrices, *IEEE Transactions on Acoustics, Speech, and Signal Processing*, **ASSP-27**, 421–423, 1979.

B. W. Dickinson, M. Morf, and T. Kailath, A Minimal Realization Algorithm for Matrix Sequences, *IEEE Transactions on Automatic Control*, **AC-19**, 31–38, 1974.

E. DuBois and A. N. Venetsanopoulos, Convolution Using a Conjugate Symmetry Property for the Generalized Discrete Fourier Transform, *IEEE Transactions on Acoustics, Speech, and Signal Processing*, **ASSP-26**, 165–170, 1978.

J. Durbin, The Fitting of Time-Series Models, *Reviews of the International Statistical Institute*, **23**, 233–244, 1960.

S. R. Dussé and B. S. Kaliski, A Cryptographic Library for the Motorola DSP56000, *Advances in Cryptology, Eurocrypt90*, I. B. Damgard, editor, 230–244, New York, Springer-Verlag, 1991.

J. O. Eklundh, A Fast Computer Method for Matrix Transposing, *IEEE Transactions on Computers*, **C-21**, 801–803, 1972.

R. M. Fano, A Heuristic Discussion of Probabilistic Decoding, *IEEE Transactions on Information Theory*, **IT-11**, no. 9, 64–74, 1963.

S. V. Fedorenko, A Method for Computation of the Discrete Fourier Transform over a Finite Field, *Problemy Peredachi Informatsii*, **42**, 81–93, 2006, English translation: *Problems of Information Transmission*, **42**, 139–151, 2006.

E. Feig and S. Winograd, Fast Algorithms for the Discrete Cosine Transform, *IEEE Transactions on Signal Processing*, **SP-40**, 2174–2193, 1992.

C. M. Fiduccia, Polynomial Evaluation via the Division Algorithm – The Fast Fourier Transform Revisited, *Proceedings of the 4th Annual ACM Symposium on the Theory of Computing*, 88–93, 1972.

G. D. Forney, Convolutional Codes III: Sequential Decoding, *Information and Control*, **25**, 267–297, 1974.

J. B. Fraleigh, *A First Course in Abstract Algebra*, 2nd edition, Reading, MA, Addison-Wesley, 1976.

B. Friedlander, M. Morf, T. Kailath, and L. Ljung, New Inversion Formulas for Matrices Classified in Terms of Their Distance from Toeplitz Matrices, *Linear Algebra and Its Application*, **27**, 31–60, 1979.

R. G. Gallager, *Information Theory and Reliable Communications*, New York, John Wiley, 1968.

R. A. Games, Complex Approximations Using Algebraic Integers, *IEEE Transactions on Information Theory*, **IT-31**, 565–579, 1985.

R. A. Games, An Algorithm for Complex Approximations in $Z[e^{2\pi i/8}]$, *IEEE Transactions on Information Theory*, **IT-32**, 603–607, 1986.

C. F. Gauss, Nachlass: Theoria interpolationis methodo nova tractata, Werke band 3, 265–327. Göttingen Königliche Gesellschaft der Wissenschaften, 1866.

I. Gertner, A New Efficient Algorithm to Compute the Two-Dimensional Discrete Fourier Transform, *IEEE Transactions on Acoustics, Speech, and Signal Processing*, **ASSP-36**, 1036–1050, 1988.

G. Goertzel, An Algorithm for the Evaluation of Finite Trigonometric Series, *American Mathematics Monthly*, **65**, 34–35, 1968.

I. J. Good, The Interaction Algorithm and Practical Fourier Analysis, *Journal of the Royal Statistics Society*, **B20**, 361–375, 1958. Addendum, **22**, 372–375, 1960.

I. J. Good, The Relationship between Two Fast Fourier Transforms, *IEEE Transactions on Computers*, **C-20**, 310–317, 1971.

D. Haccoun and M. J. Ferguson, Generalized Stack Algorithms for Decoding Convolutional Codes, *IEEE Transactions on Information Theory*, **IT-21**, 638–651, 1975.

G. H. Hardy and E. M. Wright, *The Theory of Numbers*, Oxford, Oxford University Press, 1960.

D. B. Harris, J. H. McClellan, D. S. K. Chan, and H. W. Schuessler, Vector Radix Fast Fourier Transform, *Record of the 1977 IEEE International Conference on Acoustics, Speech, and Signal Processing*, 548–551, 1977.

M. T. Heideman, D. H. Johnson, and C. S. Burrus, Gauss and the History of the FFT, *IEEE Signal Processing Magazine*, **1**, 14–21, 1984.

C. A. R. Hoare, Quicksort, *Computer Journal 5*, **5**, no. 1, 10–16, 1962. (Reprinted in Hoare and Jones: *Essays in Computing Science*, New York, Prentice Hall, 1989.)

J. E. Hopcroft and I. Musinski, Duality Applied to the Complexity of Matrix Multiplication and Other Bilinear Forms, *SIAM Journal of Computation*, **2**, 159–173, 1973.

T. S. Huang, How the Fast Fourier Transform Got Its Name, *Computer*, **3**, 15, 1971.

I. M. Jacobs and E. R. Berlekamp, A Lower Bound to the Distribution of Computations for Sequential Decoding, *IEEE Transactions on Information Theory*, **IT-13**, 167–174, 1967.

F. Jelinek, *Probabilistic Information Theory*, New York, McGraw-Hill, 1968.

F. Jelinek, A Fast Sequential Decoding Algorithm Using a Stack, *IBM Journal of Research and Development*, **13**, 675–685, 1969a.

F. Jelinek, An Upper Bound on Moments of Sequential Decoding Effort, *IEEE Transactions on Information Theory*, **IT-15**, 140–149, 1969b.

W. K. Jenkins and J. V. Krogmeier, The Design of Dual-Mode Complex Signal Processors Based on Quadratic Modular Number Codes, *IEEE Transactions on Circuits and Systems*, **CAS-34**, 354–364, 1987.

W. K. Jenkins and B. J. Leon, The Use of Residue Number Systems in the Design of Finite Impulse Response Digital Filters, *IEEE Transactions on Circuits and Systems*, **CAS-24**, 191–201, 1977.

R. Johannesson and K. Sh. Zigangirov, *Fundamentals of Convolutional Coding*, New York, IEEE Press, 1999.

H. W. Johnson and C. S. Burrus, Large DFT Modules: 11, 13, 17, 19 and 25, *Technical Report 8105*, Department of Electrical Engineering, Rice University, Houston, TX, 1981.

H. W. Johnson and C. S. Burrus, The Design of Optimal DFT Algorithms Using Dynamic Programming, *IEEE Transactions on Acoustics, Speech and Signal Processing*, **ASSP-31**, 378–387, 1983.

S. G. Johnson and M. Frigo, A Modified Split-Radix FFT with Fewer Arithmetic Operations, *IEEE Transactions on Signal Processing*, **SP-55**, 111–119, 2007.

B. S. Kaliski, The Montgomery Inverse and its Applications, *IEEE Transactions on Computers*, **C-44**, 1064–1065, 1995.

D. E. Knuth, *The Art of Computer Programming, Vol. 1: Fundamental Algorithms*, Reading, MA, Addison-Wesley, 1968.

D. P. Kolba and T. W. Parks, A Prime Factor FFT Algorithm Using High Speed Convolution, *IEEE Transactions on Acoustics, Speech, and Signal Processing*, **ASSP-25**, 281–294, 1977.

L. M. Leibowitz, A Simplified Binary Arithmetic for the Fermat Number Transform, *IEEE Transactions on Acoustics, Speech, and Signal Processing*, **ASSP-24**, 356–359, 1976.

N. Levinson, The Wiener RMS Error Criterion in Filter Design and Prediction, *Journal of Mathematical Physics*, **25**, 261–278, 1947.

B. Liu and F. Mintzer, Calculation of Narrow-Band Spectra by Direct Decimation, *IEEE Transactions on Acoustics, Speech, and Signal Processing*, **ASSP-26**, 529–534, 1978.

J. Makhoul, A Fast Cosine Transform in One and Two Dimensions, *IEEE Transaction on Acoustics, Speech, and Signal Processing*, **ASSP-28**, 27–34, 1980.

J. L. Massey, Shift-Register Synthesis and BCH Decoding, *IEEE Transactions on Information Theory*, **IT-15**, 122–127, 1969.

J. H. McClellan, Hardware Realization of a Fermat Number Transform, *IEEE Transactions on Acoustics, Speech, and Signal Processing*, **ASSP-24**, 216–225, 1976.

J. H. McClellan and C. M. Rader, *Number Theory in Digital Signal Processing*, Englewood Cliffs, NJ, Prentice-Hall, 1979.

R. Mersereau and T. C. Speake, A Unified Treatment of Cooley–Tukey Algorithms for the Evaluation of the Multidimensional DFT, *IEEE Transactions on Acoustics, Speech, and Signal Processing*, **ASSP-29**, 1011–1018, 1981.

R. L. Miller, T. K. Truong, and I. S. Reed, Efficient Program for Decoding the $(255, 223)$ Reed–Solomon Code Over $GF(2^8)$ with Both Errors and Erasures Using Transform Coding, *Proceedings of the IEEE*, **127**, 136–142, 1980.

Y. Monden and S. Arimoto, Generalized Rouche's Theorem and Its Application to Multivariate Autoregressions, *IEEE Transactions on Acoustics, Speech, and Signal Processing*, **ASSP-28**, 733–738, 1980.

P. L. Montgomery, Modular Multiplication Without Trial Division, *Mathematics of Computation*, **44**, 519–521, 1985.

M. Morf, B. W. Dickinson, T. Kailath, and A. C. O. Vieira, Efficient Solution of Covariance Equations for Linear Prediction, *IEEE Transactions on Acoustics, Speech, and Signal Processing*, **ASSP-25**, 429–433, 1977.

R. L. Morris, A Comparative Study of Time Efficient FFT and WFTA Programs for General Purpose Computers, *IEEE Transactions on Acoustics, Speech, and Signal Processing*, **ASSP-26**, 141–150, 1978.

M. J. Narasimha and A. M. Peterson, On the Computation of the Discrete Cosine Transform, *IEEE Transactions on Communications*, **COM-26**, 934–936, 1978.

H. Nawab and J. H. McClellan, Bounds on the Minimum Number of Data Transfers in WFTA and FFT Programs, *IEEE Transactions on Acoustics, Speech, and Signal Processing*, **ASSP-27**, 393–398, 1979.

H. J. Nussbaumer, Digital Filtering Using Complex Mersenne Transforms, *IBM Journal of Research and Development*, **20**, 498–504, 1976.

H. J. Nussbaumer, Digital Filtering Using Polynomial Transforms, *Electronics Letters*, **13**, 386–387, 1977.

H. J. Nussbaumer, New Algorithms for Convolution and DFT Based on Polynomial Transforms, *Proceedings of the 1978 IEEE International Conference on Acoustics, Speech, and Signal Processing*, 638–641, 1978.

H. J. Nussbaumer and P. Quandalle, Fast Computation of Discrete Fourier Transforms Using Polynomial Transforms, *IEEE Transactions on Acoustics, Speech and Signal Processing*, **ASSP-27**, 169–181, 1979.

H. J. Nussbaumer and P. Quandalle, New Polynomial Transform Algorithms for Fast DFT Computations, *Proceedings of the IEEE 1979 International Acoustics, Speech and Signal Processing Conference (1979)*, 510–513, 1979.

A. V. Oppenheim and R. W. Shafer, *Digital Signal Processing*, Englewood Cliffs, NJ, Prentice-Hall, 1975.

O. Ore, *Number Theory and its History*, New York, McGraw-Hill, 1948.

O. Panda, R. N. Pal, and B. Chatterjee, Error Analysis of Good-Winograd Algorithm Assuming Correlated Truncation Errors, *IEEE Transactions on Acoustics, Speech, and Signal Processing*, **ASSP-31**, 508–512, 1983.

T. W. Parsons, A Winograd–Fourier Transform Algorithm for Real-Valued Data, *IEEE Transactions on Acoustics, Speech, and Signal Processing*, **ASSP-27**, 398–402, 1979.

R. W. Patterson and J. H. McClellan, Fixed-Point Error Analysis of Winograd Fourier Transform Algorithms, *IEEE Transactions on Acoustics, Speech, and Signal Processing*, **ASSP-26**, 447–455, 1978.

I. Pitas and M. G. Strintzis, On the Multiplicative Complexity of Two-Dimensional Fast Convolution Methods, *Abstr. 1982 IEEE International Symposium on Information Theory*, Les Arcs, France, 1982.

J. M. Pollard, The Fast Fourier Transform in a Finite Field, *Mathematics of Computation*, **25**, 365–374, 1971.

F. P. Preparata and D. V. Sarwate, Computational Complexity of Fourier Transforms over Finite Fields, *Mathematics of Computation*, **31**, 740–751, 1977.

R. D. Preuss, Very Fast Computation of the Radix-2 Discrete Fourier Transform, *IEEE Transactions on Acoustics, Speech, and Signal Processing*, **ASSP-30**, 595–607, 1982.

J. G. Proakis and D. K. Manolakis, *Digital Signal Processing* (fourth edition), Englewood Cliffs, NJ, Prentice Hall, 2006.

L. R. Rabiner and B. Gold, *Theory and Application of Digital Signal Processing*, Englewood Cliffs, NJ, Prentice-Hall, 1975.

C. M. Rader, Discrete Fourier Transforms When the Number of Data Samples Is Prime, *Proceedings of the IEEE*, **56**, 1107–1108, 1968.

C. M. Rader, An Improved Algorithm for High Speed Autocorrelation with Applications to Spectral Estimation, *IEEE Transactions on Audio Electroacoustics*, **AU-18**, 439–441, 1970.

C. M. Rader, Discrete Convolutions via Mersenne Transforms, *IEEE Transactions on Computers*, **C-21**, 1269–1273, 1972.

C. M. Rader, Memory Management in a Viterbi Decoder, *IEEE Transactions on Communications*, **COM-29**, 1399–1401, 1981.

C. M. Rader and N. M. Brenner, A New Principle for Fast Fourier Transformation, *IEEE Transactions on Acoustics, Speech, and Signal Processing*, **ASSP-24**, 264–265, 1976.

I. S. Reed and T. K. Truong, The Use of Finite Fields to Compute Convolutions, *IEEE Transactions on Information Theory*, **IT-21**, 208–213, 1975.

B. Rice, Some Good Fields and Rings for Computing Number Theoretic Transforms, *IEEE Transactions on Acoustics, Speech, and Signal Processing*, **ASSP-27**, 432–433, 1979.

B. Rice, Winograd Convolution Algorithms overt Finite Fields, *Congressus Numberatium*, **29**, 827–857, 1980.

G. E. Rivard, Direct Fast Fourier Transform of Bivariate Functions, *IEEE Transactions on Acoustics, Speech, and Signal Processing*, **ASSP-25**, 250–252, 1977.

S. Robinson, Towards an Optimal Algorithm for Matrix Multiplication, *SIAM News*, **9**, 8, 2005.

D. V. Sarwate, Semi-Fast Fourier Transforms Over GF(2^m), *IEEE Transactions on Computers*, **C-27**, 283–284, 1978.

J. E. Savage, Sequential Decoding-The Computation Problem, *Bell System Technical Journal*, **45**, 149–175, 1966.

I. W. Selesnick and C. S. Burrus, Automatic Generations of Prime Length FFT Programs, *IEEE Transactions on Signal Processing*, **44**, 14–24, 1996.

H. F. Silverman, An Introduction to Programming the Winograd Fourier Transform Algorithm (WFTA), *IEEE Transactions on Acoustics, Speech and Signal Processing*, **ASSP-25**, 152–165, 1977.

R. C. Singleton, An Algorithm for Computing the Mixed Radix Fast Fourier Transform, *IEEE Transactions on Audio Electroacoustics*, **AU-17**, 93–103, 1969.

K. Steiglitz, *An Introduction to Discrete Systems*, New York, John Wiley, 1974.

T. G. Stockham, High Speed Convolution and Correlation, *Spring Joint Computer Conference, AFIPS Conference Proceedings*, **28**, 229–233, 1966.

G. Strang, *Linear Algebra and Its Applications*, second edition, New York, Academic Press, 1980.

V. Strassen, Gaussian Elimination is not Optimal, *Numerical Mathematics*, **13**, 354–356, 1969.

Y. Sugiyama, M. Kasahara, S. Hirasawa, and T. Namekawa, A Method for Solving Key Equations for Decoding Goppa Codes, *Information and Control*, **27**, 87–99, 1975.

N. S. Szabo and R. I. Tanaka, *Residue Arithmetic and Its Application to Computer Technology*, New York, McGraw-Hill, 1967.

L. H. Thomas, Using a Computer to Solve Problems in Physics, in *Applications of Digital Computers*, Boston, MA, Ginn and Co., 1963.

R. M. Thrall and L. Tornheim, *Vector Spaces and Matrices*, New York, John Wiley, 1957.

R. Tolmieri, M. An, and C. Lu, *Mathematics of Multidimensional Fourier Transform Algorithms*, New York, Springer-Verlag, 1993, second edition, 1997.

W. F. Trench, An Algorithm for the Inversion of Finite Toeplitz Matrices, *J. SIAM*, **12**, no. 3, 512–522, 1964.

P. V. Trifonov and S. V. Fedorenko, A Method for Fast Computation of the Fourier Transform Over a Finite Field, *Problemy Peredachi Informatsaii*, **39**, no. 3, 3–10, 2003, English translation, *Problems of Information Transmission*, **39**, 231–238, 2003.

B. L. Van der Waerden, *Modern Algebra*, two volumes, translated by F. Blum and T. J. Benac, New York, Frederick Ungar Publishing Co., 1949, 1953.

A. Vieira and T. Kailath, On Another Approach to the Schur–Cohn Criterion, *IEEE Transactions on Circuits and Systems*, **CAS-24**, no. 4, 218–220, 1977.

A. J. Viterbi, Error Bounds for Convolutional Codes and an Asymptotically Optimum Decoding Algorithm, *IEEE Transactions on Information Theory*, **IT-13**, 260–269, 1967.

J. E. Volder, The CORDIC Trigonometric Computing Technique, *IRE Transactions on Electronic Computers*, **EC-8**, no. 3, 330–334, 1959.

J. S. Walther, A Unified Algorithm for Elementary Functions, *Conference Proceedings, Spring Joint Computer Conference*, 379–385, 1971.

L. R. Welch and R. A. Scholtz, Continued Fractions and Berlekamp's Algorithm, *IEEE Transactions on Information Theory*, **IT-25**, 19–27, 1979.

R. A. Wiggins and B. A. Robinson, Recursive Solution to the Multichannel Filtering Problem, *Journal of Geophysical Research*, **70**, 1885–1891, 1965.

N. M. Wigley and G. A. Jullien, On Modulus Replication for Residue Arithmetic Computation of Complex Inner Products, *IEEE Transactions on Computers*, **C-39**, 1065–1076, 1990.

N. M. Wigley, G. A. Jullien, and D. Reaume, Large Dynamic Range Computations over Small Finite Rings, *IEEE Transactions on Computers*, **C-43**, 78–86, 1994.

J. W. J. Williams, Algorithm 232: Heapsort, *Communications of the ACM*, **7**, 347–348, 1964.

S. Winograd, A New Algorithm for Inner Product, *IEEE Transactions on Computers*, **C-17**, 693–694, 1968.

S. Winograd, On Computing the Discrete Fourier Transform, *Proceedings of the National Academy of Science USA*, **73**, 1005–1006, 1976.

S. Winograd, Some Bilinear Forms Whose Multiplicative Complexity Depends on the Field of Constants, *Mathematical Systems Theory*, **10**, 169–180, 1977.

S. Winograd, On Computing the Discrete Fourier Transform, *Mathematics of Computation*, **32**, 175–199, 1978.

S. Winograd, On the Complexity of Symmetric Filters, *Proceedings of the International Symposium on Circuits and Systems*, Tokyo, 262–265, 1979.

S. Winograd, Arithmetic Complexity of Computations, *CBMS-NSF Regional Conference Series Appl. Math*, Siam Publications 33, 1980a.

S. Winograd, Signal Processing and Complexity of Computation, *Proceedings of the International Conference on Acoustics, Speech, and Signal Processing*, **24**, 94–101, 1980b.

J. M. Wozencraft, Sequential Decoding for Reliable Communication, *1957 National IRE Convention Record*, **5**, part 2, 11–25, 1957.

J. M. Wozencraft and B. Reiffen, *Sequential Decoding*, Cambridge, MA, MIT Press, 1961.

E. L. Zapata and F. Arguëllo, Application-Specific Architecture for Fast Transforms Based on the Successive Doubling Method, *IEEE Transactions on Signal Processing*, **SP-41**, 1476–1481, 1993.

Y. Zheng, G. Bi, and A. R. Leyman, New Polynomial Transform Algorithm for Multidimensional DCT, *IEEE Transactions on Signal Processing*, **SP-48**, 2814–2821, 2000.

K. Zigangirov, Some Sequential Decoding Procedures, *Problemy Peredachi Informatsii*, **2**, 13–25, 1966.

S. Zohar, The Solution of a Toeplitz Set of Linear Equations, *Journal of the Association for Computing Machinery*, **21**, 272–276, 1974.

Index

abelian group, 22, 304
adder, 10
addition, 26
Agarwal–Cooley algorithm, 168, 171, **350**
algebraic integer, 306
algebraic number, 306
algorithm
    Agarwal–Cooley, 350
    Berlekamp–Massey, 249
    Bluestein, 91
    Cook–Toom, 148
    Cooley–Tukey, 17, 69
    cosine, 144
    Durbin, 237
    euclidean, 46, 54
    Fano, 274
    Goertzel, 85
    Good–Thomas, 17, 80
    Levinson, 232
    Montgomery, 320
    Nussbaumer–Quandalle, 411
    Preparata–Sarwate, 339
    Rader, 91, 318
    Rader–Brenner, 76
    Rader–Winograd, 97
    semifast, 328
    sequential, 270
    sine, 144
    stack, 271
    Strassen, 125
    Trench, 239
    Viterbi, 18, 267
    Winograd, 18
    Winograd convolution, 158
antidiagonal, 38
associativity, 21, 26, 35
autocorrelation, 222
autoregressive filter, 10, 15, 85, 231, 249

Bézout polynomial, 56, 188
Bézout's identity, 55
basis, 37
    of a vector space, 37

Berlekamp–Massey algorithm, 249
    accelerated, 260
Bluestein algorithm, 91, 112, 189
buffer, 119
buffer overflow, 271, 273, 277
Butler matrix, 18
butterfly
    decimation-in-frequency, 74
    decimation-in-time, 73
    two-point, 217

cancellation, 50
characteristic
    of a field, 34, 295
    of a ring, 28
Chinese remainder theorem, 58, 80, 168, 350
    for integers, 58
    for polynomials, 61
closure, 21, 26
coefficient, 49
cofactor, 39
column rank, 41, 181
column rank theorem, 182
column space, 41
commutative group, 22
commutative property, 22
commutative ring, 27
companion matrix, 187
complex multiplication, 2, 191
complex number, 31
complex rational, 31, 370
componentwise product, 35
composite, 44
congruence, 45, 51
    polynomial, 51
conjugate, 300, 329
connection polynomial, 250
constraint length, 264
continued fraction, 261
convolution, 12
    cyclic, 11
    two-dimensional, 346

convolution algorithm, 145
  Agarwal–Cooley, 350
  Cook–Toom, 148
  iterated, 168, 199, **362**
  two-dimensional, 350
  Winograd, 155
convolution theorem, 17
  polynomial ring, 375
Cook–Toom algorithm, 148, 166, 365
Cooley–Tukey FFT, 17
  decimation in frequency, 74
  decimation in time, 73
  radix-four, 78
  radix-two, 72
  two-dimensional, 387
coordinate rotation, 128
coprime
  integers, 44
  polynomials, 50
cordic algorithm, 144
correlation, 12
coset, 25
  left, 25
  right, 25
coset decomposition, 25
coset leader, 25
cosine transform, 86
  two-dimensional, 389
crosscorrelation, 222
cycle, 24
cyclic convolution, **12**, 228
  two-dimensional, 347
cyclic group, 23, 288
cyclic subgroup, 24
cyclotomic polynomial, 172, **300**, 369, 376, 406

decimating FIR filter, 213
decimation-in-frequency, 74
decimation-in-time, 73
decoding window
  Fano algorithm, 277
  Viterbi algorithm, 268
decomposition, 25
deconvolution, 231
degree, 49
derivative
  formal, 50
descendant, 119
determinant, 38
diagonal, 38
dimension, 36
  vector space, 36
direct product
  of groups, 24
direct sum, 24
  of abelian group, 24

discrepancy
  Fano algorithm, 274
  stack algorithm, 278
  Viterbi algorithm, 267
discrete cosine transform, 86
  inverse, 86, 112
  two-dimensional, 389
discrete Fourier transform, 15
distance
  euclidean, 265
  Fano, 272
  Hamming, 265, 269
distance function, 265
distributivity, 26
  scalar multiplication, 35
  vector addition, 35
divisible, 44, 49
division, 49
  of integers, 44
  of polynomials, 15, 49
division algorithm, 44
  for integers, 44
  for polynomials, 50
doubling, 116
doubly-linked list, 120
down-sampling filter, 213
Durbin algorithm, 237

element
  primitive, 34
elementary matrix, 41
elementary row operation, 41
error-control code, 231
euclidean algorithm, 46, 245
  accelerated, 130
  for polynomials, 54
  recursive, 130, 245
euclidean distance, 265
Euler's theorem, **288**
exchange matrix, 38, 64, 175, 232, 240
extended euclidean algorithm, 47
extension field, 33

Fano algorithm, 274
Fano distance, 272
fast Fourier transform
  Cooley–Tukey, 68
  Good–Thomas, 80
  Johnson–Burrus, 399
  Nussbaumer–Quandalle, 412
  Rader–Brenner, 76
  Winograd large, 395
  Winograd small, 102
feedback shift register, 250, 252
Fermat number transform, 314
Fermat prime, **315**, 331
Fermat's theorem, 288

field, 30
  characteristic, 34
  finite, 31, 295
  Galois, 31
  number, 306
  prime, 295
field of constants, 179
field of the computation, 179
filter
  autoregressive, 10
  decimating, 213
  down-sampling, 213
  finite-impulse-response, 10
  interpolating, 213
  skew-symmetric, 207
  symmetric, 207
  up-sampling, 213
filter section, 196, 200
finite field, 31, 295
finite group, 22
finite-dimensional vector space, 36
finite-impulse-response (FIR) filter , 10
finite-state machine, 262
first-in first-out (FIFO) buffer, 119
formal derivative, 50, 308
Fourier transform, 15, 64
  finite field, 328
  limited-range, 221
  punctured, 422
  two-dimensional, 384
frame, 264

Galois field, 31
gaussian elimination, 231
gaussian integer, 306
gaussian rational, 306
generalized Rader polynomial, 97
generator, 23
Goertzel algorithm, 84
Good–Thomas FFT algorithm, 17, 82, 351
greatest common divisor
  of integers, 44
  of polynomials, 50
ground field, 179
group, 21
  abelian, 22
  commutative, 22
  cyclic, 23
  finite, 22
  quotient, 23

Hamming distance, 265, 269
Horner's rule, 83

identity, 21
identity element, 21
identity matrix, 38

indeterminate, 49, 179
indirect address, 120
inner product, 36
integer, 44
  algebraic, 306
  gaussian, 306
  of a ring, 28
  prime, 44
integer ring, **44**, 293
integer ring transform, 336
interpolating FIR filter, 213
inverse, 22
  left, 27
  matrix, 38
  nonsingular, 38
  right, 27
irreducible polynomial, 49
isomorphic, 22
iterated algorithm
  convolution, 362
  filter section, 202

Klein four-group, 63
Kronecker product, 40, 354, 393

Lagrange interpolation, 58, 149, 154, 168
Lagrange theorem, 26
Laplace expansion formula, 39
last-in first-out (LIFO) buffer, 119
leader
  coset, 25
least common multiple
  of integers, 44
  of polynomials, 50
left coset, 25
left inverse, 27
Levinson algorithm, 18, 232
linear combination, 36
linear convolution, 11
linear prediction, 231
linearly dependent, 36
linearly independent, 37
linked list, 120
list, 119

marginalize, 281
matrix
  companion, 187
  exchange, 38, 232
  identity, 38
  inverse, 38, 40
  nonsingular, 38
  persymmetric, 239
  singular, 38
  square, 37
  Toeplitz, 37, 231
  transpose, 38

matrix algebra, 37
matrix exchange theorem, 175
matrix inverse, 38, 40
matrix multiplication, 37
mergesort, 121, 143
Mersenne number transform, 317
Mersenne prime, **317**, 331, 332
metric, 265
minimal polynomial, 299
minor, 39
monic polynomial, 49
Montgomery multiplication, 320
Montgomery reduction, 321
multiple
   least common, 50
multiplication, 26
   complex, 2
   matrix, 37
   Montgomery, 320
multiplier, 10

nesting, 391
nonsingular matrix, 38
null space, 41
number field, 306
number system, 8
number theory, 286
Nussbaumer–Quandalle FFT, 412, 425

one's-complement, 9, 317
optimum algorithm, 200
order, 22, 24, 65, 288
origin, 35
orthogonal, 36
orthogonal complement, 36, 41
outer product, 41
overlap, 15
overlap–add method, 197
overlap–save method, 195

parametric algorithm, 273
path sequence, 265
permutation, 38
persymmetric matrix, 239
polar transformation, 128
polynomial, 48
   Bézout, 56
   connection, 250
   cyclotomic, 172, **300**
   irreducible, 49
   minimal, 299
   monic, 49
   prime, 49, 171
   quotient, 51
   reciprocal, 207
   remainder, 51
   zero, 49

polynomial over a field, 48
polynomial ring, 48
prime, 44
prime field, 295
prime integer, 44
prime polynomial, 49, 171, 297
primitive element, 34, **304**, 330
product
   componentwise, 35
   inner, 36
   Kronecker, 40
   of groups, 24
   outer, 41
punctured FFT algorithm, 415
push-down stack, 117, 119

quadratic residue, 335
quaternion, 65, 381
queue, 119
quicksort, 121
quotient, 44
quotient group, 23
quotient polynomial, 51
quotient ring, 27
   integers, 293
   polynomials, 296

Rader polynomial, 94, 95, 103, 105, 140, 318
   generalized, 97, 107, 110, 140
Rader prime algorithm, 91, 103, 112, 189, 318, 403
   two-dimensional, 408
Rader–Brenner FFT, 76
Rader–Winograd algorithm, 97
radix, 72
radix-four Cooley–Tukey FFT, 72
   two-dimensional, 388
radix-two Cooley–Tukey FFT, 72
   two-dimensional, 387
rank, 43
rational number, 31
real number, 31
reciprocal polynomial, 207
recursive procedure, 117
relatively prime, 44
   polynomials, 50
remainder, **45**
remainder polynomial, 51
right coset, 25
right inverse, 27
ring, 26
   algebraic, 306
   characteristic, 28
   commutative, 27
   gaussian, 306
   identity, 27

integer, 28
noncommutative, 27
quotient, 293
unit, 29
ring integer, 44
ring of polynomials, 48
ring with identity, 27
row rank, 41, 181
row rank theorem, 181
row space, 41
row-echelon form, 42

scalar, 34, 49, 179
scalar multiplication, 34, 35
scaler, 10
semifast algorithm, 328
sequential algorithm, 270
shift-register stage, 10
singular matrix, 38
skew-symmetric filter, 207
source sequence, 265
span, 36
spectral analysis, 237
spectral estimation, 231
square
    of a prime field, 335
square matrix, 37
stack, 119
    push-down, 117
stack algorithm, 271
state diagram, 263
Strassen algorithm, 124, 144
string, 119
subfield, 33
subgroup, 24
    cyclic, 24
subring, 27
subspace
    vector, 35
surrogate field, 311
symmetric filter, 207

theorem
    column rank, 182
    Euler, 288
    Fermat, 288
    Lagrange, 26
    matrix exchange, 175
    row rank, 181
    unique factorization, 53
Toeplitz matrix, **37**, 143, 205, 231, 232
    symmetric, 232
totient function, **286**, 302
transcendental number, 299
transformation principle, 200
transpose, 38
transposition, 38
tree, 119
trellis, 263
Trench algorithm, 239
two's-complement, 9

unique factorization theorem, 53
unit
    of a ring, 29, 65
up-sampling filter, 213

variable, 179
vector, 34, 119
vector addition, 34, 35
vector space, 34
    finite-dimensional, 36
vector subspace, 35
Viterbi algorithm, 18, 267

Winograd convolution algorithm, 158
Winograd large FFT, 395
Winograd small fast Fourier transform, 102
Winograd small FFT, 393

zero, 27
    of a polynomial, 57
zero polynomial, 49